AF248529

IN VIVO BODY COMPOSITION STUDIES
Recent Advances

BASIC LIFE SCIENCES

Ernest H. Y. Chu, Series Editor
The University of Michigan Medical School
Ann Arbor, Michigan

Alexander Hollaender, Founding Editor

IN VIVO BODY COMPOSITION STUDIES
Recent Advances

Edited by

Seiichi Yasumura
Brookhaven National Laboratory
Upton, New York

Joan E. Harrison
Toronto General Hospital
Toronto, Ontario, Canada

Kenneth G. McNeill
University of Toronto
Toronto, Ontario, Canada

Avril D. Woodhead

and

F. Avraham Dilmanian
Brookhaven National Laboratory
Upton, New York

PLENUM PRESS • NEW YORK AND LONDON

Library of Congress Cataloging-in-Publication Data

International Symposium on In Vivo Body Composition Studies (1989 :
  University of Toronto)
    In vivo body composition studies : recent advances / edited by
  Seiichi Yasumura ... [et al.].
        p.   cm. -- (Basic life sciences ; v. 55)
    "Proceedings of the International Symposium on In Vivo Body
  Composition Studies, held June 20-23, 1989, at the University of
  Toronto, Toronto, Ontario, Canada"--T.p. verso.
    Includes bibliographical references and index.
    ISBN 0-306-43618-3
    1. Body composition--Congresses.   I. Yasumura, Seiichi.
  II. Title. III. Series.
  QP88.I53 1989
  612'.01585--dc20   90-7551
                                                            CIP

Proceedings of the International Symposium on *In Vivo* Body Composition Studies,
held June 20-23, 1989, at the University of Toronto,
Toronto, Ontario, Canada

© 1990 Plenum Press, New York
A Division of Plenum Publishing Corporation
233 Spring Street, New York, N.Y. 10013

Printed in the United States of America

PREFACE

This book is the compilation of papers presented at the International
Symposium on _in vivo_ Body Composition Studies, held at the University of
Toronto, Ontario, Canada, June 20 - 23, 1989. The purpose of this
conference was to report on advances in techniques for the _in vivo_
measurement of body composition and to present recent data on normal body
composition and changes during disease. This conference was the most
recent of several meetings on body composition studies, and follows two
successful such meetings, one at Brookhaven National Laboratory in 1986,
and at Edinburgh in 1988. The large number of excellent research papers
and posters presented at these conferences demonstrates the rapid growth
of the field and the broad interest in the subject of _in vivo_ body
composition studies. The proceedings of the Brookhaven meeting "In Vivo
Body Composition Studies", is published by The Institute of Physical
Sciences in Medicine, London. Both the Brookhaven and the current Toronto
meeting emphasized the clinical applications, together with the techniques
employed. The Edinburgh meeting placed more emphasis on the
methodological problems and design of instrumentation.

Because of the number of papers presented at the meeting it was
necessary to ask the authors from the same institution to combine their
presentations into a single paper where appropriate. The editors wish to
thank the authors for their cooperation and for graciously accepting the
minor revisions made to each manuscript.

The order of presentation of the papers at the conference forms the
basis of the order of appearance of the papers in this book, although the
titles of some of the sections have been changed to better reflect the
scope of all the papers included under each chapter heading. Also, topics
listed in the program under "Obesity, Fitness, and Aging" and "Wasting
Diseases" have been combined into a single chapter entitled, "Lean and Fat
Tissues". Papers representing poster material have been included in each
section, according to the most appropriate topic. In some cases this
choice was arbitrary, especially when the subject matter of the paper fit
several categories. Finally, a separate chapter heading: "Recent
Developments in Instrumentation" is composed of papers that should not be
construed as being exclusive in any way. They were placed in this chapter
on the basis of subjective judgement of the editors.

In addition to the high quality of the presentations, the success of
the symposium was most certainly due to the efforts of the International
Organizing Committee, the Local Arrangements Committee, and the
secretarial and technical staff. Also, the generous support of the many
companies and institutions listed below is gratefully acknowledged.

ACKNOWLEDGEMENTS

<u>Organizing Committee</u>

P. S. Allen; R. F. Code; J. E. Harrison; S. S. Krishnan; K. G. McNeill,
S. Yasumura.

<u>International Committee</u>

B. J. Allen (Australia); L. Burkinshaw (U.K.); J. Compston (U.K.);
K. J. Ellis (U.S.A.); D. Glaros (Greece); G. L. Hill (N. Zealand);
S. Mattsson (Sweden); B. D. Maziere (France); R. N. Pierson (U.S.A.);
P. Tothill (U.K.); M. Wahlqvist (Australia).

<u>Sponsors</u>

Medical Research Council of Canada
The Ontario Milk Marketing Board
Monserco Limited
Hologic, Inc.
Canberra-Packard Canada Ltd.
Sandoz Laboratories Inc.
Ross Laboratories
Rorer Canada Inc.
Fisher Scientific

<u>Exhibitors</u>

Futrex Inc.
Lunar
Seritex Inc.
Monserco Ltd.
Hologic Inc.

<u>Special thanks to</u>:

A. L. Ruggiero, Publication Secretary
K. J. Vivirito, Technical Editing
J. O'Reilly, Conference Secretary
I. Kierans, Conference Secretary
A. Hitchman, Assistant Secretary
S. Goodwin, Assistant Secretary
A. Strauss, Assistant Secretary
A. Bayley, Audio Visual
B. Hasany, Audio Visual

CONTENTS

AN INTRODUCTION - QUID SCIENDUM

K.G. McNeill and Joan E. Harrison

This conference on in vivo body composition studies celebrates the continuing fruitful marriage of physics and clinical medicine. The meeting was designed to clarify what clinicians want to know about body composition, to assess the value and limitations of current procedures used to provide this knowledge, to discuss procedures which currently exist but are yet unused, and to identify what further development in methodology is required. Lord Kelvin, late Chancellor of Glasgow University, said: "...that when you can measure what you are speaking about, and express it in numbers, you know something about it; but when you cannot measure it, when you cannot express it in numbers, your knowledge is of a meagre and unsatisfactory kind..."[1] Measurement is the solid basis on which physics builds, and a major contribution of physics to medicine is the quantitation, in vivo, of different constituents of the body.

Body composition studies primarily concern measurements of bone mass and of the various components of soft tissue. In vivo measurements of some trace elements are included,[2] as they can be useful to investigate diseases associated with trace element deficiency or toxicity.

First, it is necessary to decide which components of body composition are of clinical importance. The important component will vary, depending on the clinical problem. Where possible, the component of interest should be measured directly and in the clinically relevant part of the body. Many indirect measurements are made which are assumed to reflect the component of interest, but such indirect measurements, even if simple and precise, are of little clinical value unless they reflect reliably the component of interest. An ancestor of ours is reported to have been seen at night on all fours in his kilt crawling in the gutter under a street lamp. His explanation was that he had lost some money some 25 yards away, but was looking under the lamp because its light made search much easier there. In many cases it is much easier to make and use apparatus to measure, for example, bone or electric resistance in an easily accessible part of the body - but is it a relevant measurement? It can be so only if it be proven that the portion measured is, in fact, representative of the part of interest in the patients. It is best to measure the diagnostically important site.

In the case of bone, this measurement is now possible using partial body in vivo neutron activation analysis (IVNAA), dual photon absorptiometry (DPA), dual photon (x-based) absorptiometry (DPX) or computed tomography (CT).[3] The current problems are: firstly, that correlations of the results from different procedures are only fair (despite the fact that all the procedures measure bone mass or bone density, although in varying amounts of bone volume); secondly, that sensitivity and specificity for osteoporotic fractures is suboptimal.[4] Clearly, the problem of accuracy must be cleared up by physicists, but the second problem raises an important question: what really does one want to know? Bone strength rather than bone mass? Almost certainly, bone strength depends, not only on bone mass, but also on other factors related to the quality of bone. In that case, some other measurement of bone quality is presumably what physicians want, even though the definition of osteoporosis specifies "loss of bone." Can different modalities, such as ultrasonics, help here?[5] Or could we use CT to measure, not only mass of trabecular bone, but also its organization?

The major part of this conference focuses on measurements of soft tissue; the fat, water, and protein of the body. Although some components can be measured directly, e.g. water (TBW) by isotope dilution[6] and protein (TBP) by prompt gamma analysis for nitrogen (TBN)[7], because of the perceived risks or complexities of procedure indirect methods of assessment are widely employed.[8,9,10] These methods are based on the fact that for the "normal" adults there are relationships between the different components, for example, between TBW or TBK and lean body mass (LBM). From these facts, indirect means of estimating particular body components have been devised, based on measuring another component and using the "normal" relationship between them. Indeed, normal relationships will give body components based on weight alone.

, But, by definition, sick people are not normal and, therefore, none of the body components in patients need be related to others by the factors found in normal persons. The same will be true of the very thin and the very obese, and might be true of the old and the very young. And it is the sick, rather than the normal, in whom physicians tend to be interested from a clinical point of view. The fact that, in groups of persons, ill or well, there are good correlations between masses of body components does not necessarily mean that the two masses are directly related to one another, but is only indicative of the fact that larger persons have more of everything than smaller persons. The masses of any of the major body components will correlate well with height or weight and then, indeed, any of them will correlate with any of the others. So will quantities which vary with any one of them, for example, electrical conductance, as electricity is preferentially conducted by body fluids. As individuals deviate from the norm, for example with prematurity, aging, obesity or wasting disorders, greater variations from the average proportions of each of the components of body weight can be expected. Thus, clinicians are primarily concerned with abnormal situations in which measurements of one part of body composition cannot be expected to give data that predict reliably other components of body composition. Nor can gross properties, such as density of LBM, be trusted to be "normal" in abnormal situations. Indeed it has been extensively shown that changes in one body component (e.g. TBK) may be in the opposite direction to changes in another (e.g. TBN).[7,11] Investigators have recognized the inadequacy of these indirect measurements, even in relatively normal subjects and, as a result, some have accepted weight normalized for height (bone mass index or BMI) as sufficient, despite the fact that it is recognized that changes in BMI in an individual cannot be ascribed unequivocally to variation in body fat (TBF).[12] It is best to measure the diagnostically important components directly.

The total weight of LBM is made up of water (intra- and extra-cellular), protein, minerals, and a small quantity of glycogen. It could be useful to differentiate the intra- and extra-cellular component of body protein, as the intra-cellular protein is considered the more metabolically active fraction. Can physics help here? TBK has been used as a measure of cell mass, and thence as a measure of intra-cellular protein, but, as noted above, changes in TBK are not necessarily indicative of changes in TBP (and extra-cellular

protein is not believed to change in the short term). This change in TBK
can be attributed to change in intra-cellular potassium concentration [13] and,
therefore, in this case, measuring both TBK and TBN would be of particular
clinical significance. However, a change in TBK could also reflect a change
in intra-cellular water, which would not have the same clinical significance.
Would the measurement of intra-cellular (or, total minus extra-cellular)
water volume, in addition to TBK and TBN, now be of particular significance,
to show whether a change in TBK is due to a change in concentration or to
a change in volume? However, all of these measurements still do not give
intra-cellular protein! Can NMR help here, or can CT?

If direct measurements of a body component are done, how valid are
the results? Values for body elements obtained by NAA and chemical
analysis in cadavers[14-15] (TBN, TBCl), pigs[16-17] (TBN, TBP), and rats[6] (TBK,
TBCl, TBW), are in close agreement. Accuracy can be a few percent. Any
particular measurement has an error in precision, in addition to the error
in accuracy of the technique. Indeed, this error in precision is commonly
the important error in diagnostic work, as normally the measured quantity
is compared to the corresponding value in a "standard" person, so that
errors in accuracy cancel out.

As yet there are no established methods of measuring TBF directly,
although infrared[18] and neutron inelastic scattering[19] methods are discussed
in this conference. In principle, it is possible to measure total body fat by
subtracting of other body component masses from weight. As both accuracy
and precision errors now all contribute, the errors on the derived TBF will
be very large, typically 30% (as indeed they would be if one attempted to
measure any other body element by subtraction, for example, TBP, assuming
that TBF were known).

We have stressed, and we stress again, the necessity of direct measurement of the desired quantity at the relevant place. But, as noted
earlier, this may require equipment perceived to be complex, costly and, in
many cases, to give potentially hazardous radiation doses. As methods of
measuring TBP, TBK, TBW, and bone mineral are in routine use in many
hospitals around the world (in much simpler form than in national laboratories) we feel that the ideas of complexity and cost are more a matter of
perception than of fact. With respect to risk, physicians must, at all times,
weigh the potential hazard of any procedure against potential benefit.

Any measurement involves risk, but currently, ionising radiation ranks
high in the public's ideas of comparative potential risk (despite all evidence
to the contrary). We must be clear as to potential risks in using ionising
radiation and the potential risks of not using it. Measurement for TBP is
reported to give an effective dose equivalent to a child of 1.2 mSv. [20] This
value is translated to a detriment of $\sim 10^{-5}$, that is, a 1 in 100,000 chance in a
normal lifetime of contracting a cancer as a result of the neutron exposure.
As in the Western world the "normal" chance of contracting a lethal cancer
is about 1 in 4, the total probability is raised from 0.25 to 0.25001 - or
raising the incidence in 100,000 persons from 25,000 to 25,001. Putting this
in another way, the life shortening is, on average, about 3 hours. The
physician has to ask if the knowledge to be gained is worth 3 hours of the
patient's life - will the patient live more than a day longer, for instance, or
will the patient have a more pleasant week, as a result of treatment resulting from the TBP diagnostic test? Of course the physician must also answer
this about x-rays; in the case of an upper-GI series of x-rays, a reasonable
way of checking on reasons for malnourishment, the dose equivalent is
about four times as great. And, indeed, the probability of death or severe
injury in traffic accidents is roughly as high when a person comes to
hospital from 50 miles away as it is in a TBP procedure; properly, this

should also be taken into account when ordering any tests, whether or not they involve radioactivity.

We believe that the chapters of this book indicate the depth and breadth of on-going studies in body composition, and that the relationship between physicists and medical scientists will continue to be creative. Again we believe that a close and, indeed, much closer collaborative contact between physicists and physicians is essential to clarify many misty areas. Indeed, we believe that a much better collaboration is desirable to define more clearly what we need to know and to determine how to measure it; to define how precisely measurements are to be made and how this can be done most effectively; to discover new parameters of body composition that will increase our understanding, and to create new methods of measuring them.

REFERENCES

1. S.P. Thompson, The Life of William Thomson, Baron Kelvin of Largs, McMillan, London (1910).
2. D. Chettle, Measurements of trace elements in vivo. This volume.
3. P. Tothill, Evaluation of methods of bone mass measurement. This volume.
4. S.M. Ott, R.F. Kilcoyne, C.H. Chesnut, Comparisons among methods of measuring bone mass and relationship to severity of vertebral fractures in osteoporosis. J. Clin. Endo. and Metab. 66:501-507 (1988).
5. W.D. Evans, E.A. Jones, G.M. Owen, The in vivo precision of broadband ultrasonic attenuation. This volume.
6. S. Yasumura, D. Glaros, J. Kalef-Ezra, A.F. LoMonte, J.K. Yeh, R.I. Moore, Distribution of body water in rats. This volume.
7. K. Gaskin, Body composition in cystic fibrosis. This volume.
8. H. Lukaski, Applications of bioelectrical impedance analysis: a critical review. This volume.
9. S. Heymsfield, Dual photon absorptiometry: validation of mineral and fat measruements. This volume.
10. R.N. Pierson, High precision in vivo neutron activation analysis: a new era for compartmental analysis in body composition. This volume.
11. J.E. Harrison, K.G. McNeill, A.L. Strauss, A nitrogen index – total body protein normalized for body size – for diagnosis of protein status in health and disease. Nutr. Res. 4:209-224 (1984).
12. J.S. Garrow, Body composition for the investigation of obesity. This volume.
13. K. Jeejeebhoy, Mechanism of reduction of total body potassium in malnutrition. This volume.
14. G.S. Knight, A.H. Beddoe, S.J. Streat, G.L. Hill, Body composition of two human cadavers by neutron activation and chemical analysis. Am. J. Physiol. 250:E179-E185 (1986).
15. F.D. Moore, J. Lister, C.M. Boyden, M.R. Ball, N. Sullivan, F.I. Dagher, The skeleton as a feature of body composition. Hum. Biol. 40:135-188 (1968).
16. K.G. McNeill, J.R. Mernagh, K.N. Jeejeebhoy, S.L. Wolman, J.E. Harrison, In vivo measurements of body protein based on the determination of nitrogen by prompt gamma analysis. Am. J. Clin. Nutr. 32:1955-1961 (1979).
17. M.N. Thompson, private communication.
18. J.M. Conway, K.H. Norris, Noninvasive body composition in humans by near infrared interactance. In Ed. K.J. Ellis, S. Yasumura, W.D. Morgan, In Vivo Body Composition Studies (1986).
19. J. Kehayias, Measurement of body fat by neutron inelastic scattering: comments on installation, operation and error analysis. This volume.
20. B.J. Allen, G.M. Bailey, B.J. McGregor, The AAEC total body nitrogen facility, Proc. Fourth Australian Conf. on Nuclear Techniques of Analysis, AINSE, Lucas Heights, p.101 (1985).

INFANT BODY COMPOSITION MEASUREMENTS AS AN ASSESSMENT OF NUTRITIONAL

STATUS

Buford L. Nichols, Hwai-Ping Sheng, and Kenneth J. Ellis

USDA/ARS Children's Nutrition Research Center, Department
of Pediatrics, Baylor College of Medicine, Houston, TX

INTRODUCTION

Growth is both a determinant and an index of the unique nutritional
needs of childhood. Although growth can be defined simply as an
accumulation of nutrient stores, it actually consists of an integration
of cellular differentiation, hyperplasia, and hypertrophy, which results
in structural and functional maturation of tissues. Rapid growth in the
human occurs during the fetal, nursling, and pubertal stages of
development. Childhood, a prolonged period of relatively slower growth,
separates the nursling and pubertal stages.

The relationship between nutrient intake and growth is not
completely understood. We do know that growth results from nutrient
retention, which accounts for only a small fraction of the total diet.
When dietary intake is in excess of the threshold for growth, weight
gain occurs in proportion to the excess. When the intake exceeds a
ceiling for growth, however, the incremental growth rate is reduced as
storage occurs in fatty tissue of greater energy density.

Infants with equivalent energy intakes have different rates of
growth. The causes of this variability among normal infants are not
understood. There is, for example, the increase in apparent growth
efficiency observed in infants of obese parents, or the small percentage
of infants who actually develop clinical malnutrition despite a marginal
dietary intake. Additional research is needed to document the sources
of variation in growth efficiency and body composition observed in
normal children.

One of the most commonly used methods for nutritional assessment in
adults (and in pediatric practice) is to measure changes in body weight.
In the absence of disease or changes in exercise or activity level,
changes in body weight in the adult can be attributed to a change in
adiposity. A weight change in the infant, however, is more difficult to
interpret, because weight gain represents an increase in both the
adipose and the lean tissues. Most studies of energy and protein needs
of infants have used changes in body weight as one measure of the
nutritional efficacy of the various foods. This weight gain, however,
may be the result of accretion of fluid volume or increased lean or fat
tissue. Therefore, the composition of the weight gain is the key to our

understanding of the nutritional efficacy of the different feeding
patterns.

Another method used extensively for nutritional assessment in the
infant is a cumulative balance of nutrients. However, balance studies
have many deficiencies. They are tedious and labor-intensive, not all
routes of loss are measured (e.g., skin losses are usually ignored), and
the observation period is usually short. Furthermore, extrapolation of
short-term (48- to 72-hour) balance data to that accumulated over a week
or month requires multiplying intrinsic errors by a large factor and
does not take into consideration the changes in the body composition
over this extended period.

Our knowledge of the changes in body composition from conception
through infancy is based primarily on analyses of fetuses and infant
cadavers (Widdowson and Dickerson, 1964; Fee and Weil, 1963) and
estimates derived from indirect methods for total body water and total
body potassium measurements (Fomon et al., 1982). Ziegler et al. (1976)
and Fomon et al. (1982) compiled data from the literature to calculate
their estimates of water, protein, fat, and minerals for the "reference
fetus" and "reference infant" from which most studies on nutrient
requirements for the infants were based.

The rate of rapid growth of the infant differs from that of the
adult. Fig 1 shows body weight of the "reference fetus and infants"
increased from 600 g at 24 weeks gestation to 3500 g at birth, an
average increase of 26 g/day. At one year of age, body weight had
increased by three times the birth weight to an average of 10,000 g.
Although body length increased during this period, the increase was less
dramatic than that of body weight gain. The relative contribution of
the various chemical compartments (water, protein, and fat) to total
body weight gain is shown in Figs 2 and 3. At 24 weeks of gestation,
86% of body weight was composed of water, 9% was protein and 2% fat.
The proportion of fat increased rapidly to 14% at birth, protein
increased less rapidly to 12%, and water declined to 70%. Changes in
body composition during the first year of life were less dramatic than
those during the last trimester of gestation. Body fat continued to
increase to an average of 23% at 3 months of age, whereas the proportion
of protein remained the same; body water continued to decline to
approximately 60% by 4 months of age. In other words, protein mass
accumulated at an average rate of 3.8 g/d from 24 weeks gestation to
birth. The rate of accumulation decreased to an average of 2.0 g/d from
the fourth month through the first year of life. The peak of fat
accumulation occurred approximately one to three months postnatally
(13.7 g/d), resulting in a rapid increase in fat content in the infant
from 13% body weight at birth to 23% at 3 months of age. These data are
calculated from a few observations and should provide only the average
nutrient retentions and should only be viewed as an estimate of the
magnitude of changes in body composition during growth.

Convenient, noninvasive methods are needed for measurements of the
body content of a substance or tissue mass so that the growth or
accumulation of the nutrients in individual infants can be assessed
accurately. At present, only a few methods provide accurate and precise
in vivo measures of body composition in adults. There are even fewer
methods that can be applied to infants.

In vivo methods suitable for measurements of lean body mass (LBM)
and adipose tissue (FAT) in the infant will be reviewed in this chapter.
For such noninvasive in vivo measurements of body composition, body mass
can be considered the sum of adipose tissue and LBM or, alternatively,

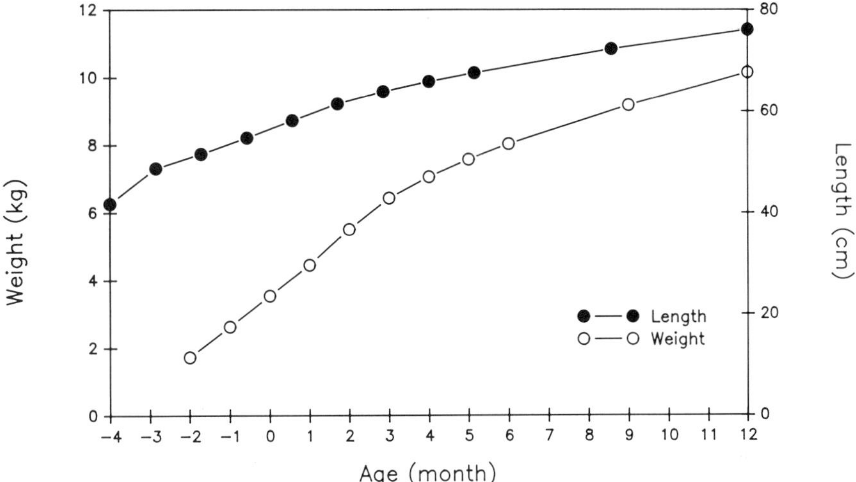

Fig 1.   Typical changes in body weight and length of the human infant
          (derived from the 'reference infant' data of Fomon et al.,
          1982).

the sum of ether-extractable fat and FFM. The terms lean body mass (LBM)
and fat-free mass (FFM) are often used interchangeably and sometimes
cause confusion among investigators. The terms would be synonymous if
adipose tissue were 100% fat. Instead, it is comprised of approximately
80 to 85% fat, 2% protein, and 13 to 18% water. The distinction between
LBM and FFM is not critical in a fairly lean subject, but is important
in an obese subject in whom the contribution of the non-fat components
of adipose tissue can be large.

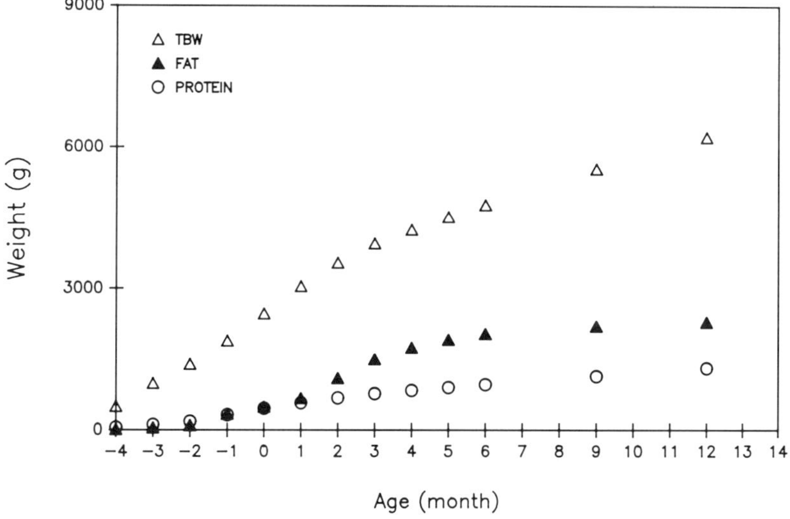

Fig 2.   Estimates of the changes in body protein, water, and fat
          during the first year for the human infant (data from Fomon et
          al., 1982).

In vivo procedures of body composition analysis have been developed
to measure the whole body or specific sites and regions from which
extrapolation to the whole body can be made. Methods range from the
simple to the complex; most make use of constants and assumptions
derived from cadaver analyses reported by Widdowson and Dickerson
(1964), Ziegler et al. (1976), and Forbes et al. (1956). Most in vivo
methods are indirect and have been validated by correlation against
another method, which also is usually indirect. Only a few studies have
reported validation using direct analysis of human cadavers; such
studies have been reported only in adults (Rudd et al., 1972; Knight et
al., 1986).

WHOLE-BODY MEASUREMENTS

Models of Body Composition

On the basis of the simplest concept, the body can be divided into
a fat compartment and a non-fat compartment which is assumed to have a
constant chemical composition: FAT and FFM. The fat-free or lean tissue
mass can be subdivided into different components depending on the
interest of the investigator. The most informative model of body
composition consists of four major body compartments: protein, water,
minerals, and fat.

Because body fat cannot be determined satisfactorily by a direct
method, it is usually determined as the difference between body weight
and fat-free mass. The first concept of fat-free body tissue was based
on a constant chemical composition. Behnke et al. (1942) later extended
this concept to lean body mass (LBM), which they defined as the "primary
energy-exchanging mass" in the body. The composition of LBM is assumed
to consist of relatively constant proportions of water, bone, and cell
mass. The assumption that LBM has a relatively constant composition has
led to the development of techniques which measure body water,
potassium, or density, as indices of the lean body mass.

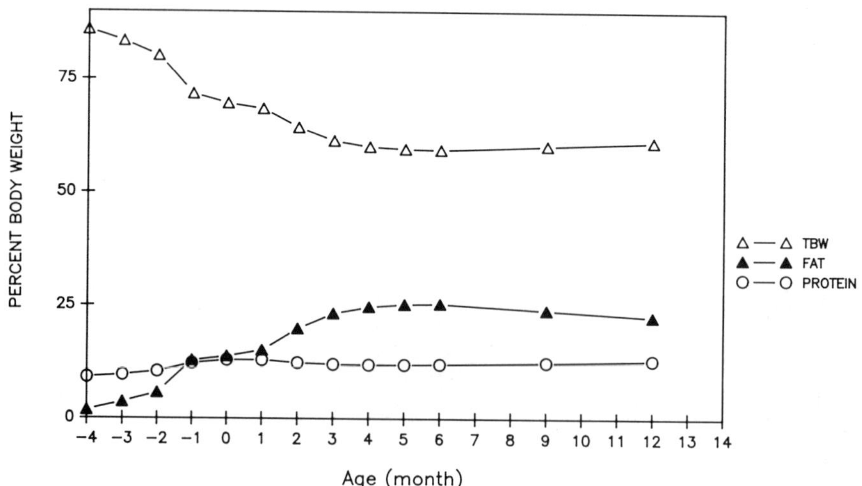

Fig 3.   Relative percentage changes in body composition during first
          year for the human infant (data from Fomon et al., 1982).

Measurement of Body Fat

Inert gases (cyclopropane, xenon, and krypton) are highly soluble
in fat but not in water. Thus, several attempts have been made to
measure body fat by the dilution principle using these gases. When
tested in small animals, the results agreed with those obtained by
carcass analysis. Two major disadvantages of the method, however, are
the necessity of a closed spirometer system (a difficult system for
studies of most infants) and long inhalation times (usually greater than
2 hr). The latter constraint is required to attain equilibrium in body
fat. Attempts to reduce the time of inhalation and then extrapolate
from the early phase have been relatively unsuccessful. A recently
developed technique based on inelastic scattering of neutrons has been
used to measure body fat in adults (Kehayias et al., 1987). Although
the radiation dose is low, one can not recommend its use in pediatric
studies at this time.

Density and Volume Measurements

Densitometric determination of the body is still considered by many
investigators as the "reference method" or the standard against which
other indirect methods are compared. The estimation of body fat from
densitometry was pioneered by Behnke et al. (1942), who reasoned that if
the densities of the body components of fat and LBM were known and if
the density of the whole body could be measured, then the proportional
masses of fat and LBM could be calculated. Although the concept is
theoretically sound, accurate data are required for the densities for
body fat and LBM. Body density is calculated as:

Body density = Body mass/body volume.

Although body weight can be measured precisely ($\pm$ 0.1 g), the
measurement of body volume does not yet yield a satisfactory level of
accuracy, especially in infants and children. Corrections for lung
volume introduce significant errors in the determination of body volume.
Even with the assurance that body volume, thus body density (weight/
volume), can be measured with great 'precision' and 'accuracy' for a
given individual, the application of the densitometric approach to
measure FFM and fat is not without errors. The values used for the
density of fat and FFM (usually 0.9 g/cc and 1.095 to 1.10 g/cc,
respectively) greatly influence the final estimates of fat and lean
tissue mass (Brozek et al., 1963). Variations in the density of body
fat have been reported at less than 2% (Pearson et al., 1968), thus
contribution to error in the estimation of body fat is small. There is,
however, evidence that the chemical makeup of FFM in adults changes with
age and may be altered significantly in several disease states (Cohn et
al., 1984), thereby altering the validity of the constant density values
used in the equations. The greatest chemical changes in the composition
of FFM occur during growth. The mean estimates for the lean tissues
change from 1.064 g/cc in infants (Fomon et al., 1982) to 1.095 for the
older children (Brozek et al., 1963). Although these differences appear
small, their impact is evident in the calculation of body fat for an
infant. The fat content would be estimated as 11% of body weight when a
density value of 1.064 is used for FFM and 23% when 1.095 is used.
Thus, the value of 1.095 for the density of FFM (derived from cadaver
analysis) must be used cautiously, and reported percentages of body fat
must be viewed with caution. The point made by Brozek et al. (1963)
more than 25 years ago in a review of the body density methods still
appears to be valid, i.e., "that no universally valid formulas for
densitometric estimation of the fat content can be offered."

To circumvent the use of a changing density value for FFM when the two-compartment model is used for infants, the densitometric method has been expanded to the four-compartment model in which the sum of the weights of the four compartments (water, fat, protein, and minerals) and the sum of the volumes of the four compartments are assumed to equal body weight and body volume, respectively:

$$W_t = W_f + W_w + W_p + W_m \qquad \text{(Eq. 1)}$$

$$V_t = V_f + V_w + V_p + V_m \qquad \text{(Eq. 2)}$$

Substituting density, $d = W/V$, into Eq 2 yields:

$$V_t = \frac{W_f}{d_f} + \frac{W_w}{d_w} + \frac{W_p}{d_p} + \frac{W_m}{d_m} \qquad \text{(Eq. 3)}$$

If one assumes that the densities are $d_f = 0.9$, $dw = 0.993$ at 37°F, $d_p = 1.34$, $dm = 3.0$, and that $W_m = 2.5\%$ body weight, then body fat can be calculated from equations 1 and 3:

$$W_f = 2.74V_T - 2.20W_T - 0.72W_W. \qquad \text{(Eq. 4)}$$

If one can measure total volume (V_T), body weight (W_T), and total body water (W_W), then an estimate of total body fat can be obtained.

Although the original, and still most widely used, physical method to measure body volume is by underwater weighing (Gnaedinger et al., 1963) or water displacement (Garn and Nolan, 1963), this approach cannot be used on infants because whole body submersion is essential with this method. Two alternate techniques, therefore, are being devised to measure body volume in infants: a modification of the air displacement method, and an acoustic method.

In theory, the volume of air displaced by an infant placed in a rigid chamber can be measured either by the helium dilution method or by measuring the pressure difference as described by Boyle's law (Taylor et al., 1985). The method is theoretically simple, but technically difficult to use in infants. For example, if a chamber volume of 30 liters is constructed and a piston is used to change the volume by 1% (0.3 liters), then a premature infant of 2-liter volume would only change the incremental pressure over that of the empty chamber by 0.76 cm of water (or a 0.073% change). The accurate measurement of such a small pressure change is a difficult task, because a 1°C temperature change from 36° to 37°C at a constant volume and an ambient pressure of 760 mm Hg alone would produce a pressure rise of 3.34 cm of water. Taylor et al. (1985) minimized this technical difficulty by developing a differential dynamic system where identical volume changes in two identical chambers are induced by two yoked pistons. Any differential pressure between the two chambers, as measured by a manometer, would result primarily from the difference in air volume between the chambers. This system would enable a resolution of 1.0% in the differential pressure measurement (instead of 0.073 %) for a 1% change in body volume. Although body volume measurements obtained by this system appear generally reasonable, widely divergent values are produced occasionally, probably because of pressure fluctuations from respiratory movement and temperature changes (Taylor et al., 1985). As the technical difficulties are resolved, the method may become particularly suited for use in infants, because corrections for residual volumes of air in the lungs and gut are not considered necessary.

Dell et al. (1987) reported the use of the pressure differential method for body volume measurements on 54 low birth weight infants who weighed between 900 and 1750 g at birth. During their hospitalization, the infants were fed one of six formulas, and were discharged when their body weight reached approximately 2.2 kg. The diets provided different amounts of protein and energy. A total of 220 volume measurements were performed on the 54 infants during this period. Each group of infants had a significant decrease in total body density as body weight increased, except for those who received supplemented human milk. The change in relative body density (g/cc per kg body weight) was also a function of the composition of the protein/energy ratio of the diet. Although all infants appeared to become fatter, those who received the highest protein:energy ratios appeared also to be leaner. Thus, it may be possible to devise diets that enable a growth rate in low birth infants that is similar to intrauterine growth.

The acoustic plethysmograph (Fig 4) is an alternate method that also measures infant body volume (Deskins et al., 1985; Sheng et al., 1987). It makes use of the Helmholtz principle that resonant frequency is inversely proportional to the volume of the resonating chamber; that is, the volume of an object placed inside a chamber of fixed volume can be calculated from the difference in resonant frequencies. The acoustic plethysmograph can be constructed and operated relatively inexpensively and can be used to measure body volume in infants. This technique also suffers from disadvantages similar to those of the air displacement technique in that there is a need to adjust for the residual volume in the lungs and gut. The acoustic method is being used in a pediatric population, and the the four-compartment equation is being evaluated with the inclusion of total body water measurement using stable isotopically labeled water (Sheng, private communication). The effect of small differences in the density estimate of the mineral content and its percentage of the total body weight appeared to introduce only a small error in the estimate of body fat.

FFM Determined by Hydrometry

The determination of FFM from total body water measurements was based on the assumption that the water content of fat-free mass is constant at a value of 73.2%. One can thus obtain a measure of FFM and

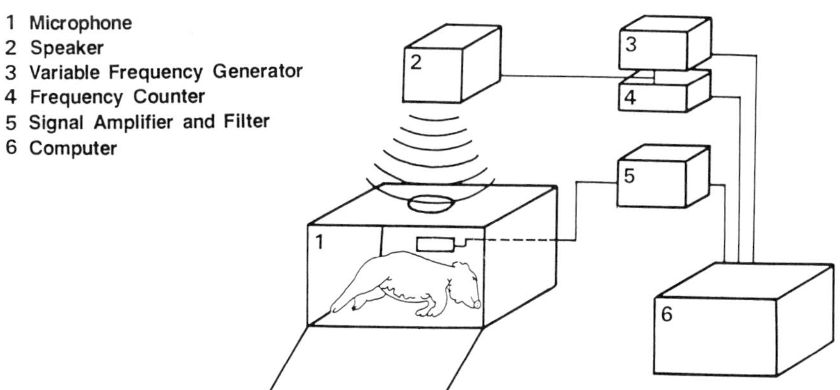

Fig 4.   Schematic diagram of the acoustic plethysmograph technique for infants (Sheng et al., 1987).

FAT from the following:

FFM = TBW/0.732,
FAT = body weight- FFM.

The use of a constant hydration factor for FFM has been questioned. In the derivation of this constant from data for several species of animals, Pace and Rathbun (1945) recognized that a constant water content (73.2%) can be applied only to adult animals. This provision has occasionally been overlooked by some investigators. Even in the mature animals, fatter animals were seen to have a higher FFM water content. For more than a half century, animals have been assumed to reach 'chemical maturity' only when their relative water content stabilizes, and the age at which stabilization occurs is species-dependent (Moulton, 1923). This concept has been challenged by Shields et al.(1963) who reported data in which a constant chemical composition in the fat-free mass of growing pigs was not evident through body weights of 150 kg. Consequently, care must be exercised when the value 73.2% is applied in the young; otherwise, the percentage of true body fat may be underestimated. TBW has been estimated to be approximately 82% of the FFM compartment in the newborn infant, to decrease during infancy, and to reach a concentration similar to that in adults at approximately 3 yr of age (Friis-Hansen, 1961). For this reason, it is difficult to estimate fat or lean tissue mass on the basis of body water alone in the infant.

Another source of error in the estimation of FFM from a body water measurement in the infant is the possible "overestimation" of body water when isotopes of deuterium or tritium are used. In the normal adult, the deuterium or tritium spaces are expected to overestimate true body water volume by not more than 2%. However, increasing evidence suggests that the tritiated water technique may overestimate the true TBW value depending on the nutritional and physiological status of the subject, especially in the case of tissues involved in rapid growth (McManus et al., 1969; Sheng and Huggins, 1979). In an infant whose growth is normal, the degree of TBW overestimation needs further investigation. In turn, this effect may alter the degree by which body fat was underestimated.

FFM Determined by Total Body Potassium

Total body potassium has been measured _in vivo_ by detecting the 1.46 Mev gamma emitted from the body by the decay of the naturally occurring radioisotope ^{40}K. Although only 0.012% of the natural K is radioactive, it can produce a sufficient signal for the accurate and precise measurement of body potassium. The high energy gamma rays emitted from ^{40}K are detected using highly sensitive whole-body counters. In most cases, FFM in adults is estimated from total body potassium (TBK) using a conversion constant of 68.1 mEq K/kg based on the data of four cadaver analyses reported by Forbes and Lewis (1956). Evidence today shows that the potassium content of the body decreases with advancing age at a rate different from that of the total FFM. The TBK/FFM ratio in adults may also differ among populations and ethnic groups (Meneely et al., 1963; Cohn et al., 1984), which could result in the derivation of an incorrect value for body fat or LBM if the possible differences are not considered.

In the newborn, body water comprises a higher percentage of body weight than in the adult. The newborn, therefore, has a lower K concentration per unit body weight than the adult. According to the calculations of Fomon et al. (1982), the K content of LBM is approximately 49 mEq/kg at birth and increases to approximately 63

mEq/kg by age 5. Burmeister and Romahn (1973) reported the TBK content
in infants under 2 yrs of age by ^{40}K counting. They derived a
conversion factor of 92.5 mEq/kg body cell mass for the newborn. Hager
et al. (1977) used a similar concentration value of 90 mEq K/kg for the
BCM in their four-compartment model of body composition in infants.

More recently, Spady et al. (1986) have measured TBK in low birth
weight infants using a whole-body counter consisting of a single 4" x 4"
x 16" NaI detector (Fig 5). Measurements were performed in 50 infants
whose ages were < 75 days and whose weights were between 1.1 and 3.6 kg.
Initial TBK levels ranged from 3 g to 6 g in these infants with a mean
increase of 1.7 mg K/g body weight. Shepherd et al. (1988) also used TBK
measurements to monitor the composition of growth in infants. Their
counter consisted of three 4" x 4" x 16" NaI detectors arranged in a
single plane. They measured changes in the body composition of
exclusively breast-fed infants and in those who received a whey-based
formula. The longitudinal study involved TBK measurements of 82 infants
from birth to age 3 months. The male infants had significantly higher
weights, greater weight gains, and more lean tissue than the female
infants. In both sexes, tissues low in K were accreted more rapidly than
lean tissue. Different rates of weight gain were noted between the
formula-fed and breast-fed infants for the first 10 days after birth;
the breast-fed infants gained more weight. For days 10 to 90, however,
the formula-fed infants gained more weight and had higher deposits of
fat. The authors also observed a higher TBK accretion in males than in
females, and by 3 months, males had a higher TBK/weight ratio than
females.

Total Body Electrical Conductivity and Impedance

Total body electrical conductivity (TOBEC) and bioelectrical
impedance analysis (BIA) have recently re-emerged as techniques to
assess body composition (Cochran et al., 1986; Fiorotto et al., 1987;
Harrison and Van Itallie, 1982; Lukaski et al., 1985; Segal et al.,

INFANT WHOLE BODY COUNTER

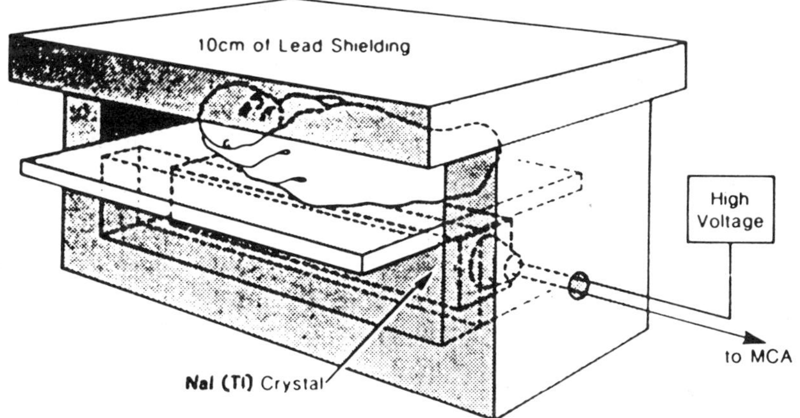

Fig 5.   Schematic diagram of a single crystal 'shadow-shield'
          whole-body counter for the measurement of ^{40}K in infants
          (Spady et al., 1986).

1985). Briefly, these two techniques are based on the principle that
lean tissue is a better conductor of electrical current than fat tissue.
Values reported for FFM in infants obtained by these techniques compare
favorably with those obtained by other indirect methods, such as
anthropometry, and hydrometry (Cochran et al., 1986; Fiorotto et al.,
1987). The TOBEC number has been calibrated with direct carcass
analysis of animals (Fiorotto et al., 1987).

Although results in children and adults show that the values for
FFM obtained by BIA compare favorably with those obtained by other
indirect methods, this technique has not been studied systematically in
the very young infant. The observations look promising and further
investigations in infants may be in order, especially before either BIA
or the TOBEC method can be used as an 'established reference method'
similar to the TBW or TBK techniques (Cohn, 1985). To date, data that
'validate' these techniques are incomplete; additional studies must be
performed.

REGIONAL MEASUREMENTS OF BODY COMPOSITION

Subcutaneous Fat Measurements

The body fat compartment has been derived from regional body
measurements. Skin-fold thicknesses are measured at specific sites of
the body, and comprise the least expensive and most frequently used
method to determine body fat content. Skin-fold calipers are used to
make the measurements. Although numerous studies have attempted to
derive a calculated fat value from subcutaneous fat thickness
measurements, these extrapolations are often inaccurate and misleading.
Some of the limitations may be difficult to overcome, even if separate
formulas are used for estimations of body fat, based on specific
population groups (Lohman, 1981). More recently, infrared interactance
has been proposed as a rapid, safe, and noninvasive method to measure
subcutaneous fat in both research and field settings (Conway et al.,
1984). Another method considered for pediatric use is ultrasonography
(Borkan et al., 1982). The technique, however, requires a nonportable
instrument which would prohibit its use in field studies.

Photon Absorptiometry

There is little information on the composition and growth of bone
in the infant, in part, because an accurate and precise method to
measure bone mineralization in vivo has not been identified. One of the
major concerns in infant nutrition, however, has been bone mineral
retention in the low birth weight infant. Whereas rapid skeletal
development normally occurs during the last trimester for the full-term
infant, this process is interrupted in the preterm infant by its early
birth. Two techniques have been developed to measure bone mineral
content (BMC) in the infant; photon absorptiometry (Gotfredsen et al.,
1986) and ultrasound (Wright, 1987). Single photon absorptiometry (SPA)
is the method most frequently used, although no single bone site has
reflected bone mineralization changes accurately in small infants. Dual
photon absorptiometry (DPA) has also been used to measure total body
bone mineral content in infants born after gestations of 31 to 42 weeks.
On the basis of DPA measurements, skeletal size was estimated to
increase fourfold during the last ten weeks of pregnancy (Gotfredsen,
1987). Infants who were small for gestational age were found to have
smaller mineral and fat masses than those who were appropriate for
gestational age. Not unexpectedly, the bone mineral content of the low
birth weight, preterm infant is low compared with that of the full-term
infant. Whether this deficit has long-term consequences, or whether

aggressive dietary intervention is sufficient to overcome the deficit
has not been determined. Although dietary intervention has been shown
to improve bone mineral content slowly, it is difficult to attain bone
mass similar to that found in age-matched infants or to achieve the
expected intrauterine mineralization rate. This 'catch-up' process may
be lengthy. Steichen et al. (1988) have shown that it took 2 years for
BMC values in the low birth weight group to parallel those of children
born full-term (Fig 6).

The initial application of NMR to body composition analyses also
appeared promising, although major advancements have not been realized.
Estimates of the regional distributions of body fat, for example, can be
obtained from NMR images (Fuller et al., 1985). Recently, Lewis et al.
(1986) reported estimates of TBW in an infant baboon using an NMR
technique. The application of NMR measurements for pediatric use must
undergo further development, because the technique offers the
possibility of body fat determinations independent of the body
composition assumptions required by both the two- or four-compartmental
models. NMR measurements also have the added advantage of using
nonionizing radiation, an important consideration in pediatric studies.

SUMMARY AND FUTURE DIRECTIONS

The impact of diet on the growth and body composition of infants is
an area of future research priority. Such studies are necessary to

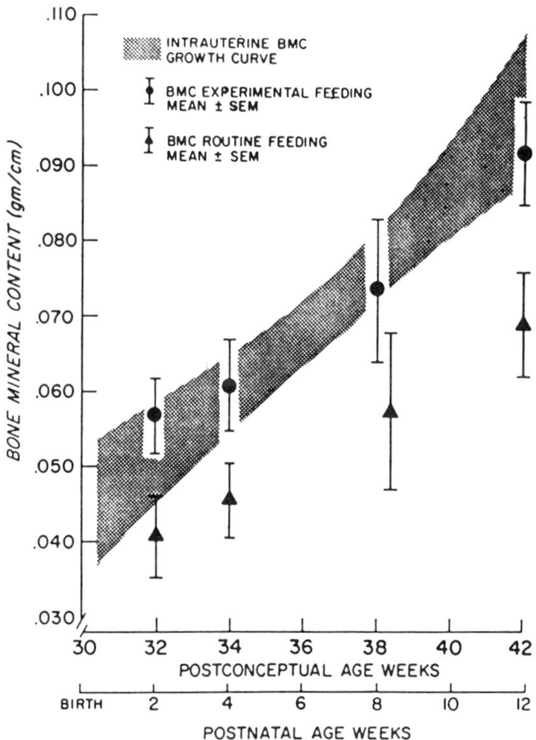

Fig 6.   BMC values for intrauterine growth, preterm infants on routine
          formula, and infants with a Ca and P supplement to the formula
          (data from Steichen et al., 1980).

enable a more precise definition of the food needs and nutritional
status of children. Early growth is known to be sensitive to both
maternal and infant diets and to vary among children. Although indirect
methods of varying degrees of complexity are available for estimates of
lean body mass and body fat, the validation of most of these methods has
been based on other indirect methods. Underwater densitometry is the
reference method used in adults. This technique, however, is not
applicable for infant measurements and is probably not the method of
choice. More rigorous calibrations by direct carcass analysis have been
performed only in a few instances. For pediatric considerations, the
final choice of any method ultimately depends on the research or
clinical objectives, the physical conditions under which the techniques
are to be used, and, the cost. At present, TBK, TBW, and BMC
measurements in the infant appear to be the most practical and
informative. Although there is a possibility of overestimating true
body water using isotopic water, this method may become a major tool in
some studies of infants. Changes in the potassium content of the lean
tissue in early age is also an important consideration. Either of these
effects can result in incorrect estimates of the lean tissue mass
derived from the two-compartment model if only a single conversion
factor is used for all ages. This consideration becomes particularly
significant when body fat determinations are also of interest. The
volume measurement by acoustic plethysmograph and the bioelectrical
techniques of BIA and TOBEC hold promise for pediatric studies, but have
not been evaluated fully in healthy or diseased infant populations.

ACKNOWLEDGMENTS

This work is a publication of the USDA/ARS Children's Nutrition Research
Center, Department of Pediatrics, Baylor College of Medicine and Texas
Children's Hospital, Houston, TX. This project has been funded with
federal funds from the USDA/ARS under Cooperative Agreement
58-7MN1-6-100. The contents of this publication do not necessarily
reflect the views or policies of the US Department of Agriculture, nor
does mention of trade names, commercial products, or organizations imply
endorsement by the US Government. We also thank ER Klein and JD Eastman
for their editorial support in the completion of the manuscript.

REFERENCES

Behnke, A. R., Feen, B. G., and Welham, W. C., 1942, The specific
        gravity of healthy men: Body weight and volume as an index to
        obesity, J. Am. Med. Assoc., 118:495.
Borkan, G. A., Hults, D. E., Cardarelli, J., and Burrows, B. A., 1982,
        Comparison of ultrasound and skinfold measurements in assessment
        of subcutaneous and total fatness, Am. J. Phys. Anthropol.,
        58:307.
Brozek, J., 1963. Body composition, Parts I and II, Ann. N.Y. Acad.
        Sci., 110:1.
Burmeister, W., and Romahn, A., 1973, Potassium content in full-term and
        premature babies: Energetics for the synthesis of body cell mass,
        in: "Current Aspects of Perinatology and Physiology of Children,"
        Linneweh, F., ed. Springer-Verlag, Berlin.
Cochran, W. J., Klish, W.J., Wong, W. W., and Klein, P. D., 1986, Total
        body electrical conductivity used to determine body composition
        in infants, Pediatr. Res., 20:561.
Cohn, S. H., 1985, How valid are bioelectric impedance measurements in
        body composition studies? Am. J. Clin. Nutr, 42:889.
Cohn, S.H., Vaswani, A. N., Yasumura, S., Yuen, K., and Ellis, K.J.,
        1984, Improved models for determination of body fat by in vivo

neutron activation, Am. J. Clin. Nutr., 40:255.

Conway, J. M., Norria, K. H., and Bodwell, C. E., 1984, A new approach
    for the estimation of body composition: Infrared interactance,
    Am. J. Clin. Nutr, 40:1123.

Dell, R., Aksoy, Y., Kashyap, S., Forsythe, M., Ramakrishnan, R.,
    Zucker, C., and Heird, W. C., 1987, Relationship between density
    and body weight in prematurely born infants receiving different
    diets. in: "In Vivo Body Composition Studies," K. J. Ellis, S.
    Yasumura, and W. D. Morgan, eds., The Institute of Physical
    Medicine, London.

Deskins, W. G., Winter, D., Sheng, H.-P., and Garza, C., 1985, Use of a
    resonating cavity to measure body volume, J. Acoust. Soc. Am.,
    77:756.

Fee, B. A. and Weil, W. B., Jr., 1963, Body composition of infants of
    diabetic mothers by direct analysis, Ann. N.Y. Acad.,
    Sci,110:869.

Fiorotto, M. L., Cochran, W. J., Funk, R. C., Sheng, H.-P., and Klish,
    W. J., 1987, Total body electrical conductivity measurements:
    Effects of body composition and geometry, Am. J. Physiol.,
    252:R794.

Fomon, S. J., Haschke, F., Ziegler, E. E., and Nelson, S. E., 1982, Body
    composition of reference children from birth to age 10 years, Am.
    J. Clin. Nutr., 35:1169.

Forbes, G. B., and Lewis, A., 1956, Total body sodium, potassium, and
    chloride in adult man, J. Clin. Invest., 35:596.

Friis-Hansen, B., 1961, Body water compartments in children, Pediatrics
    28:169.

Fuller, M. F., Foster, M. A., and Hutchison, J. M. S., 1985, Estimation
    of body fat by nuclear magnetic resonance imaging, Proc. Nutr.
    Soc., 44:108A.

Garn, S. M., and Nolan, P., Jr., 1963, A tank to measure body volume by
    water displacement (BOVOTA), Ann. N.Y. Acad. Sci., 110:91.

Gnaedinger, R. H., Reineke, E. P., Pearson, A. M., Van Huss, W. D.,
    Wessel, J. A., and Montoye, H. J., 1963, Determination of body
    density by air displacement, helium dilution, and underwater
    weighing, Ann. N.Y. Acad. Sci., 110:96.

Gotfredsen, A., Jensen, J., Borg, J., and Christiansen, C., 1986,
    Measurement of lean body mass and total body fat using dual
    photon absorptiometry, Metabolism, 35:38.

Hager A., Sjostrom L., Arividsson B., Bjorntorp P., and Smith V., 1977,
    Body fat and adipose tissue cellularity in infants: a
    longitudinal study, Metabolism, 26:607.

Harrison, G. G., and Van Itallie, T. B., 1982, Estimation of body
    composition: A new approach based on electromagnetic principles,
    Am. J. Clin. Nutr. 35:1176.

Kehayias, J.J., Ellis, K.J., Cohn, S.H., Yasumura, S., and Weinlein,
    J.H., 1987, Use of pulsed neutron generator for in vivo
    measurement of body carbon, in: "In vivo body composition
    studies," Ellis KJ, Yasumura S, and Morgan WD, eds., Institute of
    Physical Sciences in Medicine, London.

Knight, G. S., Beddoe, A. H., Streat, S. J., and Hill, G. L., 1986, Body
    composition of two human cadavers by neutron activation and
    chemical analysis, Am. J. Physiol., 250:E179.

Lewis, D. S., Rollwitz, W. L., Bertrand, H. A., and Masoro, E. J., 1986,
    Use of NMR for measurement of total body water and estimation of
    body fat, J. Appl. Physiol., 60:836.

Lohman, T. G. 1981, Skin folds and body density and their relation to
    body fatness: A review, Human Biol., 53:181.

Lukaski, H. C., Johnson, P. E., Bolonchuk, W. W., and Lykken, G. I.,
    1985, Assessment of fat-free mass using bioelectrical impedance
    measurements of the human body, Am. J. Clin. Nutr., 41:810.

McManus, W. R., Prichard, R. K., Baker, C., and Petruchenia, M. V., 1969, Estimation of water content by tritium dilution of animals subjected to rapid live weight changes, J. Agric. Sci. Cambridge, 72:31.

Meneely, G. R., Heyssel, R. M., Ball, C. O. T., Weiland, R. L., Lorimer, A. R., Constantinides, C., and Meneely, E. U., 1963, Analysis of factors affecting body composition determined from potassium content in 915 normal subjects, Ann. N.Y. Acad. Sci., 110:271.

Moulton, C. R., 1923, Age and chemical development in mammals, J. Biol. Chem., 57:79.

Pace, N., and Rathbun, E. N., 1945, Studies on body composition, III. Water and chemically contained nitrogen content in relation to fat content, J. Biol. Chem., 158:685.

Pearson, A. M., Purchas, R. W., and Reineke, E. P., 1968, Theory and potential usefulness of body density as a predictor of body composition. in: "Body Composition in Animals and Man: Proceedings of a Symposium," National Academy of Sciences, Washington, DC.

Rudd, T.G., Pailthorp, K.G., Nelp. W.B., 1972, Measurement of nonexchangeable sodium in normal man, J. Lab. Clin. Med., 80:442.

Segal, K. R., Gutin, B., Presta, E., Wang, J., and Van Itallie, T. B., 1985, Estimation of human body composition by electrical impedance methods: A comparative study, J. Appl. Physiol., 58:1565.

Sheng H.-P., Dang T., Adolph A. L. Schanler R. J., and Garza C., 1987, Infant Body Volume measurements by acoustic plethysmography, in: "In Vivo Body Composition Studies, K. J. Ellis, S. Yasumura, and W. D. Morgan, eds. The Institute of Physical Medicine, London.

Sheng, H.-P., and Huggins, R. A., 1979, A review of body composition studies with emphasis on total body water and fat, Am. J. Clin. Nutr., 32:630.

Shepherd, R. W., Oxborough, D. B., Holt, T. L., Thomas, B. J., and Thong, Y. H., 1988, Longitudinal study of the body composition of weight gain in exclusively breast-fed and intake-measured whey-based formula-fed infants to age 3 months, J. Pediatr. Gastroenterol. Nutr., 7:732.

Shields, R. G., Jr., Mahan, D. C., and Graham, P. L., 1963, Changes in swine body composition from birth to 145 kg, J. Anim. Sci., 57:43.

Spady, D. W., Filipow, L. J. Overton, T. R., and Szymanski, W. A., 1986, Measurement of total body potassium in premature infants by means of a whole-body counter, J. Pediatr. Gastroenterol. Nutr., 5:750.

Steichen, J. J., Gratton T. L., Tsang R. C., 1980, Osteopenia of prematurity: The cause and possible treatment, J. Pediatr., 96:528.

Steichen, J. J., Steichen-Asch, P. A., Tsang, R. C., 1988, Bone mineral content measurement in small infants by single-photon absorptiometry: current methodologic issues, J. Pediatr., 113:181

Taylor, A., Aksoy, Y., Scopest, J. W., du Mont, G., and Taylor, B. A., 1985, Development of an air displacement method for whole body volume measurement of infants, J. Biomed. Eng., 7:9.

Widdowson, E. M., and Dickerson, J. W. T., 1964, Chemical composition of the body, in: "Mineral Metabolism," Comar, C. E. and Bronner, F., eds. Academic Press, New York.

Wright, L.L., Glade, M.J., Gopal, J., 1987, The use of transmission ultrasonics to assess bone status in the human newborn, Pediatr. Res., 22:541.

Ziegler, E. E., O'Donnell, A. M., Nelson, S. E., and Fomon, S. J., 1976, Body composition of the reference fetus, Growth, 40:329.

BODY COMPOSITION IN CYSTIC FIBROSIS

K.J. Gaskin[*], D.L.M. Waters[*], V.L. Soutter[*], L. Baur[*],
B.J. Allen[+], N. Blagojevic[+], D. Parsons[+]

[*] The Children's Hospital, Camperdown, Sydney, Australia
[+] A.N.S.T.O., Lucas Heights, Sydney, Australia

INTRODUCTION

Over the last two decades there has been an increasing interest in the nutritional problems of patients with cystic fibrosis (CF). This situation has arisen from reports (Yassa et al., 1978; Soutter et al., 1986) indicating that stunting and wasting are common, and contrasting reports from the Hospital for Sick Children, Toronto, cystic fibrosis clinic (Corey, 1980; Corey et al., 1984), demonstrating that their patients conform to the normal distribution for height and, to a lesser extent, weight percentiles. Moreover, the maintenance of normal nutritional status and growth may have considerably enhanced the prognosis of patients at the Toronto clinic (Corey, 1980) in comparison to other clinics. Consequently, a number of studies have been performed to define the origin of CF malnutrition, its deleterious effect on body composition, growth and lung function, and the possible benefits of nutritional rehabilitation programs. Dietary intake analyses have revealed that most CF patients consume large amounts of protein similar to controls (Soutter et al., 1986). In contrast, due to the almost universal policy of the provision of a low fat diet, a persistent energy deficit has occurred (Roy et al., 1984). The latter has been compounded by the effects of anorexia due to lung, gut and liver disease and drug administration, and possibly excessive energy expenditure related to lung disease, recurrent lung infections and the underlying disease process.

Regarding the possible deleterious effects of the energy deficit in CF on body composition, studies have focused on defining body fat content, lean body mass and muscle mass (Miller et al., 1982; Shepherd et al.,1983; Pencharz et al., 1984). The interest in defining muscle mass has been a special focus as, obviously, a deficit of this compartment, specifically the respiratory muscles, could contribute to a deterioration in lung disease. A study of underweight adolescent CF patients with normal linear growth demonstrated a reduction in body fat mass, but normal total body nitrogen (TBN) content (Pencharz et al., 1984). In a more chronically undernourished group of patients with evidence of stunting and wasting, deficits of body fat mass, lean body mass, total body potassium (TBK) and muscle mass were demonstrated (Miller et al., 1982; Shepherd et al., 1983). However, the numbers in these studies were small, and at least in the latter two, only age and sex matched control subjects were used, whereas controls matched for size (height) may well have demonstrated less significant deficits in lean body and muscle masses.

Larger studies are thus required to determine the effect of malnutrition on lean body and muscle mass to confirm the above results. More precise studies are also required to assess the effect of malnutrition on protein status, and if a deficit exists, can this be reversed with aggressive nutritional intervention? The latter also raises the question of how best to assess protein nutritional status in those patients. Although total body potassium measurements had been used by two Units to assess lean body mass during either malnutrition or nutritional rehabilitation, is this an appropriate investigation considering the underlying electrolyte secretion abnormalities in CF or the electrolyte disturbances in patients with malnutrition? With the advent of a body composition unit at A.N.S.T.O. at Lucas Heights in Sydney, we have commenced an in-depth investigation to address these issues. Our results on protein deposition during nutritional rehabilitation are presented later in this book. The current study will focus on the abnormalities of body composition which we have found amongst our CF population.

METHODS

Patients

CF patients and their parents were recruited for this study from the CF Clinic at The Children's Hospital, Sydney. Malnutrition was defined as height or weight for age less than the third centile, less than 90% of expected weight for height, or with a measured linear growth velocity of less than 4.0cm/year.

Normal subjects with no family history of CF were recruited from a normal school, from friends of the patients, or from the Hospital staff.

Anthropometry

Height was measured with a Harpenden stadiometer, weight with an electronic scale, and four skinfolds (triceps, biceps, subscapular and suprailiac) with Harpenden skinfold calipers. Body fat was estimated for prepubertal children from Brook (1971), and in pubertal subjects from Durnin and Rahaman (1967). Fat free mass (FFM) was estimated from the equation: FFM = Body Mass minus Body Fat Mass.

24-Hour Urinary Creatinine Excretion

Study subjects were prescribed a flesh free diet for four days. During the fourth day a timed 24-hour urine collection was performed. Urinary creatinine was determined by automated analysis using the classic Jaffe reaction and muscle mass calculated from the equation: 1 gm of creatinine excreted per 24 hours is derived from 18.6 kg of muscle according to Picou et al., (1976).

Total Body Nitrogen (TBN)

TBN was measured at Lucas Heights by in vivo prompt gamma neutron activation using a ^{252}Cf neutron source, as described previously (Allen et al., 1987).

Total Body Potassium (TBK)

TBK was determined from the measurement of ^{40}K in a whole body counter at Lucas Heights. Measurements are made over 40 minutes. The facility has been standardized and calibrated against child-sized phantoms with known potassium content.

Table 1. Weight, Fat Content, FFM and Muscle Mass in CF Patients and
         Controls

n	CFMal 37	N1 37	CFWN 18	N2 18
Height (cms)	136.8 ±16.6	135.4 ±13.2	142.4 ±15.4	138.9 ±10.9
Weight (kgs)	27.59 f ±8.45	32.46 ±8.51	36.58 ±11.02	34.09 ±7.25
Fat (kgs)	3.84 § ±2.13	8.17 ±4.32	7.92 ±3.86	8.45 ±3.94
FFM (kgs)	23.87 ±6.01	24.07 ±4.92	28.22 ±7.55	25.07 ±4.49
Muscle Mass (kgs)	8.83 * ±4.26	10.75 ±4.86	10.69 ±3.90	11.22 ±2.88

* p <0.01, f p <0.002,   § p <0.001

All participating subjects or their parents gave written informed
consent for the study which was approved by The Children's Hospital Ethics
Committee. Results are presented as mean ± S.D.

RESULTS

<u>Assessment of Fat Free Mass (FFM) and Muscle Mass in CF Patients and
Controls</u>

Average weight, fat mass, FFM and muscle mass were determined in 37
malnourished CF patients (CFMal) and compared with 37 height and sex
matched healthy normal children (N1). The same parameters were determined
in 18 well-nourished CF patients (CFWN) and compared with 18 height and sex
matched healthy controls (N2). The results are summarized in Table 1. CFMal
weighed significantly less than their controls (p (<0.002) and had
significantly less body fat (p <0.001) and muscle mass (p <0.01). Total FFM
was similar, however, and non-muscle fat free mass significantly increased
in CFMal (p <0.02), as shown in Fig. 1. Average weight, FFM, muscle mass
and fat content were not significantly different in CFWN compared with
their controls, as shown in Table 1. However, the average FFM was
marginally greater and non-muscle FFM again significantly greater (p <0.01)
in the CFWN (cf Fig. 1).

<u>TBK and TBN Values</u>

TBK and TBN measurements were performed consecutively in each of 31
malnourished CF patients, 9 well-nourished CF patients and 8 normal
controls. There were no significant differences in age or sex of the
subjects studied. Nearly all patients were prepubertal, and as there were
no significant differences in TBK or TBN between sexes in these three
groups, the values have been analyzed collectively. The preliminary results
are summarized in Table 2. There was no significant difference in age
between the three groups, although there was a trend for the CFWN to be
younger and the normal controls older. Average TBK values did not differ
significantly between the three groups. However, in relation to age
individual TBK estimations in CFWN were, in general, in excess of those

Table 2. TBK, TBN and K/N Ratios in CF Patients and Controls

n	CFMal 31	CFWN 9	Controls 8
Age (yrs)	12.19 ±3.58	11.29 ±2.68	13.02 ±2.58
TBK (mmoles)	1780 ±473	1909 ±532	1927 ±328
TBN (gms)	884 ±315	1029 ±367	1249 * ±254
K/N	2.06 ±0.34	1.92 ±0.22	1.57 f ±0.16

* Value significantly greater p <001 than CFMal
f " "   less   p <0.001 than CFMal or CFWN

observed in CFMal, which were surprisingly similar in distribution to those of the normal controls. The average TBN was significantly greater in the normal controls compared to CFMal (p <0.01), but not different from CFWN. K/N was significantly elevated in both the CF groups vs normal controls (p <0.001).

BODY FAT, FAT FREE MASS AND MUSCLE MASS IN CF AND CONTROL PATIENTS

The elevated K/N ratios observed in the CF patients could reflect either an increase in TBK, a decrease in TBN, or both. A total protein deficit,

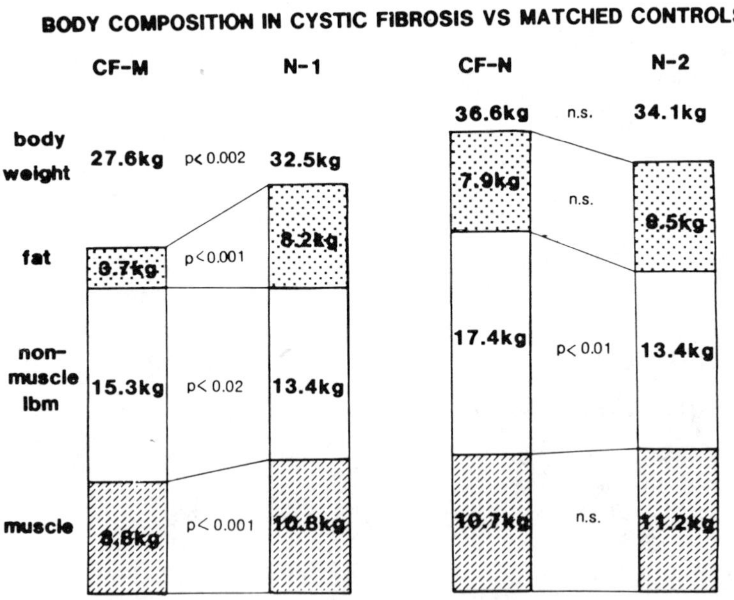

Fig. 1. CFMal and CFWN demonstrated a significant increase in non-muscle FFM vs their controls (p <0.02 and p <0.01, respectively).

for instance, could explain the elevated K/N ratio in the CFMal group.
However, a protein deficit does not explain the elevated ratio in CFWN
whose age-related nitrogen values were similar in distribution to the
normal controls and in excess of those observed in CFMal. Furthermore, if a
protein deficit was the key factor, then protein repletion should be
associated with normalization of the ratio. In this regard, 21 patients had
repeat TBK, TBN measurements at an average 8 months after the commencement
of a nocturnal nutritional supplementation programme, and 14 of these
patients had a third measurement after a further 12 months of supplementation. As shown in Table 3, the average K/N improved significantly with
time, and a number of values were within the normal range. However, even
after nearly 2 years of nutritional rehabilitation the rates were still
elevated compared to controls.

To investigate further the possibility that the elevated ratio was
related to an increase in TBK, results from a group of obligate
heterozygotes, i.e. 15 mothers and 10 fathers of children with proven CF,
were compared with those of 13 comparably aged normal females and 10 males,
respectively. The results are shown in Table 4. The TBK and TBN results of
the heterozygotes were not significantly different from their normal male
or female counterparts. However, CF mothers had a significantly greater K/N
ratio than normal females ($p < 0.001$). The CF fathers also had a higher
ratio, but the difference with normal males did not reach significance at
the .05 level.

Table 3. K/N Ratios in CFMal During Nutritional Rehabilitation

Time (months)	0	8	20
K (mmoles)	1771 ±443	1858 ±477	1905 ±547
N (g)	912 ±311	1057 ±314	1121 ±366
K/N	2.07 ±0.37	1.83 ±0.21	1.75 * ±0.19

* $p < 0.05$ when compared with controls
  $p < 0.001$ when compared with Time 0

DISCUSSION

The current study demonstrates the deleterious effect of malnutrition on
body composition in CF patients. In a large group of malnourished patients,
they had significant deficits in body weight, fat and muscle mass in
comparison with height and sex matched normal controls. These results
confirm the previous findings in smaller studies with age and sex matched
controls (Miller et al., 1982; Shepherd et al., 1983; Pencharz et al.,
1984) and indicate that the previously demonstrated reduction in muscle
mass, as measured by creatinine excretion, was not simply related to the
shorter stature of the malnourished patients participating in these
studies. The results both in the present study and the previous ones are,
however, based on average group data of non-invasive measurements with a
degree of overlap between the groups. Considering the latter, and the

Table 4. TBK, TBN and K/N Ratios in Obligate CF Heterozygotes and Controls

| | Males | | Females | |
	CF	Control	CF	Control
TBK (mmoles)	3671 ±287	3869 ±360	2609 ±351	2463 ±298
TBN (g)	2220 ±214	2422 ±309	1556 ±193	1674 ±250
K/N	1.71 ±0.11	1.61 ±0.13	1.68 ±0.08	1.48 * ±0.11

* significantly less than CF p <0.001

inherent variability of the techniques involved, it is difficult to use any of these tests to precisely define nutritional status in the individual patient.

To overcome these problems TBN and TBK measurements were used in the study to define precisely the effect of malnutrition on the body composition of CF patients. TBN values in malnourished patients were clearly lower than in well-nourished subjects, with virtually no overlap of age-related values. Although, as yet, these results are preliminary, with a relatively small group of well-nourished subjects they do serve to emphasize that the malnourished subjects were protein deficient. The results also indicate that with a larger number of control values the precise deficit in protein will be able to be quantified, thus producing a clinical research tool that will potentially have widespread application to the investigation and management of malnutrition in children. In contrast, TBK estimations appeared less reliable in precisely predicting the patient's nutritional status as there was a considerable overlap between the malnourished CF patients and normal controls. The results were, in fact, similar to the FFM values which were nearly identical in both the malnourished CF patients and their height matched controls. In malnutrition one would have anticipated finding depleted TBK and FFM. That the latter did not occur suggests that there is an anomaly of body composition in this group of malnourished patients. The present results, however, also suggest that there is a similar anomaly in well-nourished CF patients who appear to have a mild expansion of their FFM compartment and elevation of their TBK values in relation to age. Further evidence supporting the presence of abnormal body composition in CF was provided by the significant elevation of K/N ratios in CF patients (either well or malnourished) and CF heterozygotes compared to their respective controls. From these preliminary results there thus appears to be a generalized anomaly of body composition in CF patients related to both a variable increase in potassium and a variable deficit in protein.

The present study does not provide an explanation for these findings. The results as yet require confirmation with larger groups of well-nourished CF patients, controls and heterozygotes. If confirmed, the abnormal body composition of CF patients may relate to the underlying electrolyte transport problems. Currently, in secretory tissues this is defined as a defect in chloride secretion (Cook and Young, 1988) which may be accompanied by intracellular potassium retention. Whatever the ultimate explanation is, the present study suggests that TBK measurements in malnourished CF patients, to determine nutritional status, specifically the presence of proteins deficits, or to guide protein repletion during nutritional supplementation, should be interpreted with caution. Further

investigation of TBK measurements are, however, warranted to define the
anomaly as this may help to unravel the pathophysiology of this disease.

Overall, besides the recognition of an abnormal composition in CF, the
present study emphasizes that malnutrition can cause a variable and
sometimes severe depletion in body fat, muscle and protein. The
significance of these findings remains to be determined, and in particular,
further work is required to assess their effect on lung function and
survival, and whether correction of these deficits will alter this
prognosis.

REFERENCES

Allen, B. J., Blagojevic, N., Parsons, D., Gaskin, K. J., Soutter, V. L.,
     Waters, D. L. M., Allman, M., Stewart, P., and Tiller, D., 1987, _In
     vivo_ determination of protein in malnourished patients, _In_: "_In Vivo_
     Body Composition Studies", K. J. Ellis, S. Yasumura, and W. D.
     Morgan, eds., The Institute of Physical Sciences in Medicine,
     London.
Brook, C. G. D., 1971, Determination of body composition of children from
     skinfold measurements, _Arch. Dis. Child._, 46:182.
Cook, D. I., and Young, J. A., 1988, Cystic Fibrosis: A disorder of
     biological signal transduction, _J. Gastro. and Hepatol._, 3:645.
Corey, M., 1980, Longitudinal studies in cystic fibrosis, _In_: "Perspectives
     in Cystic Fibrosis", J. M. Sturgess, ed., Canadian Cystic Fibrosis
     Foundation, Toronto.
Corey, M., Gaskin, K. J., Durie, P. R., Levison, H., and Forstner, G. G.,
     1984, Improved prognosis in CF patients with normal fat absorption,
     J. Pediatr. Gastro. and Nutr., 3:S, 99.
Durnin, J. B. G. A., and Rahaman, M. M., 1967, The measurement of the
     amount of fat in the human body from measurements of skinfold
     thickness, _Br. J. Nutr._, 21:681.
Miller, M., Ward, L., Thomas, B. J., Cooksley, W.G.E., and Shepherd, R. W.,
     1982, Altered body composition and muscle protein degradation in
     children with cystic fibrosis, _Amer. J. Clin. Nutr._, 2:439.
Pencharz, P., Hill, R., Archibald, E., Levy, L., and Newth, C., 1984,
     Energy needs and nutritional rehabilitation in undernourished
     adolescents and young adult patients with cystic fibrosis, _J.
     Pediatr. Gastro. and Nutr._, 3:S, 147.
Picou, D., Reed, P. J., Jackson, A., and Poulter, N., 1976, The measurement
     of muscle mass in children using ISN creatine, _Pediatr. Res._,
     10:184.
Roy, C. C., Darling, P., and Weber, A. M., 1984, A rational approach to
     meeting macro- and micronutrient needs in cystic fibrosis, _J.
     Pediatr. Gastro. and Nutr._, 3:S, 154.
Shepherd, R. W., Thomas, B. J., Bennett, D., Cooksley, W. G. E., and
     Ward, L. C., 1983, Changes in body composition and muscle protein
     degradation during nutritional supplementation in nutritionally
     growth retarded children with cystic fibrosis, _J. Pediatr. Gastro.
     and Nutr._, 2:439.
Soutter, V. L., Kristidis, P., Gruca, M. A., and Gaskin, K. J., 1986,
     Chronic undernutrition/growth retardation in cystic fibrosis,
     Clinics in Gastroenterology, 15:137.
Yassa, J. G., Prosser, R., and Dodge, J.A., 1978, Effects of an artificial
     diet on growth of patients with cystic fibrosis, _Arch. Dis. Child._,
     53:777.

BODY COMPOSITION STUDIES IN CYSTIC FIBROSIS AND MYELOMENINGOCELE

B.J. Thomas,[1] R.W. Shepherd,[2,3] T.L. Holt,[3] K. Shepherd,[2]
R. Greer,[3] and G.J. Cleghorn[2,3]

[1]Department of Physics, Queensland University of Technology,
GPO Box 2434, [2]Department of Gastroenterology and Nutrition,
Royal Children's Hospital, [3]Department of Child Health,
Brisbane 4001 Australia

INTRODUCTION

One of the major considerations of improving the management of
many chronic diseases has been the realisation of the importance of
nutrition, and, in children, the maintenance of normal growth. Cystic
Fibrosis (CF) and Myelomeningocele (MMC) are two such disease states in
which nutritional status has a significant effect on morbidity.

With improving mean age of survival, in both disease states, the
importance of ensuring normal nutrition and growth from an early age and
recognising at-risk groups or individuals has become very important
(Shepherd et al., 1984; Lloyd, 1979). Nutritional problems are common
in CF subjects and may affect growth and the course of pulmonary disease
(Shepherd et al., 1986). In CF, malnutrition is a major factor
adversely affecting survival. In MMC, the occurrence of obesity, which
may result in considerable morbidity, has been suggested to be related
to relative immobility and intake of nutrients excessive for the
individual's needs (Guy, 1978). However, there has been little research
on body composition during growth of MMC subjects, and only limited
research on factors determining the occurrence of obesity.

This paper reports the results of studies of body composition of
CF and MMF subjects from infancy to adulthood. (CF study: n = 161, 1
month - 17 year. MMC study : n = 59, 4 month - 29 year.) The study of
CF subjects investigated the potential role of total body potassium
(TBK) measurements in the management of this disease and, in particular,
in the identification of at-risk individuals and in the assessment of
response to nutritional therapy. The study of MMC subjects investigated
the nature of altered body composition in relation to growth, age, sex,
level of lesion, and the degree of mobility.

TECHNIQUES

We used several techniques to assess nutritional status, including
clinical and anthropometric measurements, biochemical assessment of, for
example, protein metabolism, and determination of body composition.
Only the results of body composition studies are reported here.

Total body potassium (TBK) is used as a measure of body cell
mass - the vital work-performing and, in children, growing body
compartment. TBK is measured with a shadow-shield whole-body monitor.

The whole-body monitor incorporates three 40 cm x 10 cm x 10 cm NaI(T1) scintillation detectors in a shielded environment, provided by 10 cm thick selected steel. Subjects to a maximum height of 60 cm, are measured while stationary beneath the detectors. Subjects whose height exceeds 60 cm are 'scanned' on a moving bed which passes beneath the detectors. The measurement typically takes 40 minutes, which results in a standard deviation of 3-8% depending primarily on the amount of TBK in the subject.

Normal TBK values for age and sex were established by pooling results obtained in our laboratory (n = 145) with data published by other groups. [Novak, 1973, (n = 63); Pierson et al, 1974, (n = 1420)]. We also used the functional relationship

$$TBK = a . \quad W^{1/3}H^2 \quad ....(1)$$

to predict TBK in relation to weight (W) and height (H), (a is an empirically determined constant). Similar predictive formulae were suggested previously by for example, Mann et al., (1974).

Anthropometric measurements of body mass and length, were also used in this work. Total body water, determined by (deuterium) isotope dilution was measured for the MMC group.

RESULTS

The Cystic Fibrosis (CF) Study

As shown in Fig. 1, the CF population of the Brisbane Clinic had

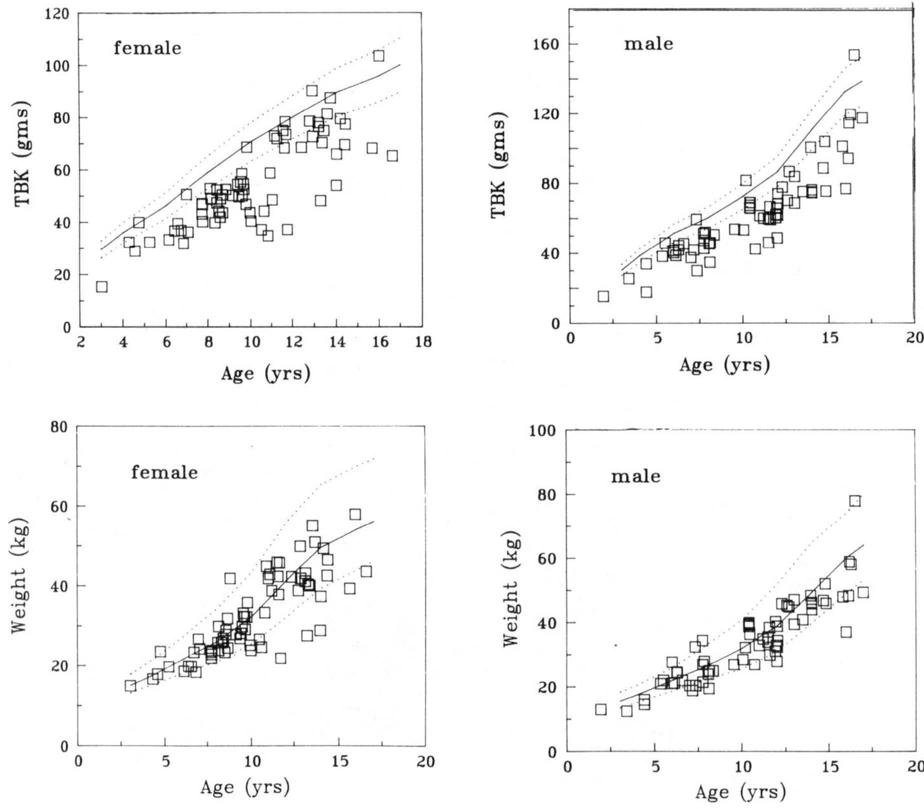

Fig. 1   Total body potassium and weight for CF males and
          females from 3-17 years. The solid lines are
          the normal values of TBK or weight expected by
          age and sex. The dotted lines are ± 1 SD of
          these normal values.

significantly lower TBK compared with that expected for age and sex,
with approximately one-third of the subjects having TBK below two
standard deviations from the mean normal TBK for age and sex. These TBK
values tended to be relatively lower with increasing age. However, TBK
for the group in the age range 3-17 years, did not differ significantly
from values predicted for their weight and height using formula (1)
although, in some individuals, TBK was markedly lower than expected.
Similarly, the weight for age of the CF group did not differ
significantly from expected values, although in individual cases marked
variations were observed (Fig. 1).

CF infants were found to be depleted in TBK at 1 month of age.
However, the infants responded to nutritional therapy and by 2 years had
TBK in the normal range. Long-term nutritional therapy also was shown
to increase the TBK of a group of nine malnourished adolescent CF
subjects. In these subjects, the mean TBK increased from 85% of the
value predicted by weight and height to 98% after 6-12 months of
nutritional therapy.

The therapy (intra-gastric feeding) supplied a minimum of 120% of
the recommended daily allowance of protein and energy. Most subjects
showed a corresponding improvement in clinical condition.

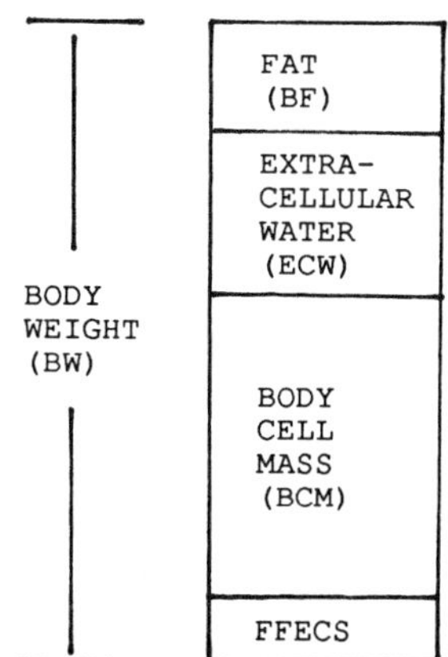

Calculations

$$BF = BW - (FFECS+BCM+ECW)$$
$$ECW = TBW - ICW$$
$$ICW = 0.75 \times BCM$$
$$BCM = 8.33 \times TBK/1000$$
$$FFECS = 0.12 \times BW_{norm}$$

TBW = Total Body Water
ICW = Intracellular Water
FFECS = Fat-Free Extracellular Solids
BW_{norm} is the normal body weight for height

Fig. 2.   The four-compartment model of body composition as derived
           from measured body weight, total body water, and total body
           potassium (Bruce et al., 1989).

The Myelomeningocele (MMC) Study

A four-compartment model of body composition as described by Bruce
et al (1989) was used in this study (Fig.2). The four compartments are
body fat, extracellular water, body cell mass (BCM), and fat-free
extracellular solids. These compartments are derived from measured TBK,
total body water, and body weight, and the calculation assumes that the
constants (eg. that intracellular water is 75% of BCM) are indeed
constants even in these conditions.

Body composition of the MMC group (Fig. 3), as indicated by this four-compartment model, was relatively normal in the first 4 years of life, but after this there was progressive relative depletion of body cell mass (TBK) and total body water, with relative excess of extracellular water and adipose deposition. These changes correlated with relatively slower weight gain and growth with age as compared with reference values.

Fig. 4 shows that this increasing relative depletion of body cell mass (TBK) with age also becomes more pronounced relative to body weight. However, again up to 4 years of age, total body potassium relative to weight was usually within the normal range for both males and females. A similar pattern was observed with the changes in total body water. Although total body water was relatively depleted with increasing age compared with normal data, there was a relative excess of extracellular water as a percentage of body weight.

Total body fat expressed as a percentage of body weight was significantly greater than that expected for age and sex in higher lesions, but not significantly greater in low lesions (Fig. 5). When the percentage body fat was plotted against a method of ambulation (Fig. 6), only those in the non-ambulatory group, that is, wheelchair bound, had significantly greater percent body fat than expected for age and sex. Other groups who were more ambulatory, whether they walked with or without aids, or were pre-ambulatory, did not show this significant difference.

DISCUSSION

Body composition measurements are of value in the clinical evaluation of subjects with diseases affecting nutrition. The measurement of TBK, a non-invasive technique, and body water by (deuterium) isotope dilution are well-suited to assessing children with chronic disease. However, the efficacy of the information obtained needs to be considered for particular disease states to obtain maximum clinical value from a measurement. A knowledge of normal TBK for age and sex, as well as relative to weight and height, is necessary for such studies.

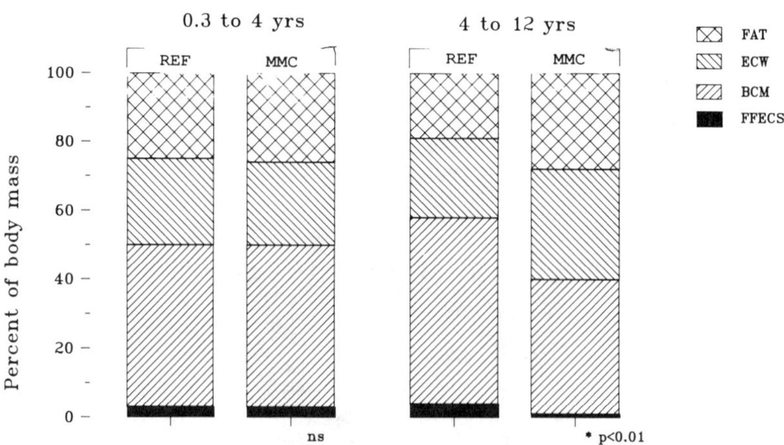

Fig. 3.   Body Composition (% body mass) vs Age myelomeningocele (MMC) vs Reference (REF) for age/sex.

The present study has considered the application of body composition techniques in CF and MMC. In summary, this study has shown that the functional cell mass of young CF subjects is often limited in its growth and that nutritional support needs to be assessed in many children even from quite a young age. The results from the nutritional therapy group suggest that this deficit of functional cell mass is potentially correctable. As a result of this study we would also see merits in measuring other body compartments in CF, such as body fat and water, as this would give a clearer picture of the whole nutritional abnormality.

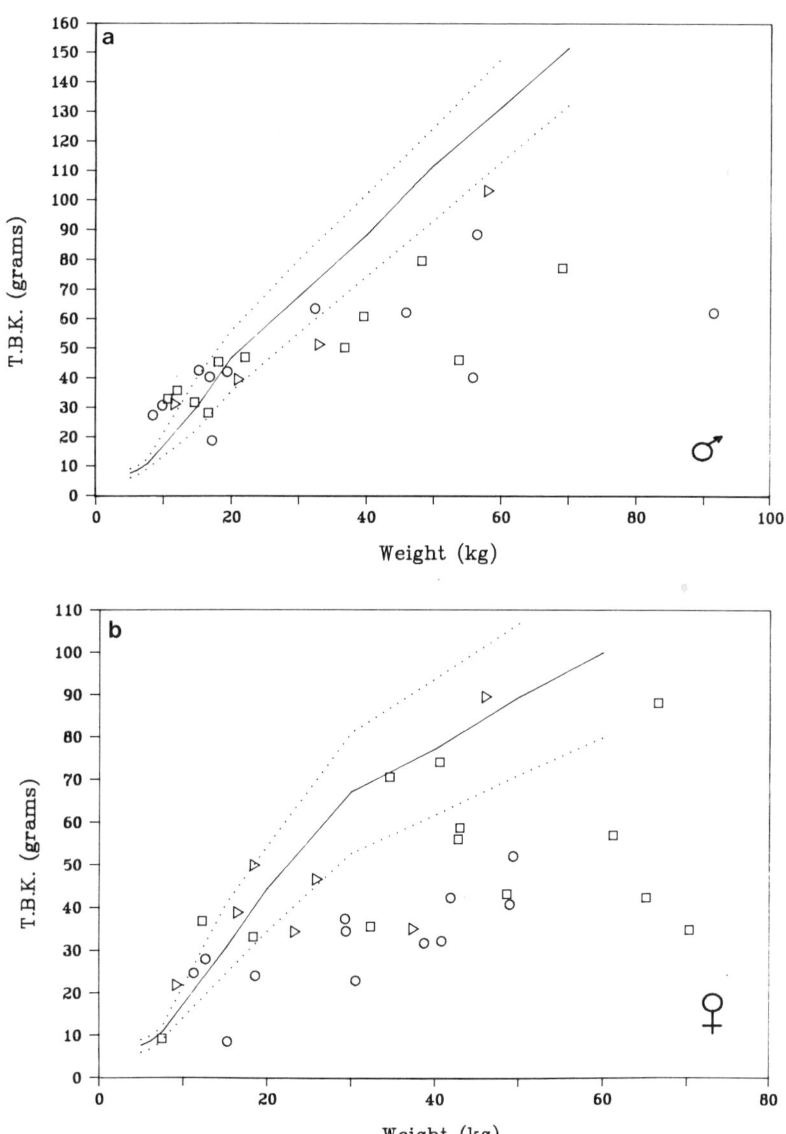

Fig. 4   Total body potassium of MMC group (a) females and (b) males plotted against weight compared with normal data (mean; solid line ± 1 SD; dotted lines) o high lesions, ▯ mid lesions, Δ low lesions.

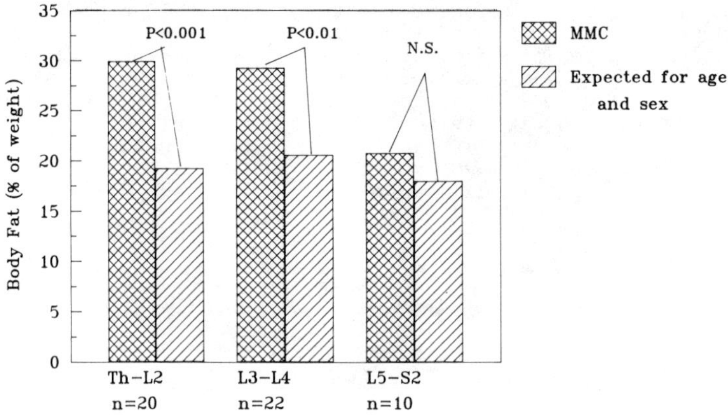

Fig. 5   Total body fat as a % of body weight
           vs the level of lesion. (T1-L2 =
           high lesion, L3-L4 = medium lesion,
           L5-S2 = low lesion).

These investigations show that in MMC body composition can be relatively normal in younger children. However, beyond the age of three or four years, myelomeningocele is increasingly associated with relative depletion of body cell mass and total body water, and there is maldistribution of body water with an expansion of extracellular water and an excess of adipose tissue. These changes, particularly those of the excess adipose tissue, were more pronounced in those with high lesions and less pronounced in those who remained ambulatory.

We speculate that the maldistribution of body water and depletion of body cell and lean mass may result in important metabolic and nutritional maladaptation during stress. This finding needs further evaluation in the context of the frequent surgical procedures necessary in MMC subjects. Finally, the progression of change with age and decreasing ambulatory activity suggests that nutritional intervention and mobility programmes should be started early and that such intervention warrants further study.

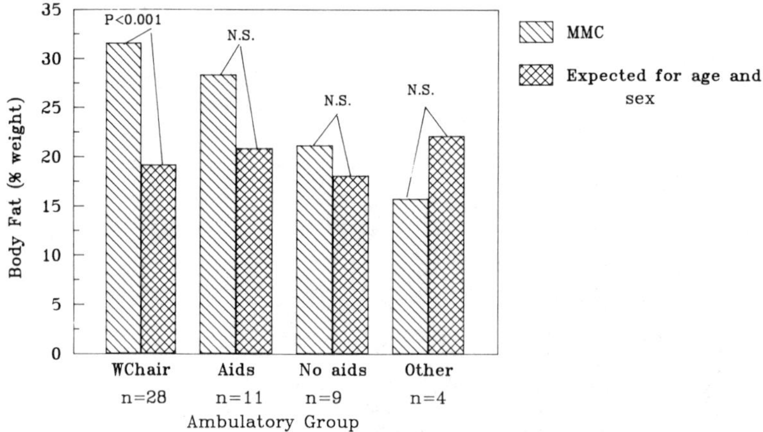

Fig. 6   Total body fat as a % of body weight plotted
           against method of ambulation.

ACKNOWLEDGMENTS

       We acknowledge the technical help of M Boyd and N Hille and we are
grateful for the cooperation of P Francis, Director, Cystic Fibrosis
Clinic, Royal Children's Hospital. Financial support for these studies
was provided by the National Health and Medical Research Council of
Australia and the Australian Cystic Fibrosis Association. Our thanks to
S Looker for typing this manuscript.

REFERENCES

Bruce, A., Andersson, B., Arvidsson, B., and Isaksson, B., 1989, Body
      composition. Prediction of normal body potassium, body water and
      body fat in adults on the basis of body height, body weight and
      age. Scand.J.Clin.Lab.Invest. 40:461.
Guy, R., 1978, The growth of physically handicapped children with
      emphasis on appetite and activity, Publ.Hlth. 92:145(London).
Lloyd, F.K., 1979, Dietary problems associated with the care of
      chronically sick children, J.Hum.Nutr. 33:133.
Mann, M.D., Bowie, M., and Hansen, J., 1974, Total body potassium
      estimation in young children. The interpretation of results,
      Pediatr.Res. 8:879.
Novak, L.P., 1973, Total body potassium during the first year of life
      determined by whole body counting of 40-K, J.Nucl.Med. 14:550.
Pierson, R.N., Lin, D.H.Y., Phillips, R.A., 1974, Total body potassium
      in health: effects of age, sex, height and fat, Am.J.Physiol.
      226:206.
Shepherd, R.W., Holt, T.L., Thomas, B.J., Ward, L.C., Isles, A., and
      Francis, P.J., 1984, Malnutrition in Cystic Fibrosis: the nature
      of the nutritional deficit and optimal management, Nutr.Abstr.Rev.
      54:1009.
Shepherd, R.W., Holt, T.L., Thomas, B.J., Isles, A., Francis, P.J., and
      Ward, L.C., 1986, Nutritional rehabilitation in Cystic Fibrosis:
      controlled studies of effects on nutritional growth retardation,
      body protein turnover and the cause of pulmonary disease,
      J.Pediatr. 109:788.

BODY COMPARTMENT CHANGES IN SICK CHILDREN

Paul B. Pencharz, Nachum Vaisman, Maria Azcue, and
Virginia A. Stallings

The Hospital for Sick Children, Toronto, Canada

INTRODUCTION

The body composition of children changes with growth and development
(Pencharz, 1985). It is also affected by disease and the nutritional
status of the child. Our group has been particularly interested in the
changes due to disease and nutritional perturbations (Pencharz, 1988).
A powerful motivation for this interest has come from our clinical
responsibilities in practising nutritional medicine. Being able to
assess the body composition of children is a key component of assessing
and monitoring their nutritional status. Traditionally, paediatricians
have measured anthropometric variables like weight and height. By
taking serial measurements it was possible to monitor not only the
weight of the child but also to measure growth rates. These
measurements, although useful, are not sufficient in themselves. It is
necessary to be able to measure changes in the lean body mass of
children in response to disease and treatment. Further, it is necessary
to be able to monitor body cell mass (i.e. intracellular mass) as
related to extracellular mass. The two components when summed make up
lean body mass. Within body cell mass there are important constituents
like protein, water, and the major intracellular cation potassium. The
ratio of these constituents do vary from tissue to tissue, but are
relatively constant in a healthy individual of a given gender and age.

In Toronto we are fortunate in having access to many of the
techniques that are available to measure body composition in living
human beings. We have sought to employ these techniques selectively in
an effort to understand the effects of disease and dietary intervention
on the body composition of children.

TECHNIQUES

Anthropometry

At a clinical level we find careful serial measurements of height
(or in children under two years of age) and weight to be extremely
useful. We also find expressing weight as a percent of ideal weight for
height to be of great help in assessing a child's status (Moore et al,
1985). We supplement the height and weight measurements with triceps

skinfold and mid arm circumference measurements (Pencharz, 1988). There
are normative values for triceps skinfold, mid arm circumference, mid
arm muscle area (and circumference) for North American children. The
mid arm muscle parameters give a useful clinical index of lean body mass
while the triceps skinfold is a reflection of adiposity. While these
measurements are useful clinically, there are situations in which
forearm parameters are not representative of the body as a whole.
Principally, on a research basis, we use the four skinfolds proposed by
Durnin (1967) and Brook (1971), namely triceps, biceps, subscapular, and
supra-iliac. The sum of these four skinfolds has been related to body
density, and hence it is possible to calculate body fat and fat free
body mass. We have consistently used this approach in a number of
studies and found a surprisingly good correlation between fat free body
mass and other measurements of lean body mass.

Total Body Potassium and Total Body Nitrogen

Our studies in this area were done in the Medical Physics Laboratory
at the Toronto General Hospital. Total body potassium is measured by
whole body counting of the radiation given off by naturally occurring
potassium 40 in the body (Harrison et al, 1975; Flynn et al, 1972).
Total body nitrogen is measured by in vivo neutron activation employing
a prompt gamma emission (Mernagh et al, 1977).

Measurement of Body Water Spaces

We prefer to measure total body water (TBW) using ^{18}O labelled
water. There is some isotope exchange across the bi-carbonate pool with
the result that the ^{18}O space slightly overestimates total body water
(Schoeller et al, 1980).

We measure extracellular water (ECW) as the corrected bromide
space. We give unlabelled sodium bromide and then measure serum bromide
after equilibration has taken place (Vaisman, 1987). After a series of
pilot studies we established that equilibration has taken place by three
hours after administration of the $H_2^{18}O$ and sodium bromide. We
showed that it is satisfactory to give the sodium bromide orally
(Vaisman, 1987), and therefore we give both labels orally. An initial
blood sample is taken followed by administration of the labels and then
a second blood sample is taken at three hours. Bromide concentration is
measured by in vitro neutron activation and ^{18}O enrichment by isotope
ratio mass spectrometry (Schoeller et al, 1980; Vaisman et al, 1987).

Once total body water and extracellular water have been determined,
intracellular water is calculated by the difference between the two.

Bioelectrical Impedance Analysis

It has been recognized for some time that the behaviour of a small
electrical current within the body is affected by the total amount of
water in the body and its distribution within and without cells. Two
related but distinct approaches have been devised to measure body
composition based on the electrical properties of the human body, namely
Bioelectrical Impedance Analysis (Khaled et al, 1988) and Total Body
Electrical Conductivity (TOBEC) (Cochran, 1988). From our point of view
the attraction of BIA is that the instrumentation is portable and can be
brought to the patient's bedside. We recently embarked on a series of
studies in which we are trying to validate the use of BIA to measure not
only total body water but also extracellular water. These validation
studies involve simultaneous isotope measurements, using $H_2^{18}O$ and

Table 1. Characteristics of the Subjects*

(Mean; SD; Range)

Age (yr)	Height (cm)	Weight (kg)	Resistance (Ohms)	Reactance (Ohms)
27.4	164.6	59.5	608	64.0
± 2.8	± 8.0	±10.1	±109	± 8.7
26–31	145–172	34.6–68.2	471–810	54–81

* n = 9; 7 females and 2 males

sodium bromide dilution studies, combined with measurements of bioelectrical impedance. We were able to validate the technique in healthy adults and we are able to estimate both total body and extracellular water spaces. Table 1 shows the characteristics of the subjects and Table 2 shows a comparison of ECW and TBW determined by BIA and by isotope dilution. We are actively studying the validity of the technique in sick children.

*TBW(BIA) was calculated from Resistance using the following equations:

provided by the manufacturer (RJL System)

$$TBW = 0.3963 \ \frac{H^2}{R} + 0.143W + 8.4 \ (Males)$$

$$TBW = 0.3821 \ \frac{H^2}{R} + 0.1052W + 8.3148 \ (Females)$$

*TBW(BIA) was highly correlated with TBW($H_2^{18}O$), producing the equation:

$$TBW(H_2^{18}O) = 1.13 \times TBW(BIA) - 4.67$$
$$(r^2 = 0.985; \ SEE = 1.225)$$

**ECW(BIA) was correlated with ECW(Br) using TBW(BIA)/Reactance as the independent variable, to produce the following equations:

$$ECW(Br) = 16.97 \times (TBW(BIA)/Reactance) + 4.86$$
$$(r^2 = 0.9195; \ SEE = 1.070)$$
From this equation, ECW can be calculated from TBW(BIA) and reactance.

STUDIES

Effect of Weight Reduction on Body Composition of Obese Adolescents

Seventeen obese adolescents, mean percent overweight 158 ± 16%, were treated with a high protein, low carbohydrate, low calorie diet for about 3 months (Archibald, et al, 1983), and then followed for one year (Archibald et al, 1988). The diet provided approximately 880 kcal per day and 2.5 g of protein per kg ideal body weight per day. We determined body composition by using the four skinfolds. Total body potassium and total body nitrogen were measured before the diet, after

Table 2 Body Water Spaces

(Mean; SD; Ranges)			
TBW (L)		ECW (L)	
$H_2^{18}O$	BIA*	Bromide	BIA**
32.6	33.0	13.8	13.8
± 6.7	± 5.8	± 2.5	± 2.3
20.1–43.9	21.9–41.6	7.9–16.6	9.06–16.6

the 3 months on the diet and then were followed up at 12 months. After 3 months on the diet the children were put on a weight maintenance diet for age. Twelve of the subjects returned for follow-up at one year. During the 3 months on the diet, all the subjects lost weight and there was a significant fall in total body potassium (Archibald et al, 1983). There was a parallel fall in fat free body mass. However, total body nitrogen did not fall significantly.

When we started the study, we were not allowed to use total body nitrogen because of the radiation exposure involved. After studying the first 10 patients and showing a significant fall in total body potassium we were able to argue successfully that it was also then necessary to see if there was a drop in total body protein, as measured by total body nitrogen. Thus, we only have measurements of total body nitrogen on 7 of the 17 subjects. Fortunately, 6 of these 7 subjects did return for follow-up at 12 months. Thus we have parallel measurements of total body potassium and nitrogen before the diet, at 3 months, and at 12 months. The subgroup of 6 subjects was representative of the 17 in terms of changes in body fat, fat free body mass and total body potassium, all of which fell significantly during the 3 months on the diet. Over the succeeding 9 months the subjects regained an average of 6.5 kg but were still 7.8 kg below their initial body weight. Associated with this increase in body weight, total body potassium increased slightly. By way of contrast the measurements of the subjects' total body nitrogen showed a small decrease at 3 months, that was not statistically significant, but total body nitrogen continued to fall so that at 12 months it was significantly less (P < 0.001) than the initial value (14). Normal total body potassium for adolescents can be predicted from sex and height. However, measurement of total body nitrogen is relatively a new method and there are no generally accepted normative data. The University of Toronto Medical Physics Laboratory has, however, developed a method of predicting the approximate expected total body nitrogen, based on the subject's size (height and arm span). As a percent predicted, the initial potassium measurements were 121% of predicted and the nitrogen measurements 119% predicted. At the end of the year both nitrogen and potassium were 102% predicted. The greater fall in potassium in response to the very low calorie diet did result in a significant reduction in the potassium to nitrogen ratio. At one year, however, this ratio had returned to the same value of approximately 70 g potassium per kg of body nitrogen.

In these studies of obese subjects undergoing weight reduction there is a significant fall in the K/N ratio during rapid weight loss. Once the diet is discontinued, however, there is a restoration of the K/N ratio to the previous value. Until quite recently it was thought by many that for a given sex and age the whole body K/N ratio was constant. Our own studies during weight reduction and during refeeding have challenged that notion (Archibald et al, 1988). Nonetheless, these follow-up studies suggested that in a steady state situation the concept of a fixed K/N ratio may still be valid.

Effects of Refeeding on the Body Composition of Subjects with Cystic Fibrosis

With advancing lung disease, patients with cystic fibrosis frequently develop a negative energy balance as a result of a combination of an inadequate food intake and increased energy expenditure (Levy et al, 1985). We sought to refeed these undernourished patients by using night time gastrostomy feeds. We measured the response to refeeding using anthropometry, measurements of total body potassium, nitrogen, and body water spaces (Levy et al, 1985). In response to refeeding we showed a significant increase in total body potassium within 10 days to 2 weeks of initiating refeeding (Pencharz, 1984). Changes in total body nitrogen were slower, of less magnitude, and only became statistically significant after some months of refeeding. Increases in fat free body mass paralleled those in total body potassium. Since many of our subjects had advanced lung disease and evidence of right ventricular hypertrophy, we were concerned that the overnight gastrostomy feeding might result in fluid overload with a concomitant expansion of extracellular space. Our measurements did show an increase in total body water; however, all of this increase could be accounted for by an increase in intracellular water. There was no significant increase in extracellular water.

The Effects of Refeeding on the Body Composition of Adolescent Females with Anorexia Nervosa

The changes in body composition during refeeding in adolescent patients with anorexia nervosa were measured using anthropometry, total body potassium, and bromide space (Vaisman, 1988). We observed that the extracellular water increased significantly by approximately two litres (16%). Extracellular water subsequently fell both in absolute terms and as a percent of body weight so that at follow-up extracellular water was 25.3% of body weight, a reduction of approximately 22% of the admission value. Our findings of an expanded extracellular water in under-nourished patients confirms other work. Similarly, its restoration to a lower percentage of body weight after refeeding was well known. We found for the first time the pattern of an expansion in extracellular water during the initial phase of refeeding. It is important to mention that at no stage did any of the subjects show clinical edema. There was a progressive, steady increase in total body potassium and intracellular water throughout refeeding. Similarly there was a parallel increase in fat free body mass and body fat mass.

The Effects of Renal Transplantation on Body Composition (Vaisman, et al, 1988)

Rapid weight gain has been observed following kidney transplantation. The composition of this weight gain, however, was not known. We studied eight patients (seven female and one male), age 3-7.5 years, who underwent renal transplantation. Body composition measurements included anthropometry, total body potassium, total body water and extracellular water. Since admission of a patient for renal transplantation is at an "on-call" basis, we were only able to carry out anthropometric measurements and bromide measurements of extracellular water before

renal transplantation. We have repeated these same measurements at 1
1/2 months post transplantation. At 3-6 months post transplantation we
added total body water and potassium to the anthropometric and bromide
measurements.

Following transplantation there was an increase in body weight that
took place principally between 1 1/2 to 3 months post transplantation.
This was associated with an increase in body fat as measured by triceps
skinfolds. There was a progressive and steady increase in mid arm
muscle circumference throughout the six months following transplantation.
Extracellular water was increased before transplantation ($32.9 \pm 6.5\%$ of
body weight) and returned to normal ($27.3 \pm 8.6\%$) three months after
transplantation ($P < 0.01$). Between 3 to 6 months there was an increase
in total body potassium ($P < 0.08$) but no change in intracellular water
($11.4 \text{ L} \pm 5.9\%$ compared with $11.6 \pm 6.2\%$).

Weight as a percent of ideal weight for height increased from $96 \pm$
13.8 to $116.3 \pm 13\%$ ($P < 0.01$). Our body composition data suggests
that this was accompanied by a gain in fat mass in the first three
months and a subsequent increase in lean body mass in the next 3
months. Weight gain following kidney transplantation in children is not
due to water retention. Glucocorticoids are used in high doses in the
first few months following renal transplantation and can perhaps account
for the increase in fat mass. The steroids have, however, been tapered
down to lower levels by 3 months, which may account for increase in lean
body mass between 3 to 6 months.

SUMMARY AND CONCLUSIONS

Our studies showed that there are important changes in body
composition in children in response to dieting and refeeding.
Similarly, we showed changes in body composition in response to renal
transplantation, which would be expected to restore more normal
homeostasis, but in its early phases changes in body composition are
complicated by the use of high doses of drugs, notably steroid.

Some of the changes in body composition can be predicted from
changes in weight but the majority cannot. Relatively simple
measurements like the four skinfolds as a means of determining body fat
are surprisingly accurate, except in extremely obese subjects.
Similarly, the measurements of fat free body mass are useful. However,
this measurement does not discriminate changes within fat free body mass
in the proportions of body cell mass versus extracellular mass.
Similarly, none of these measurements provide any information regarding
the composition of body cell mass with regard to its potassium and
protein content (as reflected by total body nitrogen).

The main drawback with total body nitrogen measurements is the
radiation exposure. Thus, we have had to limit the inclusion of total
body nitrogen measurements. It has been estimated that two total body
nitrogen measurements provide a gonadal dose of radiation roughly
equivalent to that experienced by an individual living in a large North
American city, that means 30 to 50 mREM per each total body nitrogen
measurement. It becomes ethically possible then to carry out nitrogen
measurements where it can be argued that their measurement is essential
in monitoring the safety and well being of the subject; or alternatively
if the subject's life expectancy is limited. Thus, we could argue in
the obese subjects due to the large and dramatic falls in total body
potassium that there was a safety reason to measure total body nitrogen

on two occasions within a year. Similarly, with the patients with
cystic fibrosis their longevity is limited and since refeeding them was
at that stage experimental, we could successfully argue for the
measurements of total body nitrogen as well as potassium. A further
drawback in the potassium and nitrogen measurements is that the patient
must be mobile enough to be taken to the Medical Physics Laboratory. On
the other hand, anthropometric measurement, body water spaces, and
bioelectrical impedance analysis can all be carried out at the patient's
bedside. The drawback of measuring body water spaces using isotope
dilution is a necessity for blood sampling. The cost of isotopically
labelled water has been falling and there are now a variety of methods
for measuring isotopic abundance. If bioelectrical impedance analysis
can be validated in sick children as well as in healthy adults, it will
be an important clinical diagnostic tool as well as an investigation
tool in assessing the nutritional status of sick children and monitoring
their response to treatment.

ACKNOWLEDGEMENTS

We are grateful to Dr. Joan Harrison and Professor Ken McNeill
(Medical Physical Laboratory at TGH) for their scientific support in
total body potassium measurements.

REFERENCES

Archibald, E.H., Harrison, J.E., and Pencharz, P.B., 1983, Effect of a
    weight reducing high protein diet on the body composition of obese
    adolescents, Am. J. Dis. Child, 137:658.
Archibald, E.H., Stallings, V., Pencharz, P.B., Harrison J.E., and
    Bell, L.E., 1988, One-year follow-up of weight, total body potassium
    and total body nitrogen in obese adolescents treated with the
    protein sparing modified fast, Am. J. Clin. Nutr., 48:91.
Brook, C.G.D., 1971, Determination of body composition of children from
    skinfold measurements, Arch. Dis. Child, 46:182.
Cochran, W.J., Wong, W.W., Fiorotto, M.L., Sheng, H.P., Klein, P.D., and
    Klish, W.J., 1988, Total body water estimated by measuring
    total-body electrical conductivity, Am. J. Clin. Nutr., 48:946.
Durnin, J.V.G.A., and Rahama, M.M., 1967, The assessment of the amount
    of fat in the human body from measurements of skinfold thickness,
    Br. J. Nutr., 21:681.
Flynn, M.A., Woodruff, C., Clark, J., and Chase G., 1972, Total body
    potassium in normal children, Pediatr. Res., 6:239.
Harrison, J.E., Williams, C., Watts J., and McNeill, K.G., 1975, A bone
    calcium index base on partial body calcium measurements by in vivo
    neutron activation analysis, J. Nucl. Med., 16:116.
Khaled, M.A., McCutcheon, M.J., Reddy, S., Pearman, P.L., Hunter, G.R.,
    and Weinsier, R.L., 1988, Electrical impedance in assessing human
    body composition: the BIA method, Am. J. Clin. Nutr., 47:789.
Levy, L.D., Durie, P.R., Pencharz, P.B., and Corey, M.L., 1985, The
    effects of long-term nutritional rehabilitation on body composition
    and clinical status in malnourished children and adolescents with
    cystic fibrosis, J. Pediatr., 107(2):225.
Mernagh, J.R., Harrison, J.E., and McNeill, K.G., 1977, In vivo
    determination of nitrogen using Pu-Be sources, Phys. Med. Biol.,
    22:831.
Moore, D.J., Durie, P.R., Forstner, G.G., and Pencharz, P.B., 1985, The
    assessment of nutritional status in children, Nutr. Res., 5:797.

Pencharz, P.B., 1985, Body composition and growth. _in_: "Nutrition in Pediatrics. Basic Science and Clinical Application", A. Walker, ed. Little, Brown & Co., Massachusetts.

Pencharz, P.B., 1988, Identifying the patient at nutritional risks. _J. Can. Diet. Assn._, 49:108.

Pencharz, P.B., Hill, R., Archibald, E., Levy, L., and Newth, C., 1984, Energy needs and nutritional rehabilitation in undernourished adolescents and young adult patients with cystic fibrosis, _J. Pediatr. Gastroenterol. Nutr._, 3(S1):S147.

Schoeller, D.A., van Santen, E., Peterson, D.W., Dietz,Jaspan, W., J., and Klein, P.D., 1980, Total body water measurement in humans with ^{18}O- and ^{2}H-labeled water, _Am. J. Clin. Nutr._, 33:2686.

Vaisman, N., Corey, M., Rossi, M.F., Goldberg, E., and Pencharz, P.B., 1988, Changes in body composition during refeeding of patients with anorexia nervosa, _J. Pediatr._, 113:925.

Vaisman, N., Pencharz, P.B., Geary, D., and Harrison, J., 1988, Changes in body composition in children following kidney transplantation, _Nephron_, 50:282.

Vaisman, N., Pencharz P.B., Koren G., and Johnson J.K., 1987, Comparison of oral and intravenous administration of sodium bromide for extracellular water measurements, _Am. J. Clin. Nutr._, 46:1.

MINERAL STATUS IN PRETERM INFANTS AS MEASURED BY SINGLE PHOTON

ABSORPTIOMETRY

Richard J. Schanler, Steven Abrams[1], and Hwai-Ping Sheng

USDA/ARS Children's Nutrition Research Center and Section
of Neonatology, Department of Pediatrics, Baylor College
of Medicine, Houston, TX, and [1]The National Institutes of
Health, Bethesda, MD

INTRODUCTION

An optimal mineral status is difficult to achieve in preterm
infants. Multiple nutritional regimens have been used in an attempt to
provide sufficient calcium and phosphorus, but have failed generally
because of problems with mineral salt solubility. Moreover, when near
optimal calcium and phosphorus concentrations are provided in the milk,
poor bioavailability of the mineral salt preparations limits appropriate
absorption and retention of the minerals (Schanler and Garza, 1988;
Schanler et al., 1988). Metabolic bone disease, which occurs as a
consequence of inadequate mineral retention, is reported most often in
preterm infants fed human milk (Brooke and Lucas, 1985). Preterm infants
fed their mothers' unfortified milk have demonstrated postnatal retention
of calcium (Ca) and phosphorus (P) at rates significantly less than
estimates of intrauterine mineral accretion (Atkinson, et al., 1983; Rowe
et al., 1984). These lowered postnatal Ca and P retentions are of
concern. In similar populations of infants, decreased bone mineral
content, decreased serum P concentration, increased serum alkaline
phosphatase activity, and radiologically determined fractures and rickets
also have been reported (Brooke and Lucas, 1985; Atkinson et al, 1983;
Rowe et al., 1984; Eek et al., 1957; Gross, 1983; Greer and McCormick,
1988). The conclusion derived from these investigations is that mother's
milk is an inadequate source of Ca and P for preterm infants and that
mineral mixtures are needed as fortifiers to achieve more appropriate
intakes of Ca and P (Schanler and Garza, 1987).

The objective of our series of studies in fortified human milk-fed
preterm infants was to develop methods by which estimates of bone mineral
content could be applied to the usual assessments of mineral homeostasis,
Ca and P balance, and serum mineral indices.

METHODS

We enrolled healthy preterm infants from the neonatal nurseries of
Texas Children's Hospital. The infants were between 27 and 30 weeks
gestation and were free from major congenital abnormalities and prolonged
medical illnesses. All infants had to achieve a transition from total
intravenous nutrition to enteral nutrition by 3 weeks of life. Informed

consent was obtained from mothers who wished to provide their milk for their infants, and all protocols were approved by the institutional review boards for human research.

The clinical studies were conducted in two phases. The aim of phase one was to determine the relationships between diet, Ca and P balance, and bone mineral content. Study infants were fed a mixture of their mother's milk and one of three fortifiers by nasogastric tube. Group A received their mother's milk mixed with pasteurized, freeze-dried donor human milk, Ca lactate, and phosphate salts. Group B received their mother's milk mixed (1:1 wt/wt) with either of two commercial formulas designed for preterm infants (Similac Special Care or Similac Natural Care, Ross Laboratories, Columbus, OH) and Group C received their mother's milk mixed (4 g/dL) with a commercial powder (Enfamil Human Milk Fortifier, Mead Johnson Nutritionals, Evansville, IN). A fourth group (group D) was studied similarly, but was fed formula designed for preterm infants (Preemie SMA, Wyeth Laboratories, Philadelphia, PA). The duration of the study extended from the onset of enteral feeding to the age at which a body weight of 1800 g was achieved (total duration approximately 6 weeks). Breastfeeding was subsequently introduced in groups A, B, and C.

The outcome of human milk-fed infants, enrolled in phase one, was examined during phase two, the interval between hospital discharge and their first birthday. During phase two, mothers had to decide whether to continue breastfeeding (group HM) or to feed a commercial formula (group F).

The feeding regimen for all infants consisted of the same parenteral nutrient mixture until complete enteral nutrition was achieved. Milk intakes were adjusted to 150 ml/kg daily. Glucose polymers were used to adjust the gross energy composition of the milks to 81 kcal/dl. The range in protein concentrations was 2.2 to 2.4 g/100 kcal. The milk Ca and P concentrations (mg/100 kcal) were 124 ± 19 and 65 ± 6 in group A, 108 ± 9, 58 ± 4 in group B, 134 ± 10, 69 ± 8 in group C, and 109 ± 10, 60 ± 4 in group D, respectively. The milks for groups B and C were prepared according to manufacturers' instructions, and group D formula was ready-to-feed. Multivitamins, including vitamin D (800 IU/d), were administered daily. After hospital discharge, groups HM and F were encouraged to adhere to their original feeding choices. Group F received a variety of commercial formulas during the year (Abrams et al., 1988, 1989). Infants in these groups received multivitamins, including 400 IU vitamin D, daily.

During phase one, anthropometric measurements were performed weekly, a 96-h nutritional balance was conducted at weeks 5-6 to assess N, Ca, P, and Mg absorptions and retentions and metabolizable energy, and at weeks 2 and 6, bone mineral content and serum Ca, P, and alkaline phosphatase activity were measured. During phase two, measurements of bone mineral content and skinfold thicknesses were conducted at 10, 16, 25, and 52 weeks.

Body weight was measured on an electronic scale at the same time each day. Recumbent length was measured using a rigid head support and a movable foot board. Bone mineral content (BMC) was measured by single photon absorptiometry (SP-2 scanner, Lunar Radiation Corp., Madison, WI). The scanner contains a collimated detector and an ^{125}I source which moves along a 1-cm-width path of forearm. The detector measures the attenuated counts and the measurement is proportional to the amount of bone mineral (mg/cm) in the path. Bone width was also measured by this technique. The mid-radius site was located (half the distance between the ulna

styloid and the olecranon). The forearm was wrapped in a flexible, soft
tissue-equivalent material for each scan. The average value of 2-4 scans
of the mid-portion of the radius was computed. The coefficient of
variation within an individual was less than 4%. The methods for the
analyses of Ca, P, Mg, and nitrogen in milk, urine, and feces and energy
in milk and feces have been reported previously (Schanler et al., 1985).
Retention was calculated as the difference between nutrient intake and
the sum of urinary and fecal losses.

Statistical analyses were performed by ANOVA and Student's t test.
Relationships between variables were tested by multiple linear
regression. Repeated measures ANOVA was used to analyze longitudinal BMC
data. Data are presented as means ± SEM.

RESULTS

Phase One

Twenty-nine infants were evaluated. Their average birth weight was
1.09 kg (range 0.84 to 1.36 kg) and their gestational age was 28 weeks
(range 27 to 31 weeks). Anthropometric measurements were determined
during the interval from the age at which birth weight was regained
(approximately 15 to 19 days) to the arbitrary end point of the study,
the achievement of a body weight of 1.8 kg (approximately 55 days). Body
weight gain differed significantly among groups during this interval:
group A 22 ± 0.7, group B 15 ± 1.4,   group C 19 ± 0.5, and group D
23 ± 0.8 g/kg/d (p < 0.001). The increment in recumbent length was
similar among groups, averaging 1.1 ± 0.64 cm/wk. The changes in weight
gain and length were similar to or greater than intrauterine estimates
(Ziegler et al., 1976; User and McLean, 1969).

Data from the measurements of calcium and phosphorus balance and
serum indices of mineral status are depicted in Table 1. Significant
differences were noted among groups for the retentions and absorptions of
Ca and P, but the intrauterine accretions of Ca (120 mg/kg/d) and P
(65 mg/kg/d) were not achieved in any group. Nitrogen retention (average
320 ± 6 mg/kg/d) and metabolizable energy intake (average
109 ± 2 kcal/kg/d) were appropriate for growing preterm infants (Ziegler
et al., 1976).

Differences among groups were noted for BMC measurements at 5-6
weeks (Table 1). There were no differences among groups in the changes
in BMC from age 1-2 weeks to the time of the balance study (5-6 wk). BMC
and net Ca retention were correlated significantly (r = 0.44, p = 0.019).
The relationships between BMC and net retention of P, nitrogen,
magnesium, or metabolizable energy were not significant. There were no
relationships at 5-6 weeks between BMC and age, sex, race, body weight,
recumbent length, Ca and P intakes, or serum indices of mineral status.

Phase Two

Thirty-one infants were enrolled in the longitudinal study at
hospital discharge (approximately 8 weeks). Eleven infants in group HM
(birth weight 1.1 ± 0.1 kg, gestational age 28 ± 1 wk) received their
mothers' milk until 25 weeks postnatal age and 20 infants in group F
(birth weight 1.1 ± 0.2 kg, gestational age 29 ± 1 wk) received a variety
of commercial formulas. Body weight, recumbent length, and midradius
bone width measurements were similar between groups at 10, 16, 25, and 52
weeks. Significant differences between groups were noted for BMC (Fig. 1)
after the 10-week follow-up visit. The calculated weekly changes in BMC

Table 1. Ca and P Balance Study, Serum Indices of Mineral Status, and
         Bone Mineral Content of Left Midradius in Preterm Infants

| | Study Groups | | | |
Characteristic	A (n=7)	B (n=6)	C (n=9)	D (n=7)
Postnatal age (wk)	6 ± 0.5	5 ± 0.6	6 ± 0.5	6 ± 0.2
Ca intake (mg/kg/d)	132 ± 3	123 ± 4	134 ± 3	116 ± 4
Ca retention (mg/kg/d)*	99 ± 7	61 ± 10	47 ± 8	66 ± 8
Ca absorption (%)*	81 ± 5	54 ± 7	39 ± 6	61 ± 7
P intake (mg/kg/d)	70 ± 3	66 ± 3	70 ± 2	65 ± 3
P retention (mg/kg/d)*	58 ± 4	46 ± 3	39 ± 2	44 ± 5
P absorption (%)*	98 ± 0.4	90 ± 2	63 ± 3	74 ± 4
Serum Ca (mg/dL)	9.7 ± 0.1	9.9 ± 0.3	9.4 ± 0.1	10.2 ± 0.1
Serum P (mg/dL)	5.8 ± 0.3	6.8 ± 0.2	6.0 ± 0.2	5.7 ± 0.2
Serum alkaline phosphatase activity (IU/L)	340 ± 18	301 ± 7	345 ± 28	343 ± 22
BMC (mg/cm)*	35 ± 5	19 ± 2	26 ± 1	32 ± 2
Bone width (mm)	3.7 ± 0.2	2.8 ± 0.04	3.2 ± 0.1	3.5 ± 0.1

M ± SEM
*p < 0.01 by ANOVA

during the intervals from 10–16, 16–25, and 10–52 weeks differed
significantly among groups. There were no differences in the weekly
change in BMC in the interval from 25–52 weeks.

BMC measurements at 10, 16, 25, and 52 weeks were not correlated
with either age, sex, or race. However, 78% of the infants were white.
BMC correlated with body weight at 10, 16, 25, and 52 weeks (r = 0.53 to
0.62, p < 0.01) and with recumbent length at 16, 25, and 52 weeks (r =
0.47 to 0.52, p < 0.02). Use of weight or length as covariates did not
change the relationship between BMC and diet.

DISCUSSION

Osteopenia, or delayed bone mineralization, also called metabolic bone
disease of prematurity, is a frequent occurrence in preterm infants (Brooke
and Lucas, 1985; Steichen et al., 1980). We have demonstrated that the
evaluation of bone mineralization in preterm infants can be accomplished
using single photon absorptiometry. The method is especially applicable for
longitudinal assessments.

We observed that diet had the greatest influence on BMC measurements.
We found that anthropometric indices, sex, and race had little or no
influence on that relationship. Koo et al. (1988) reported that body weight
and length were major factors influencing BMC, but the authors did not
evaluate the role of diet. When we compared the infants studied in Houston
with those in Cincinnati, accounting for differences in the site of
measurement on the forearm, the infants in group HM at 16 and 25 weeks had
BMC measurements far below those in infants with fractures and/or rickets at
14 and 27 weeks in the study of Koo et al. (1988).

Chan et al.(1986) also measured post–hospitalization BMC in larger
preterm infants (group mean birth weights, 1.5 and 1.7 kg) during the first
year of life who were fed either human milk or formula beginning early in
their hospitalization. The authors found no differences between groups at

42 weeks in distal one-third radius BMC, but the groups differed significantly at 47 and 56 weeks. This finding suggests that if larger preterm infants are fed human milk, they may encounter bone mineralization delays later in the first year of life.

The most common cause of osteopenia of prematurity is substrate (Ca and P) deficiency (Brooke and Lucas, 1985). We have demonstrated that an additional aspect to be considered is Ca and P bioavailability. Despite similar Ca and P intakes during their hospitalization, those infants absorbing and retaining more Ca had greater BMC measurements. The positive relationship between net Ca retention and BMC suggests that BMC measurements may be useful for clinical monitoring of high risk infants fed a variety of diets.

ACKNOWLEDGMENTS

This work is a publication of the USDA/ARS Children's Nutrition Research Center, Department of Pediatrics, Baylor College of Medicine and Texas Children's Hospital, Houston, TX. This project has been funded in part with federal funds from the U.S. Department of Agriculture, Agricultural Research Service under Cooperative Agreement number

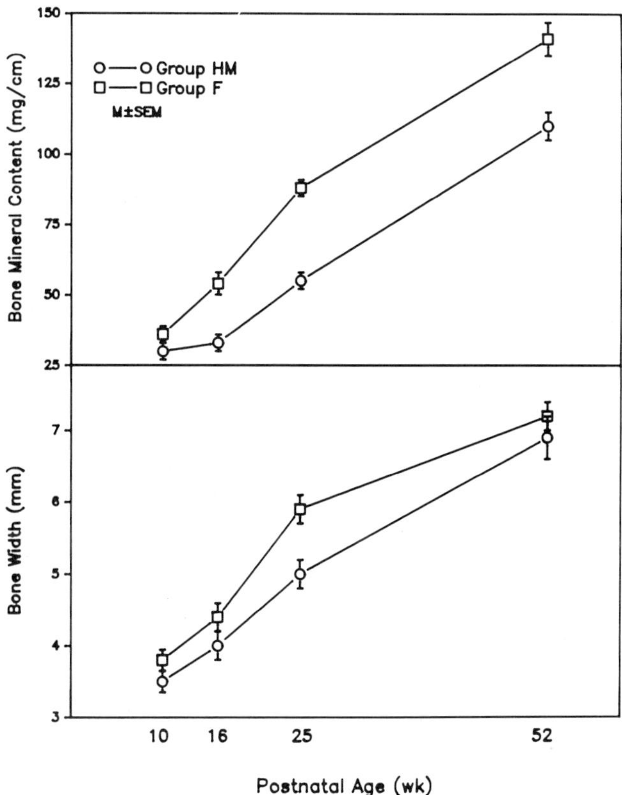

Fig. 1. Bone mineral content vs postnatal age in infants fed either human milk (group HM) or commercial formula (group F). Significant differences by repeated measures ANOVA were observed for the diet group ($p = 0.0008$), age ($p < 0.0001$), and for the interaction of diet and age ($p = 0.0107$). At 10 weeks there were no differences between groups in BMC. There were no significant differences between diet groups in bone width.

REFERENCES

Abrams, S.A., Schanler, R.J., Tsang, R.C., and Garza, C., 1989, Bone
    mineralization in former very low birth weight infants fed human milk
    or commercial formula: one year follow-up observation, J. Pediatr.,
    114:1041.
Abrams, S.A., Schanler, R.J., and Garza, C., 1988, Bone mineralization in
    former very low birth weight infants fed either human milk or
    commercial formula, J. Pediatr., 112:956.
Atkinson, S.A., Radde, I.C., and Anderson, G.H., 1983, Macromineral balances
    in premature infants fed their own mothers' milk or formula, J.
    Pediatr., 102:99.
Brooke, O.G., and Lucas, A., 1985, Metabolic bone disease in preterm
    infants, Arch. Dis. Child., 60:682.
Chan, G.M., Mileur. L., and Hansen, J.W., 1986, Effects of increased calcium
    and phosphorus formulas and human milk on bone mineralization in
    preterm infants, J. Pediatr. Gastroenterol. Nutr., 5:444.
Eek, S., Gabrielsen, L.H., and Halvorsen, S., 1957, Prematurity and rickets
    Pediatrics, 20:63.
Greer, F.R., and McCormick, A., 1988, Improved bone mineralization and
    growth in premature infants fed fortified own mother's milk, J.
    Pediatr., 112:961.
Gross, S.J., 1983, Growth and biochemical response of preterm infants fed
    human milk or modified infant formula, New Engl. J. Med., 308:237.
Koo, W.W.K., Sherman, R., Succop, P., Oestrich, A.E., Tsang, R.C.,
    Krug-Wispe, S.K., and Steichen, J.J., 1988, Sequential bone mineral
    content in small preterm infants with and without fractures and
    rickets, J. Bone Min. Res., 3:193.
Rowe, J., Rowe, D., Horak, E., Spackman, T., Saltzman, R., Robinson, S.,
    Philipps, A., and Raye, J., 1984, Hypophosphatemia and hypercalciuria
    in small premature infants fed human milk; evidence for inadequate
    dietary phosphorus, J. Pediatr., 104:112.
Schanler, R.J., and Garza, C., 1988, Bioavailability of calcium and
    phosphorus in human milk fortifiers and formula for very low birth
    weight infants, J. Pediatr., 113:95.
Schanler, R.J., Abrams, S.A., and Garza, C., 1988, Mineral balance studies
    in very low birth weight infants fed human milk, J. Pediatr.,
    113:230.
Schanler, R.J., and Garza, C., 1987, Improved mineral balance in very low
    birth weight infants fed fortified human milk, J. Pediatr., 112:452.
Schanler, R.J., Garza, C., and Nichols, B.L., 1985, Fortified mothers' milk
    for very low birth weight infants: results of growth and nutrient
    balance studies, J. Pediatr., 107:437.
Steichen, J.J., Gratton, T.L., and Tsang, R.C., 1980, Osteopenia of
    prematurity: the cause and possible treatment, J. Pediatr., 96:528.
Usher, R., and McLean, F., 1969, Intrauterine growth of live-born Caucasian
    infants at sea level: standards obtained from measurement in 7
    dimensions of infants born between 25 and 44 weeks of gestation,
    Pediatrics, 74:901.
Ziegler, E.E., O'Donnell, A.M., Nelson, S.E., and Fomon, S.J., 1976, Body
    composition of the reference fetus, Growth, 40:329.

BIOELECTRICAL IMPEDANCE INDICES IN PROTEIN-ENERGY MALNOURISHED CHILDREN AS

AN INDICATOR OF TOTAL BODY WATER STATUS

Carolina Vettorazzi, Susana Molina, Carlos Grazioso, Manolo
Mazariegos, Mei-Ling Siu and Noel W. Solomons

Center for Studies of Sensory Impairment, Aging and
Metabolism (CeSSIAM), Research Branch of the National
Committee for the Blind and Deaf of Guatemala, Guatemala, CA

INTRODUCTION

In Guatemala, as in many developing countries, various forms of
protein-energy malnutrition (PEM) -- kwashiorkor (edematous), marasmus
(starvation), and marasmic-kwashiorkor (mixed) -- still prevail. The
loss of fat mass and of lean body mass attendant to severe malnutrition
can cause a change in the distribution of water and solids within the
body. Both the percentage of total body water and of extracellular
water in malnourished children is above normal even when there is not a
clinical edema (Viteri, 1981).

Recovery from PEM represents a situation of differential and rapid
change in body-water compartments. In kwashiorkor, in which there is
initially an overhydrated state, a relative loss of water through diuresis
can be seen; in marasmus, early recovery involves the repletion of both
fat and lean tissues in varying proportions, but an overall net gain of
total body water.

Bioelectrical impedance analysis (BIA) for estimating body composition is based on the differential conductivity of electricity in adipose
(non-ionic) tissue and lean (ionic) tissue (Segal et al., 1985). Nomogramic relationships predictive of total body water have been derived from
the measurements of the vector components of impedance (Z) -- resistance
(R) and reactance (Xc) -- for adults (Hoffer et al., 1969; Segal et al.,
1985; Kushner and Schoeller, 1986), and from children as young as 7 years
of age (Cordain et al., 1988). We earlier reported procedures for placing
electrodes and positioning patients that improved the stability of the
measurement in children under 4 years of age (Barillas-Mury et al., 1986).
The applicability of BIA to the assessment of body composition of younger
children still presents problems; however, a first approximation would be
to determine whether the indices of BIA respond in the expected and
appropriate manner to changes in total body water. Previously, we reported
the expected relation between rehydration of diarrheal children and a rise
in R (Molina et al., 1987). In the present study, we followed, longitudinally, children with severe PEM over the course of recuperation to determine whether the expected changes in body composition would be reflected
in the changes observed in BIA indices.

MATERIALS AND METHODS

Subjects

We enrolled 34 children (20 males and 14 females) hospitalized with
severe protein-energy malnutrition, whose ages ranged from 2 to 60 months.
The diagnosis of the type of malnutrition was based on the McLaren classification (McLaren and Walter, 1983). Accordingly, there were 11 children with kwashiorkor, 12 with marasmus, and 11 with mixed malnutrition.

Measurements

Body weight and impedance were determined every day, and length was
determined every week, from 1 to 4 weeks. The subjects' length was
measured in supine position, to the nearest 0.5 cm with a horizontal
stadiometer (infantometer). Nude weight was determined on a SECA infant
balance, and reported to the nearest 0.1 kg. The indices of bioelectrical
impedance were determined on a BIA-103 tetrapolar body composition analyzer (RJL Systems, Detroit, MI), and expressed in ohms. BIA determinations were obtained with the subjects lying supine, with legs separated
to avoid limb contact, and the arms placed at 45° to the trunk. Electrodes were always placed on the right-side limbs. On the upper extremity, the sensor electrode was placed on the ventral side at the wrist,
and the signal electrode 6 cm higher on the ventral forearm. On the
lower limb, the sensor electrode was placed on the dorsal ankle, and the
signal electrode 6 cm higher on the shin.

RESULTS

Table 1 shows the mean and standard deviation of R and Xc at the
beginning and end of the study. The R increased by some 50% in the
kwashiorkor group, whereas it was +27% in the mixed group, and -7% in
the marasmatic group.

Table 2 shows the percentage of the children who showed increases
or decreases in R, Xc, and weight during recovery from their various
forms of severe PEM. One-hundred per cent of the children in the 12
marasmatic group gained weight, whereas 8 (73%) of the kwashiorkor children lost weight due to diuresis, and only 3 (27%) had turned the corner,
and showed a net gain by the time of discharge. The values for resistance increased in 100% of the kwashiorkor group over the course of
rehabilitation. In the marasmatic children, whereas 100% increased their
weight, 75% showed a decrement in R, and only 25% increased their
resistance.

Figures 1 and 2 show the change in percent of weight and resistance,
respectively, of all patients with PEM. The mean longitudinal change in
R for the kwashiorkor children was +167 ± 91 ohms (+ 47%) over 15 ± 7
days (range: 6 to 29 days); the mean change of R in marasmatics over an
average period of 17 ± 7 days (range: 7 to 29 days) was -42 ± 84 (-7%),
and the mean change for R in the mixed group over 15 ± 1 days (range: 13
to 19) was +124 ± 173 (24%).

So striking were the absolute differences in resistance in the kwashiorkor, as compared to marasmus, that we determined the ability of
this index in the differential diagnostic classification of the two malnutrition syndromes. For example, if 500 ohm is used as the cut-off
level, we achieve a sensitivity of 91%, a specificity of 67%, and a

Table 1. The Values of Resistance and Reactance and their
Change During Recovery in Children with
Kwashiorkor, Marasmus, and Mixed Protein-Energy
Malnutrition

		Rs	Re	%	Xcs	Xce	%
Kwashiorkor	X	388	568	+50	31	36	+18
	SD	94	105	–	6	6	–
Marasmic	X	553	511	–7	38	39	+ 5
	SD	3	87	–	5	9	–
Mixed	X	519	633	+27	31	42	+47
	SD	120	141	–	11	11	–

s = start   e = end

Table 2. Relative Changes for Weight, Resistance, and
Reactance over the Observation Interval of Six to
Twenty-Nine Days compared with Condition on
Admission

	Weight		R		Xc	
	I	D	I	D	I	D
Kwashiorkor	27	73	100	0	100	0
Marasmic	100	0	75	25	58	12
Mixed	64	36	64	36	82	18

I = Increased
D = Decreased

predictive value of a positive test of 71% for the diagnosis of kwashiorkor, based on the R value. This power also holds for children
with mixed (marasmic-kwashiorkor) type. An R value of over 500 ohms
only would have a sensitivity of 91%, a specificity of 55%, and a
positive predictive value of 67% in differentiating the children with
pure kwashiorkor from those with the mixed type.

DISCUSSION

In previous studies in healthy young children, we established the
best positions for the electrodes to ensure we had the best reproducibility (Barillas-Mury et al., 1986). Several factors in sick children
can influence the readings of impedance components, such as crying, moving,
and having wet diapers. However, it is possible to obtain stable and reliable data in sick children (Molina et al., 1987). Here we observed a
coefficient of variation of 7 to 12% in malnourished children (marasmics
and the mixed type) during observations over 7 days with daily measurements. This suggests that when changes in interstitial fluid volume are
minimal, the placement of electrodes can allow for a stable, reproducible

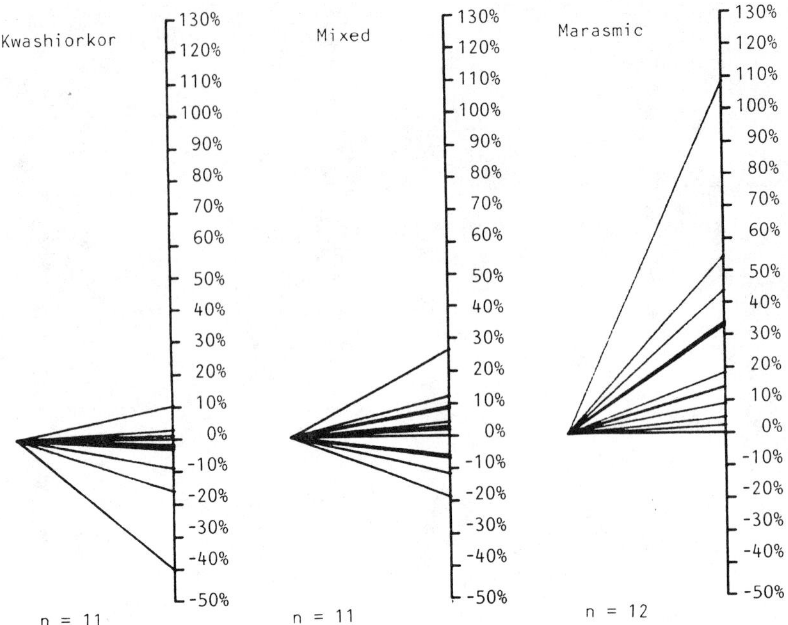

Fig. 1.   Change in weight from admission to discharge in children with P.E.M., interval ranged from 6 to 29 days of observation.

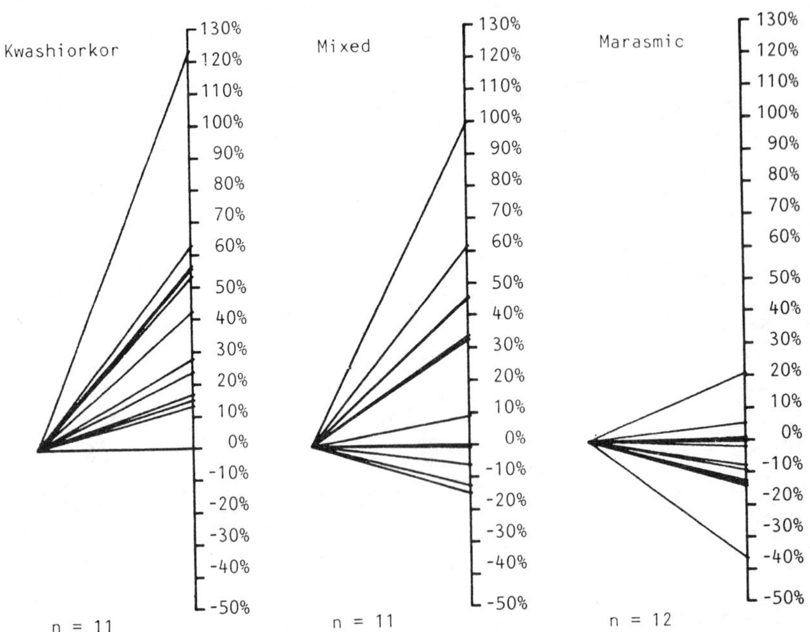

Fig. 2.   Change in resistance from admission to discharge in children with P.E.M. for the same observation period.

recording of impedance over time. In children with kwashiorkor, the CV for
R varied between 5 and 29% within a subject. Here we expected changes of a
greater magnitude, as edema fluid was being diuresed. However, given the
experience in the other malnourished children, neither placement of electrodes nor other factors affecting the measurement, per se, are likely to
explain the serial change in R values.

Children with manifest kwashiorkor are overhydrated. Over the course
of nutritional recuperation, the expected behavior in the edematous subjects
is a loss of interstitial water, accompanied by a rise in the resistance.
After the diuresis of the edema, children often have relatively well-pre-
served reserves of fatty tissue. This would explain the slightly greater
average R values in the formerly kwashiorkor and mixed-malnutrition subjects,
compared to the leaner marasmic children, after diuresis (Table 2).

The recovery in the marasmatics involves a repletion of both lean and
fat mass. One-hundred per cent of the marasmic subjects gained weight over
the period of observation. At the beginning of the study, they had a mean
R value of 558 ohms, and 511 ohms at the end. This figure is similar to
that in a study of 100 healthy Guatemalan toddlers, who had a mean R of
508 ± 58 ohms (Molina and Santizo; personal communication, 1987).

Morgan and her colleagues (1989), working at the Tropical Metabolism
Research Unit in Jamaica, with edematous and non-edematous malnourished
children, applied BIA during recovery and obtained similar findings to
those reported here. Although there is no nomogramic relationship of BIA
indices to body composition in the acute stage of clinical PEM, the
expected responses to differential changes in body water are clearly detectable and demonstrable.

REFERENCES

Barillas-Mury, C., Vettorazzi, C., Molina, S., and Pineda, O., 1986, Expe-
    rience with bioelectrical impedance analysis in young children: Source
    of variability, in: In Vivo Body Composition Studies, K. J. Ellis, S.
    Yasumura, and W. D. Morgan, eds., Bocardo press Limited, Oxford, pp 87.
Cordain, L., Whicher, R., and Johnson, J. E., 1988, Body composition
    determination in children using bioelectrical impedance. Growth
    Development & Aging, 52:37.
Hoffer, E.C., Meador, C., and Simpson, D.C., 1969, Correlation of whole-
    body impedance with total body water volume. J. Appl. Physiol., 27:531.
Kushner, R.F., and Schoeller, D.A., 1986, Estimation of total body water by
    bioelectrical impedance analysis. Am. J. Clin. Nutr., 44:417.
McLaren, D.S., and Walter, R., 1983, Classification of nutritional status
    in early childhood. Lancet 2:146.
Molina, S., Arango, T., Pineda, O., and Solomons, N.W., 1987, Response of
    bioelectrical impedance analysis (BIA) indices to rehydration therapy
    in severe infantile diarrhea. Am. J. Clin. Nutr. 45: abstract, pp 837.
Morgan, P., Golden, B., Bocage, C., and Golden, M., 1989, Prediction of
    total body water and body composition from bioelectrical impedance
    measures in Jamaican infants during recovery from severe malnutrition.
    FASEG Jr., 3:A4804.
Segal, K.R., Gutin, B., Presta, E., Wang, J., and Van Itallie, T.B., 1985,
    Estimation of human body composition by electrical impedance methods:
    a comparative study. J. Appl. Physiol., 58(5):1565.
Viteri, F.E., 1981, Primary protein-energy malnutrition: Clinical bio-
    chemical and metabolic changes, in: Textbook of Pediatric Nutrition,
    R. M. Suskind, ed., Raven Press, New York, pp 189.

ANTHROPOMETRY AND BIOELECTRICAL IMPEDANCE ANALYSIS IN NEWBORNS WITH

INTRAUTERINE GROWTH RETARDATION

Carlos Grazioso, Susana Molina, Maria Claudia Santizo,
Manolo Mazariegos, Mei-Ling Siu, Carolina Vettorazzi, and
Noel W. Solomons

Center for Studies of Sensory Impairment, Aging and
Metabolism (CeSSIAM), Research Branch of the National
Committee for the Blind and Deaf of Guatemala, Guatemala
City, Guatemala, C.A.

INTRODUCTION

Children with low-birth-weight (LBW), defined as birthweight less
than 2500 g, represent an important fraction of births in developing
nations (Villar and Belizán, 1982). Intrauterine growth retardation
(IUGR), defined as having birthweight for gestational age below the 10th
percentile, is a form of LBW ascribed to a nutritional deficiency or
infectious illnesses during pregnancy (Battaglia and Lubschenko, 1967).
IUGR can be chronic (symmetrical), i.e. persisting throughout pregnancy,
or acute (asymmetrical), i.e. occurring in the latter phase of pregnancy.
Both conditions are differentiated based on the ponderostatural index
(PSI) (Lubschenko et al., 1966). The nature of the IUGR has implications
for their post-natal prognosis for survival and development (Lubschenko et
al., 1966).

Improving our understanding of the phenomenon of intrauterine growth
faltering could be aided by better descriptive and mechanistic studies on
body composition of these children. Their small size, however, presents
both technical and theoretical limitations on the precision and accuracy
of measurements, both for conventional anthropometry and for the newer
indirect indices of body compartments, such as bioelectrical impedance
analysis (BIA). Many investigators have shied away from the application of
BIA to small children based on the potential influence of movement,
irritability, micturition and inconsistency of electrode placement. Our
Center has shown that these issues can be overcome in preschool children,
weighing 4 to 8 kg (Barillas et al., 1987). Accepting that measuring
children as small as 2 kg will involve some degree of measurement error,
we sought to evaluate just how well anthropometric measures and
determinations of BIA indices would perform in a series of 100 LBW
children with IUGR.

SUBJECTS, MATERIALS AND METHODS

Standardization

During two and a half months, February to April, 1987, three
examiners (C.G., M.C.S., and S.M.) standardized themselves by taking each

of the anthropometric measurements and BIA readings independently in the LBW and normal newborns. When the improvement in interobserver correlations reached a plateau and leveled off, it was considered that the maximum correspondence had been achieved.

Subjects

One-hundred newborns with a birthweight of <2500 g, and a gestational age of $\geq$37 weeks (as estimated by the same observer using the Ballard Score System) (Ballard et al., 1977), were enrolled during the months of May to August, 1987 in the Neonatal Service of the "San Juan de Dios" General Hospital in Guatemala City. All reported measurements were made within the first 24 h of life.

Anthropometry

Weight was measured on a Suspension Balance (Salter Type, Itac Corporation, Maryland, USA), and recorded to the nearest 50 g. Length was measured with an infantometer (Pedobaby, Nestle Corporation, Vevey, Switzerland). The recumbent, supine length was recorded to the nearest 0.1 cm. Cephalic, thoracic, hip and brachial circumferences were measured to the nearest 0.1 cm using the Inser-tape (Ross insertion tape, Ross Laboratories, Columbus, Ohio). The tricipital, subscapular and abdominal skinfold thicknesses were estimated with a Lange Caliper (Lange, Cambridge, MD, USA), and recorded to the nearest mm.

The derivative indicators calculated from these measurements included the midarm cross-sectional area, the midarm fat area, and the midarm muscle area, as well as the ratio of brachial and head circumference. Additional ratios included each of the circumferences divided by the length.

Bioelectric Impedance Analysis

BIA measurements -- Resistance (R) and Reactance (Xc) -- were recorded using a BIA-101 unit (RJL Systems, Detroit, MI, USA). The child was placed in a supine position, with only the right arm and leg extended and uncovered. The diapers were dry. Electrodes were placed on the upper extremities with the signal electrode immediately above the flexture of the wrist on the volar surface, and the sensor electrode was placed 6 cm more proximally. On the lower extremity the electrodes were placed on the anterior tibial surface, the distal in the intermaleolar line (immediately above), and the proximal 6 cm above. The child was comforted and pacified, and when a consistent, unvarying reading for R and Xc was registered on the digital meter, it was recorded.

Statistical Analyses

Comparisons between mean values for symmetrical and asymmetrical IUGR neonates were made using the Student's t test. The Pearson produced-moment correlation coefficient was used as the index of association between measures in the same subject.

RESULTS

Descriptive Statistics

We measured 71 girls and 29 boys, ranging in weight from 1660 to 2500 g with a mean of 2200 $\pm$ 190 g. Mean length was 45.2 $\pm$ 1.5 cm. Head circumference averaged 32.2 $\pm$ 1.0 cm. Mid-arm circumference was 8.9 $\pm$ 0.6 cm. Tricipital skinfold thickness averaged 2.7 $\pm$ 0.7 mm. The values for R ranged from 295 to 630 ohms (mean: 448 $\pm$ 51 ohms) and values for Xc ranged from 22 to 71 ohms (mean: 43 $\pm$ 8 ohms).

Table 1. Correlations Among Variables of Anthropometry and BIA

	Variable 1	Variable 2	"r"
Weight			
	Weight	Brachial circumference	+ 0.66
	Weight	Cephalic circumference	+ 0.65
	Weight	Length	+ 0.71
Skinfolds			
	Abdominal	Subscapular	+ 0.68
	Abdominal	Triceps	+ 0.60
	Tricipital	Subscapular	+ 0.74
	Skinfolds	Weight	+ 0.39-0.44
Indices			
	Ponderostatural index	Brachial circum/length	+ 0.66
	Ponderostatural index	Cephalic circum/length	+ 0.74
	Ponderostatural index	Thigh circum/length	+ 0.53
BIA			
	Weight	Resistance	- 0.23
	Weight	$Height^2$/Resistance	+ 0. 52

(p <0.05)

As classified by PSI, 11 of the subjects were asymmetrical and 89 were symmetrical. Significant inter-group differences were detected by conventional anthropometry with respect to several measures of length, circumference, and skinfolds (data not shown). Resistance averaged 494 ± 53 ohms in the former group and 442 ± 48 ohms in the latter group (p<0.05). The Xc mean was 43 ohms in both groups.

<u>Intervariable Correlations</u>

Within the limits of measurement errors, we found statistically significant bivariate correlations for a host of pairs. Important examples are shown in Table 1. The expected relationships between weight and other gross indices of body size (length, circumferences) were found. Weight was also significantly correlated with each of the skinfold thicknesses. The midarm fat area was not only correlated with the tricipital skinfold (from which it was derived), but also with distant skinfolds on the back and abdomen. Weight was also correlated with each of the indices of subcutaneous fat. As has been found in adults, weight is inversely related to R, and positively correlated with $length^2/R$.

DISCUSSION

Experimentalists are often intimidated by errors in measurement, failing to accept the reality that all measures have error. The relative magnitude of error, however, increases inherently as the measured value approximates that of the integral units. Thus, very small newborns may have been avoided as subjects for conventional anthropometry and have been ignored since the advent of BIA. We feel that it is not the existence of measurement error, per se, but rather the understanding of its nature and magnitude, that is important for scientific inquiry. Accepting that the precision for obtaining individual and group values in newborns weighing less than 2500 g would be less than that reported in adults, our concern

was whether or not the measurements in LBW infants show significant
stability to reveal biologically important inter-group distinctions and
within-subject associations.

Despite the small size and stature of our population, correlations
between a large number of pairs of conventional anthropometric
measurements produced the expected associations with a strong level of
significance (Table 1). That BIA measurements were relatively precise can
be argued from their ability to discriminate between two forms of IUGR,
despite a small size, and from the correlation of resistance with weight
as reported by others (Lukaski et al., 1985; Kushner and Schoeller, 1986).
A commonly used transformation has been height^2/R (Lukaski et al., 1985;
Kushner and Schoeller, 1986; Segal et al., 1985). As in the experience of
Lukaski et al. (1985) (r = +0.86) and Kushner and Schoeller (1986)
(r = +0.77), weight was positively and significantly correlated with Ht^2/R
(r = +0.52).

CONCLUSIONS

Despite their small body size, measurements of anthropometric and BIA
indices can be made in low-birth-weight infants with intrauterine growth
retardation that are sufficiently reliable to provide the biologically
expected intercorrelations. The mean value for the BIA index for
resistance, differs across two classes of IUGR, suggesting a role for
studies of intergroup phenomena. We conclude that, with sufficient
attention to detail and inter-observer standardization, conventional
anthropometry and BIA should be applicable to cross-sectional and
longitudinal studies in newborns, even the smallest of the neonatal
population.

ACKNOWLEDGEMENTS

The authors are grateful to the RJL Systems Co., for furnishing the
materials used in this study. We appreciate the cooperation of the medical
and paramedical personnel of the Neonatal Unit "Dr. Carlos Cossich", of
the "San Juan de Dios" Hospital.

REFERENCES

Ballard, J., Kazmaier, K., and Driver, M., 1977, A simplified assessment
        of the gestational age, Pediatr. Res., 11:374.
Barillas, C., Vettorazzi, C., Molina, S., and Pineda, O., 1987, Experience
        with BIA in young children: sources of variability, in: "In Vivo
        Body Composition Studies", Bocardo Press Limited, Oxford, London.
Battaglia, F., and Lubschenko, L., 1967, A practical classification of
        newborn infants by weight and gestational age, J. Pediatr., 71:159.
Kushner, R. F., and Schoeller, D. A., 1986, Estimation of total body water
        by bioelectrical impedance analysis, Am.J. Clin. Nutr., 44:417.
Lubschenko, L., Hansman, C., and Boyd, E., 1966, Intrauterine growth in
        length and head circumference as estimated from live birth at
        gestational ages from 26 to 42 weeks, Pediatrics, 37:403.
Lukaski, H. C., Johnson, P.E., Bolonchuck, W.W., and Lykken, G. I., 1985,
        Assessment of fat-free mass using bioelectrical impedance
        measurement of the human body, Am J. Clin. Nutr., 41:810.
Segal, K. R., Lutin, B., Presta, B., Wang, J., and Van Itallie, T. B.,
        1985, Estimation of body composition by electrical impedance
        methods: a comparative study, J. Appl. Physiol., 58:1565.
Villar, J., and Belizán, J., 1982, The relative contribution of
        prematurity and fetal growth retardation to low birth weight in
        developing countries, Am J. Obstet. Gynecol.,143:793.

EXTRACELLULAR WATER ESTIMATED BY THE BROMIDE DILUTION METHOD

FROM SAMPLES OF URINE, SALIVA, AND PLASMA

J.C. Morkeberg, H.-P. Sheng, and W.W. Wong

USDA/ARS Children's Nutrition Research Center
Department of Pediatrics, Baylor College of Medicine
Houston, Texas 77030, USA

INTRODUCTION

Bromide (Br) dilution has been used by many investigators to
estimate extracellular water (ECW) (Cheek et al., 1984; Brans et al.,
1984; Schaffer et al., 1987). The use of this method in infants,
however, is often limited by the need to take multiple blood samples.
Our present studies were designed to 1) evaluate the accuracy and
precision of Br measurements in micro-volumes of body fluids using ion
chromatography, and 2) investigate urine and saliva as alternatives to
plasma for the estimation of Br dilution space.

METHODS

We spiked pooled plasma samples with Br in a range of 1 to 10 µg
Br/mL plasma. The samples were diluted (1:10) and ultrafiltered; each
sample was analyzed six times. Ion chromatography (Waters) was used to
separate Br from other anions and detection was at an ultraviolet
wavelength of 210 nm. Accuracy was validated by a gravimetric method
(Wong et al., 1989). We analyzed urine and saliva samples according to
the method of Wong et al., with some modifications: the sample
preparations were filtered again; and a different anion column, a higher
temperature, a higher concentration of eluent, and a longer runtime were
used in the ion chromatography procedure.

We determined corrected Br space (CBS) in ten normal, healthy
males. The means $\pm$ SD for age, weight, and height were 27.5 $\pm$ 3.5 years,
76.7 $\pm$ 9.0 kg, and 179.0 $\pm$ 4.4 cm, respectively. After baseline samples
of plasma, saliva, and urine were obtained from each subject, Br was
administered orally at a dose of 25 mg Br/kg body weight. Samples were
then collected at 4, 6, 7, 8, and 24 hours after the oral dose. Br
concentration was measured by ion chromatography, chloride (Cl)
concentration was measured on a Buchler chloridometer, and water content
was determined gravimetrically. The protocol was approved by the
Institutional Review Board for Human Research of Baylor College of
Medicine, and informed consent was obtained from all subjects.

CALCULATIONS

Plasma Samples

CBS was first calculated from the standard equation using plasma concentrations of Br:

$$CBS = \frac{[Br]_i * volume(mL)}{[Br]_p} * 0.90 * 0.95 \qquad (Eq.1)$$

where $[Br]_i$ is the concentration of Br in the administered Br solution, $[Br]_p$ is the concentration of Br in plasma water obtained either by extrapolation of the 4- to 8-h samples to time zero or from the 4-h sample alone. The factor of 0.90 is a correction for the intracellular content of Br, and 0.95 is the Donnan equilibrium correction.

Urine Samples

CBS was calculated from urine samples using a modification of the above standard equation:

$$CBS = \frac{[Br]_i/[Cl]_i * volume(mL)}{[Br]_u/[Cl]_u} * K_u * 0.90 * 0.95 \qquad (Eq.2)$$

where $[Br]_u$ and $[Cl]_u$ are the Br and Cl concentrations in urine; $[Cl]_i$ is the Cl concentration in the administered Br solution, which is prepared in a NaCl solution that contains a [Cl] equivalent to that of plasma. Equation (2) assumes a constant relationship, K_u, between the [Br]:[Cl] ratio in urine and the [Br]:[Cl] ratio in plasma:

$$K_u = \frac{[Br]_u/[Cl]_u}{[Br]_p/[Cl]_p} \qquad (Eq.3)$$

Saliva Samples

CBS was calculated from saliva samples using Equations 2 and 3; in the calculations, Br concentration in saliva was substituted for that in urine.

RESULTS

Reproducibility of the Br determination in plasma using ion chromatography was $\pm$ 0.06 µg/mL (1SD) with an accuracy of -0.15 ± 0.36 µg/mL (mean $\pm$ 1SD). Sample volumes of 50 µL were sufficient for accurate analysis of plasma Br.

The mean [Br]:[Cl] ratio in urine and the mean [Br]:[Cl] ratio in plasma declined over the first 8 hours, from 0.0112 to 0.0093 ($p < 0.05$) and 0.0162 to 0.0146 ($p < 0.01$), respectively. A large variation was observed in the saliva [Br]:[Cl] ratio with a range from 0.088 to 0.096. The mean value of K_u, the urine [Br]/[Cl] over plasma [Br]/[Cl] for the

10 subjects, was relatively constant between 4 h and 8 h (regression analysis, $p > 0.9$) (Fig. 1): 0.66 $\pm$ 0.02 (mean $\pm$ 1SD). A larger variation was observed in the mean K-ratio for saliva: $K_s = 6.13 \pm 0.50$ (Fig. 1).

    CBS was calculated from 3 different methods (Table 1). A
significant difference was found between CBS calculated from plasma Br
concentrations obtained by extrapolation of 4- to 8-h plasma to time
zero and CBS calculated from a 4-h plasma sample alone, (p < 0.05,
paired t-test): a single 4-h sample tended to give a higher value. The
mean difference and 95% confidence limits were 5.6% ± 14.8%. CBS
calculated from the 4-h urine sample compared with that calculated from
plasma (Table 1) yielded no significant differences. The 95% confidence
limits were 9.4% ± 40.4%.

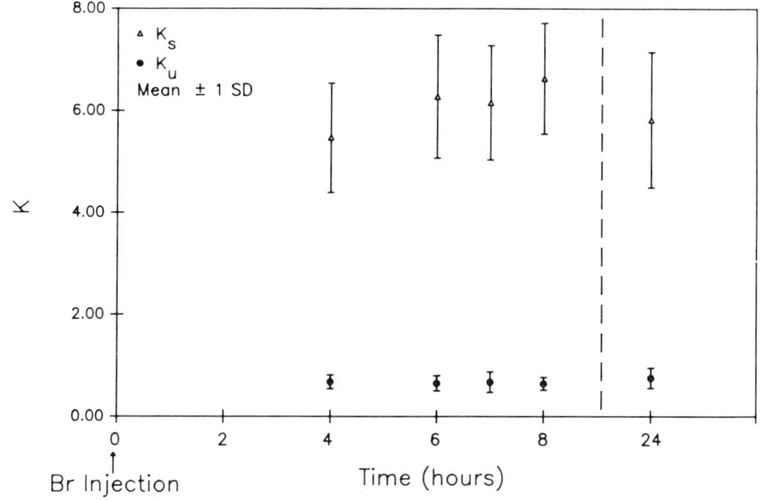

Fig. 1. Changes in K_s (saliva) and K_u (urine) over time.

DISCUSSION

    Miller and Cappon (1984) reported an anion-exchange chromatographic
method for the determination of bromide in serum. The accuracy of the
technique, however, was not evaluated. We found good accuracy and
precision for the measurement of Br in micro samples of plasma by ion
chromatography. The minimum detection limit of the assay is 0.20 µg
Br/mL which allows Br doses as low as 10 µg Br/kg body weight. This is
approximately a fourfold reduction in dose when compared to amounts
presently used in infant studies (Cheek et al., 1984; Schaffer et al.,
1987; Drongowski et al., 1982). Analyses of urine and saliva samples by
this method were also possible, but modifications were required in
sample preparation and ion chromatography procedures.

    The mean CBS measured in plasma in our young, male subjects was
232 ± 23 mL/kg body weight, a value similar to that reported by others
(Vaisman et al., 1987). We did not detect a significant difference in
the mean CBS calculated from urine compared with that obtained from
plasma. The increased variability in the measurements of CBS in urine,.

Table 1. Extracellular Water Estimated by Bromide Dilution

Subject	CBS[a]	CBS[b]	CBS[c]
	(mL/kg body weight)		
1	241	241	264
2	201	232	303
3	275	299	247
4	219	235	218
5	208	229	240
6	246	253	193
7	214	254	289
8	264	243	306
9	215	225	202
10	241	243	323
Mean	232	246	258
± 1 SD	23	21	45

[a] Corrected Br space from plasma Br concentrations obtained by extrapolation of 4–8 h plasma to time zero
[b] From plasma Br concentrations obtained a 4–h plasma sample alone
[c] From 4–h urine [Br]:[Cl] ratio and a K_u value of 0.66

however, may have made it impossible to detect such a difference. The large SD of K_s for saliva did not permit a reasonable estimate of CBS.

Our findings demonstrate that the ion–chromatographic method for the determination of Br is accurate and allows for micro samples of body fluids. Urine, but not saliva, may be used as an alternative to plasma for the determination of CBS.

ACKNOWLEDGMENTS

This work is a publication of the USDA/ARS Children's Nutrition Research Center, Department of Pediatrics, Baylor College of Medicine, Houston, TX. The project has been funded by the USDA/ARS under Cooperative Agreement number 58–7MN1–6–100. The contents of this publication do not necessarily reflect the views or policies of the U.S. Department of Agriculture, nor does mention of trade names, commercial products, or organizations imply endorsement by the U.S. Government. We gratefully acknowledge fruitful discussions with Dr. K.J. Ellis throughout the study.

REFERENCES

Brans, Y.W., Cassady, G., and Cheek, D.B., 1984, To the editor: response, Pediatr. Res., 18:393.
Cheek, D.B., Wishart, J., MacLennan, A.H., and Haslam R., 1984, Cell hydration in the normally grown, the premature and the low weight for gestational age infant, Early Hum. Dev., 10:75.
Drongowski, R.A., Coran, A.G., and Wesley, J.R., 1982, Modification of the serum bromide assay for the measurement of extracellular fluid volume in small subjects, J. Surg. Res., 33:423.
Miller, M.E., and Cappon, C.J., 1984, Anion–exchange chromatographic determination of bromide in serum, Clin. Chem., 30:781.

Schaffer, S.G., Bradt, S.K., Meade, V.M., and Hall, R.T., 1987,
    Extracellular fluid volume changes in very low birth weight
    infants during first two postnatal months, J. Pediatr.,
    11:124.
Vaisman, N., Pencharz, P.B., Koren, G., and Johnson, J.K., 1987,
    Comparison of oral and intravenous administration of sodium
    bromide for extracellular water measurements, Am. J. Clin. Nutr.,
    46:1.
Wong, W.W., Sheng, H.P., Morkeberg, J.C., Kosanovich, J.L., Clarke,
    L.L., and Klein, P.D., 1989, Measurement of extracellular water
    volume by bromide ion chromatography, Am. J. Clin. Nutr., in
    press.

POST-MENOPAUSAL OSTEOPOROSIS AND

MEASUREMENTS OF BODY COMPOSITION

John A Kanis and Jean Aaron

Dept. Human Metabolism and Clinical Biochemistry
University of Sheffield Medical School
Beech Hill Road, Sheffield S10 2RX, UK

INTRODUCTION

Many techniques have been developed to measure indices
of bone mass, often at regional sites (Table 1). Their use
has become an important component in the diagnosis of
osteoporosis, the study of its natural history and the
effects of therapy. This paper reviews briefly the place
of such measurements in the evaluation of osteoporosis,
with particular emphasis on the assumptions made and the
manner in which these might temper our interpretation of
the information which the techniques provide.

WHAT IS OSTEOPOROSIS?

Osteoporosis is a disorder in which there is a
diminution of bone mass without detectable changes in the
ratio of mineralised to non-mineralised matrix. This
definition distinguishes the condition from osteomalacia
where the proportion of osteoid to calcified bone is
increased. For this reason, measurements of bone mineral
content or density provide estimates of skeletal mass in
osteoporosis provided that coexistant osteomalacia is not
present. Since bone mass is one of the major determinants
of the compressive and torsional strength of bone, if loss
is sufficient, fracture will occur more easily. It has
thus become fashionable to consider osteoporosis
exclusively in terms of the amount of bone present or in
terms of bone mineral content. On this basis, osteoporosis
is commonly defined as a decrease in bone density which
renders the skeleton more liable to fracture.

Definition of Osteoporosis

Definitions of osteoporosis are somewhat arbitrary,
depending on whether it is the clinical end result
(fracture) which is described, or the process which gives
rise to fracture. From the point of view of a disease
process, a decrease in bone density is clearly a major
factor which gives rise to the increased risk of fracture.

Table 1. Some Techniques Used in the Assessment of Bone Loss

Technique and Site		Coefficient of Variation (%)
A. Metabolic balance		5-15
B. Serial determination of:		
1. Neutron activation	Hand	2-4
	Spine	2-10
	Whole body	5-10
2. Single photon absorptiometry	Appendicular cortex	1-4
	Appendicular spongy	3-5
	Axial	3-5
	CAT	3-5
3. Dual photon absorptiometry	Appendicular cortex	2-5
	Appendicular spongy	3-5
	Axial cortex	3-5
4. X-ray	DEXA spine	1-2
	Metacarpal width	1-2
	Vertebral morphometry	2-5
	CAT	5-15
5. Bone histology	Ilium	20-30
6. Ultrasound attenuation	Heel	4-6

There is a compelling relationship between the amount
of calcium or matrix in bone and the risk of fracture. The
lower the bone density, the lower its ability to withstand
compressive forces (Bartley et al, 1966), and the greater
the risk from fracture (Jensen et al, 1982; Cooper et al,
1987). Several prospective studies have now shown that
measurements of bone mineral content or density at a
variety of skeletal sites can provide an estimate of the
risk from future fracture (Wasnich et al, 1987; Ross et al,
1988; Hui et al, 1988). These relationships have markedly
influenced the ways in which osteoporosis have been
defined.

Thus, osteoporosis has been defined as a decrease in
bone density below that of a normal population, either age-
matched or the young adult population (Nordin, 1987). An
alternative has been to characterise the bone density of an
osteoporotic population and to define osteoporosis in terms
of bone density below a fracture threshold at which a
patient is at risk from fracture (Odvina et al, 1988).

Irrespective of the approach used and the site of
measurement, there is an overlap between the bone mineral
content or density of populations with and without
fracture. Indeed, the risk of osteoporotic fracture is
stochastic and increases progressively as bone density
decreases. Moreover, as we review later, factors other
than the amount of bone contribute to fracture risk. Thus,
bone mineral measurements may give an index of the risk but
do not capture all elements of that risk, which vary with
the techniques used, age, cause of bone loss and site of
interest.

For these reasons osteoporosis might be more usefully
defined as a clinical disorder rather than a process. In
these terms, osteoporosis represents the occurrence of a
fragility fracture and a decrease in bone density
(osteopenia) a major risk factor. There are useful
analogies to be drawn with other diseases where major risk
factors have been identified (Table 2). For the purpose of
this review, osteoporosis is considered as a disease rather
than a process.

Table 2. The Distinction Between Risk Factors and Disease

Disease	Risk factor	Clinical expression
Coronary artery disease	Hypercholesterolemia	Myocardial infarction
Cardiovascular accident	Hypertension	Stroke
Osteoporosis	Osteopenia	Fracture

DIAGNOSIS OF OSTEOPOROSIS

     Can measurements of bone mass be used to provide a
diagnostic tool? A number of studies examined the
sensitivity and specificity of bone mass measurements to
discriminate osteoporotic patients with fracture from
populations without fracture (Jensen et al, 1982; Odvina et
al, 1988; Ott et al, 1987). In general, both sensitivity
and specificity are improved by measuring sites of
biological relevance. Thus, measurements of vertebral bone
mineral content have a higher predictive value for the
detection of spinal osteoporosis than measurements at the
wrist. Conversely greater predictability is obtained for
the detection of wrist fractures by measurements at this
site than by measurements at the spine (Eastell et al,
1989).

     The real value of this type of intelligence is,
however, limited. The diagnosis of osteoporosis (the
detection of fracture) is clinically obvious with a
predictive value much greater than can be obtained with
density measurements. The strength of such measurements
can, in a diagnostic sense, only be in their ability to
provide an estimate of current fracture risk irrespective
of whether an osteoporotic fracture is present or absent.
They can also provide information whether a fracture where
present was associated with a reduced bone density. Both
types of information are diagnostically useful but do not,
and should not be used to detect osteoporotic fracture.

     Within this framework, the value of physical
measurements depends critically on their accuracy - for
example, the ability of bone mineral content to predict ash
weight content of the site examined. The accuracy of
various techniques varies from 1-20%. These figures have
to be considered alongside the variance in measurements of
the population to be examined, which ranges from 10-50%,

depending upon the techniques used for measurement and any
normalisation procedures applied (Kanis et al, 1983). It
is evident, therefore, that even techniques with an
apparently acceptable accuracy (eg 5-10% for dual photon
absorptiometry) cannot define fracture risk with accuracy
where the population variance for photon absorptiometry is
in the order of 20%. Notwithstanding, absorptiometric
techniques provide the best index currently available and
are of greater predictive value than clinical risk factors
identified thus far for osteoporotic fracture (eg smoking,
body weight, life-style). Nevertheless, the development of
techniques with greater accuracy would be of significant
value for the diagnostic use of bone mineral measurements.

THE STUDY OF BONE LOSS AND INTERVENTION

     Different considerations apply to the use of bone mass
measurements to study the natural history of osteoporosis
and the effects of treatment. They concern the precision
of the methodology (see Table 1), and in particular the
biology of osteoporotic bone loss, which is briefly
reviewed.

<u>The Organisation and Turnover of Bone Tissue</u>

     The manner in which osteoporosis arises is, in large
part, determined by changes in skeletal metabolism and
architecture. The skeleton is comprised of compact and
cancellous (trabecular) bone. In the healthy adult, bone
mass is neither increasing nor decreasing, but there is
considerable turnover of bone, and 95% of skeletal turnover
in the adult is accounted for by remodelling (Parfitt,
1982). The remodelling process comprises a discrete series
of cellular events on bone which have been well
characterised morphologically. These events occur
predominantly on surfaces of bone. The surface of
trabecular bone is greater than that of cortical bone, even
though trabecular bone may occupy only 25% of total
skeletal mass in health and only 10% in an osteoporotic
patient. Because of the high surface to volume ratio of
trabecular bone tissue, disorders of bone remodelling more
commonly affect trabecular sites earlier and more floridly
in the disease process.

     Remodelling activity is important for skeletal
strength in vivo (Frost, 1960). Clinical and experimental
studies both suggest that if the rate of remodelling is
decreased substantially, the risk of spontaneous fractures
will increase, not due to a decrease in skeletal mass, but
due to the inability of the skeleton to undergo self-repair
(Kanis, 1984).

     At the start of the remodelling sequence osteoclasts
assemble or differentiate together to excavate a resorption
cavity. At the completion of this phase osteoclasts
disappear, and several days later bone-forming cells
(osteoblasts) are attracted principally to sites of
previous resorption. This sequence of events (Fig 1)
permits the self repair of bone and is one of the

mechanisms for preserving both skeletal mass and
architecture.

The term 'coupling' describes the attraction of
osteoblasts to sites of previous resorption, but the basis
of this process is ill-understood. Clearly, it allows a
moiety of bone to self-repair. The osteoporotic process
may involve uncoupling of bone formation to previous
resorption (eg neoplastic bone disease), but more commonly
a decrease in bone mass is due to an imbalance between the
amount of mineral and matrix removed and subsequently
incorporated into each resorption cavity, so that skeletal
mass decreases progressively. In postmenopausal
osteoporosis and many other types of osteoporosis, the
imbalance between the amount of bone resorbed and that
formed at each remodelling site is due to a decrease in the
functional capacity of osteoblasts or to a decrease in the
number of osteoblasts recruited to resorption sites.

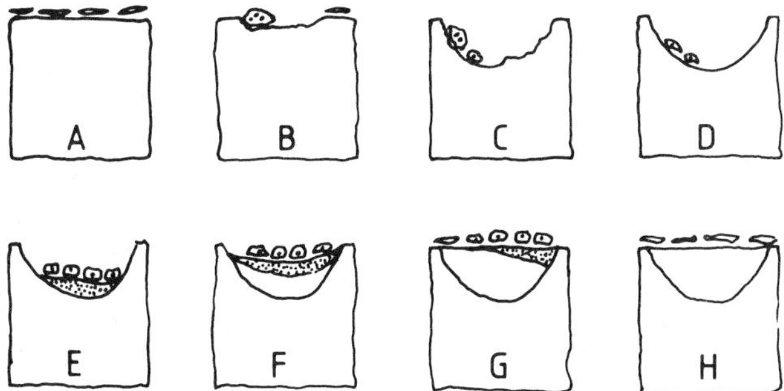

Fig. 1 Steps in the bone remodelling sequence of
       trabecular bone. Early in the remodelling
       sequence osteoclasts are attracted to a
       quiescent bone surface (A) and excavate a
       resorption cavity (B,C). Mononuclear cells
       smooth off the resorption cavity (D) which is a
       subsequent site for the attraction of
       osteoblasts which synthesise an osteoid matrix
       (E). Continuous new matrix synthesis (F) is
       followed by calcification (G) of the newly
       formed bone. When complete, lining cells once
       more overlie the trabecular surface (H).

Irrespective of the mechanism, a finite deficit of
bone is the end result of each remodelling sequence. If
bone turnover is increased, then the number of bone
remodelling units extant at any one time also increases.
If the imbalance at each site remains constant then the
result of increasing bone turnover will be to amplify the
rate of bone loss (Fig. 2). There is now good evidence
that oestrogen deficiency not only induces a focal

imbalance at remodelling sites, but also increases the remodelling rate of bone. In this way bone loss is accelerated.

The implications for the measurement of bone loss is that the rate of loss will depend upon the site measured and is clearly greater at trabecular sites than at cortical bone sites. Even within trabecular bone tissue there is a great heterogeneity in rates of bone remodelling, so that rates of bone loss similarly differ. As we review later, the ability to detect a change depends not only on the change but also on the precision of technique. The choice of technique should be determined in part by these biological considerations.

<u>Therapeutic Modulation of Bone Turnover</u>

An increasing number of drugs are used in the treatment of established osteoporosis (Table 3). Their role is reviewed in detail elsewhere (Kanis, 1984). In general, the inhibitors of bone turnover (also used in prophylaxis) appear to decrease the rate of bone loss, but do not always prevent bone loss entirely. The reason for this relates to their effect on bone remodelling. Thus, at each remodelling site a finite volume of bone is resorbed, and in osteoporosis a somewhat lesser amount is formed. In terms of calcium transport, approximately 6 mmol of calcium is resorbed daily and 5 mmol put back by bone formation in the million or so remodelling sites. This imbalance gives rise to a net deficit of 1 mmol of calcium per day (equivalent to a bone loss of approximately 1% per annum). When bone turnover alone is decreased, the number of remodelling sites also decreases, but the imbalance between formation and resorption at each remodelling site persists.

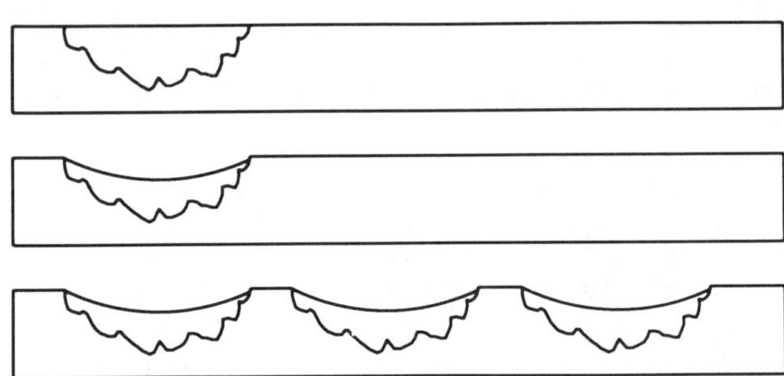

Fig. 2 Schematic representation of a trabecular bone
        surface to illustrate the effect of balance
        and remodelling on the rate of bone loss. The
        top panel shows the infilling of a resorption
        bay with an equal volume of new bone. In
        osteoporosis less bone is deposited in
        resorption cavities (centre). If bone
        turnover is increased without altering this
        balance (lower panel), the rate of trabecular
        bone loss will increase in proportion to the
        increment in bone turnover.

Thus, if bone turnover is decreased by 50%, bone loss would
be reduced from 1% to 0.5% per annum. Whereas a decrease
in the rate of bone loss may have clinical dividends, such
regimes will not reverse a high risk of refracture in a
patient who has already sustained an osteoporotic fracture
and has a bone mass below a given fracture threshold.

Table 3. Some Drugs Used For Osteoporosis

Inhibitors of bone turnover
Oestrogens
Calcitonins
Progestins
Diphosphonates
Anabolic steroids
Calcium
Thiazides
Stimulators of bone formation
Fluorides
Sequential calcitonin and phosphate
Parathyroid hormone
Uncertain efficacy or action
Vitamin D derivatives
Other sequential and combination treatments

In contrast to this account a number of observations
have shown that the inhibitors of bone turnover may
increase skeletal mass in osteoporosis. Indeed, treatment
is associated with an increase in skeletal mass, but this
is modest (2-10% depending on the site measured) and is
sometimes ill-sustained. The reason for the transient
increase in bone mass is that the agents used to decrease
bone turnover are generally inhibitors of bone resorption,
which decrease the activation of new remodelling sites.
Early during treatment, bone formation will continue at
previously existing remodelling sites and bone mass will
increase transiently (Fig. 3). Since bone turnover is a
slow process, bone mass may increase for up to 3 years
before a new steady state is achieved (Parfitt, 1980). The
occurrence of this transient state probably explains why
almost no treatments advocated for osteoporosis have been
shown to be ineffective! These considerations imply that
if the long-term consequences of treatment are to be
predicted in patients with established osteoporosis, they
must be studied for at least 3-5 years (Kanis et al, 1983).
It is quite misleading to study the effects of a drug for 1
year and, from the increment in bone density observed,
assume that the increments in skeletal mass would be
similar in later years.

<u>Spatial Organisation of Trabecular Tissue</u>

The continual imbalance between formation and resorption would be expected to result in a progressive decrease in the width of trabecular bone elements. A decrease in trabecular width occurs in many forms of osteoporosis, but it is clear that in postmenopausal osteoporosis there is not only a decrease in trabecular width, but also a marked loss of trabecular elements themselves (Parfitt et al, 1983; Aaron et al, 1987). Indeed, there is little evidence for trabecular thinning, so that the loss of trabecular elements must be, in part, a result of the generation of resorption cavities which transect or perforate trabecular structures.

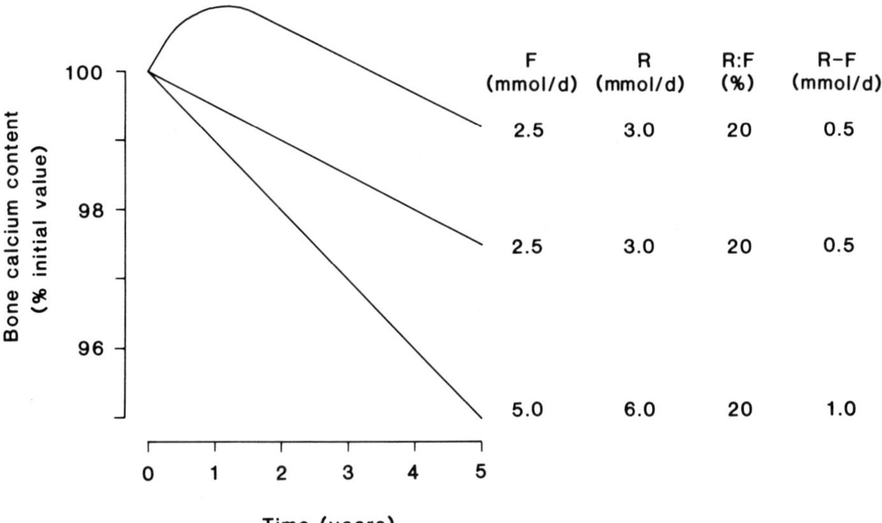

Fig. 3 Effects of altering bone turnover in osteoporosis. The relationship between bone mineral content with time is shown for a patient losing 1% of bone mass per year due to an imbalance of bone at remodelling sites. In terms of calcium fluxes more calcium is resorbed (R; 6mmol/day) than formed (F; 5mmol/day) - a net deficit of 1 mmol/day representing a skeletal loss of 1% per annum. If bone turnover is halved without altering the imbalance between the amounts formed and resorbed then the rate of bone loss is halved. The administration of inhibitors of bone resorption permits, however, continued formation at preexisting resorption sites so that bone mineral content increases to infill this resorption space. When bone formation decreases to match the prevailing rate of bone resorption, bone loss will occur once more, albeit at a slower rate than before treatment.

The disruption of trabecular architecture characteristic of
postmenopausal vertebral osteoporosis has important
implications for structural strength. The selective
destruction of cross bracing elements lead to failure of
the structure out of proportion to the amount of material
removed. Even though remaining trabeculae in
postmenopausal osteoporosis can thicken, evident even
radiographically, this cannot compensate adequately and
prevent skeletal failure. Similarly, therapeutic
manipulation of the remodelling process might be expected
to thicken trabecular structures without necessarily
restoring trabecular continuity (Fig 4). The majority of
techniques used to measure bone density do not capture
information on trabecular continuity. This suggests that
changes in bone mass, whether due to natural history or to
therapeutic intervention, cannot be interpreted reliably as
changing fracture susceptibility without additional
skeletal information.

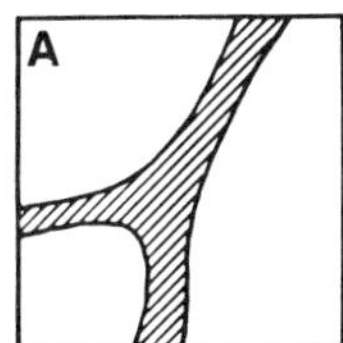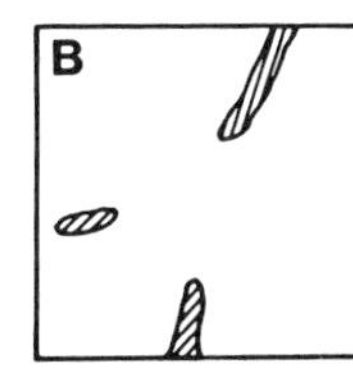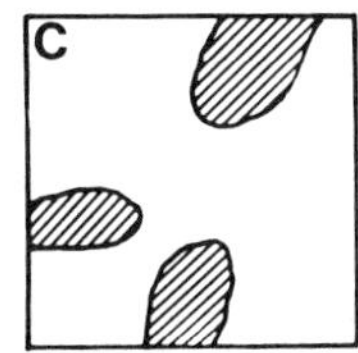

Fig. 4 Schematic representation of trabecular bone
showing normal trabecular architecture (A) and
the thinning and discontinuity of trabecular
elements during post-menopausal osteoporosis
(B). The surface deposition of bone by
anabolic agents may thicken remnant structures
without necessarily restoring trabecular
continuity.

As in the case of the diagnostic use of density
measurements, the site of measurement is important in that
changes at one site may not be reflected by parallel
changes elsewhere. For example, the therapeutic use of
fluoride induces marked increases in trabecular bone volume
over several years of treatment, but has less effect on
cortical bone (Kanis and Meunier, 1984). Like all
currently used treatments, however, fluoride alters bone
mass by affecting bone remodelling and at each remodelling
site, the resorption cavity is filled with a greater volume
of new bone (Erikson, 1986). It is to be expected then
that trabecular width would increase, but its effect on
fracture rate in established osteoporosis is likely to be
less than would be predicted from the increment in bone
density, unless the drug were also to have effects on
trabecular continuity.

The less marked effects of anabolic agents on cortical bone relate in part to bone remodelling. At cortical sites where bone is compact, the resorption cavity is excavated as a cutting cone. The tunnel cannot be overfilled due to the physical constraints of the surrounding compact bone. In contrast, the marrow cavity at trabecular sites does not constrain the overfilling of resorption bays.

Many problems in the interpretation of changes in bone density in established osteoporosis could be overcome by the development of techniques which measured the connectivity of trabecular bone. At present this can be measured reliably only by invasive techniques (Aaron et al, 1987). X-rays at the hip provide a semiquantitative index in the form of the Singh score (Cooper et al, 1987), which are of value in population studies. It has been suggested

Table 4. Pathogenesis of Osteoporotic Fracture

Skeletal
Bone mass
Spatial organisation of bone
Turnover of bone
Extraskeletal
Falls - frequency and severity
Response to trauma - neuromuscular coordination

that broadband ultrasound attenuation of bone provides an index of skeletal competence (Langton, 1984). We have recently shown that this technique discriminates patients with and without osteoporosis at least as well as measurements of bone mineral density (McCloskey et al, 1990a). Moreover, it captures some aspects of true bone density not dependent on mineral density (McCloskey et al, 1990b), suggesting the partial dependence of ultrasound attenuation on the intrinsic trabecular architecture of cancellous bone.

Skeletal Versus Extraskeletal Factors

The skeletal abnormalities which increase the risk of osteoporosis are summarised in Table 4. Bone density measurements alone do not capture all information relating to fracture risk. Several studies, including some prospective data, have now shown that age is an independent factor. Thus, for any given bone density, the risk of fracture is greater in the elderly (Hui et al, 1988; Ross et al, 1988b). A factor of probable importance is the role of falls which increase with advancing age (Cummings, 1985). The osteopaenic skeleton can be likened to a fragile vase. The vase will remain intact forever if left on display, but if it is tipped over it will certainly break. The role of falls differs with different types of fracture as does the importance of protective neuromuscular defences (Table 5).

<u>Assessment of Bone Loss and Gain</u>

There are several additional biological factors which determine the applicability of bone mass measurements. It is important to be aware of the nature of the measurements made and their physiological significance. For example, measurements of cortical width do not give an indication of bone density, and bone density measurements do not necessarily give an indication of the stress required to fracture a bone. A less dense but larger wrist may be less liable to fracture than a denser, but smaller wrist. Indeed, the techniques used to normalise data are at best arbitrary and are based on assumptions which are difficult to test.

These considerations are at first sight less relevant to the assessment of bone loss or gain where repeated measurements are made in the same individual. However, if

Table 5. Relative Inportance of Skeletal and
Extraskeletal Factors in Osteoporosis

Type of fracture	Role of falls	Role of bone
Colles	+++	+
Femoral neck	++	++
Vertebral	+	+++

bone width alters, if fat distribution changes or extraskeletal calcification occurs due to interventions, apparent changes are liable to misinterpretation. For example, the apparent anabolic effects of decadurabolin on skeletal density is partly artefactual, and the apparent increment observed in bone density is due in part to a significant decrease in fatty tissue surrounding bone (Hassager et al, 1989). Conversely, a component, perhaps 50%, of the apparent bone loss after oophorectomy assessed by CAT is attributable to an increase in marrow fat.

The precision or reproducibility of the technique used is also a critical factor in determining whether or not a change has occurred. The conditions under which reproducibility is assessed are important. For example, the precision of single photon absorptiometry is less than 1% when bone mineral content of a standard is determined on the same day. But the measurement of greater clinical significance is that derived in real people where the reproducibility is halved (>2%). In osteoporotic patients with a decreased bone mineral content the coefficient of variation exceeds 3% in most laboratories. The importance of determining the coefficient of variation applicable to working conditions in real patients is obvious since the probability of detecting a difference between two measurements (when a difference truly exists) depends upon the confidence that can be placed on the single (or

repeated) measurement. Techniques for improving the power
of measurements to detect changes in individuals and in
populations, reviewed elsewhere (Kanis et al, 1983), can
considerably enhance the value of density and bone mass
measurements.

CONCLUSIONS

Bone is a complex tissue. Physical studies of the
characteristics of bone have largely been confined to its
mechanical properties in vitro, or to the measurement of
the amount of bone present. Whereas the amount of bone
present is clearly an important factor which determines
structural strength, it is now clear that spatial
organisation of bone and its turnover as well as
extraskeletal factors are also important for its structural
integrity in vivo. Physical measurements of bone density
or mineral content do not capture these determinants of
skeletal strength and the interpretation of bone mass
measurements should take account of these biological
considerations as well as statistical and technical
factors.

The development of new techniques to assess other
determinants of bone strength will form an important
component for our further understanding of osteoporosis and
the rational development and assessment of therapeutic
regimens.

ACKNOWLEDGEMENTS

We are grateful to the Medical Research Council and to
Rorer Central Research for their support of our work.

REFERENCES

Aaron, J.E., Makins, N.B., and Sagreiya, K., 1987, The
    microanatomy of trabecular bone loss in normal ageing
    men and women, Clin Orthop, 215:260.
Bartley, M.H., Arnold, J.S., Haslam, R.K., and Webster,
    S.S.J., 1966, The relationship of bone strength and bone
    quantity in health, disease and ageing, J Gerontol,
    21:517.
Cooper, C., Barker, D.J.P., Morris, J., and Briggs, R.S.,
    1987, Osteoporosis falls and age in fracture of the
    proximal femur, Br Med J, 295:13.
Cummings, S.R., 1985, Are patients with hip fracture more
    osteoporotic, Am J Med, 78:487.
Eastell, R., Wahner, H.W., O'Fallon, M. Amadio, P.C.,
    Melton, L.J., and Riggs, B.L., 1989, Unequal decrease in
    bone density of lumbar spine and ultradistal radius in
    Colles' and vertebral fracture syndromes, J Clin Invest,
    83:168.
Eriksen, E.F., 1986, Normal and pathological remodelling of
    human trabecular bone: three-dimensional reconstruction
    of the remodelling sequence in normals and metabolic
    bone disease, Endocrine Rev, 7:379.
Frost, H.M., 1960, Presence of microscopic cracks in vivo in
    bone, Henry Ford Hospital Bulletin, 8:25.

Hassager, C., Borg, J., and Christiansen, C., 1989,
    Measurement of the subcutaneous fat in the distal
    forearm by single photon absorptiometry, <u>Metabolism</u>,
    35:159.
Hui, S.L., Slemenda, C.S., and Johnston, C.C., 1988, Age and
    bone mass as predictors of fracture in a prospective
    study, <u>J Clin Invest</u>, 81:1804.
Jensen, G.F., Christiansen, C., Boesen, J., Hagedus, V.,
    Transbol, I., 1982, Epidemiology of postmenopausal
    spinal and long bone fractures, <u>Clin Orthop Rel Res</u>,
    166:75.
Kanis, J.A., and Meunier, P.J., 1984, Should we use fluoride
    to treat osteoporosis?, <u>Quart J Med</u>, 53:145.
Kanis, J.A., Caulin, F., and Russell, R.G.H., 1983, Problems
    in the design of clinical trials in osteoporosis. <u>in</u>:
    "Osteoporosis: A Multidisciplinary Problem", St. J.
    Dixon A., R.G.G. Russell, T.C.B. Stamp, eds., Royal
    Society of Medicine Int Cong Symp Series 55:205.
Langton, C.M., 1984, The measurement of broadband ultrasound
    attenuation in cancellous bone (PhD Thesis), Hull,
    University of Hull.
McCloskey, E.V., Murray, S.A., Miller, C., Charlesworth, D.,
    Tindale, W., O'Doherty, D.P., Bickerstaff, D.R.,
    Hamdy, N.A.T., and Kanis, J.A., 1990a, Broadband
    ultrasound attenuation in the os calcis: relationship to
    bone mineral at other skeletal sites, <u>Clin Sci</u>, 78 (in
    press).
Nordin, B.E.C., 1987, The definition and diagnosis of
    osteoporosis, <u>Calcif Tissue Int</u>, 40:57.
Odvina, C.V., Wergedal, J.E., Libanati, C.R., Schulz, F.E.,
    and Baylink, D.J., 1988, Relationship between trabecular
    vertebral bone density and fractures: a quantitative
    definition of spinal osteoporosis, <u>Metabolism</u>, 37:221.
Ott, S.M., Kileoyne, R.F., and Chesnut, C.H., 1987, Ability of
    four different techniques of measuring bone mass to
    diagnose vertebral fractures in postmenopausal women, <u>J
    Bone Mineral Res</u>, 2:201.
Parfitt, A.M., 1980, Morphological basis of bone mineral
    measurements: transient and steady state effects of
    treatment in osteoporosis, <u>Min Electrolyte Metab</u>, 4:273.
Parfitt, A.M., 1982, The physiological and clinical
    significance of bone histomorphometric data. <u>in</u>: "Bone
    Histomorphometry. Techniques and Interpretation", R.
    Recker, ed. CRC Press, Boca Ratton.
Parfitt, A.M., Mathews, C.H.E., Villaneuva, R.A.,
    Kleerkoper, M., Frame, B., and Rao, D.S., 1983,
    Relationship between surface volume and thickness of
    iliac trabecular bone in aging and osteoporosis, <u>J Clin
    Invest</u>, 72:1396.
Ross, P.D., Wasnich, R.D., and Vogel, J.M., 1988a, Detection
    of prefracture spinal osteoporosis using bone mineral
    absorptiometry, <u>J Bone Mineral Res</u>, 3:1.
Ross, P.D., Wasnich, R.D., MacLean, C.J., Hagino, R., and
    Vogel, J.M., 1988b, A model for estimating the potential
    costs and saving of osteoporosis prevention strategies,
    <u>Bone</u>, 9:337.
Wasnich, R.D., Ross, P.D., Heibrun, L.K., and Vogel, J.M.,
    1987, Selection of the optimal site for fracture risk
    prediction, <u>Clin Orthop Rel Res</u>, 216:262.

LONGITUDINAL STUDY OF TOTAL BODY CALCIUM MEASUREMENTS IN
PATIENTS WITH INFLAMMATORY BOWEL DISEASE : CORRELATIONS WITH
QUANTITATIVE CT AND SINGLE PHOTON ABSORPTIOMETRY

J.E. Compston (1), S.J.S. Ryde (4), R.J. Motley (2)
E.O. Crawley (3), W.D. Evans (3), W.D. Morgan (4)

Department of Pathology (1), Gastroenterology (2) and
Medical Physics (3), University Hospital of Wales,
Cardiff, and Department of Medical Physics and Clinical
Engineering (4), Singleton Hospital, Swansea, U.K.

INTRODUCTION

Osteoporosis is known to occur in association with inflammatory bowel disease (Compston et al., 1987) and recently we
demonstrated rapid rates of spinal trabecular bone loss in
these patients (Motley et al., 1988). The pathogenesis of bone
loss is likely to be multifactorial, with corticosteroid therapy, ovarian deficiency, malabsorption and malnutrition all
playing a role. However, at present we do not know whether
bone loss in these patients is generalised or occurs only at
certain sites. In this study we measured longitudinal changes
in total body calcium in a group of patients with inflammatory
bowel disease and compared these with changes in spinal trabecular bone mineral density and radial cortical bone mineral
content.

PATIENTS

We studied eleven patients, 5 female, 6 male, aged 25 to
59 years. Eight had small bowel Crohn's disease, one had large
bowel Crohn's disease, one had Crohn's disease affecting both
small and large bowel and one had ulcerative colitis. All
patients with small bowel Crohn's disease had undergone one or
more small intestinal resections. Eight of the 11 patients
were receiving corticosteroid therapy.

MEASUREMENTS

Total body calcium was measured using a californium-252
based _in vivo_ neutron activation analysis instrument (Ryde et
al., 1987). This method utilises an on-line prompt-gamma reaction rather than the delayed method of counting induced Ca-49
activity. Total body measurements were achieved by scanning

Table 1.   Mean Annual Percentage Changes in the 11 Individual
           Patients in Total Body Calcium (TBCa) and Spinal
           Trabecular Bone Mineral Density (BMD).

PATIENT NO.	SEX	AGE (YRS)	% CHANGE IN TBCa	% CHANGE IN SPINAL BMD
1	F	44	-5.1	-2.6
2	M	44	-11.9	-5.5
3	F	39	-10.4	-4.6
4	F	28	-1.1	+2.4
5	M	45	-5.1	+4.9
6	M	59	+24.3	-2.5
7	M	25	-6.2	-5.0
8	M	42	-7.7	-2.1
9	F	51	-2.4	-11.0
10	F	39	-15.2	-4.8
11	M	49	+0.5	-4.6

the subject from shoulder to knee across a vertical collimated
neutron beam from a single 200 ug (4GBq) Cf-252 source. High
resolution gamma-ray spectra are obtained from two heavily
shielded HPGe detectors positioned above the subject. The _in
vitro_ precision of measurement is 4 to 5%.

Spinal trabecular bone mineral density was measured in the
first three lumbar vertebrae by quantitative computed tomography (QCT), using a Philips 350 X-ray CT scanner, by a modification of the method of Cann and Genant (Cann and Genant, 1980).
Scans were made through the middle of each vertebra, using a
slice thickness of 6 mm, field of view of 400 mm and tube
voltage of 120 kV. The _in vivo_ precision of this technique in
our hands is 2.3%.

Bone mineral content in the forearm was measured using
single photon absorptiometry (SPA) at a point one-third of the
distance between the styloid process of the radius and the
lateral epicondyle of the humerus. The _in vivo_ precision of
this technique in our hands is 1.7%.

Measurements of total body calcium were carried out over a
period of 2.3 years, the mean length of follow-up being 21.3
months and 2 or 3 measurements per patient were made. QCT and
SPA measurements were carried out over 2-5 years; the mean
length of follow-up was 38.1 months and the mean number of
measurements in each patient was 3.4 (range 2 to 5).

RESULTS

Changes in total body calcium and spinal trabecular bone
mineral density in individual patients are shown in Table 1.
The mean annual change in total body calcium in the group as a
whole was -6.5% (P<0.02). Only one patient showed significant
increase in total body calcium; the rise of 24.3% in a 59 year
old man is difficult to explain and further measurements are
planned to check the validity of this high value.

Reduction in spinal trabecular bone mineral density was
also seen in the majority of patients, the median annual rate
of bone loss in the lumbar vertebrae being 4.4%. Only small
changes were seen in the radial bone mineral content and the
median change found, (+0.7%/annum) does not represent a significant change.

DISCUSSION

Our results demonstrate large reductions in total body
calcium in some patients with inflammatory bowel disease over
the course of one to two years, the mean annual change in the
group being -6.5% (P<0.02). Most of the patients had Crohn's
disease of the small bowel with previous intestinal resection
and were being treated with corticosteroids; in the small
number studied, it was not possible accurately to define risk
factors although corticosteroid therapy and malabsorption seem
likely candidates. However, a large decrease in total body
calcium was also seen in one 42 year old man with non-
corticosteroid treated ulcerative colitis, indicating that
other factors must also contribute.

Spinal trabecular bone mineral density also showed a large
decrease in some patients and in the group as a whole the
annual rate of trabecular bone loss at this site was 4.4%,
higher than would be expected in the general population but
less than that observed in total body calcium. This might
indicate preferential cortical bone loss in these patients or
alternatively could be explained by the comparatively low
precision of the technique used for measurement of total body
calcium or influence on its measurement by fluctuations in body
weight. Also, QCT measurements were made over a longer period
of time, increasing the accuracy of the calculated mean annual
changes. At the radial site, no significant loss could be
detected in the mean follow-up period of 38 months; nevertheless, the large changes in total body calcium indicate that
bone loss in this group of patients is not localised to the
axial skeleton but represents a more generalised phenomenon.
The magnitude of the losses observed together with the presence
of severe clinical manifestations of osteoporosis in some
relatively young patients (Compston et al., 1987) emphasises
the need for further studies on the prevention and treatment of
bone loss in patients with inflammatory bowel disease.

ACKNOWLEDGEMENTS

We are grateful to the Medical Research Council for their
support in the development of the total body calcium measurement technique.

REFERENCES

Cann, C.E., and Genant, H.K., 1980, Precise measurement of
     vertebral mineral content using computed tomography,
     J. Comput. Assoc. Tomogr., 4:493.

Compston, J.E., Judd, D., Crawley, E.O., Evans, W.D., Evans, C.,
     Church, H.A., Reid, E.M., and Rhodes, J. 1987, Osteo-
     porosis in patients with inflammatory bowel disease, Gut,
     28:410.

Motley, R.J., Crawley, E.O., Evans, C., Rhodes, J., and
     Compston, J.E., 1988, Increased rate of spinal trabecular
     bone loss in patients with inflammatory bowel disease,
     Gut, 29:1332.

Ryde, S.J.S., Morgan, W.D., Sivyer, A., Evans, C.J., and Dutton,
     J., 1987, A clinical instrument for multielement _in vivo_
     analysis by prompt, delayed, and cyclic neutron activation
     using Cf-252, _Phys. Med. Biol._, 32:1257

THE RELATIONSHIP BETWEEN SPINAL TRABECULAR BONE MINERAL CONTENT AND

ILIAC CREST TRABECULAR BONE VOLUME

C.D.P. Wright[1], E.O. Crawley[2], W.D. Evans[2], N.J. Garrahan[1],
R.W.E. Mellish,[1], P.I. Croucher[1], and J.E. Compston[1]

Department of Pathology[1], University of Wales College of
Medicine and Medical Physics[2], University Hospital of
Wales, Cardiff, CF4 4XN, UK

INTRODUCTION

Osteoporosis is characterised by bone loss, the distribution of
which is heterogenous. Measurement of bone mineral density at clinically
relevant sites, for example, the vertebrae and femoral neck can be made
using techniques such as quantitative computed tomography (QCT) and dual
photon absorptiometry (DPA). Assessment of trabecular bone volume (TBV)
in iliac crest biopsies has also been used as a diagnostic test for
osteoporosis; however, this may not be representative of other skeletal
sites. Previous studies of the relationship between spinal trabecular
bone mineral density (BMD) and iliac crest TBV have produced conflicting
results, some reporting a strong positive correlation (Bikle et al.,
1985; Torres et al., 1986; Delmas et al., 1986) between the two
measurements, and another failing to find any significant relationship
(Ott et al., 1988). The aim of the present study was to investigate
the correlation between spinal trabecular BMD and iliac crest TBV in
several different patient groups.

MATERIALS AND METHODS

Eighty four patients were studied; 23 (9 males) aged 12-78 years
(mean 54) with primary osteoporosis, 19 (8 males) aged 21-78 years (mean
43) with osteoporosis secondary to inflammatory bowel disease and 42
(15 males), aged 35-71 (mean 52) with rheumatoid arthritis. In the first
two patient groups, osteoporosis was defined as a spinal trabecular
BMD $>$ 2SD below the age and sex-matched normal mean value with or without
one or more vertebral crush fractures. The patients with secondary
osteoporosis had either Crohn's disease or ulcerative colitis and some
were receiving corticosteroid therapy. The remaining patients comprised
a group with rheumatoid arthritis (classical or definite) (Compston
et al., 1988), who were receiving treatment with a variety of non-
steroidal anti-inflammatory agents and with a second line drug (gold,
penicillamine or hydroxychloroquine). None of these patients had
received systemic steroid therapy and none was severely disabled.

Spinal trabecular BMD was measured by single energy CT on a
Philips 350 X-ray CT scanner using a modification of the method by Cann
and Genant, 1980. The results were corrected for the contribution of
non-mineral components using calculated values of estimated soft
tissue corrections based on published tissue composition data (Crawley
et al., 1988). The _in vivo_ precision of this method in our hands is 2.3%.

Full thickness iliac crest biopsies were obtained with a 6 or 8 mm internal diameter Bordier trephine, 2.5 cm below and behind the anterior superior iliac spine. TBV was measured in an average of 40 fields using an IBAS II image analyzer (Kontron, West Germany) at a magnification of x207. Correlations were examined using linear regression analysis or the Spearman rank correlation test and significance was assessed using a two-tailed Student's t test.

RESULTS

As shown in Fig. 1, when all 84 patients were grouped together, there was a significant positive correlation between spinal BMD and iliac crest TBV (r=0.60, P<0.001). When the results for the three patient groups were considered separately, differences emerged. In patients with primary osteoporosis, no correlation was found between the two measurements (r=0.07, P=NS).

In contrast, those with osteoporosis secondary to inflammatory bowel disease showed a significant correlation between spinal trabecular BMD and iliac crest TBV (r=0.65, P<0.01). When the patient with a TBV of 25% and QCT of 220 mg/cm^3 was excluded from the analysis, the correlation, although reduced, remained significant (r=0.503, P<0.05). In patients with rheumatoid arthritis, nearly all of whom had normal spinal BMD values, no significant correlation was found between spinal trabecular BMD and iliac crest TBV (r=0.19, P=NS).

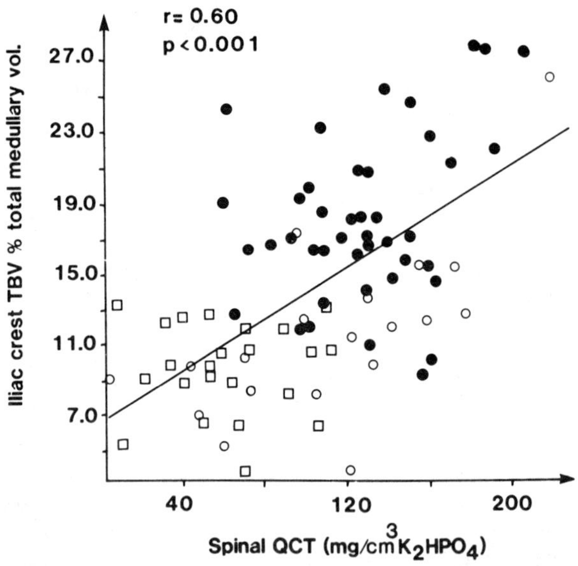

Fig. 1. Correlation between spinal bone mineral density assessed by QCT and iliac crest TBV in all three patient groups (□ = primary osteoporosis, O = inflammatory bowel disease, ● = rheumatoid arthritis). The regression equation is TBV = 6.71 + 0.072 QCT and the 95% confidence limits are 0.401-0.697.
Lower limit for QCT is 90 mg/cm^3 K$_2$HPO4 and for TBV is 14% total medullary volume.

DISCUSSION

Our results demonstrate that the relationship between trabecular BMD
in the spine and iliac crest differs according to the underlying disease
process, a correlation being found in patients with osteoporosis
secondary to inflammatory bowel disease, whilst no such relationship was
evident in patients with primary osteoporosis or rheumatoid arthritis.
These differences may thus reflect variations in the pattern of bone loss
in different types of osteoporosis, bone loss being more generalized in
secondary than in primary osteoporosis. This is further demonstrated by
the overlap in QCT values between rheumatoid arthritis and primary
osteoporotic patients with widely differing TBV values. Another possible
explanation for the poor correlation between the values obtained at the
two skeletal sites in some patient groups could be the measurement
variance associated with histomorphometric assessment of iliac crest TBV
and to a lesser extent measurements made with QCT.

Our results thus confirm previous data demonstrating that assessment
of iliac crest TBV is not an accurate diagnostic technique in
osteoporosis as defined in this paper (Chavassieux et al., 1985, Compston
et al., 1986). Nevertheless, iliac crest biopsy is of value in
determining pathophysiological mechanisms of bone loss, and is also
sometimes indicated to exclude other forms of metabolic bone disease.
The variation in relationship between spinal trabecular BMD and iliac
crest TBV in different patient groups indicates possible different
distributions of bone loss between these groups, a finding which has
interesting pathophysiological implications.

ACKNOWLEDGEMENTS

We would like to thank the Cancer Research Campaign (London) and the
Welsh Office for their financial support.

REFERENCES

Bikle, D.D., Genant, K.H., Cann, C., Recker, R.R., Halloran, B.P.,
    Strewler, G.J., 1985, Bone disease in alcohol abuse, <u>Ann Intern
    Med,</u> 103:42.
Cann, C.E., and Genant, H.K., 1980, Precise measurement of vertebral
    mineral content using computed tomography, <u>J. Comput Assist Tomogr,</u>
    4:493.
Chavassieux, P.C., Arlot, M.E., and Meunier, P.J., 1985, Intersample
    variation in bone histomorphometry: comparison between parameter
    values measured on two contiguous transiliac bone biopsies, <u>Calcif
    Tissue Int,</u> 37:345.
Compston, J.E., Vedi, S., and Stellon, A.J., 1986, Inter-observer and
    intra-observer variation in bone histomorphometry, <u>Calcif Tissue
    Int,</u> 38:67.
Compston, J.E., Crawley, E.O., Evans, C., O'Sullivan, M.M., 1988, Spinal
    trabecular bone mineral content in patients with non-steroid
    treated rheumatoid arthritis, <u>Ann Rheum Dis,</u> 47:660.
Crawley, E.O., Evans, W.D., and Owen, G.M., 1988, A theoretical analysis
    of the accuracy of single energy CT bone mineral measurements, <u>Phys
    Med Biol,</u> 33(10):1113.
Delmas, P.D., Fontanges, E., Duboeuf, F., Boivin, G., Chavassieux, P.,
    and Meunier, P.J., 1986, Comparison of bone mass measured by
    histomorphometry on iliac biopsy and dual photon absorptiometry of
    the lumbar spine, <u>Bone,</u> 9:209.

Ott, S.M., Kilcoyne, R.F., and Chesnut, C.C., III, 1988, Comparison among methods of measuring bone mass and relationship to severity of vertebral fractures in osteoporosis, <u>J Clin Endocrinol Metab</u>, 66(3):501.

Torres, A., Lorenzo, V., and Gonzales-Posada, J.M., 1986, Comparison of histomorphometry and computerized tomography of the spine in quantitating trabecular bone in renal osteodystrophy, <u>Nephron</u>, 44:282.

CLINICAL STUDIES ON OSTEOPOROSIS

J.E. Harrison, K.G. McNeill, S.S. Krishnan, T.A. Bayley,
F. Budden, R. Josse, T.M. Murray, W.C. Sturtridge, C. Müller,
N. Patt, A. Strauss, S. Goodwin

Toronto General Hospital, the Bone and Mineral Group
University of Toronto, Toronto, Canada.

Osteoporosis is, by definition, a reduction in bone mass with remaining bone of normal composition. The related loss in bone strength results in fractures with little or no abnormal force. Since a valid measurement of low bone mass or osteopenia was not widely available until recently, the diagnosis of osteoporosis was based on evidence of vertebral fractures, together with generalized osteopenia, based on the radiological assessment of spinal radiographs. A vertebral fracture causes partial or complete compression (anterior or mid-vertebral compression or loss in height of the full width of the vertebrae). In the absence of low bone mass, however, these vertebral deformities would not be considered osteoporotic fractures. A vertebral deformity without osteopenia may be attributed, for example, to an old traumatic fracture, to genetic anomaly or to such pathological bone disease as tumour. With aging, we observe an increasing incidence of mid-thoracic anterior wedge deformities without radiological evidence of osteopenia. These age related deformities are not considered by radiologists to be vertebral fractures and usually the radiographs are reported as normal or normal for age. The radiological assessment of osteopenia is subjective, however, and a quantitative measurement of bone mass is required to provide a more reliable estimate of the degree of osteopenia associated with vertebral deformities. In addition, the quantitative measurement of bone mass should detect significant osteopenia with increased risk of fracture in patients without vertebral deformity.

At Toronto, neutron activation analysis (NAA) has been used for more than 15 years to measure the total bone mineral mass in the central third of the skeleton.[1] The facility has been substantially upgraded over the past year. With the current facility, the subject is exposed to a flux of neutrons within a heavily shielded room. The PuBe sources (110 Ci) are moved from a shielded position into hollow tubes to provide bilateral neutron exposure over a 60 cm length of trunk and upper thighs. After a 20 minute exposure, the subject is moved to the whole body counter where sensitivity for gamma counting is optimized by use of 8 extruded detectors, 40 x 10 x 10 cm^3, 4 above and 4 below the subject, providing near 4π geometry. The signal from each detector is entered into a separate section of the multichannel analyser by use of two mixer routers, each with 4 inputs. The ^{49}Ca count is reported as an index, CaBI, which is the ratio of the patient's count to the mean value for

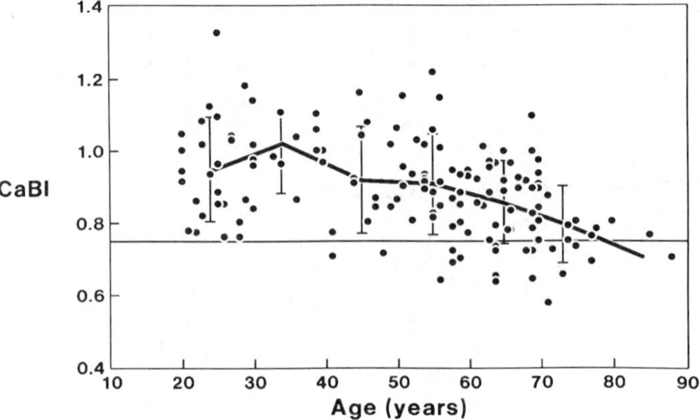

Fig. 1. CaBI values for female volunteers plotted as a
function of age. The mean values for each decade
are shown and the vertical error bars represent
1SD of the mean. The line at 0.75 represents the
lower limit for adults <40 yrs of age.

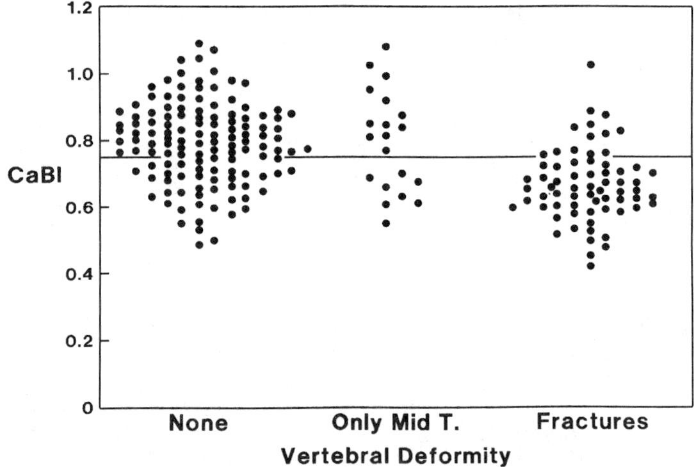

Fig. 2. CaBI values on women investigated for postmenopausal osteoporosis. The line at CaBI = 0.75 represents the lower limit for normal premenopausal
women. The group called "fractures" are those
with deformities in distal thoracic or lumbar vertebrae. Only Mid T represents subjects with deformity confined to the mid thoracic vertebrae.

volunteers (<55 years) of the same size, based on maximum height and arm span.[2] CaBI data in healthy, female volunteers are shown as a function of age in Fig. 1. Women under 40 years have mean CaBI near unity and values range ± 25% from unity, with a lower limit of 0.75. In agreement with all reported data, the mean CaBI decreases with age. With our cross sectional data, the mean decrease in CaBI was 5.5% per decade after the age of 50.

CaBI values have been related to spinal radiograph data obtained on patients investigated for possible postmenopausal osteoporosis. (Fig. 2) Women with no evidence of vertebral deformity on spinal radiographs had a mean CaBI of 0.80 ±(0.13)(1SD). About a third of these patients had CaBI values < 0.75. Patients with at least one deformity in the distal thoracic or lumbar vertebra had significantly lower CaBI values (mean CaBI=0.66 ±[0.10][1SD]), providing confirmation of osteopenia and supporting a diagnosis of osteoporotic fractures. Twenty patients had only 1 or 2 deformities, confined to the mid thoracic vertebrae, and these subjects had the same range of CaBI values (mean CaBI 0.81, ±[0.15][1SD]) as those without deformity. This CaBI data confirms the radiological evaluation that most of these mid-thoracic deformities are not osteopenic fractures. Possibly, with aging, some mid thoracic deformity can occur as a result of molding to prolonged poor posture.

Osteopenia is also observed in patients with no radiological evidence of vertebral fracture. Presumably, those osteopenic patients have increased risk of fractures but some other factors, in addition to osteopenia, might be required to be present before fractures occur. The osteopenic patients of the three groups in Fig. 2, those with CaBI values < 0.75, differed in age. (Table 1) Osteopenic patients with no vertebral fractures or deformity were, on average, 61 years of age, or 8 years younger than the patients with vertebral deformity. Perhaps, with aging, fractures occur as a result of loss in muscle tone and muscle strength allowing, during normal physical activity, abnormal force on osteopenic bone.

We have had the opportunity to follow more than half of these patients over a 4-year period. (Table 1) Of the patients who had osteopenia but initially no vertebral deformity, only 3 (or 12.5%) had a vertebral fracture during the 4-year follow-up period. Of the 5 with only mid thoracic deformity, 1 (20%) developed a vertebral fracture. In contrast, more than half the patients with distal thoracic or lumbar deformities initially developed further fractures over the subsequent 4 years.

The data of Table 1 were obtained on patients treated with fluoride over the 4 years. Over this period, the CaBI values increased substantially, on average, from 0.65 to 0.82.[3] (Fig. 3) The observed mean increase in CaBI of 0.17 (0.65 to 0.87) represents half the estimated average loss in bone mass by osteoporosis, as indicated by the pretreatment reduction in CaBI of 0.35 from the normal value of 1.0 to the observed value of 0.65. This improvement in bone mass is reflected in the radiological assessment of increased vertebral bone density on spinal radiographs. (Fig. 4)

However, the improvement in CaBI was variable. (Fig. 5) About 20% of the patients showed no change in CaBI, while a few doubled their bone mass in the area of measurement. We sought reasons for this variable response. The change in CaBI was weakly but significantly inversely correlated to initial CaBI (CaBI$_0$) (r=-0.32, p<0.05). The change in CaBI was not significantly correlated to the individual variations in fluoride dose, but showed strong correlations to serum fluoride (r=0.50, p<0.01)

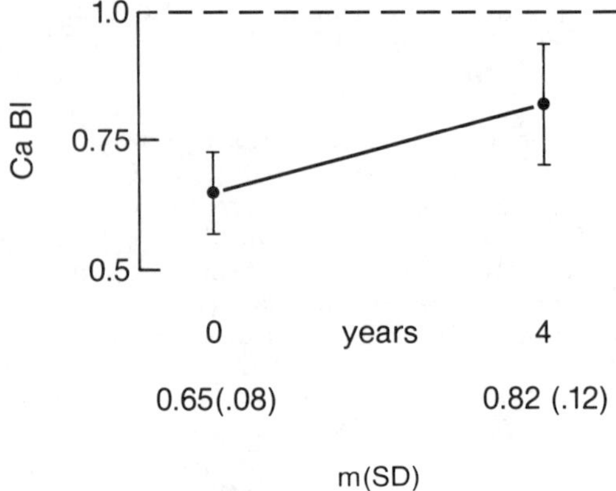

Fig. 3. Mean CaBI values on 61 patients with postmenopausal osteoporosis prior to and after 4 years of continuous treatment with sodium fluoride 20 mg b.i.d., elemental calcium 1 g/day and vitamin D_2 50,000 I.U. every 2 wk. The error bars represent 1 SD. The dotted line at CaBI value 1.0 represents the mean value for normal premenopausal women.

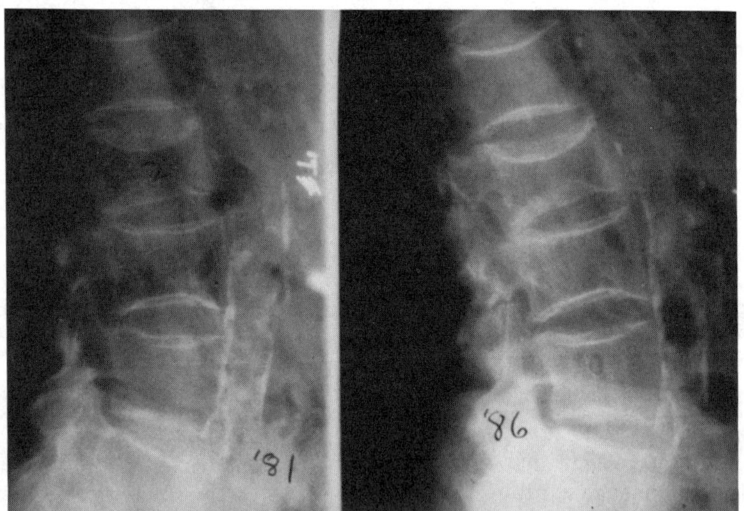

Fig. 4. Lateral lumbar spine radiographs. The radiographs were taken prior to (left) and after (right) 4 years on fluoride treatment and show increased bone mineral density over the 4 year interval.

Table 1. Osteopenic Patients

	No.	CaBI M(SD)	Age, yr M(SD)	4 yr Follow-up No.	New Vertebral Fractures No. (%)
No vertebral deformity	41	0.65(0.06)	61(9)	24	3 (12.5)
Mid thoracic deformity	7	0.64(0.06)	70(8)*	5	1 (20)
Distal thoracic and lumbar deformities	58	0.63(0.07)	69(7)*	30	17 (57)

* Significant difference from patients with no vertebral deformity: p<0.001

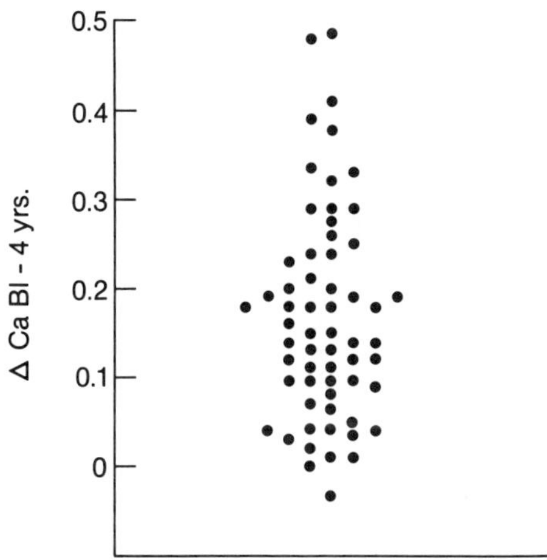

Fig. 5. The change in CaBI values (Δ CaBI) over 4 years on fluoride treatment, for each of the 61 patients. Δ CaBI values are the differences between initial and final values calculated from the linear regression for all measurements (5–8) carried out on each subject over the 4 years.

and to bone fluoride (r=0.52, p<0.01). Serum and bone fluoride depend
not only on fluoride dose but also on fractional intestinal absorption and
on renal excretion. Serum fluoride was significantly correlated to loss in
renal function as measured by reduction in the rate of creatinine
clearance (r=0.52, p<0.01).

Since osteoporotic fractures are due to loss of bone mass and related
loss in bone strength, effective treatment requires restoration of low
bone mass to normal. Fluoride is the only treatment shown to increase
substantially bone mass over many years and the increase in bone mass
occurs in the vertebrae where fractures commonly occur. Thus, with
fluoride treatment, improvement in bone strength and reduction in the
risk of further vertebral fractures would be anticipated, but, as yet, a
reduction in fractures has not been established. In our study, the
number of new vertebral fractures ($V\#_4$) did not correlate to change in
CaBI or to final CaBI after 4 years of treatment. The number of new
vertebral fractures correlated inversely to initial CaBI ($CaBI_0$)(r=-0.37,
p<0.01) and directly to the number of vertebral fractures at onset of
treatment ($V\#_0$)(r=0.40, p<0.01). The number of new vertebral fractures
also correlated to age (r=0.33) but partial correlations showed significance to $V\#_0$ and $CaBI_0$ (r=0.50), while the addition of age did not increase
significance.

In conclusion, the CaBI provides a test for the diagnosis of osteopenia and, just as important, a test to exclude osteopenia in symptomatic
patients suspected of osteoporosis but who are found to have normal
CaBI values. The CaBI also demonstrates changes in bone mass with
progression of disease or in response to treatment. The CaBI has shown,
for example, that fluoride treatment can reverse osteoporotic bone loss.
Although improvement in bone strength would be expected with the
increase in bone mass, a reduction in the risk of further fractures still
needs to be proven. If increase in bone strength does, in fact, not go
with increased bone mass, the physician will require a non-traumatic way
of measuring bone strength.

ACKNOWLEDGEMENTS: We thank S. Lyn and E. Bogdanovic for technical
assistance. The work was supported, in part, by the Canadian Geriatrics
Research Society and by the Auxiliary of the Toronto General Hospital.

REFERENCES

1. K.G. McNeill, B.J. Thomas, W.C. Sturtridge, J.E. Harrison, In vivo
    neutron activation analysis for calcium in man. J. Nucl. Med.
    14:502-506 (1973).
2. J.E. Harrison, C. Williams, J. Watts, K.G. McNeill, A bone calcium index
    based on partial body calcium measurements by in vivo activation
    analysis. J. Nucl. Med. 16:116-122 (1975).
3. T.M. Murray, J.E. Harrison, T.A. Bayley, R.G. Josse, W.C. Sturtridge, R.
    Chow, F. Budden, L. Laurier, K.P.H. Pritzker, R. Kandel, R. Vieth, A.
    Strauss, S. Goodwin, Fluoride treatment of postmenopausal osteopo-
    rosis: age, renal function and other clinical factors in the osteogenic
    response, J. Bone Miner. Res., In press.

ALTERATIONS OF BONE MINERALS IN UREMIC PATIENTS AND RENAL GRAFT RECIPIENTS

John Kalef-Ezra, Konstantinos Siamopoulos,
Apostolos Karantanas, John Xatzikonstantinou,
George Sferopoulos, Seiichi Yasumura, and
Dimitrios Glaros

Medical Physics Laboratory, Department of Internal
Medicine and Radiology Laboratory, Medical School
University of Ioannina, Ioannina, Greece, GR 451.10

INTRODUCTION

Renal osteodystrophy is a common complication in chronic renal failure
(CRF). A successful kidney transplantation alleviates most of the complications of the disease. However, the influence of kidney transplantation
on the status of bone is not clear. In this study, the bone mineral
status in such patients was evaluated by applying two techniques:
Quantitative Computed Tomography (QCT) for assessing spinal trabecular
bone density, and Partial Body Neutron Activation Analysis (PBNAA) for
measuring the phosphorus content of the hands.

METHODS

Patients

One hundred and thirteen CRF patients (aged 25y to 70y) and 17 renal
graft recipients (RGR) (aged 25y to 70y) were studied. Among the CRF
group, 37 were non-dialyzed (NDCRF) patients, 62 were receiving chronic
haemodialysis (HD), and 14 were on continuous peritoneal dialysis (CAPD).
All subjects belonged to these subgroups for at least 4 months. Half of
the patients were examined with both techniques.

Quantitative Computed Tomography (QCT)

A modified version of the Cann-Genant single KV_p technique (Cann and
Genant, 1980) was applied, using a Siemens DR-2 scanner. The linear
attenuation coefficient that corresponds to each voxel reflects the
electron density in this volume element. The trabecular bone equivalent
density (TBED) of the T_{12}, L_1, L_2, L_3 vertebral bodies was expressed as the
density of a K_2HPO_4-solution that exhibits an identical linear attenuation
coefficient. Reference solutions of K_2HPO_4 with concentrations 0, 50,
100, 150, 200 mg cm^{-3}, positioned under the subject's back, were used for
calibration. Whereas a correction factor for field non-uniformity was
experimentally determined and applied to the measurements, no correction
was applied for the inhomogeneity of the organic content in the trabecular
bone, i.e. the influence of fat and collagen.

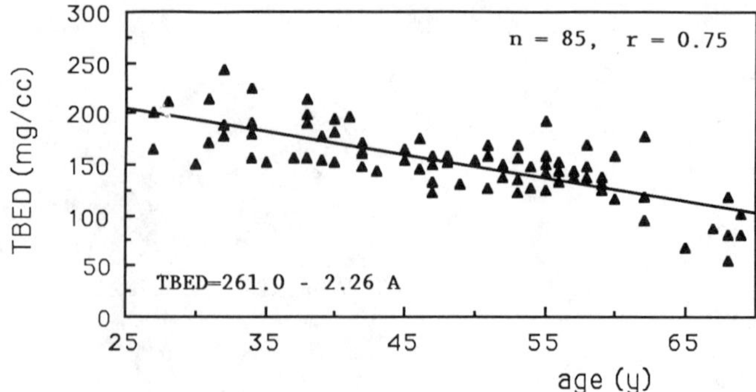

Fig. 1.   Dependence of trabecular bone equivalent density (TBED) on
age for normal men.

Repeated in vivo measurements of 45 subjects asymptomatic for osteoporosis (30y to 65y) showed that the coefficient of variation depends on
age, and ranges from 1.5% to 3.6%.   Thermoluminescent dosimetry indicated
that the imparted energy for each examination (4 sections and a scout
view) is 9 mJ and the effective dose equivalent is 370 μSv.   The
corresponding fatal risk factor of 5×10^{-6} is lower than that of the
routine radiological examination of the lumbar spine.

Data on 220 Greek subjects asymptomatic for osteoporosis, aged 30y to
65y, served as the reference line (Figs. 1,2).

<u>Partial Body Neutron Activation Analysis (PBNAA)</u>

PBNAA was applied for measuring hand bone phosphorus (HBP).   The
technique is based on the detection of the delayed γ-rays emitted from
^{28}Al produced via the ^{31}P(n,α)^{28}Al fast neutron reaction (Glaros et al.,
1987).

The irradiation unit consists of two ^{241}Am-Be sources with a neutron
emission rate of 7×10^6 s^{-1} each.   A 3 mm-thick copper layer around the
sources in the irradiation position shields the low energy photons.   The
subject grasps the copper shielding and the sources are pulled from the

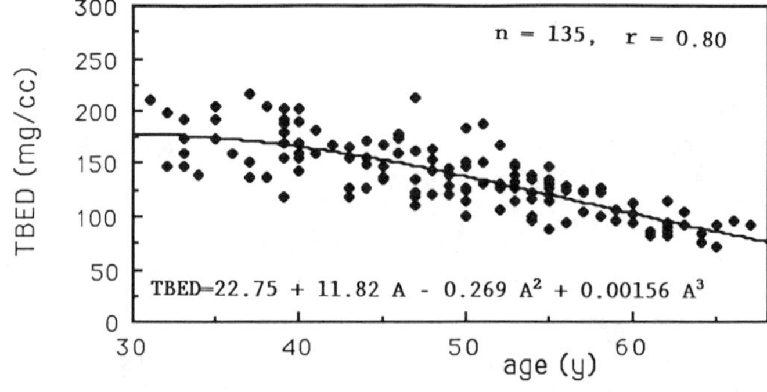

Fig. 2.   Dependence of trabecular bone equivalent density (TBED) on
age for normal women.

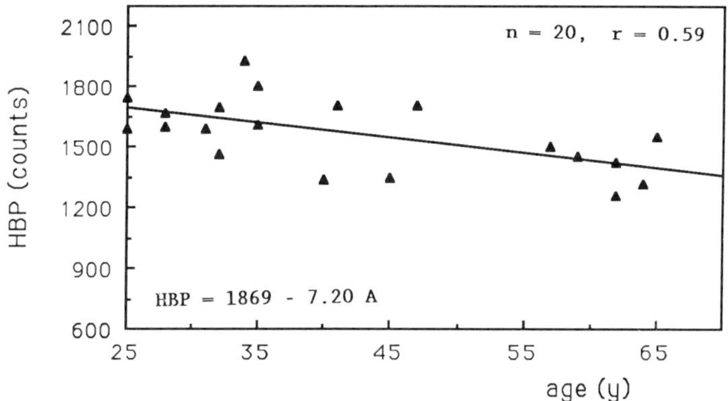

Fig. 3. Dependence of hand bone phosphorus (HBP) on age for normal men.

underground storage position to the irradiation position. The 4-min irradiation is followed by a 4-min counting of the ^{28}Al induced activity. There is a 0.5-min lag time between irradiation and counting. The 1.78 MeV-photons emitted by ^{28}Al ($T_{1/2}$ = 2.24 min) are counted with the hands sandwiched in contact with two cylindrical 20.3 cm x 10.0 cm NaI(Tl) detectors (Glaros et al., 1987). Background radiation is reduced by the detector shielding of 6cm thick lead. Relative measurements were used and no attempt was made to correlate quantitatively the ^{28}Al counts with the phosphorus mass, due to the potential presence of large systematic errors.

Repeated in vivo measurements in 30 subjects showed that the coefficient of variation ranges from 2.5% to 3.0%, for persons asymptomatic for osteoporosis, aged 30 to 65 years, respectively. Dosimetric measurements with foil activation detectors (Al, In, and Au), track detectors, and thermoluminescent dosimeters (TLD-700) showed that the absorbed dose at any location of the skin palm in contact with the copper shielding does not exceed 2 mGy. Eighty percent of this dose is due to fast neutrons (Q=13.5) and the rest is due to photons. Therefore, the dose equivalent at any location of the skin palm, does not exceed 22 mSv. Dose equivalent to the surface of the thorax does not exceed 0.2 mSv.

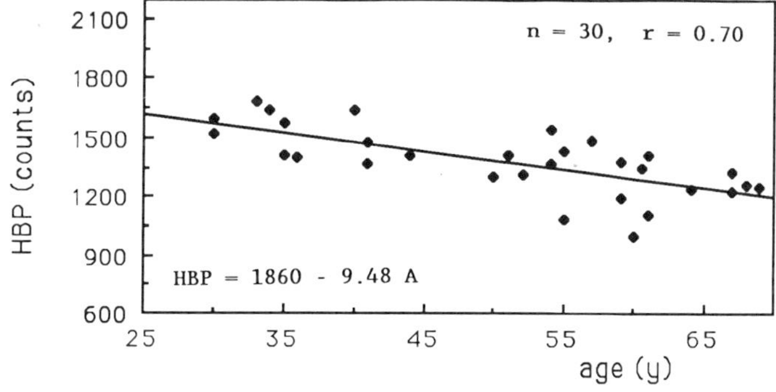

Fig. 4. Dependence of hand bone phosphorus (HBP) on age for normal women.

Table 1. Trabecular Bone Equivalent Density (TBED) and Hand Bone
Phosphorus (HBP) Relative to Matched Controls in Uremic
Patients

Group	n	relative TBED	CV*	n	relative HBP	CV*
NDCRF	19	1.05 ± 0.06	24%	23	0.99 ± 0.03	14%
HD	47	0.82 ± 0.03	22%	50	0.91 ± 0.02	16%
CAPD	6	0.79 ± 0.09	28%	13	0.93 ± 0.04	16%
RGR	15	0.55 ± 0.03	24%	13	0.94 ± 0.04	14%

* C.V. Coefficient of variation
NDCRF: non-dialyzed chronic renal failure patients; HD: chronic
hemodialysis patients; CAPD: continuous peritoneal dialysis patients; RGR:
renal graft recipients

Data on 50 normal subjects asymptomatic for osteoporosis, aged 25y to
70y, served as the reference line (Figs. 3 and 4).

RESULTS

Table 1 shows the TBED and HBP values of uremic subjects compared with
controls matched for sex and age.

The TBED on 11 patients with osteosclerosis (Rugger-Jersey) are not
included in Table 1. These patients revealed a remarkable inhomogeneity
in the vertebral trabecular bone area, resulting in a meaningless measurement. The HBP values in osteosclerotic HD patients, however, were on
average 81% ± 12% (x ± 1 SD) of the values found in non-osteosclerotic HD
patients.

The NDCRF patients showed no statistically significant differences in
either TBED or HBP values from those of the controls.

The hand phosphorus content of RGR (6 months to 13 years after
transplantation) was similar to that of normals. However, a significant
reduction (55% on average) of the TBED was found in the same group. In
four RGR, studied before and four months after grafting, a 37% ± 2%
($\bar{x}$ ± 1 SD) reduction of TBED was demonstrated. Repeated measurements of
HBP in 4 RGR one year later showed no significant change (-1.3% ± 4.5%).

A weak correlation was found between TBED and HBP relative values for
24 patients undergoing long-term haemodialysis (more than two years).

TBED measurements were repeated after 6 months in 36 HD patients and
an average TBED reduction of 6% ± 2% (x ± 1 SEM) was found. HBP
measurements were repeated after twelve months in 23 HD patients and an
average reduction of 4% ± 1.7% (x ± 1 SEM) was observed.

DISCUSSION

Severe chronic renal failure is associated with alterations of bone
mineral metabolism (Harrison et al., 1977). These alterations contribute
to the well-known manifestations of renal osteodystrophy, such as osteitis

fibrosa, osteomalacia, osteosclerosis and osteoporosis. These histological manifestations may occur either exclusively, or concurrently, in the
same subject. However, these lesions do not appear to the same extent at
various parts of the skeleton. There is some evidence that the earliest
and most severe manifestations of the disease may occur in bones with high
trabecular content and the hands. Thus, these sites appear to be approximate for the study of renal osteodystrophy.

The TBED and HBP values of the NDCRF patients were not statistically
different from those of the controls. However, a significant reduction in
TBED values in dialyzed patients was found, and the majority of the
subjects showed slightly lower HBP values than those of the matched
controls. Repeated measurements on HD subjects confirmed these findings.
Catto et al. (1973), applying PBNAA in thirteen HD patients, found a high
rate of calcium loss in the hands (11% loss in 0.5y).

Eastell et al. (1985) found no significant change in total body Ca,
seventeen months after transplantation. In the present study, whereas the
RGR showed only a minor reduction in HBP, the TBED was remarkably reduced.
The TBED reduction seems to take place shortly after the transplantation.

These findings reflect the complexity of bone tissue disorders
associated with chronic renal failure. Long-term investigation of the
same subjects with various in vivo techniques, will hopefully, provide a
better understanding of bone alterations in the human body.

REFERENCES

Cann, C., and Genant, H. K., 1980, Precise measurement of vertebral
        mineral content using computed tomography, J. Comput. Assist.
        Tomogr., 4:493.
Catto, R., McIntosh, J., MacDonald, A., and Mc.Leod, M., 1973,
        Haemodialysis therapy and changes in skeletal calcium, The
        Lancet, 1:1150.
Eastell, R., Kennedy, N. S. J., Smith, M. A., Tothill, P., and
        Anderton, J. L., 1985, Changes in total body calcium after renal
        transplantation: Effects of low dose steroid regime. Nephron,
        40:139.
Glaros, D., Xatzikonstantinou, J., Leontiou, J., and Kalef-Ezra, J.,
        1987, A partial body activation analysis technique for the
        measurement of phosphorus in bone, in: "In Vivo Body Composition
        Studies", K. Ellis, S. Yasumura, W. D. Morgan, eds., The
        Institute of Physical Sciences in Medicine, London.
Harrison, J. E., McNeill, K. G., Meema, H. E., Oreopoulos, D., Rabinovich,
        S., Fenton, S., and Wilson, D. R., 1977, Partial body calcium
        measurements on patients with renal failure, Metabolism, 26:255.

BODY COMPOSITION STUDIES IN PREMENOPAUSAL HEALTHY WOMEN

R. Lindsay, S. Himmelstein, B.S. Herrington and
F. Cosman

Regional Bone Center, Helen Hayes Hospital
W. Haverstraw, NY 10993, USA

INTRODUCTION

Dual photon absorptiometry (DPA) is a potentially
exciting tool in studies of body composition. It allows
direct measurement of total bone mass and lean body mass as
well as fat mass that can then be calculated from all pixels
of a total body scan that do not contain bone. Thus, a
semi-direct measurement of fat mass is obtained.
Importantly, DPA or its modern x-ray equivalent can be
applied to a much broader population including the aged and
other individuals who have disorders that would preclude
their participation in underwater weighing. We have
evaluated the technique in a relatively homogeneous
population of premenopausal women and here report the
preliminary results and problems perceived thus far.

METHODS

Normal premenopausal women were recruited from our
previous normal study of regional bone mass (Lindsay et al,
1986). All were healthy young women with regular menses,
who were not taking any medication likely to affect mineral
metabolism or weight control. Specific criteria were
established prior to recruitment and inclusion or exclusion
of each individual was decided by a single observer (BSH)
based on a standard questionnaire.

Body composition was determined by dual photon
absorptiometry using a Norland 2600 (Norland Corp, Fort
Atkinson, Wisconsin). In the original experiments the
software supplied by the manufacturer was used. (In later
experiments performed in conjunction with the Body
Composition Unit, St. Lukes-Roosevelt Medical Center, New
York, calibration was performed against a standard
consisting of soft tissue (beef) and bone of known
composition. In addition, 12 individuals were measured on
the Lunar DP-4 at St. Lukes, and our own Norland 2600.)
Regional bone mass was estimated in the lumbar spine and
femoral neck also by DPA using a Lunar Radiation DP-3, with

corrections for variability in source as we have previously
described (Lindsay et al, 1987). Bone mass in the radius
was estimated by single photon absorptiometry using a
Nuclear Data 1100, which estimates bone mass proximally
from, and distally to, the 5 cm gap between the radius and
ulna.

Body composition was also determined by estimation of
body mass index from height and weight (wt/ht^2), and from
skinfold thickness at four standard sites using the formula
of Durnin and Womersley to calculate fat mass from the sum
of the skinfold measurements. Finally, bioelectrical
impedance measurements were obtained on all women using a
commercially available device (RJL BIA 101, Detroit, MI).

All data were assembled on a microcomputer and
statistical analyses were performed by standard techniques.
Comparison between means was performed by a two-tailed
t-test after testing for normal distribution. Linear
regression analyses were performed using the least squares
method.

Table 1. Characteristics of the Population

	Mean	SD	Range
Age	39.95	9.51	17 – 53
Height (cm)	163.14	7.74	137.16 – 185.42
Weight (kg)	63.17	9.97	43.18 – 89.55
Lumbar spine (g/cm^2)	1.19	0.16	0.80 – 1.58
Femoral neck (g/cm^2)	0.85	0.13	0.59 – 1.23
Ward's triangle (g/cm^2)	0.76	0.16	0.49 – 1.36
Trochanter (g/cm^2)	0.70	0.09	0.52 – 0.89
Distal radius (g/cm)	36.08	5.63	19.80 – 51.60
Proximal radius (g/cm)	37.06	4.96	22.60 – 51.40
TBBM (g)	2599.40	59.24	1700.50 – 4074.50

RESULTS

The characteristics of the population are detailed in
Table 1.

Bone Mass

Mean total body bone mass (TBBM) was 2599 ± 59 gm.
TBBM did not decline with age in this population (Fig. 1),
but was significantly related to height, weight, and percent
fat measured by all of the techniques used. TBBM
constitutes a variable percentage of lean body mass ranging
from 4-10%. The proportion of lean body mass occupied by
the skeleton did not change with age in the premenopausal
group, but declined dramatically in a separate group of

postmenopausal women. Regional measurements of bone mass
also did not decline with age in premenopausal women, with
the exception of bone mineral density in the femoral neck,
particularly the region of Ward's triangle (r=-0.35).

<u>Fat Measurements</u>

     Percentage body fat was calculated from four separate
sets of measurements.

     Fat measured by DPA was higher (39.67 ± 12.46%) than by
all other techniques. Percentage fat was lowest using
height and weight to calculate body mass index (BMI).
Measurements by all techniques correlated highly with each

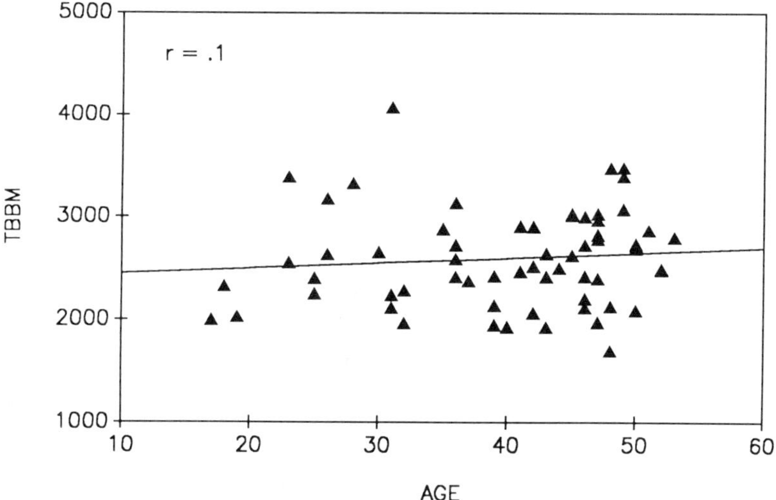

Fig. 1.   Relationship between TBBM and age in premenopausal
          women.

other with correlation coefficients ranging from 0.7 - 0.9.
Body fat increased significantly with age with all
techniques, except BMI, but the increments were highly
variable; DPA showed a 3% increase per decade with only a
nonsignificant increase of less than 1% per decade using
BMI.

     The difference in measurements of mean fat (up to 10%)
and the greater variance observed between the DPA results
and other techniques were caused by at least two phenomena.
First, there appeared to be inadequate DPA calibration
resulting in underestimation of body fat at values under 20%
but overestimation at fat values in excess of this. Thus,
all techniques equated at approximately 20% fat.

Standardization using beef of known fat content against a
second DPA technique (using a Lunar Radiation DP-4 already
standardized) corrected this problem, and reduced mean fat
percent by DPA to $32.86 \pm 0.92\%$. However, this figure was
still significantly greater than that recorded by other
techniques. The differences in fat measurement from DPA to
all other techniques were correlated positively with the
ratio TBBM/LBM. This suggested that the assumption that the
skeleton occupies a constant fraction of lean mass, required
for underwater weighing against which the other techniques
are standardized, creates the problem.

<u>Lean Body Mass</u>

     Lean body mass (kg) did not change with age or with
increasing % fat. Although weight, which appears to be an
important determinant of TBBM, increases with age, this is

Table 2.   Percent Fat in Premenopausal Women
           Estimated by 4 Separate Techniques (see text)

Characteristics of the population % body fat		
DPA	32.86 ± 0.92	
BIA	29.43 ± 0.76	$P < .005$
Skinfolds	30.51 ± 0.81	$p = .05$
BMI	29.14 ± 0.70	$p < .02$

totally due to changes in fat content. Thus fat/LBM
increased from 39.8 ± 4.6 in 20 year old individuals to 54.1
$\pm 2.3\%$ in those 40 years or older.

DISCUSSION

     Dual photon absorptiometry is a useful technique for
estimation of body composition giving direct estimate of
total body bone mineral and semi-direct measurement of body
fat and lean body mass. Like the data from Gallagher et al(1987)
TBBM did not change with age in premenopausal women while
regional measurements in the hip tended to decline. The
fall in mass was, however, quite small, an estimated
reduction of .2 gm/cm in the region of interest, which is a
small part of the skeleton (.089% of total skeletal area),
and may account for the failure to detect an age-related
decline in TBBM. Alternatively, small increments in
cortical bone mass, undetectable in the relatively small
areas of interest evaluated, could easily offset the
apparent reduction in Ward's triangle, an area consisting of
both cancellous and cortical bone. Indeed, increases in
cortical bone during this period of life are suggested by
other cross-sectional data.

Measurement of body fat by our method (Norland DPA)
recorded values considerably higher than those estimated by
other techniques. Without adequate calibration, however,
the results may be inaccurate, even though there are
excellent correlations among the methods. After
calibration, significant improvements in accuracy were
obtained and the variance between the DPA fat measurements
was significantly reduced. The further differences in fat
between the DPA results and the other techniques could not
be explained on the basis of calibration. Closer
examination of the data confirmed the suggestion of others
(Mazess et al, 1984) that the differences in fat
measurements among DPA and other techniques are related to
the fraction of lean body mass occupied by the skeleton.
This suggests that the two compartmental model required by
the underwater weighing techniques, considered by many to be
the gold standard for body composition, is inadequate.
Admittedly, the errors introduced are not particularly
large. However, the major advantage of the DPA technique
and its more modern derivative, dual energy x-ray
absorptiometry, is the capability to allow investigation of
patients who could not undergo underwater weighing, such as
those with spinal cord injury or cerebro-vascular accident,
and in whom alterations in skeletal size may be important.

SUMMARY

Body composition studies of healthy premenopausal women
were performed using dual photon absorptiometry and the
results compared with fat estimates obtained from skinfold
measurements, body mass index, and bioelectrical impedance.
Total body bone mineral did not decline with age nor did the
fraction of lean body mass occupied by the skeleton, which
ranged from 4-10%. DPA values for fat were significantly
higher than those obtained by other techniques, in part
because of calibration but also because of apparent
inadequate correction for the fraction of lean body mass
occupied by bone in the assumptions made for the other
techniques.

REFERENCES

Gallagher, J. C., Goldgar, D., and Moy, A., 1987, Total
        bone calcium in normal women: effect of age and
        menopause status, J Bone Min Res., 2:491.
Lindsay, R., Williams, S., Dempster, D., McMahon, D., and
        Tohme, J., 1989, Bone mineral density changes with
        age, body size, and crush fractures, J Bone Min
        Res., 1:121.
Lindsay, R., Fey, C., and Haboubi, A., 1987, Dual photon
        absorptiometric measurements of bone mineral density
        increase with source life, Calcif Tis., 41:293.
Mazess, R. B., Peppler, W. W., and Gibbons, M., 1984, Total
        body composition by dual-photon (^{153}Gd)
        absorptiometry, Am J Clin N., 40:834.

TOTAL AND REGIONAL BONE MASS IN

HEALTHY AND OSTEOPOROTIC WOMEN

Anders Gotfredsen, Christian Hassager, and
Claus Christiansen

Department of Clinical Chemistry, Glostrup
Hospital, University of Copenhagen, Denmark

INTRODUCTION

Osteoporosis is characterized by a reduced bone density,
which because of a reduced compressive strength leads to the
typical non-traumatic fractures. Although reliable measurements of bone mass have been made for two decades in at least
20 different centers over the world, there is still no definite agreement on the patterns of postmenopausal and osteoporotic bone loss. There are obvious reasons for this controversy: 1) small sample sizes, 2) differences in measuring
equipment, and 3) insufficient precision. However, differences
in choice of statistics and interpretation of the data also
play a role. On the Department of Clinical Chemistry at
Glostrup Hospital we have made absorptiometric bone measurements for the last 18 years. The aims of the work presented
here have been: 1) to study the pattern of postmenopausal bone
loss, 2) to investigate regional differences in bone loss, 3)
to study the pattern of osteoporotic bone deficit, and 4) to
compare the diagnostic ability of different methods.

METHODS

Total body bone mineral (TBBM) and density (TBBD) were
measured according to Gotfredsen et al. (1984a+b) using dual
photon absorptiometry (DPA). Bone mineral content (BMC) and
density (BMD) values were calculated in six regions of the
total skeleton: head, arms, chest, spine, pelvis, and legs.

Lumbar BMD was also measured by DPA (Nilas et al., 1986).

Forearm BMC was measured by single photon absorptiometry
(SPA) according to Christiansen et al. (1975).

MATERIAL

I. 135 healthy women of which 56 were postmenopausal
(Gotfredsen et al., 1987; 1989a; Nilas et al., 1988).

II. 104 healthy, recently postmenopausal women who were
followed during a year receiving either estrogen (n = 52) or
placebo (n = 52) treatment (Gotfredsen et al., 1986a+b).

III. 118 postmenopausal women with osteoporotic fractures: 45 Colles' fracture, 46 vertebral fracture, and 27 hip
fracture (Gotfredsen et al., 1989a+b; Nilas et al., 1986;
1987).

STATISTICS

In order to standardize different variables for inter-
comparisons z-scores were calculated. The z-scores are the
deviation of the individual values of a variable from the mean
value of a reference group divided by the standard deviation
of the variable in that reference group. ROC analysis (receiver operation characteristic analysis) was used in order to
compare the diagnostic value of different variables. In ROC
analysis one plots the true positive fraction (sensitivity)
against the false positive fraction (one minus specificity).
The method with the greatest diagnostic ability has a ROC
curve which is closest to the upper left corner of the
diagram. Other statistical measures included paired and
unpaired students t-tests, analysis of variance, as well as
simple linear, multiple, and curvilinear regressions.

RESULTS

We tested several different regressions for the relationship between bone mass and age (linear, quadratic, cubic,
combined linear-exponential). In the postmenopausal women age
was adjusted to the duration of the menopause. The best fit to
the data regarding all regions and areas (TBBM, TBBD, regional
BMD's lumbar BMD, forearm BMC) was achieved using the combined
premenopausal linear + postmenopausal exponential regression,
which in all instances yielded the smallest residual mean
square (Gotfredsen et al., 1987; 1989a; Nilas et al., 1988;
Thomsen et al., 1986). The combined linear + exponential
regression of TBBD on menopausal adjusted age gave the following: r = 0.56, SEE = 7.8%. No significant premenopausal bone
loss was found. Looking at TBBM, the postmenopausal exponential loss was 4%, 0.44% and 0.03% per year at 1, 10 and 20
years , respectively, after the menopause. The overall loss in
TBBM from age 20 to age 80 was approximately 20%. Looking at
the regional BMDs from the total measurement (head, arms,
spine, chest, pelvis, legs), there were no significant
departures from the regression derived from TBBD, and no
significant inter-regional differences.

In our longitudinal study we found that estrogen pre-
served bone mass in all regions and areas as well as in the
total skeleton, whereas the 52 recently postmenopausal women
who were treated with placebo lost bone from all regions and
from the total skeleton. The loss from TBBM was 5% during the
year of the study, and there were no significant differences
between the six regions of the TBBM (possibly due to a type II
error) (Gotfredsen et al., 1986a+b). The changes in lumbar BMD

and forearm BMC were similar to those of the TBBM and the six regional BMCs. However, the ratio of change/precision-error was larger for the forearm BMC compared to spinal BMD (2 versus 0.5).

There was a large overlap between healthy postmenopausal women and postmenopausal women with all three fracture types (Colles', vertebral, hip), regarding both TBBD and all other indices of bone mass (Table 1) (Gotfredsen et al., 1989a+b; Nilas et al., 1986; 1987). We therefore chose to calculate z-scores and ROC plots relative to healthy premenopausal women. This is also the most logical according to the well-known stronger relation between bone mass and fracture incidence compared to that between age and fracture incidence (Melton et al., 1986). Looking at the regional BMDs from the total measurement, we found a clear generalization of the bone deficit (Table 1). However, there were preferential regional deficits in the regions affected by or close to fracture. For instance, the largest deficit in BMD value was that of the legs in the hip fracture patients (72% of healthy premenopausal women). ROC analysis of a comparison between TBBD, lumbar BMD, and forearm BMC revealed the same tendency regarding all fracture types: TBBD had curves that were closest to the upper left corner, and forearm BMC had ROC curves in-between those of TBBD and lumbar BMD. Therefore TBBD had the greatest diagnostic ability, and lumbar BMD by DPA the smallest (Gotfredsen et al., 1989b). Fig. 1 shows the ROC plot for femoral neck fracture.

DISCUSSION

The present data favor the view of a generalization of the bone loss both in the normal menopause and in osteoporosis. This is opposed to the commonly held view that there should exist a specific type of osteoporosis characterized by an early accelerated loss of bone from trabecular regions leading to spinal fractures at an early age. We did, however, find some regional differences in the fracture patients: there was a preferential bone loss in the regions affected by or close to the fracture, but no disproportionate loss in trabecular regions.

It is intriguing that we did not find a higher diagnostic ability of lumbar BMD in vertebral fracture patients than TBBD or forearm BMC, as others have found such a relationship (Eastell et al., 1989). We feel that part of the explanation must be the insufficient precision of lumbar BMD by DPA. With the advent of high-precision methods such as dual energy X-ray absorptiometry (DEXA) (Mazess et al., 1989) a much greater reliability of lumbar measurements will be expected.

CONCLUSIONS

1) Premenopausal bone loss is negligible. 2) Postmenopausal bone loss begins at the menopause and follows an exponential pattern. 3) Postmenopausal bone loss is generalized.

Table 1. BMD Values of Six Regions of the Total Skeleton in
          Five Different Groups of Healthy Women or Women With
          Osteoporotic Fractures. Values Are Given as Mean
          (SD).

	Head	Arms	Chest	Spine	Pelvis	Legs
Premenopausal women, 21-54 y.	4.3 (0.6)	1.4 (0.1)	1.1 (0.1)	1.8 (0.2)	1.9 (0.2)	2.1 (0.2)
Postmenopausal women, 45-77 y.	3.5 (0.5)	1.2 (0.2)	1.0 (0.1)	1.5 (0.3)	1.6 (0.2)	1.9 (0.2)
Colles' fract., 55-77 y.	3.3 (0.5)	1.2 (0.1)	0.9 (0.1)	1.4 (0.2)	1.4 (0.2)	1.7 (0.2)
Spinal fract., 56-75 y.	3.1 (0.6)	1.2 (0.1)	0.9 (0.1)	1.3 (0.2)	1.4 (0.2)	1.7 (0.2)
Hip fract., 42-86 y.	3.1 (0.5)	1.1 (0.1)	0.9 (0.1)	1.4 (0.2)	1.3 (0.1)	1.5 (0.2)

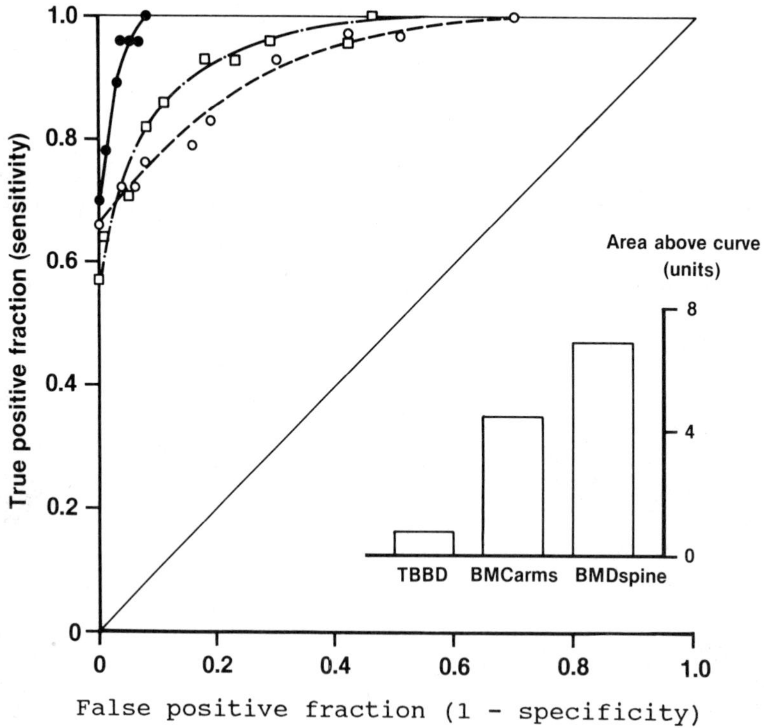

Fig. 1. ROC diagram of a comparison between TBBD (●),
forearm BMC (☐), and lumbar BMD (O) in hip fracture.

4) Osteoporotic bone loss is generalized. 5) In osteoporosis
there are some regional differences depending on the fracture
localization. The bone deficit is largest in the regions
affected by (or adjacent to) fracture. 6) The regional
differences in osteoporosis may be explained purely on the
basis of biological variation. The variation may include total
trabecular/cortical ratio, local trabecular/cortical ratio,
total amount of integral bone, local amount of integral bone,
rate of trabecular bone loss, rate of cortical bone loss, time
of onset of trabecular bone loss, and time of onset of
cortical bone loss. 7) TBBD has better diagnostic ability in
osteoporotic fracture patients than local lumbar or forearm
measurements. The forearm BMC lies in-between in our study. 8)
Local peripheral measurements may be used to screen for
osteoporotic fracture risk although they may only inaccurately
predict the axial bone mass.

REFERENCES

Christiansen, C., Rødbro, P., and Jensen, H., 1975, Bone
    mineral content in the forearm measured by photon ab-
    sorptiometry: principles and reliability, Scand J Clin
    Lab Invest, 35: 323.
Eastell, R., Wahner, H.W., O'Fallon, W.M., Amadio, P.C.,
    Melton, L.J. III, and Riggs, B.L., Unequal decrease in
    bone density of lumbar spine and ultradistal radius in
    Colles' and vertebral fracture syndromes, J Clin Invest,
    83: 168.
Gotfredsen, A., Borg, J., Christiansen, C., and Mazess, R.B.,
    1984a, Total body bone mineral in vivo by dual photon
    absorptiometry. I. Measurement procedures, Clin Physiol,
    4: 343.
Gotfredsen, A., Borg, J., Christiansen, C., and Mazess, R.B.,
    1984b, Total body bone mineral in vivo by dual photon
    absorptiometry. II. Accuracy, Clin Physiol, 4: 357.
Gotfredsen, A., Riis, B.J., and Christiansen, C., 1986a, Total
    and local bone mineral during estrogen treatment: a
    placebo controlled trial, Bone and Mineral, 1: 167.
Gotfredsen, A., Nilas, L., Riis, B.J., Thomsen, K., and
    Christiansen, C., 1986b, Bone changes occurring sponta-
    neously and caused by oestrogen in early postmenopausal
    women: a local or generalised phenomenon?, Br Med J, 292:
    1098.
Gotfredsen, A., Hadberg, A., Nilas, L., and Christiansen, C.,
    1987, Total body bone mineral in healthy adults, J Lab
    Clin Med, 110: 362.
Gotfredsen, A., Pødenphant, J., Hadberg, A., Nilas, L., and
    Christiansen, C., 1989a, Regional bone mass in healthy
    and osteoporotic women. A cross sectional study, Scand J
    Clin Lab Invest, in press.
Gotfredsen, A., Pødenphant, J., Nilas, L., and Christiansen,
    C., 1989b, Discriminative ability of total body bone
    mineral measured by dual photon absorptiometry, Scand J
    Clin Lab Invest, 49: 125.
Mazess, R.B., Collick, B., Trempe, J., Barden, H., and Hanson,
    J., 1989, Performance evaluation of a dual-energy X-ray
    bone densitometer, Calcif Tissue Int, 44: 228.
Melton, L.J. III, Wahner, H.W., Richelson, L.S., O'Fallon,
    W.M., and Riggs, B.L., 1986, Osteoporosis and the risk of
    hip fracture, Am J Epidemiol, 124: 254.

Nilas, L., Gotfredsen, A., Riis, B.J., and Christiansen, C., 1986, The diagnostic validity of local and total bone mineral measurements in postmenopausal osteoporosis and osteoarthritis, Clin Endocrinol 25: 711.

Nilas, L., Pødenphant, J., Riis, B.J., Gotfredsen, A., and Christiansen, C., 1987, Usefulness of regional bone measurements in patients with osteoporotic fractures of the spine and distal forearm, J Nucl Med, 28: 960.

Nilas, L., Gotfredsen, A., Hadberg, A., and Christiansen, C., 1988, Age-related bone loss in women evaluated by the single and dual photon technique, Bone and Mineral, 4:95.

Thomsen, K., Gotfredsen, A., and Christiansen, C., 1986, Is postmenopausal bone loss an age-related phenomenon? Calcif Tissue Int, 39: 123.

EVALUATION OF METHODS

OF BONE MASS MEASUREMENT

Peter Tothill

Department of Medical Physics and Medical Engineering
Western General Hospital
Edinburgh EH4 2XU, U.K.

INTRODUCTION

Quantitative assessment of bone mineral is necessary for an understanding of those factors which influence the development of osteoporosis and the investigation of possible remedial measures. Most
studies have been upon groups, rather than individuals, and substantial
overlap between osteoporotic and normal subjects and between different
disease and treatment groups has been shown, in spite of significant
differences of mean values. However, it is now being argued that there
is a case for screening individuals to identify subjects with a bone
mass low enough to suggest a high risk of vertebral or femoral fracture,
so that prophylactic measures may be recommended. Accuracy is clearly
important for all measurements, especially when comparisons between
different centres are used; for longitudinal studies precision is
paramount. Techniques must also carry a negligible risk, from radiation
for example. The most appropriate part of the skeleton for a particular
purpose must be investigated. These factors will all be considered
here.

Bone mineral measurements have been carried out for some 30 years
by a variety of methods. All have yielded useful results and could
still do so. To some extent the attention paid to each in this review
will reflect its current popularity. The author has recently published
a more extensive review with the same scope (Tothill, 1989).

RADIOGRAMMETRY

Radiogrammetry was an early quantitative technique. It relies on
linear measurements of X-ray films of cortical bone taken under
standardised conditions and is simple and inexpensive to perform. The
method was introduced by Barnett and Nordin (1960) and Virtama and
Mähönen (1960). In spite of giving information only about the
dimensions of bones and not their porosity, it was successful in
demonstrating age-related bone loss in normal women (Barnett and Nordin,
1960; Meema, 1962) and the value of oestrogen treatment in reducing
that loss. Bones measured have included the radius, humerus, femur,
clavicle and tibia, but the site most commonly used is the midshaft of a
metacarpal.

Assessments of accuracy have been few, but when measurements of metacarpal cortical area have been compared with bone ash per unit length on post mortem specimens, a correlation coefficient of about 0.8 has been found (Shimmins et al., 1972; Exton-Smith et al., 1969). Precision is easier to assess and coefficients of variation ranging from 3 to 10% have been reported. Absorbed radiation dose to the hand is small, perhaps 100 μGy per exposure, with an effective dose equivalent well below 10 μSv.

RADIOGRAPHIC PHOTODENSITOMETRY

This technique depends on measuring the optical density of X-ray films of bones and is the oldest quantitative method of assessing bone mineral. The exposure of a reference wedge alongside the bone is neces-sary to minimise problems of spatial and temporal variations in kilovol-tage, exposure, film characteristics, processing and soft tissue thick-ness. Metacarpal, radius, ulna, femur and tibia have been studied, but most success has come from the phalanges. Many substances have been used for the reference wedge, aluminium being the commonest, results usually being expressed as an equivalent thickness of the wedge material.

Single spot measurements may be made with a photodensitometer. Reproducibility may be enhanced by linear or area density plots using a scanning instrument (Colbert and Bachtell, 1981). Although the method has largely been superseded by direct photon absorptiometry, which does not suffer from so many dubious assumptions, there may be a place for it where space and accessibility are limited, for example in the dual-energy technique with digital analysis of the films applied by Hawkes et al., (1987) to neonatal measurements.

The problems of determining "in vivo" accuracy are the same as those for radiogrammetry, but Shimmins et al., (1972) found a rather better correlation coefficient against bone ash of 0.88. Radiation dose is the same as for radiogrammetry.

PHOTON ABSORPTIOMETRY

This is the most widely practised technique and was introduced to overcome the problems associated with radiographic densitometry. Usually a gamma ray source and coupled scintillation detector scanning across the area of interest eliminate temporal, spatial, film and processing non-uniformities and minimise the effect of scatter.

Single Photon Absorptiometry (SPA)

SPA was introduced by Cameron and Sorensen (1963). The amount of bone mineral in the tissue traversed by the well-collimated gamma-ray beam is derived from the attenuation through bone-plus-soft-tissue rela-tive to that through soft tissue alone. The overall thickness must be constant, usually achieved by immersing the limb in water. Although measurements of the femur, humerus, finger, calcaneous and whole hand or foot have been made, most attention has been paid to the forearm and commercial instruments are available to measure the distal radius or ulna. I-125 is used as a source, its only drawback being a rather short half-life of 60 days.

Reproducible positioning of the scan imposes the biggest limitation on precision. When manual location was used, a site in the mid-shaft

where the bone mineral content does not change rapidly with position was
usually chosen. Here the bone is almost entirely cortical. A more
distal site is to be preferred, to include trabecular bone; the necessary precision and objectivity can be achieved by automatic site selection based on the separation of the radius and ulna (Christiansen
et al., 1975). Scanning distal to the point where the separation is
6 mm can interrogate bone with up to half its content trabecular (Awbrey
et al., 1984; Nilas et al., 1985). With such techniques a precision of
1% can be obtained.

Accuracy may be compromised by a non-uniform thickness of fat, with
attenuation characteristics different from water or lean soft tissue.
In some equipment the computer program assumes the fat to be a uniform
shell around the bone and makes a correction. True in vivo determinations of accuracy have not been made. Radiation dose in SPA is very
low and to a small volume of tissue, giving an effective dose equivalent
no more than 1 μSv.

Dual Photon Absorptiometry (DPA)

In some parts of the body, it is impossible to achieve a uniform
overall thickness. Even if the trunk could be immersed in a water bath,
errors would arise from the unknown and variable amounts of gas in the
gut. Different soft tissue thicknesses can be allowed for by the simultaneous measurement of the transmission of gamma rays of two different
energies, allowing measurement in any part of the body, but particularly
the important sites of lumbar spine and hip. Theoretical studies of
optimum photon energies (Watt 1975; Smith et al., 1983) have shown
that for the trunk Gd153 provides in a single radionuclide a combination
of energies (44 and 100 keV) that is close to ideal. This is the source
most commonly used in commercially-available DPA equipment. Drawbacks
are the cost and half-life of 240 days. A combination of Am 241 and
Cs137 is cheaper and longer lived, but leads to somewhat poorer statistical variance and geometrical resolution (Smith et al., 1983).

All modern DPA equipment includes computing facilities which make
necessary corrections and calculations and produce a bone mineral image.
Areas of interest over the bone and background are selected by the oper-
ator, perhaps assisted by the computer, and bone mineral in a number of
vertebrae printed out. Division by the length or area of spine considered provides some normalisation for size and reduces the variation due
to difficulties in defining intervertebral spaces.

Measurements of the proximal femur have been developed later than
those of the spine, although the same apparatus can be used. The latest
DPA development is the introduction of an X-ray generator instead of
gamma radiation sources. The necessary pairs of effective energies can
be obtained either by K-edge filtering, using cerium or samarium (Mazess
et al., 1989), or rapidly switching the generator potential (Wahner
et al., 1988). The advantages are a higher intensity, and therefore
faster scan (10 minutes instead of 20 for the lumbar spine), improved
spatial resolution, and therefore easier identification of vertebrae,
and better precision (1% instead of 2% for a gamma-ray source). In
osteoporotic or obese patients, precision can be markedly worse.

The theory of DPA requires that there are only two components
present — bone and soft tissue of uniform composition. In practice fat
forms a third component with attenuation characteristics different from
water, muscle or most organs. A uniform layer does not matter, but
Tothill et al., (1989) and Farrell and Webber (1989) have shown from CT

scans that fat is distributed very non-uniformly in the region of the
lumbar spine. Errors of spine bone mineral of up to 10% can be
introduced. An error can also be introduced by fat within the
vertebral bone marrow.

Total body bone mineral can be measured by DPA (Peppler and Mazess
1981). Instrumental problems are greater owing to the wide range of
count rates and non-uniform fat distribution introduces even more
errors. However, total body bone mineral measured by absorptiometry
correlated highly with total body calcium measured by neutron activation
analysis (Mazess et al., 1981).

As with SPA, the radiation dose for DPA is low, the effective dose
equivalent for part-body examinations being only a few μSv.

QUANTITATIVE COMPUTED TOMOGRAPHY

In computed tomography (CT) a thin transverse slice through the
body is imaged; in appropriate circumstances the image can be quanti-
tated to give a measure of bone mineral density. It offers the possibility of measuring trabecular bone independently of surrounding cortical
bone. Developments have concentrated in two directions: the construction of special equipment using a radionuclide source for measurements of
the forearm and the use of X-ray CT machines installed for general
radiology to measure vertebral bone mineral density.

Forearm CT Scanner

A dedicated forearm scanner was first described by Ruegsegger
et al., (1976). I-125 is used as the photon source, which is mounted in
a gantry with the NaI scintillation detector; a linear scan is
performed at each of 48 angular positions. Computer reconstruction
leads to an image in which a region of interest in the trabecular bone
of the distal ulnar is selected. Hosie and Smith (1986) made further
developments to give higher precision (better than 1% instead of 2%) and
a 2, rather than 5 minute scan. The local absorbed dose is well below
1 mGy.

X-ray CT of the Spine

Capital and running costs of an X-ray CT system are so high that an
installation solely for bone mineral measurement could not be contemp-
lated. Nevertheless, the possibility of measuring trabecular bone
density in the vertebrae has attracted a lot of attention. A group in
San Francisco has pioneered this development (Genant et al., 1981; Cann,
1988).

A lateral plane projection scan is necessary for precise selection
of slice position through the centres of the vertebrae. The slice must
be perpendicular to the axis of the vertebra, calling for a tilting
gantry or an angled reconstruction. Regions of interest within the
vertebral bodies are selected; circular, elliptical, rectangular or
manually drawn areas just inside the cortex have been used. Automatic
choice of region to minimise subjectivity and to enhance precision has
been introduced (Kalender et al., 1987).

The relationship between the observed CT number and the true attenuation coefficient is subject to short and long-term variation, so that
it is necessary to scan a calibration standard simultaneously with the

patient. The standard placed under the patient by Cann and Genant
(1980) is crescent-shaped and contains tubes filled with solutions of
dipotassium hydrogen phosphate, ethanol and glycerol plus water. More
recently, simpler standards with fewer components, based on suspensions
of calcium hydroxyapatite in plastic have been adopted and are
commercially available (Kalender et al., 1987). Comparison between the
Hounsfield numbers of the trabecular region of the vertebral bodies and
the standard allows the bone mineral density to be expressed in terms of
the equivalent concentration of the material of the standard.

The biggest error in single X-ray potential CT systems is due to
fat within the bone marrow; errors up to 30% may be introduced (Mazess,
1983). The inaccuracy can largely be overcome by carrying out scans at
two different potentials; typically 80 and 120 kVp are used. Kalender
et al. (1987) then claim an accuracy of 1%. However, precision is
thereby sacrificed, falling below the 2% obtainable with a single-energy
scan.

Quite a wide range of radiation doses has been quoted for CT, with
values as high as 40 mGy for a dual-energy measurement (Kalender et al.,
1987). Lower dose measurements are possible and Cann (1981) has achieved 1 mGy. Several slices are usually examined but each involves only a
small part of the body. The effective dose equivalent may be only
0.2 mSv for females and less for men as their gonads receive a lower
dose.

PHOTON SCATTERING METHODS

These also allow the possibility of measuring trabecular bone
without interference from the surrounding cortex although only at
peripheral sites, notably the heel. Only a small proportion of photons
interact by coherent scattering and most attention has been paid to
Compton scattering, although both have been considered together by
some authors.

Compton Scattering Only Techniques

The volume examined is defined by the intersection of a narrow
photon beam and the projection of the bore of the collimator of a detector, usually a scintillation counter. The angle between the primary and
scattered beams is related to the change in energy on scattering. Both
beams are subject to attenuation and corrections are necessary. Garnett
et al. (1973) used two gamma-ray sources and measured transmitted as
well as scattered intensities. Energies of around 100 keV are best for
dose reduction, but these favour multiple scattering, which introduces
errors. Short-term precision in the os calcis is quoted as 1.6%
(Webber, 1976) with an effective dose equivalent of perhaps 1 μSv.

Combined Coherent/Compton Scattering Method

Coherent scattering is strongly dependent on atomic number, offer-
ing good sensitivity for the determination of bone mineral. Problems of
attenuation and geometry can be minimised by expressing results as a
ratio of coherent to Compton scatter intensities, using a semi-conductor
detector to provide the necessary energy resolution. The method was
introduced by Puumaleinen et al. (1976) and further developed and
applied by Shukla et al. (1986). Both groups used Am241 as the source.
For measurements of cadaver calcanei Shukla et al. quote an accuracy of
5%, reproducibility of 3% and an absorbed dose of 3 mGy.

The application of IVNAA to bone mineral measurement relies on the fact that 99% of the calcium in the body is in the skeleton, for elemental analysis is involved. When the body is irradiated with neutrons, many of its constituent elements become radioactive and can be identified and quantified by examining the characteristic gamma-ray emission. The most useful reaction is $Ca48(n,\gamma)Ca49$. Ca48 is only 0.18% abundant, but has a reasonable cross-section. The half-life of Ca49 is only 8.8 minutes. Although the reaction is thermal, it is necessary to irradiate with a fast neutron beam, with thermalisation in the body and possibly a pre-moderator to obtain sufficient uniformity of thermal neutron flux. There have been reviews of part-body (Maziere, 1981; McNeill and Harrison, 1981) and whole-body (Cohn, 1981) IVNAA for bone mineral determination.

Part-body IVNAA

Radionuclide neutron sources suffice. Large NaI scintillation crystals are required to maximise sensitivity, particularly as the main gamma-ray has a high energy of 3.1 MeV. Sites examined include the hand, forearm and spine. The dose equivalent received in part-body IVNAA is of the order of 50 mSv; while this has been considered acceptable (the effective dose equivalent is probably below 10 mSv), it is much higher than in any of the other techniques considered. Although useful clinical results have been obtained, I believe that part-body IVNAA is no longer justified and should be replaced by photon absorptiometry, for example. This stricture does not apply to systems that examine the whole trunk (McNeill et al., 1973) for they are more analogous to whole-body measurements.

Whole-body IVNAA

Total-body IVNAA can measure several elements simultaneously and until quite recently was the only way to assess total bone mineral. Cyclotrons have been used as neutron sources by Nelp et al. (1970), Spinks et al. (1977) and Kennedy et al. (1982). The patient stands in a special enclosure at a distance of several metres, with rotation to give bilateral irradiation. Mean incident neutron energy is typically between 4 and 8 MeV.

Fourteen MeV neutron generators are cheaper and justify dedicated IVNAA use (Boddy et al., 1973). The greatest convenience and reliability come from the use of radionuclide neutron sources. Multiple Pu238,Be sources have been described by Cohn et al. (1972) and McNeill et al. (1973).

The high sensitivity whole-body counter can be of the steel room or shadow-shield type and should contain the largest detectors that can be afforded. Rapid transfer of the subject between source and detector is required.

Variable output sources call for intensity monitoring, including perhaps simultaneous activation of a standard. Corrections for body size are required to obtain an absolute measurement.

Accuracies based on phantom measurements of about 5% are claimed. Assessments of precision are mostly based on repeated measurements of anthropomorphic phantoms. For most systems the figure is about 2%. It is possible to assess precision from measurements on patients repeated at intervals for clinical purposes. The IVNAA system in Edinburgh, which had a precision of 1.8% with phantoms, was shown to have an _in vivo_ precision for calcium of 2.9% (Nicoll et al., 1987).

The dose equivalent associated with IVNAA has ranged from about 3 to 20 mSv, mostly using a quality factor of 10. However, it has recently been proposed that the quality factor for neutrons should be doubled.

An interesting recent development has been the introduction of a prompt-gamma IVNAA method of measuring calcium (Ryde et al., 1987). A Cf252 source is used and hyperpure germanium detectors count the high energy gamma-rays during the irradiation while the patient passes through the sensitive volume. The precision and dose equivalent compare favourably with the performance of the delayed gamma technique.

ULTRASOUND MEASUREMENTS OF BONE

This method has not yet received widespread validation, but the results are sufficiently encouraging to warrant consideration, particularly as no ionising radiation is used (Langton et al., 1984, Palmer and Langton 1987). So far only the os calcis has been examined. The system consists of a water tank containing two broadband ultrasonic transducers, one acting as a transmitter, the other as a receiver, at a fixed separation, together with a computer-interfaced electronic generation and detection unit. A short burst of ultrasound is passed through the heel, the frequency varying from 200 to 1000 kHz. The amplitude spectrum is compared with that from water alone to give a plot of attenuation in the os calcis against frequency. The slope of the linear portion of this graph is taken to characterise the bone. The attenuation is related to both the amount of bone in the path of the ultrasound and to the trabecular structure. A reproducibility of about 3.5% is achieved (Poll et al., 1986).

The technique has been validated by a study of a population of 940 elderly women (Miller and Porter, 1987). In spite of a large variance (CV 43%) age-related bone loss was demonstrated. More importantly, the ultrasound result was linked to the incidence of hip fracture.

COMPARISONS AND CONCLUSIONS

There have been many comparisons of measurements by different techniques and/or at different sites. A necessary comparison is of a new technique with a better-established yardstick; an example already mentioned is whole-body bone mineral by DPA versus calcium by IVNAA (Mazess et al., 1981). More commonly the aim has been to examine how well an easy or cheap measurement at one site can be used to predict the bone mineral at another more difficult, but perhaps more important site. Almost invariably the answer is not well enough. Typically the correlation coefficient, r, is between 0.6 and 0.7. Such a correlation may be highly significant statistically, but is quite inadequate for the prediction of one parameter from the other. If bone mineral of the spine is the parameter of interest, it should be measured directly and similarly with any other site.

A correlation of importance that is well established is that between bone mineral content and compressive bone strength, with r>0.9. This reinforces the rationale for bone mineral measurements, but is not the ultimate consideration. Other factors than compressive bone strength influence fracture incidence. To examine the relationship between a bone mineral measurement at a particular site and the

likelihood of an osteoporosis-induced fracture occurring, it is
necessary to screen a large population, which is then followed over a
period of years to record fractures. The most comprehensive such study
has been undertaken by Wasnich et al. (1985), who used photon
absorptiometry to measure bone mineral in the os calcis, distal and
proximal radius and lumbar spine and then followed the 1098 women (most
pre-menopausal) for six years. In this period 39 new fractures occurred
and it was found that fracture incidence increased with diminishing bone
mineral at all sites. The os calcis was the best predictor of non-spine
fracture risk and, perhaps surprisingly, also the best indicator of
spine fracture risk. There is a need for more prospective studies like
this.

I believe that it is not possible to recommend a single best technique of bone mineral measurement and many questions remain unanswered.

REFERENCES

Awbrey, B. J., Jacobson, P. C., and Grubb, S. A., 1984, Bone density in
    women: A modified procedure for measurement of distal radial dens-
    ity, J. Orthop. Res., 2:314.
Barnett, E., and Nordin, B. E. C., 1960, The radiological diagnosis of
    osteoporosis: a new approach, Clin. Radiol, 11:166.
Boddy, K., Holloway, I., and Elliot, A., 1973, A simple facility for
    total body in vivo activation analysis, Int. J. Appl. Radiat.
    Isot., 24:428.
Cameron, J. R., and Sorensen J., 1963, Measurement of bone mineral in
    vivo: An improved method, Science, 142:230.
Cann, C. E., 1981, Low dose CT scanning for quantitative spinal mineral
    analysis, Radiology, 140:813.
Cann, C. E., 1988, Quantitative CT for determination of bone mineral
    density: a review, Radiology, 166:509.
Cann, C. E., and Genant, H. K., 1980, Precise measurement of vertebral
    mineral content using computed tomography, J. Comput Assist.
    Tomogr., 4:493.
Christiansen, C., Rödbro, P., and Jensen, H., 1975, Bone mineral content
    in the forearm measured by photon absorptiometry; Principles and
    reliability, Scand. J. Clin. Lab. Invest., 35:325.
Cohn, S. H., 1981, Total body neutron activation, in: "Non-invasive
    Measurements of Bone Mass and their Clinical Application" S. H. Cohn
    ed., CRC Press, Boca Raton, Florida.
Cohn, S. H., Shukla, K. K., Dombrowski, C. S., and Fairchild, R. G.,
    1972, Design and calibration of a "broad-beam" [238]Pu,Be neutron
    source for total-body neutron activation analysis, J. Nucl. Med.,
    13:487.
Colbert, C., and Bachtell, R. S., 1981, Radiographic absorptiometry
    (photodensitometry), in: "Non-invasive Measurements of Bone Mass
    and their Clinical Application" S. H. Cohn, ed., CRC Press, Boca
    Raton, Florida.
Exton-Smith, A. N., Millard, P. H., Payne, P. R., and Wheeler, E. F.,
    1969, Method for measuring quantity of bone, Lancet, 2:1153.
Farrell, T. J., and Webber, C. E., 1989, The error due to fat inhomogen-
    eity in lumbar spine bone mineral mass measurements, Clin. Phys.
    Physiol. Meas., 10:57.
Garnett, E. S., Kennett, T. J., Kenyon, D. B., and Webber C. E., 1973, A
    photon scattering technique for the measurement of absolute bone
    density in man, Radiology, 106:209.
Genant, H. K., Boyd, D., Rosenfeld, D., Abols, Y., and Cann, C. E.,
    1981, Computed tomography, in: "Non-invasive Measurements of Bone

Mass and their Clinical Application" S. H. Cohn, ed., CRC Press,
    Boca Raton, Florida.
Hosie, C. J., and Smith, D. A. S., 1986, Precision of measurement of
    bone density with a special purpose computed tomography scanner,
    Br. J. Radiol., 59:345.
Kalender, W. A., Klotz, E., and Suess, C., 1987, Vertebral bone mineral
    analysis: an integrated approach with CT, Radiology, 164:419.
Kennedy, N. S. J., Eastell, R., Ferrington, C. M., Simpson, J. D.,
    Smith, M. A., Strong, J. A., and Tothill, P., 1982, Total body
    neutron activation analysis of calcium: calibration and
    normalisation, Phys. Med. Biol., 27:697.
Langton, C. M., Palmer, S. B., and Porter, R. W., 1984, The measurement
    of broadband ultrasonic attenuation in cancellous bone, Eng. Med.,
    13:89.
Lyon, A. J., Hawkes, D. J., Doran, M., McIntosh, N., and Chan, S., 1989,
    Bone Mineralisation in pre-term infants measured by dual energy
    radiographic densitometry, Arch. Dis. Childhood, 64:919.

Mazess, R. B., 1983, Errors in measuring trabecular bone by computed
    tomography due to marrow and bone composition, Calif. Tissue Int.,
    35:148.
Mazess, R. B., Peppler, W. W., Chesnut, C. H., Nelp, W. B., Cohn, S. H.,
    and Zanzi, I., 1981, Total body bone mineral and lean body mass by
    dual photon absorptiometry, II Comparison with total body calcium
    by neutron activation anlysis, Calcif. Tissue Int., 33:361.
Mazess, R. B., Sorenson, J. A., Hanson, J. A., and Collick, B. D., 1989,
    Dual-photon X-ray absorptiometry, in: "Proceedings of the
    Conference on Osteoporosis and Bone Mineral Measurement" Bath,
    April 1988, E. F. J. Ring, W. D. Evans, and A. S. Dixon, eds.,
    Institute of Physical Sciences in Medicine, York.
Maziere, B., 1981, Partial body neutron activation - hand, in: "Non-
    invasive Measurements of Bone Mass and their Clinical Application"
    S. H. Cohn, ed., CRC Press, Boca Raton, Florida.
McNeill, K. G., and Harrison, J. E., 1981, Partial body neutron
    activation - truncal, in: "Non-invasive Measurements of Bone Mass
    and their Clinical Application" S. H. Cohn, ed., CRC Press, Boca
    Raton, Florida.
McNeill, K. G., Thomas, B. J., Sturtridge, W. C., and Harrison, J. E.,
    1973, in vivo neutron activation analysis for calcium in man, J.
    Nucl. Med., 14:502.
Meema, H. E., 1962, The occurrence of cortical bone atrophy in old age
    and in osteoporosis, J. Can. Assoc. Radiol., 13:27.
Miller, C. G., and Porter, R. W., 1987, Broadband ultrasonic attenuation
    as a predictive index for hip fracture in the elderly, in: "Ultra-
    sonic Studies of Bone" S. B. Palmer and C. M. Langton, eds., Insti-
    tute of Physics, Bristol.
Nelp, W. B., Palmer, H. E., Murano, R., Pailthorp, K., Gervas, M. H.,
    Rich, C., Williams, J., Rudd, T. G., and Denny, J. D., 1970, Measu-
    rements of total body calcium (bone mass) in vivo with the use of
    total body neutron activation analysis, J. Lab. Clin. Med., 76:151.
Nicoll, J. J., Tothill, P., Smith, M. A., Reid, D., Kennedy, N. S. J.,
    and Nuki, G., 1987, in vivo precision of total body calcium and
    sodium measurements by neutron activation analysis, Phys. Med.
    Biol., 32:243.
Nilas, L., Borg, J., Gotfredsen, A., and Christiansen, C., 1985, Compar-
    ison of single and dual-photon absorptiometry in postmenopausal
    bone mineral loss, J. Nucl. Med., 26:1257.
Palmer, S. B., and Langton, C. M. eds., 1987, "Ultrasonic Studies of
    Bone" Institute of Physics, Bristol.
Peppler, W. W., and Mazess, R. B., 1981, Total body mineral and lean
    body mass by dual-photon absorptiometry, I. Theory and measurement
    procedure, Calcif. Tissue Int., 33:353.

Poll, V., Cooper, C., and Crawley, M. I. D., 1986, Broadband ultrasonic attenuation in the os calcis and single photon absorptiometry in the distal forearm: a comparative study, _Clin. Phys. Physiol. Meas._, 7:375.

Puumaleinen, P., Uimarihuhta, A., Alhava, E. M., and Olkkonen, H., 1976, A new photon scattering method for bone mineral density measurements, _Radiology_, 120:723.

Ruegsegger, P., Elsasser, V., Anliker, M., Grehm, H., Kind, H., and Prader, A., 1976, Quantification of bone mineralisation using computed tomography, _Radiology_, 121:93.

Ryde, S. J. S., Morgan, W. D., Sivyer, A., Evans, C. J., and Dutton, J., 1987, A clinical instrument for multi-element _in vivo_ analysis by prompt, delayed and cyclic neutron activation using ^{252}Cf, _Phys. Med. Biol._, 32:1257.

Shimmins, J., Anderson, J. B., Smith, D. A., and Aitken, M., 1972, The accuracy and reproducibility of bone mineral measurements "_in vivo_"; (a) The measurement of metacarpal mineralisation using an X-ray generator, _Clin. Radiol._, 23:42.

Shukla, S. S., Leichter, I., Karellas, A., Craven, J. D., and Greenfield, M. A., 1986, Trabecular bone mineral density measurement _in vivo_: use of the ratio of coherent to Compton-scattered photons in the calcaneous, _Radiology_, 158:695.

Smith, M. A., Sutton, D., and Tothill, P., 1983, Comparison between ^{153}Gd and ^{241}Am, ^{137}Cs for dual-photon absorptiometry of the spine, _Phys. Med. Biol._, 28:709.

Spinks, T. J., Bewley, D. K., Ranicar, A. S. O., and Joplin, G. F., 1977, Measurement of total body calcium in bone disease, _J. Radioanal. Chem._, 37:345.

Tothill, P., Pye, D. W., and Teper, J., 1989, The influence of extra-skeletal fat on the accuracy of dual-photon absorptiometry of the spine, _in_: "Proceedings of the Conference on Osteoporosis and Bone Mineral Measurement" Bath, April 1988, E. F. J. Ring, W. D. Evans, and A. S. Dixon, eds., Institute of Physical Sciences in Medicine, York.

Tothill, P., 1989, Methods of bone mineral measurement, _Phys. Med. Biol._, 34:543.

Virtama, P., and Mähönen, H., 1960, Thickness of the cortical layer as an estimate of mineral content of human finger bones, _Br. J. Radiol._, 33:60.

Wahner, H., Morin, R., Dunn, W., Brown, M., and Riggs, B., 1988, Dual energy radiography for bone mineral analysis of the lumbar spine, _J. Nucl. Med._, 29(Suppl.):855.

Wasnich, R. D., Ross, P. D., Heilbrun, L. K., and Vogel, J. M., 1985, Prediction of postmenopausal fracture risks with use of bone mineral measurements, _Am. J. Obstet. Gynecol._, 153:745.

Watt, D. E., 1975, Optimum photon energies for the measurement of bone mineral and fat fractions, _Br. J. Radiol._, 48:265.

Webber, C. E., 1976, Experience with photon scattering measurements of bone density, _Am. J. Roentgenol._, 126:1280.

DETERMINATION OF BONE MINERAL CONTENT IN THE HEEL BONE BY

DUAL PHOTON ABSORPTIOMETRY

R. Jonson, L.-G. Månsson, Å. Rundren,[*]
and J. Szücs

Department of Radiation Physics and [*]Geriatrics,
University of Göteborg, S-413 45 Göteborg, Sweden

INTRODUCTION

A dual-photon absorptiometry (DPA) system for the
determination of the bone mineral content _in vivo_ in the os
calcis is described. This system has the advantage over
existing DPA methods (Tothill, 1989) in that it is possible
to measure the bone mineral content accurately in the
presence of fat without the use of three different photon
energies (Jonson et al., 1988). This is accomplished by
knowing the total thickness of the heel at the measuring
point.

MATERIALS AND METHODS

The radiation source consists of a 125-I source (mean
photon energy 27.4 keV), activity 3.7 GBq, and a 57-Co source
(photon energy 122 keV), activity 370 MBq. The sources are
mounted in a brass holder which can be slid along runners in
a holder of lead. They are placed alternatively under a 140
mm long lead collimator with a 10 mm diameter circular hole.

The detector is a 2.5 cm x 2.5 cm NaI(Tl) detector
connected, via a linear amplifier, to a dual-counter-timer.
A portable computer is used to control the counter and to
calculate the bone mineral values. Fig. 1 is a schematic
diagram of the apparatus.

The patient measurement is carried out as a stationary
measurement with the source and detector collimators in close
contact with the skin of the heel. The point of measurement
is placed 4 cm from the sole of the foot and 3.5 cm from the
ankle (Shukla et al., 1987). This point is situated in the
centre and at the even part of the calcaneus. After the
patient measurement the unattenuated count rate is determined
from measurement of a phantom composed of 30 mm perspex and 5
mm aluminum. The thickness of the heel at the measuring
point is determined by taking the distance between the source
and detector collimators. An equation system containing two

Advances in In Vivo Body Composition Studies
Edited by S. Yasumura _et al.,_ Plenum Press, New York, 1990

exponentials and the total thickness of the heel at the
measuring point is solved for the bone mineral content
(Rundgren et al., 1984; Jonson et al., 1989).

The resulting bone mineral value is presented as bone
mineral equivalent length, B_1, in units of μm. By dividing
this value by the width of the heel at the measuring point,
one gets a dimensionless quantity in arbitrary units (a.u.).
By multiplying these quantities by the crystal density of
hydroxyapatite, 3.2 g/cm^3, the bone mineral value can be
expressed in g/cm^2 or g/cm^3, respectively.

The time for the patient measurement is about 2 minutes
and for the phantom also 2 minutes, which gives a total
measurement time of 4 minutes. The effective dose equivalent
to the patient will be less than 10 μSv.

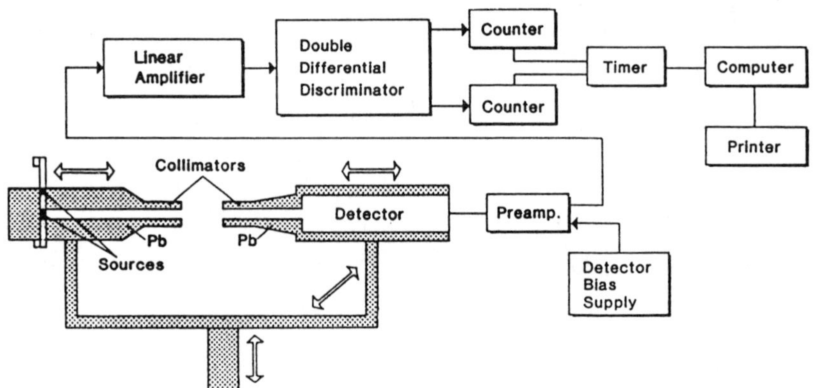

Fig. 1.   Schematic diagram of apparatus used for bone
          mineral measurements. The source and detector
          are movable in all directions.

RESULTS AND DISCUSSION

When 200 measurements were carried out on a phantom
composed of 3 mm Al and 4 cm water over a period of 2 years,
the precision was found to be 1.2% (coefficient of
variation). The precision for measurements of humans was
determined from repeated measurements of 14 healthy
volunteers over a period of 3 months and found to be 1.8%.

Simultaneously performed measurements of the heel bone
and spine by DPA showed a high correlation for the bone
mineral content in these two sites (Jonson et al., 1989).
Wasnich et al. (1985) also showed that measurement of the
bone mineral content in the heel bone could be used for
prediction of spine fracture risk.

REFERENCES

Jonson, R., Månsson, L.-G., Rundgren, Å., Szücs, J., 1989,
    Dual photon absorptiometry for determination of bone
    mineral content in the calcaneus with correction for
    fat, Unpublished results.
Jonson, R. Roos, B., and Hanson, T., 1988, Triple photon
    energy absorptiometry in the measurement of bone
    mineral, <u>Acta Radiological</u>, 28(4):461.
Rundgren, Å., Eklund, S., and Jonson, R., 1984, Bone mineral
    content in 70- and 75-year-old men and women: an
    analysis of some anthropometric background factors,
    <u>Age and Ageing</u>, 13:6.
Shukla, S.S., Leu, M.Y., Tighe, T., Krutoff, B., Craven,
    J.D., and Greenfield, M.A., 1987, A study of the
    homogeneity of the trabecular bone mineral density in
    the calcaneus, <u>Med. Phys.</u>, 14(4):687.
Tothill, P., 1989, Methods of bone mineral measurements,
    <u>Phys. Med. Biol.</u>, 34(5):543.
Wasnich, R.D., Ross, P.D., Heilbrun, L.K., and Vogel, J.M.,
    1985, Prediction of postmenopausal fracture risk with
    bone mineral measurements, <u>Am J. Obstet. Gynecol.</u>,
    153:745.

A FACILITY FOR THE _IN VIVO_ MEASUREMENT

OF Ca AND P CONTENT IN THE HUMAN HAND

Dimitrios Glaros,[2] John Kalef-Ezra,[2]
John Xatzikonstantinou,[2] Antony LoMonte,[1]
and Seiichi Yasumura[1]

[1]Medical Department, Brookhaven National Laboratory, Upton,
NY 11973, USA
[2]Medical Physics Laboratory, Medical School, University of
Ioannina, Ioannina, Greece, GR 451.10

INTRODUCTION

A facility was constructed for the _in vivo_ multi-elemental analysis
of either the human hand or the whole body of small animals. The
facility was installed in a room (area $20m^2$) supported on a 30-cm thick
concrete floor. Adjacent to the irradiation room are the counting room,
secretarial offices, and corridors. Therefore, attention was given to a
compact, light and efficient radiation protection shielding. Because
similar conditions exist in hospitals, an identical neutron activation
facility can be installed and used for in hospital applications.

IRRADIATION UNIT

Human hands or small animals can be bilaterally irradiated in the
8 cm wide, 10 cm high, and 30 cm deep irradiation cell. Both the neutron
spectrum and the neutron fluence rate can be modified according to the
application. The irradiation unit houses two kinds of neutron sources,
^{252}Cf and ^{238}Pu/Be. The ^{238}Pu/Be source is used for the measurement of
phosphorus (^{31}P(n,α)^{28}Al), whereas the ^{252}Cf is used for the analysis of
elements through thermal neutron capture reactions. An inner tank,
surrounding the sources and the irradiation cell, can be filled with
different moderating fluids such as air, H_2O or D_2O. Also, the skin to
source distance can be selected by varying the distance between the
sources and the irradiation cell; there is an option of distances: 8, 12,
and 20 cm from the midline of the irradiation cell.

Six ^{252}Cf sources with a total weight 430 μg were positioned inside
two cylinders parallel to the long axis of the irradiation cell. The
sources were arranged in such a way so that each cylinder simulates an
"extended" source (215 μg) about 20 cm long. The extended sources slide
within tubes, located inside a large water tank, that extends from a
shielded storage area to the irradiation positions on both sides of the
irradiation cell. The 740 GBq ^{238}PuBe source, located in a separate tube,
can be pulled from its storage location to a position 40 cm out of the

tank for the fast neutron irradiation. This allows the determination of phosphorus through the $^{31}P(n,\alpha)^{28}Al$ fast neutron reaction (Glaros et al., 1987).

The shielding material surrounding the source, starting from the innermost layer, includes water, polyethylene, boronated-polyethylene, polyethylene-lead bricks, and concrete bricks. The total thickness in any direction exceeds 1 m. During irradiations with the ^{252}Cf sources, the equivalent dose rate has been reduced to levels below 6 μSv/h immediately adjacent to the shielding, and to less than 0.5 μSv/h in areas of public access. With the sources at the storage, the dose rate is less than 2 μSv/h adjacent to the shielding and 0.2 μSv/h at areas of public access.

ACTIVATION (UNIFORMITY-DOSIMETRY)

Bare or cadmium-covered, indium gold foils were used for the determination of the fast and the thermal neutron fluence, Φ_{th} in air, and in phantoms. With the inner tank filled with water and the sources 16 cm apart, the maximum variation of the thermal neutron fluence from the mean value in a lucite phantom (5 cm wide, 7 cm high, and 16 cm long) was 5%, 6%, and 12% for the width, height, and length, respectively. The variation of the thermal neutron fluence increased slightly when the H_2O was replaced with D_2O and even more when the inner tank was empty.

The dose rate free in air from fast neutrons ($\dot{D}_n$) and photons ($\dot{D}_\gamma$) was assessed using a pair of Far West Technology ionization chambers (a IC-17 "tissue equivalent" chamber and a graphite/CO_2 IC-17A chamber) covered with metallic ^6Li. The photon exposure rates, $\dot{X}_{TE}$ and $\dot{X}_G$, that lead to currents equal to those collected in the ^{252}Cf field, are related to the dose rate in soft tissue by the equations:

$$\dot{X}_{TE} = 1.036 \ \dot{D}_\gamma + 0.968 \ \dot{D}_n \tag{1}$$

$$\dot{X}_G = 1.037 \ \dot{D}_\gamma + 0.099 \ \dot{D}_n \tag{2}$$

The correction factors involved in the determination of $\dot{X}_{TE}$ and $\dot{X}_G$ are discussed by Kalef-Ezra et al. (in press).

As an alternative to the graphite chamber, ^7Li enriched LiF thermoluminescent dosimeters (TLD-700) were used. The photon dose, TL_{eq}, that leads to thermoluminescent signals equal to the signal induced by the ^{252}Cf irradiation is given by the equation:

$$TL_{eq} = 0.84 \ \dot{D}_\gamma + 0.025 \ \dot{D}_n + k \ \Phi_{th} \tag{3}$$

where the thermal neutron sensitivity (k) was found to be k = 0.7 μGy ^{60}Co γ-equivalent for a thermal neutron fluence 10^{10} m^{-2}. Therefore, combining relations (1) and (3), the $\dot{D}_\gamma$ and $\dot{D}_n$ were estimated.

With air present in the inner tank and the ^{252}Cf sources 16 cm apart, the fast neutron dose rate at the center of the irradiation cell is 1.17 mGy/min and the photon dose rate is 0.45 mGy/min. Applying Q=16 for fast neutrons, the equivalent dose rate is 20 mSv/min. The use of D_2O, and even more with H_2O as moderating material, reduced significantly the fast neutron dose rate with minor influence in the photon dose rate. Preliminary dosimetric measurements with the inner tank filled with water indicated dose equivalent rates below 10 mSv/min.

MEASUREMENT OF INDUCED ACTIVITY

The counting system is located in a room adjacent to the irradiation
unit. This allows us to use only a 30-second delay time between the end
of activation and the start of counting. The counting system consists of
four rectangular NaI(Tl) detectors (10 cm x 10 cm x 40cm) shielded with
8-cm thick Pb bricks and 1 mm of Cu. The detectors arranged in a cross
setup geometry allow the hand or the small animal to be measured in a
geometry approximating 4π. Standard nuclear electronic units are used in
combination with a multichannel analyzer (nucleus/IBM PC).

PERFORMANCE OF THE SYSTEM

Rats in stainless steel containers were irradiated for 20 minutes
with the inner tank containing H_2O and the ^{252}Cf sources located 16 cm
apart. The rats then were counted for 20 minutes in lucite containers.
Yasumura et al. (this volume) found that the coefficient of variation for
Ca, Na, Cl and K is less than 3%, with a 60-min counting time for ^{40}K.
For rat phantoms containing a solution of 2 g of calcium, the accuracy
and reproducibility of the system was about 2%. Similar results were
obtained in a human hand phantom containing a solution of Ca.

In conclusion, partial body neutron activation analysis provides a
non-invasive means for measuring bone mineral mass at sites of high
clinical importance (Catto et al., 1973; Maziere et al., 1979; Ebifegha
et al., 1986; Glaros, 1987, Kalef-Ezra, this volume). The facility for
multi-elemental analysis presented in this study, is simple, precise, has
low installation and maintenance costs and can be easily used in a
hospital environment for routine measurements.

ACKNOWLEDGEMENTS

This research was supported in part by the U.S. Department of Energy
under Contract DE-AC02-76CH00016.

REFERENCES

Catto, G., McIntosh, J., MacLeod, M., 1973, Partial body neutron
        activation analysis _in vivo_: A new approach to the investigation
        of metabolic diseases, _Phys. Med. Biol._, 18:508.
Ebifegha, M., Harrison, J., McNeill, K., Krishnan, S., Sengabi, K.,
        1986, A system for _in vivo_ measurement of bone calcium by local
        neutron activation of the hand, _Appl. Radiat. Isot._, 37:159.
Glaros, D., Xatzikonstantinou, J., Leodiou, J., and Kalef-Ezra, J., 1987,
        A partial-body activation analysis technique for the measurement
        of phosphorus in bone, _in_: "In Vivo Composition Studies",
        K.J. Ellis, S. Yasumura, W. Morgan, eds., The Institute of
        Physical Sciences in Medicine, London.
Kalef-Ezra, J., Saraf, S.K., Fairchild, R.G., Laster, B., Fiarman, S.,
        and Ramsey, E., Epithermal beam development at the BMRR:
        Dosimetric evaluation, Proc. of the Workshop on Neutron Beam
        Design, Development and Performance for Neutron Capture Therapy,
        Boston, April 1989, Plenum Press, New York, in press.
Maziere, B., Kuntz, D., Comar, D., and Ryckewaert, A., 1979, _In vivo_
        analysis of bone calcium by local neutron activation of the hand:
        Results in normal and osteoporotic subjects, _J. Nucl. Med._, 20:85.

LOCAL BODY COMPOSITION MEASUREMENTS BY NMR

R. Fraser Code and Kenneth G. McNeill

Department of Physics
University of Toronto
Toronto, Ontario, CANADA M5S 1A7

INTRODUCTION

The recent emergence of in vivo Magnetic Resonance Imaging (Partain et
al., 1983) and Spectroscopy (Bottomley, 1989) as clinical and research diagnostic tools has made available new types of information on the local composition of the human body. Since these methods detect signals from the weak
nuclear magnetism of materials, they are limited to observing elements which
have both strong nuclear magnetic moments and relatively high abundances in
the tissue under study. The common proton, which has one of the strongest of
nuclear magnetic moments, is very easy to detect by in vivo magnetic resonance. It is also very significant that the times for re-magnetization (T_1)
and coherent dephasing (T_2) depend on the microscopic molecular environments
of the protons. These properties have led to the development of proton
Magnetic Resonance Imaging techniques capable of providing detailed three-
dimensional views of the soft tissue regions of the body, where conventional
X-ray tomography has limited use because of poor image contrast.

Although in vivo magnetic resonance imaging for humans is often referred
to as a "whole body" technique, what is usually obtainable from this procedure is a cross-sectional image over major body regions, usually not bigger
than the length of the spine. In cases where the particular body components
of "soft" (or liquid-like) tissues have a strong and distinctive nuclear
magnetic resonance signal, it is possible to modify the magnetic resonance
methods of an imager to obtain local quantitative body composition measurements. Some examples of soft tissue analysis based on signals from protons
(^{1}H), as well as the nuclei ^{13}C, ^{23}Na and ^{31}P, will be reviewed in the next
section. For nuclei in "hard" (or solid) tissues such as ^{19}F in bone mineral,
it is not possible to construct a magnetic resonance image of their
distribution because the resonance signals in a solid environment decay much
faster than the time normally required to encode positional information by
means of applied magnetic field gradients (Partain et al., 1983). The major
emphasis of this paper will be focused on the development of nuclear
magnetic resonance procedures which have enabled in vivo measurements of the
mass of accumulated bone fluorides in human index fingers.

IN VIVO NMR ANALYSIS OF SOFT TISSUES

The major portions of soft tissues consist of water-based solutions or

oily-type lipids. Both of these tissue components contain large numbers of
protons. Because of the "chemical shift" effect produced by the diamagnetism
of electrons surrounding a particular nucleus (Partain et al., 1983), in a
very uniform external magnetic field the NMR resonance frequency of water
protons differs by about 3 to 4 parts per million (ppm) from the resonance
frequency of lipid protons, whereas the full width at half height of both
resonances is about 1.5 ppm. Consequently, the observed liquid-phase proton
magnetic resonance spectrum separates into two fully-resolved "peaks" for
water and lipid molecules. Each of these spectral components can be selectively excited by pulsed magnetic resonance techniques (Dixon, 1984). In
addition to the conventional magnetic resonance image showing both water and
lipid protons, selective proton rf excitation can produce separate images of
the water and of the lipid content of soft tissues. This is often useful for
the purposes of tumour identification (Bloem et al., 1988).

With regard to other nuclei of interest in NMR body composition measurements, there is sufficient signal from natural abundance ^{13}C in subcutaneous
human adipose tissue to allow a quantitative in vivo measurement to be made
of its linoleic acid content (Moonen et al., 1988). This measurement was
done by chemical shift NMR spectroscopy, and was in quantitative agreement
with the results of tissue biopsies. Furthermore, the measurement was
sensitive enough to establish a reduced content of linoleic acid in the
adipose tissue of cystic fibrosis patients as compared to normal controls.

In vivo magnetic resonance images of the ^{23}Na nucleus in various organs
of the human body have been reported (Ra et al., 1988). In this case several
overlapping sodium components were observed, and contrast between them was
enhanced by using a short spin-spin relaxation time imaging algorithm. Calibrated ^{23}Na content measurements were not obtained, and image detail was poor
compared to proton images. However, the results are significant because of
the feasibility of using magnetic resonance to distinguish between intracellular and extracellular Na reservoirs, as well as to discriminate between
normal and pathological tissue.

There is also a report of non-invasive magnetic resonance measurements
of absolute metabolite concentrations in soft tissues. A spectroscopic technique using a double-tuned surface coil for both ^{1}H and ^{31}P nuclei has been
shown to yield ^{31}P metabolite concentrations in rats that are consistent with
in vitro values established by chemical analysis (Tofts, 1988).

Extending subcutaneous absolute concentration measurements to large
organs located in the interior of the human body may not be easy in certain
experimental configurations because of the absorption of rf energy by tissue
(Roeschmann, 1987; Surowiec et al., 1987) and variations in the sensitivity
of rf coils with position (Brey and Narayana, 1988). In addition, the difficulty of calibrating quantitative measurements in a reliable way, even for
^{1}H resonances, is quite severe in cases where different tissue components
contribute to the same magnetic resonance signal.

If the above difficulties could be overcome, then, in principle, proton
chemical shift imaging techniques (Morris, 1986) could be used to construct
an integrated measure of the water and lipid components over selected
regions of the body. By using local calibration techniques in conjunction
with NMR mapping of the entire body, it would seem possible to obtain by NMR
a measure for the total body water and the total body lipids which could be
then compared with the results of other measurement techniques. This has not
yet been done, largely because of the complexity and expense of the task.

IN VIVO NMR ANALYSIS OF HARD TISSUES

Usually, no contributions are seen from bone mineral to the magnetic resonance image of the body. Regions of compact bone are commonly displayed as black areas in the images taken of the musculoskeletal system. Bone mineral contains several nuclei which have a significant magnetic moment: ^{1}H, ^{31}P, and the important substitutional impurity ^{19}F. Since their environment in bone mineral is mostly static, the magnetic dipole-dipole interactions between neighbouring magnetic nuclei strongly dephase each other, causing their resonance signals to decay by spin-spin relaxation processes in a few hundred microseconds or less. Since it takes several milliseconds or more to encode the spatial distribution information by the gradient coils of a magnetic resonance imager, signals from the resonant nuclei in bone mineral are consequently lost from the image.

If imaging is dispensed with, and pulse NMR techniques suitable for solid samples are used (Gerstein and Dybowski, 1985), then it is possible to obtain in vitro signals from ^{1}H, ^{31}P and ^{19}F in bone samples (Funduk et al., 1984; Roufosse et al., 1984; Ebifegha et al., 1986). However, this is not so straightforward for the first two of these three nuclei in the in vivo case. There is a large solid-like background signal from protons in the collagen of the bone matrix that is approximately 20 times greater than the signal from the OH group protons in hydroxyapatite, the major component of bone mineral. In addition, the narrow line proton signal from soft tissues may be several hundred times larger than the wide line proton signal from the OH groups in hydroxyapatite. The solid ^{31}P signal from hydroxyapatite also is mixed with the high resolution chemical shift peaks from ^{31}P containing metabolites. In some parts of the body, such as the finger, this ^{31}P signal may be more easily isolated from the soft tissue resonances.

The situation for the in vivo detection of ^{19}F signals from bone mineral is more favourable than for ^{1}H and ^{31}P because there are no significant amounts of fluorides in the blood and soft tissues compared to the amount of fluorides that accumulate in bone mineral. Fluorides are incorporated in bone mineral as substitutional impurities for the OH groups of hydroxy-apatite. The specific ratio of fluorides in soft tissue to fluorides in bone is not available for humans, but environmental studies have shown that this ratio in fish is approximately 1:70 (Christensen, 1987). Initially it was shown by us that ^{19}F nuclei were detectable by in vivo NMR methods (Ebifegha et al., 1987) in the index finger bones of patients who also showed elevated levels of bone fluorides in pelvic biopsy samples. These levels were a result of sodium fluoride therapy for osteoporosis (Budden et al., 1988). More recent work in our laboratory using an improved "split-ring" resonator has shown that fluoride signals can be unambiguously detected in adult subjects who have no history of elevated bone fluoride levels (Code et al., in press). In our in vivo experiments, the localization of the ^{19}F NMR signal to the index finger is a result of the small size of the rf resonator that excites and detects the ^{19}F signal, and the relatively small volume of the homogeneous field region of our electromagnet.

IN VIVO MEASUREMENT OF ACCUMULATED BONE FLUORIDES

In our experiments, we assume that the NMR signals from all bone fluor-ides have exactly the same basic characteristics (i.e. the same longitudinal and transverse relaxation times T_1 and T_2). This is equivalent to assuming that all fluorides are excited and detected with uniform efficiency regard-less of their microscopic environment. The amount of fluoride in the index finger is determined by comparison with a powdered rat bone calibrator whose fluoride impurity content was previously measured by in vitro neutron activation analysis (Ebifegha et al., 1986).

The main difference in the NMR technique for bone fluorides compared to soft tissue is the requirement for a more intense pulsed rf excitation field. We use a rotating magnetic field at 27 MHz which has an amplitude of approximately 8 Gauss, which excites the fluorine spins in approximately 7.5 microseconds, and requires an rf power of approximately 5 Watts per cc of resonator volume. (Our finger resonator has a volume of approximately 30 cc). The peak rf electric fields to which the finger is exposed are less than 4 kV/m, significantly less than the NATO limit of 100 kV/m (Polk and Postow, 1986). We estimate that the mean thermal heating of the finger is less than 50 milliwatts per kilogram in our experiment, lower than the recommended limit of approximately 1000 milliwatts per kilogram. Our measurement of bone fluorides in human index fingers has been approved by the Human Subjects Review Committee of the University of Toronto.

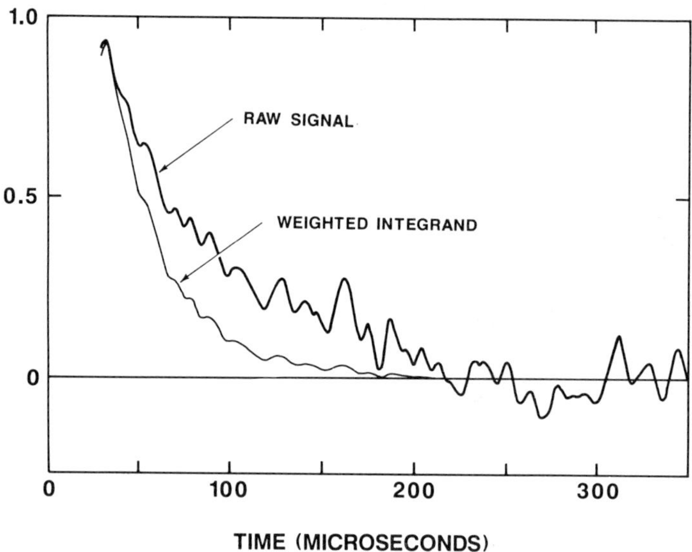

Fig. 1. ^{19}F NMR signal from an index finger containing approximately 2.5 mg of F. An observation time of 100 minutes (2000 scans) was used.

A typical ^{19}F NMR signal from an index finger is shown in Fig. 1. The initial amplitude of the ^{19}F free induction decay signal (which is observable after an instrumental dead time of approximately 35 microseconds) contains the information on the fluoride content. The second curve in Fig. 1 (labelled "weighted integrand") is the product of the observed finger signal times a smooth weighting function, which is a "least squares" fit to the rat bone calibrator NMR signal. The area under the weighted curve is taken as a measure of the NMR signal amplitude. This measurement procedure enables an "average signal amplitude" to be established over an observational interval of approximately 40 microseconds, which is about 15 times longer than the spectrometer's response time.

Our experiments measure the fluorides contained within a 6.5 cm length of an index finger, provided that amount is greater than approximately 0.5 mg. We cannot yet determine the bone mineral content of the observed portion

Table 1. Comparison of Standard Errors in the Mean of NMR Measurements of Accumulated Bone Fluorides in Index Fingers as a Function of Observing Time.

Observing Time T (Minutes)	No. of Fingers Measured	Standard Error in Mean (mg F)	S.E.M. x Sqrt(T) (Normalized)
40	4	0.17	0.75
50	4	0.20	1.0
160	4	0.11	1.0
310	2	0.12	1.5

of the finger, but we can measure the total calcium content in the hand by localized in vivo neutron activation analysis (Ebifegha et al., 1986a). From this we obtain the bone mineral mass of the hand and then estimate the bone mineral contents of the finger by anthropomorphic measurements to determine an approximate bone fluoride concentration in the finger (Code et al., in press). We are now developing a ^{31}P resonance method to measure the mineral mass of bone in the portion of the index finger observed by NMR. Modified versions of single photon absorptiometry in the hand (Nicoll et al., 1987) may also be capable of quantitating the bone mass in portions of the finger.

SENSITIVITY VERSUS OBSERVING TIME

Because of the particular in vivo characteristics of ^{19}F nuclear magnetism in solid bone mineral, the intensity of the ^{19}F resonance signal can be directly measured for only about 35 microseconds out of every 3 seconds. The sensitivity of our 27 MHz spectrometer is fairly close to the thermal noise limit. If the experimental noise were always randomly distributed with constant spectral power, then the product of the standard error in the mean of an ensemble of measurements times the square root of the number of individual observations in the ensemble should be a constant. As can be seen below in the fourth column of Table 1, this product is not truly constant. (Note that the time T in the caption of the fourth column is proportional to the total number of observations in the ensemble of repetitions.) This suggests that the ultimate sensitivity of our measurement is determined by excess noise, such as might arise from rf interference or motions of the finger inside the resonator.

The individual observing periods for the index finger bone F measurements shown in Table 1 were each less than 10 minutes long, and have been repeated as necessary to make up the total observing time T. The fluoride contents over the 6.5 cm length of finger under study in this experiment varied from 0.38 to 2.72 mg F. None of the participating adult volunteers had a history of abnormal exposure to fluorides.

We have found that a convenient measurement session for most volunteers consists of three 10-minute observing periods separated by a few minutes to stretch and restore circulation to the hand, which is kept as motionless as possible during each 10-minute period. Under these circumstances, we can expect a standard error in the mean of the three measurements of the index finger bone fluoride content to be 0.25 mg F. Hence, upon repeat measurements we would expect that an observed change of 0.75 mg F would be statistically significant at a $p < 0.05$ level (using a Student's t distribution test).

SIGNIFICANCE OF BONE F MEASUREMENTS IN FINGERS

Osteoporosis is a disease state in which the total amount of bone is reduced, but the existing bone quality is good, rather than diseased. Fluoride therapy for established osteoporosis has been found to be beneficial in cases where the fluorine has been taken up by the bone (Harrison and McNeill, 1988). Testing for this uptake has relied heretofore on invasive bone biopsies and subsequent in vitro testing for fluorine (Mernagh et al., 1977). These bone samples have normally been taken from the iliac crest (the site of the biopsy is important because the two different types of bone in general lose and gain mass at different rates). Therefore, demonstrated loss of trabecular bone may not be reflected (at least not immediately) in loss of cortical bone. Conversely, increased mineralisation in cortical bone may not be indicative of corresponding uptake in trabecular bone.

It follows from the above that, without further investigation, the measurement of fluorine in the finger bones (which may be dominantly cortical) may not be taken as indicative of F uptake in the critical trabecular bone.

The distribution of fluorine in rat bone has been measured in the laboratories of the DSIR in New Zealand (Coote and Vickridge, 1988). The rats had been given drinking water spiked with sodium fluoride for times and at concentrations found appropriate in earlier experiments (Harrison et al., 1984). After sacrifice, thin sections of femurs and vertebrae were exposed to a proton beam whose position could be so accurately controlled that spatial resolution was 20 micrometers. By the (proton, gamma) reaction, characteristic 6 MeV gamma rays were produced by interaction of protons with fluorine nuclei. Variations in intensity of these gamma rays as the proton beam was scanned across the bone sample gave variations of the concentration of F from point to point in the bone. Simultaneously, characteristic X-rays were detected after their stimulation by interaction of protons with calcium nuclei. The relative proportions of F and Ca at different places in the bone - and therefore in different types of bone - could be estimated.

In general, as the proton beam scanned across the section of femur or vertebra, a ring of high Ca content (the outer ring of cortical bone) was apparent with, inside the ring, spikes of high Ca content corresponding to the positions of trabecular bone; in between the trabeculae, in the marrow, there were regions of low Ca concentration. A similar picture was shown by the F gamma rays, with the relative concentration of F depending on the length of time that the rat had been drinking spiked water and on the rate of uptake. The important point is, however, that in the cases tested, uptake of F by cortical bone was always accompanied by uptake by trabecular bone.

Insofar as results on rats can be taken to be applicable to humans, these results indicate that a finding of uptake of F by cortical bone (for instance, by bones in the finger) will be indicative of uptake by trabecular bone, for instance, of the vertebrae. In osteoporotic patients, a positive increase of F in the index finger by NMR may reasonably be taken as evidence of uptake by the critical bone regions which cannot be observed directly by solid state NMR.

DISCUSSION

The motivation for our developing a local NMR technique for measuring bone fluoride contents in index fingers was to enable repeat measurements for establishing a rate of accumulation of bone fluorides in patients being treated for osteoporosis by dietary supplements of fluorides. The

sensitivity we have achieved appears sufficient to enable feasibility
studies to begin. Since the ratio of trabecular to cortical bone may be
different in the index finger than in iliac crest biopsy material,
correlation studies of the long-term changes in bone F from the pelvis and
from the index finger must also be done.

It is recognized that a higher mineral turnover rate exists in
trabecular bone (Leichert et al., 1987). Indirect evidence from the free
induction decay shape of our NMR experiments (Code et al., in press)
indicates that more fluorides may accumulate in the ends of the finger bones
rather than the sidewalls than had previously been expected. As stated
above, the retention of fluorides in the cortical portions of fingers seems
to be a significant measure of its retention in other parts of the skeleton.
The sample volume of finger bone observed by our NMR technique is much
larger than usually analyzed by invasive biopsy methods. Therefore, our
finger fluoride measurements may be fairly representative of the bone
fluoride retention processes throughout the skeleton.

The non-invasive nature of NMR bone fluoride measurements also makes it
suitable for screening populations for excessive burdens of accumulated
fluorides, either from industrial and environmental exposure or as a result
of various metabolic diseases. Over the long term, in vivo studies of the
fluoride contents of living bones may help to answer the larger question of
the overall risks and benefits of fluorides to the human body.

In summary, magnetic resonance spectroscopy enables new types of local
body composition studies to be made by non-invasive means. Some of these
measurements are still in the research stage, while others (especially those
involving ^{31}P chemical shift spectra) are now being introduced in the
clinical laboratory. During the past decade, magnetic resonance imaging
techniques have revolutionized the practise of radiology. The future impact
of localized magnetic resonance spectroscopy on in vivo composition studies
of the human body should be equally profound.

REFERENCES

Bloem, J. L., Taminau, A. H. M., and Bloem, R. M., 1988, MRI of
    musculoskeletal disease, in: "Essentials of Clinical MRI," T. H. M.
    Falke, ed., Martinus Nijhoff, Dordrecht.
Bottomley, P. A., 1989, Human in vivo NMR spectroscopy in diagnostic
    medicine: clinical tool or research probe?, Radiology, 170:1.
Brey, W. W. and Narayana, P. A., 1988, Correction for intensity falloff in
    surface coil magnetic resonance imaging, Med. Phys., 15:241.
Budden, F. H., Bayley, T. A., Harrison, J. E., Josse, R. G., Murray, T. M.,
    Sturtridge, W. C., Kandel, R., Veith, R., Strauss, A. L., and Goodwin,
    S., 1988, The effect of fluoride treatment on bone histology for
    postmenopausal osteoporosis depends on adequate fluoride absorption and
    retention, J. Bone Min. Res., 3:127.
Christensen, B., 1987, Uptake and release of fluoride in Arctic char,
    Environ. Toxicol. Chem., 6:529.
Code, R. F., Harrison, J. E., McNeill, K. G., and Szyjkowski M., In vivo ^{19}F
    spin relaxation in index finger bones, Magn. Reson. Med., (in press).
Coote, G. E., and Vickridge, I. C., 1988, Application of a nuclear
    microprobe to the study of calcified tissues, Nucl. Instr. and Meth. in
    Phys. Res., B30:393.
Dixon, T., 1984, Simple spectroscopic imaging, Radiology, 153:189.
Ebifegha, M. E., Code, R. F., McNeill, K. G., and Szyjkowski, M., 1986,
    Nuclear magnetic resonance determination of fluorine content and
    relaxation times in bone powder, Can. J. Phys., 64:282.

Ebifegha, M. E., Harrison, J. E., McNeill, K. G., Krishnan, S. S., and Ssengabi, J., 1986a, A system for in vivo measurement of bone calcium by local neutron activation of the hand, Int. J. Appl. Radiat. Isotop., Part A, 37:159.

Ebifegha, M. E., Code, R. F., Harrison, J. E., McNeill, K. G., and Szyjkowski, M., 1987, In vivo analysis of bone fluoride content via NMR, Phys. Med. Biol., 32:439.

Funduk, N., Kydon, D. W., Schreiner, L. J., Peemoeller, H., Miljkovic, L., and Pintar M. M., 1984, Composition and relaxation of the proton magnetization of human enamel and its contribution to the tooth NMR image, Magn. Reson. Med., 1:66.

Gerstein, B. C., and Dybowski, C. R., 1985, "Transient Techniques in NMR of Solids: An Introduction to Theory and Practise," Academic Press, Orlando.

Harrison, J. E., Hitchman, A. J. W., Hasany, S. A., Hitchman, A., and Tam C. S., 1984, The effect of diet calcium on fluoride toxicity in growing rats, Can. J. Physiol. Pharmacol., 62:259.

Harrison, J. E., and McNeill, K. G., 1988, The skeletal system, in: "Nutrition and Metabolism in Patient Care," J. M. Kinney, K. M. Jeejeebhoy, G. L. Hill, and O. E. Owen, eds., W.B. Saunders, Philadelphia.

Leichert, I., Bivas, A., Giveon, A., Margulies, J. Y., and Weinreb, A., 1987, The relative significance of trabecular and cortical bone density as a diagnostic index for osteoporosis, Phys. Med. Biol., 32:1167.

Mernagh, J. R., Harrison, J. E., Hancock, R., and McNeill, K. G., 1977, Measurement of fluoride in bone, Int. J. Appl. Radiat. Isot., 28:581.

Moonen, C. T. W., Dimand, R. J., and Cox, K. L., 1988, The non-invasive determination of linoleic acid content of human adipose tissue by natural abundance ^{13}C nuclear magnetic resonance, Magn. Reson. in Med., 6:140.

Morris, P. G., 1986, "Nuclear Magnetic Resonance Imaging in Medicine and Biology," Clarendon Press, Oxford.

Nicoll, J. J., Smith, M. A., Reid, D., Law, E., Brown, N., Tothill P., and Nuki, G., 1987, Measurement of hand bone mineral content using single-photon absorptiometry, Phys. Med. Biol., 32:607.

Partain, C. L., James, A. E. Jr., Rollo, F. D., and Price, R. R., eds., 1983, "Nuclear Magnetic Resonance (NMR) Imaging," W.B. Saunders, Philadelphia.

Polk, C., and Postow, E., eds., 1986, "CRC Handbook of Biological Effects of Electromagnetic Fields," CRC Press, Boca Raton.

Ra, J. B., Hilal, S. K., Oh, C. H., and Mun, I. K., 1988, In vivo magnetic resonance imaging of sodium in the human body, Magn. Reson. in Med., 7:11.

Roeschmann, P., 1987, Radiofrequency penetration and absorption in the human body: Limitations to high-field whole-body nuclear magnetic resonance imaging, Med. Phys., 14:922.

Roufosse, A. H., Aue, W. P., Roberts, J. E., Glimcher, M. J., and Griffin, R. G., 1984, Investigation of the mineral phases of bone by solid state ^{31}P magic angle sample spinning nuclear magnetic resonance, Biochemistry, 23:6115.

Surowiec, A., Stuchly, S. S., Eidus, L., and Swarup, A., 1987, In vitro dielectric properties of human tissues at radiofrequencies, Phys. Med. Biol., 32:615.

Tofts, P. S., 1988, The noninvasive measurement of absolute metabolite concentrations in vivo using surface-coil NMR spectroscopy, J. Magn. Reson., 80:84.

A CLINICAL APPROACH TO BODY COMPOSITION IN WASTING

Mark Wahlqvist and Sharon Marks

Department of Medicine, Monash University

Prince Henry's Hospital, Melbourne, Australia

INTRODUCTION

The term wasting is used by physicians as a clinical statement about loss of muscle mass and not even lean body mass in its entirety. Whatever the clinician may think, it is not possible to make, for example, an assessment about reduced liver nitrogen or protein in conjunction with protein energy malnutrition (P.E.M.) in the face of associated changes of fatty liver. Where there is sufficient evidence of wasting by inspection or by scrutiny of a weight chart, the diagnosis is often made without further confirmatory investigations. In patients with oedema or obesity the wasting is less overt and the need for further assessment arises.

Clinicians are interested in a precise, non-invasive anatomical definition of body composition. The combination of this need with that for a clinical assessment of nutritional status and metabolic state has become a challenge for technologists and, in particular, physicists. The opportunity to directly measure body composition has encouraged the clinician to ask for measures of nutrient intactness and distribution, in addition to anatomical and metabolic parameters of body composition.

As far as human nutrition is concerned, with body composition studies we are just beginning to move beyond the state we were at earlier this century with proximate food analysis for macronutrients, with little ability to measure micronutrients, let alone trace elements. Other than in a few centres, body composition studies have been limited to research projects. The development of hospital-based laboratories has enabled body composition measurements to be an integral part of a clinical assessment. Facilities such as our own (Stroud et al., this volume) have been established at relatively low cost and make routine studies more feasible. To be of practical value to the treating physician, results need to be reproducible and valid, even in the presence of oedema, obesity or osteoporosis. It is particularly in these clinical settings that indirect studies have failed to measure body compartments accurately in the wasted patient. This should not be a surprise, as many of the basic assumptions used are for "the reference man" and are not applicable to malnourished patients, or to the sick in general.

Recognition of the value of simultaneous assessments of body composition is seen with new developments such as the Dual-energy X-ray Absorptiometer (DEXA) (Rigg, 1988) which measures bone mineral density as well as directly measuring the percentage body fat. Although validation

studies with In Vivo Neutron Activation Analysis (IVNAA) or other
techniques are not available, DEXA is a modality which may be of great
clinical benefit.

The economical practice of clinical medicine must consider cost,
time, and space, as well as convenience. There is a need for inexpensive,
portable instruments which are more accessible to the population most at
risk. These populations are those in developing countries, as well as the
underprivileged in our own societies.

Table 1. Types of Wasting

A. WASTING ALONE

Primary Malnutrition

- developing countries with limited food supply
- general community in developed countries
  (educationally and socioeconomically disadvantaged)
- eating disorders
- hospital or institutional settings

Chronic Wasting Disease

- Cardiac failure
- Infectious disease
- Liver failure
- Malabsorption
- Malignancy
- Neurological disorders
- Obstructive lung disease
- Renal failure

B. WASTING ASSOCIATED WITH OBESITY   As for

C. WASTING ASSOCIATED WITH OEDEMA   Wasting

D. WASTING ASSOCIATED WITH OSTEOPOROSIS   Alone

WASTING TYPES

Wasting is a consequence of protein-energy malnutrition (P.E.M.)
when the body's need for protein and energy fuels or both cannot be
satisfied by diet. Wasting can occur on its own, as in primary malnutrition where there is inadequate food intake, but wasting is also seen in
chronic disease (Table 1). When wasting is seen in association with
either obesity, oedema or osteoporosis it also occurs in the setting of
either primary malnutrition or with chronic wasting diseases.

Primary Malnutrition

The extent of P.E.M. related to inadequate intake is illustrated by
the World Health Organization's estimate of 300 million children world-
wide with growth retardation related to malnutrition (WHO, Document WHA
36/1983/7). The emphasis has recently begun to shift from childhood to
adult life with the recognition of an increasingly aging population in
developing countries. It has been estimated that the number of people
aged over 65 in developing countries will climb dramatically from 129
million in 1980 to well over 220 million by the year 2000. This is a far

greater increment than the 167 million projected for developed countries,
where in 1980 there were also 129 million elderly people (Andrews et al.,
1986). With this rate of elderly population growth, the prevalence of
chronic disease and wasting is likely to escalate and the need will arise
for more sensitive low-cost technology to detect milder degrees of malnutrition.

In industrialized societies primary malnutrition is also seen,
especially in the aged, the institutionalized and the underprivileged,
who are at particular risk of inadequate food intake. Even in a hospital
setting, wasting due to primary malnutrition occurs, for example in
post-operative patients, unable to return to their usual dietary intake.
Studies using body composition techniques are attempting to relate surgical risk to pre-operative nutritional status and weight loss (Windsor and
Hill, 1988).

Table 2. Changes in Weight, Body Composition after 1 Year of
          Intervention

	Exercisers (n-47)	Dieters (n=42)	Controls (n-42)
Total wt (kg)	-4.0*	-7.2*	0.6
Non-fat body mass (kg)	0.1	-1.3*	0.8
Fat body mass (kg)	-4.1*	-5.9*	-0.3

* p<0.001 vs controls. Data from Wood et al, 1988

<u>Chronic Wasting Disease</u>

The wasting associated with chronic disease cannot be attributed
solely to lack of adequate intake, as in primary malnutrition (Table 1).
Cardiac cachexia is a term used to describe end-stage cardiac failure
where although food intake may be reduced due to breathlessness, many
other factors contribute to the severe wasting seen. In patients with
cancer, factors such as tumor necrosis factor (TNF) and cachectin have
been implicated in the wasting. TNF possibly plays a significant role in
the pathogenesis of alcoholic liver disease (McClain and Cohen, 1989) and
may account for some of the wasting seen. Lean body mass as calculated
from total body nitrogen (IVNAA) was reduced in 13 patients with chronic
obstructive pulmonary disease who were all underweight. The Nitrogen
Index in each patient was calculated before dietary supplementation. This
index is the measured nitrogen divided by the predicted nitrogen, as
calculated by the following equation:

$$\text{Predicted Nitrogen} = \text{Constant} \times \text{HAS}^{2.6}$$

where the constant=0.42 for women and 0.50 for men. HAS is the mean of
height and armspan in meters (Harrison et al., 1984). In these 13 patients the Nitrogen Index ranged from 0.4-0.7 compared with the normal
range of 0.8-1.2. In addition to the reduction in fat mass seen by
anthropometric techniques, a reduction in body protein, was able to be
demonstrated.

Reduction in lean body mass and, therefore, muscle mass, will diminish the patient's functional capacity. Nutritional supplementation may increase lean body mass and aid diaphragmatic and respiratory muscle function, although in a recent study this was disputed. Otte et al.(1989) showed a weight gain in patients given nutritional supplementation. However, they did not show a significant improvement in other indices of well-being. This study did not have the benefit of direct body compositional measures for lean body mass and relied upon anthropometry to document changes in body protein. More direct methods of measuring total body protein such as I.V.N.A.A. are needed to determine whether weight changes are related to increased body protein, water or fat.

The changes in body composition in chronic disease are often complicated by the underlying pathology. A common example in industrialized societies is the vasculopath with risk factors like smoking. The vascular disease, which may be calcific, is associated with outcomes of immobility due to ischemic heart disease or to peripheral vascular disease. Cardiac decompensation and deformity related to stroke are additional factors which, because of their effect on body water, protein or fat, need consideration.

Table 3. The Wellcome Classification

Weight (percentage and standard*)	Oedema	
	Present	Absent
80-60	Kwashiorkor	Underweight
< 60	Marasmic kwashiorkor	Marasmus

* Standard - fiftieth centile Boston value

In Association with Obesity

Although it seems a contradiction in terms, wasting can occur in obese patients. First, although muscle mass and bone mass are usually increased in association with obesity, they can be too low for the level of obesity. Second, account needs to be taken of compartment changes other than fat, such as in muscle and bone. Third, sick obese patients are at greater risk than their less obese counterparts through underrecognition of reduced body protein, and cardiac decompensation with fluid overload. Garrow et al. (1988) recently reported an increased risk of post-operative morbidity in obese patients which was due to the occurrence of wound infection. This complication may be related to reduced non-fat mass in the obese patients and not simply to mechanical factors.

Wood et al (1988) assessed the loss of non-fat body mass in overweight men randomized to diet, exercise or to a control group (Table 2). Dieters lost approximately 1.3 kg of non fat-body mass, in addition to losing approximately 5.9 kg of fat body mass. Patients who exercised did not significantly alter their non-fat body mass despite a reduction in fat body mass.

In Association with Oedema

Alleyne et al. (1977) adopted the Wellcome classification (1970) in their book **Protein-Energy Malnutrition**. This classification divides

P.E.M. into four categories, based on the presence or absence of oedema
and describes three clinical types of severe malnutrition - kwashiorkor,
marasmus, and marasmic kwashiorkor. The recognition of lesser degrees or
earlier accumulation of excess tissue fluid would be of great clinical
value and would increase the understanding of the different types of
P.E.M. The Wellcome classification (Table 3) is also based on body weight
deficit and while using it as a simplistic classification, Alleyne et al
acknowledged the inherent difficulties. Oedema may be associated with
cardiac, renal, or liver disease and its presence may maintain the body
weight at a normal level despite severe wasting. Thus, the diagnosis of
malnutrition may not be suspected. Body composition studies with considerations of total body water and lean body mass as separate, measurable
compartments allow for earlier recognition and classification.

In Association with Osteoporosis

Anorexia nervosa is one situation where loss of lean body mass is
found with osteoporosis (see discussion of body composition in patient
J.B. below). The Diagnostic and Statistical Manual (DSM-111) Criteria for
Eating Disorders (APA, 1987) included loss of more than 25% of original
body weight as a prerequisite for the diagnosis of Anorexia Nervosa. This
diagnosis has been revised (DSM-111-R) (1987) and is now worded as
"...weight loss leading to maintenance of body weight 15% below that
expected." Neither description includes the loss of lean body mass or
bone density or fluctuations in body water which may obscure the
diagnosis.

Table 4.   Limitations of Body Mass Index (B.M.I.)

- Presumes sedentary western life-style
- Presumes frame size
- Does not allow for effects of aging on bone density
  and height
- Makes no allowance for fat distribution
- Presumes normal body water

MEASUREMENT OF WASTING DISORDERS

One of the most basic anthropometric measures is that of body weight
and the calculation of the Body Mass Index $[wt(kg)/ht^2(m)]$ as described
by Quetelet (1969) provides a reasonable guide to the contribution of fat
to body composition in healthy subjects. But there are many circumstances
in which these measurements cannot adequately assess body fatness (Table
4).

The most crucial of these is where the fluid balance is deranged, as
occurs in cardiac failure as well as in renal and liver disease. In these
patients, even assessment of skinfold thickness is hindered by subcutaneous oedema. Morbid obesity also poses large technical problems with
skinfolds too large for precise measurement. Osteoporosis, while not necessarily associated with a reduction in lean body mass or body fatness,
may reduce body weight and so factitiously reduce B.M.I. Unfortunately,
few patients fit the "normal man" who is usually young, healthy, male and
weighs 70kg. Amputees (whose missing limb will alter their weight
without necessarily changing the measured height) and patients with
collapsed vertebrae from osteoporosis are two such examples. Where more

precision than can be expected from anthropometric measures is needed, further body composition studies are required.

Lukaski (1987) recently reviewed the available methods of assessing human body composition. In grading the precision of the various techniques at measuring non-fat mass and fat mass, he addresses the question of "...which existing method best meets the objectives of the proposed research". IVNAA was the only method to achieve a similar degree of precision as densitometry (underwater weighing). Coward et al. (1988) considered IVNAA to be "...the most significant development..." so far. By incorporating Burkinshaw's idea (Burkinshaw, 1985) of measuring multiple components of body mass and then deriving estimates of unmeasured components, IVNAA can give useful clinical information. In patients with wasting, however, many of the assumptions break down and more direct methods of measuring total body water and body fat are needed.

Table 5.   Compartment Disorders of Interest in Wasting

---

1.   Lean Body Mass
   • Metabolic mass - energy expenditure and requirement
   • Cellular function - eg. cardiac mechanics, skeletal muscle
     function

2.   Fat
   • Energy stores
   • Drug distribution

3.   Water
   • Water overload or reduction
   • Distribution of nutrients and drugs

4.   Bone
   • Bone strength

CLINICAL INDICATIONS FOR BODY COMPOSITION STUDIES IN WASTING

Indications for body composition studies in wasting include the assessment of many complex chronic diseases where it is not sufficient to monitor only one component or a compartment in isolation. Wasting is rarely limited to the lean body mass and measures of body fat and bone mineral density are needed for completeness. Adequate documentation of the extent of the wasting gives a clear picture of the severity of the disease and, with sequential studies, the rapidity of progress of the disease can be assessed.

The compartments of interest vary with individual diseases (Table 5). Lean body mass may be important as a measure of metabolic mass when attempting to find the cause for wasting. Body composition studies allow earlier recognition of disorders such as the wasting that can occur in obese people or low-energy diets.

Any change in body composition, either reduction in lean body mass or increases in total body water will alter drug distribution. These changes can occur during growth spurts as in childhood and adolescence, as part of a disease process such as congestive cardiac failure or during normal aging. Steen (1988) showed that the most important cause of a decrease in body weight during the eighth decade of life is a reduced amount of extracellular water; this is considered to be part of normal

Table 6. Body Composition on a 24-year Old Man (D.C.) with Chronic
Renal Failure and Chronic Pancreatitis - Before and After
Pancreatic Enzyme Replacement

	Sept 88	May 89	Healthy range (if 1.62 m tall)
Weight (kg)	42.1	44.9	53-66
Height (m)	1.62	1.62	
BMI	16	17.1	20-25
IVNAA			
Total Body Protein (kg)	4.26	5.9	
Nitrogen Index	0.37	0.5	0.8-1.2
Impedance			
Total Body Water (%)	66(28L)	64(29L)	64-70(33-46L)
Lean Mass (%)	78(33kg)	78(35kg)	75-85(40-56kg)
Fat Mass (%)	22(9kg)	22(10kg)	15-25(8-16.5kg)

aging. However, any change in body water whether due to aging or to
illness can drastically alter the volume of distribution of a drug and
drug dosages need to be adjusted accordingly. Until a precise measure is
readily available, dosages will be assessed with a large amount of
guesswork (Gilman et al., 1985).

<u>Clinical Examples of Wasting</u>

The following two patients, one with chronic renal failure (D.C.)
and one with anorexia (J.B.) highlight some of the complexities of
studies in chronic diseases associated with wasting.

Patient D.C. is a 24-year old man with the problems of chronic renal
failure due to focal glomerulosclerosis, chronic pancreatitis, poor pubertal development, fluid and electrolyte disturbances, and renal bone
disease. IVNAA showed a reduced total body protein (Table 6) in September
1988. His malabsorption secondary to chronic pancreatitis was then
treated with pancreatic enzyme replacement. The Nitrogen Index (usually
between 0.8 and 1.2) increased from 0.37 to 0.50 after eight months
treatment. Impedance measurements showed a gain in total body water of
1 litre, in lean body mass of 2 kilogram, and in fat mass of one kilogram. During this time his weight increased from 42.1 kg to 44.9 kg. In
patients with end-stage renal failure who depend on either hemo- or
peritoneal dialysis, weight changes are usually attributed to fluid
retention. In this situation, it was invaluable to document an increased
lean body mass and, hence, prevent attempts to reduce his weight to the
"normal" level.

Patient J.B. is a 32-year old woman who has had anorexia nervosa
with amenorrhoea for 15 years. In this eating disorder, a reduction in
lean body mass and total body water as well as in bone mineral contribute
to the large weight loss often seen. Osteoporosis is a particular threat
in view of the poor dietary intake of calcium and protein together with
the associated amenorrhoea and oestrogen deficiency state.

The patient J.B. had body composition studies using IVNAA,
Bioelectrical Impedance, and DEXA (Table 7).

Table 7.   Body Composition Studies in a 32-year Old Woman (J.B.)
          with Anorexia Nervosa

	June 89	Healthy range (if 1.57 m tall)
Weight (kg)	34.5	50-61.5
Height (m)	1.57	
B.M.I.	14.0	20-25
IVNAA		
Total Body Protein (kg)	4.65	
Nitrogen Index	0.55	0.8-1.2
Impedance		
Total Body Water (%)	70(24L)	55-60(27-37L)
Lean Mass (%)	85(29kg)	70-80(35-49kg)
Fat Mass (%)	15(5kg)	20-30(10-18kg)
DEXA - Total Body		
Bone mineral content (kg)	1.1386	
Bone mineral density (g/cm^2)	0.467	0.8-1.49/sq cm
Soft tissue mass (kg)	33.29	
Lean mass (kg)	29.2	
Fat content (%)	12.5(4.1kg)	20-30

Her body fat is reduced to 12.5% (by DEXA) which is below the
healthy range of 20-30% for women. In addition, her Total Body Protein
and calculated Nitrogen Index are far below that expected for her age and
height. If her lean body mass (LBM) is estimated using the equation
quoted by Forbes (1987):

$$LBM \ (kg) = \frac{total \ N \ (g)}{33}$$

an estimate of 22.54 kg is arrived at. This value is well below that
measured with impedance and DEXA, and reflects the difficulties in using
assumptions to derive unmeasured compartments. This raises yet again the
question of which method should be used and also which constant is
appropriate for a given clinical situation.

The reduced bone density does, however, place her at increased risk
of fractures: indeed, she has already suffered two fractured ribs from
minimal trauma. Longitudinal studies on this patient are needed to assess
the rapidity of bone mineral loss and any possible improvement with
therapy.

FUTURE DIRECTIONS

Hill and Beddoe (1988) quoted Beneke, 1878 "Nothing is measured with
greater error than the human body". This is no longer true. In 1989, we

140

have seen the advent of new technology such as Magnetic Resonance Spectroscopy (Chance and Veech, 1988), IVNAA and DEXA, together with total body water and potassium measures. Today, the ability to make simultaneous measurements of body composition is critical and is the way to the future. DEXA is a technique that allows several measures to be taken; however, the technique must be validated.

The possibility of measuring individual nutrients such as magnesium, zinc, essential fatty acids (Moonen et al., 1988), vitamins, or even total body cholesterol becomes more realistic as the technology improves. The accurate, direct measures of body compositional variables currently available make body composition studies a more integral part of contemporary metabolic medicine. Ultimately, by the introduction of low cost, portable units, the technology will become more and more a part of everyday clinical and public health practice.

ACKNOWLEDGEMENTS

The authors wish to acknowledge the help given by various members of the Body Composition Unit at Prince Henry's Hospital, Melbourne. These include D.B. Stroud, B.J.G. Strauss, J.R. Lambert, all of who have contributed with both patient data and advice for this paper. Thanks are also due to Wendy Yu for typing of the manuscript.

REFERENCES

Alleyne, G. A. O., Hay, R. W., Picou, D. I., Starfield, J. P., and Whitehead, R. G., 1977, in: "Protein-Energy Malnutrition", Edward Arnold Ltd.
American Psychiatric Association, 1980, Diagnostic and Statistical Manual of Mental Disorders, ed. 3, Washington D.C., APA.
Andrews, G. R., Esterman, A. J., Braunack-Mayer, A. J., and Rungie, C.M., 1986, "Aging in the Western Pacific," World Health Organization, Manila.
Burkinshaw, L., 1985, Measurements of human body composition in vivo, Prog Med Rad Phys, 2:113.
Chance, B., and Veech, R. L., 1988, Phosphorus magnetic resonance spectroscopy as a probe of nutritional state, in: "Nutrition and Metabolism in Patient care," J. M. Kinney, K. N. Jeejeebhoy, G. L. Hill, and O. E. Owen, eds., W. B. Saunders Company.
Coward, W. A., Parkinson, S. A., and Murgatroyd, P., 1988, Body composition measurements for nutrition research, Nutr Res Rev, 1:115.
DSM-111-R, 1987, in "Diagnostic and Statistical Manual of Mental Disorders", ed. 3 Revised, American Psychiatric Association, Washington.
Forbes, G. B., "Human Body Composition", Springer Verlag, New York 1987.
Garrow, J. S., Hastings, E. J., Cox, A. G., North, W. R. S., Gibson, M. Thomas, T. M., and Meade, T. W., 1988, Obesity and post- operative complications of abdominal operation, BMJ, 297 (6642):181.
Gilman, A., Goodman, L. S., Rall, J. W., and Murad, F., 1985, "The Pharmacological basis of therapeutics", 7th ed, MacMillan Publishing Company, New York.
Harrison, J. E., McNeill, K. G., and Strauss, A. L., 1984, Nitrogen Index - total body protein normalized for body size, Nutr Res, 4:209.
Hill, G. L., and Beddoe, A. H., 1988, Dimension of the human body and its compartments, in: "Nutrition and Metabolism in Patient Care," J. M. Kinney, K. N. Jeejeebhoy, G. L. Hill, and O. E. Owen, eds., W. B. Saunders Company.

Lukaski, H. C., 1987, Methods for the assessment of human body
        composition: traditional and new, Am J Clin Nutr, 46:537.
McClain, C. J., and Cohen, D. A., 1989, Increased tumor necrosis factor
        production by monocytes in alcoholic hepatitis, Hepatology, 9
        (3):349.
Moonen, C. T. W., Dimand, R. J., and Cox, K. L., 1988, The non-invasive
        determination of linoleic acid content of human adipose tissue by
        natural abundance carbon-13 nuclear magnetic resonance, Magn Reson
        Med 6:140 (1988).
Otte, K. E., Ahlberg, P., D'Amore, F., and Stellfeld, M., 1989, Nutrition
        repletion in malnourished patients with emphysema, JPEN 13:2, 152.
Quetelet, L. A., 1969, Physique Sociale, Brussels C., Musquardt, 2:92.
Rigg, B. L., 1988, Bone densitometry and clinical decision making in
        osteoporosis, Ann Intern Med, 108:293.
Steen, B., 1988, Body composition and ageing, Nutr Rev, 46 (2):45.
Stroud, D. B., Borovnicar, D. J., Lambert, J. R., McNeill, K. G.,
        Marks, S. J., Rainer, H. C., Rassool, R. P., Strauss, B. J. G.,
        Tai, E. H., Thompson, M. N., Wahlqvist, M. L., Watson, B. A., and
        Wright, C. M., Clinical studies of total body nitrogen in an
        Australian hospital, this volume.
Wellcome Trust Working Party, 1970, Classification of Infantile
        Malnutrition. Lancet, ii:302.
WHO: Infant and Young Child Nutrition. Report by the Director
        General to the World Health Assembly, May 1983 (Document WHA
        36/1983/7).
Windsor, J. A., and Hill, G. L., 1988, Weight loss with physiologic
        impairment: a basic indicator of surgical risk, Ann Surg, 20:290.
Wood, P. D., Stefanick, M. L., Dreon, D. M., Frey-Hewitt, B. Garay, S.C.,
        Albers, J. J., Vranizan, K. M., Ellsworth, N. M., Terry, R. B.,
        and Haskell, W. L., 1988, Changes in plasma lipids and
        lipoproteins in overweight men during weight loss through dieting
        as compared with exercise, NEJM 319 (18):1173.

MECHANISM OF REDUCTION OF TOTAL BODY POTASSIUM IN MALNUTRITION

K.N. Jeejeebhoy

Department of Medicine

University of Toronto, Toronto, Canada

INTRODUCTION

Lean Body Mass (LBM) consists of tissues which are metabolically active and are responsible for locomotion. In addition, circulation, respiration and metabolism, are all dependent upon the functional activity of the LBM. It is not surprising, therefore, that in nutritional studies an estimation of LBM becomes of great importance. LBM is the main reservoir of body protein and nitrogen. Unfortunately, direct measurement of total body nitrogen (TBN) was not possible until recently: consequently, indirect methods were used to estimate LBM. Analytical data suggested that in healthy muscle the ratio of intracellular potassium to nitrogen is closely controlled at 3 mM K^+ to 1 g nitrogen. This ratio was extrapolated to the whole body measurements by using total body potassium (TBK) as an index of body cell mass (BCM) (Forbes and Hursh, 1963) and calculating LBM from these data. Later prompt gamma analysis (PGA) became available to measure total body nitrogen (TBN) (Mernagh et al., 1977); using these techniques, and also carcass analysis, the ratio of TBK to TBN in normals and in a chronically wasted individual (Knight et al., 1986) was found to be 1.6 mM K^+/g nitrogen.

Subsequent studies indicated that the tight relationship between TBK and TBN was not maintained during the refeeding of wasted individuals who showed a rapid gain in TBK but little change in TBN over 3 weeks (Jeejeebhoy et al., 1982). In this paper, data are presented which are consistent with the hypothesis that the mass of TBK is determined by cell energetics altering the cellular membrane potential and so influencing the intra- and inter-cellular distribution of potassium.

RESULTS

Relationship of TBK to TBN

Earlier chemical analysis showed that human soft tissues are composed of 3% nitrogen and 0.2% potassium. Therefore, the potassium to nitrogen ratio of such tissues should be 1.7 mM K^+/g nitrogen. For muscle, this ratio is 3.05 mM K^+/g nitrogen.

Recently we showed that the relationship of TBK measured by counting naturally occurring ^{40}K, and TBN measured by prompt gamma analysis (Jeejeebhoy et al., 1982) was 1.6 mM K^+/g nitrogen. This figure corresponds closely to the data obtained by tissue analysis (Knight et al., 1986).

Advances in In Vivo Body Composition Studies
Edited by S. Yasumura *et al.,* Plenum Press, New York, 1990

Table 1. Effect of Refeeding on Patients with Anorexia

Weeks refed	TBN Kg	TBK mM	CHI %	K/TBN ratio
0	1.20	1590	49.7	1.31
4	1.33	1897	64.9	1.42
8	1.37	2103	67.0	1.53

TBN = Total body nitrogen
TBK = Total body potassium
CHI=creatine height index.

Effect of Nutritional Support on TBK and TBN:

The effect of nutritional support can be studied in several situations, including oral refeeding of starving anorexia nervosa patients, and in malnourished and cancer patients receiving total parenteral nutrition (TPN). Table 1 gives the data for anorexia patients: it shows that the baseline ratio of TBK/TBN is 1.31 mM of K^+/g N, which is well below the normal ratio of 1.6 mM/g N. With refeeding, the ratio approaches normality. Anorexic patients were relatively K^+ depleted and refeeding increased K^+ to a disproportionately greater extent than nitrogen. In patients with anorexia, eating a normal diet for 8 weeks resulted in an increase of only 10% in TBN, whereas TBK rose 30% during the same time. In patients with cancer similar findings were noted (Shike et al., 1984). Patients with small cell cancer of the lung were randomized into groups, one receiving TPN for a month, the second eating an _ad lib_ diet (control). The TBN showed no differences. However, the TBK increased significantly in patients receiving TPN, whereas it fell in controls.

Studies of intracellular potassium, phosphogens, free ADP, free Mg^{2+}, free energy change of ATP hydrolysis and pH:

Nutritional manipulation: Rats were randomly allocated to an _ad lib_ fed control group or a hypocaloric group fed 25% of the calories eaten by their pair-fed controls. The measurements were made after the hypocaloric animals had lost 20% of their initial weight over about one week.

Intracellular potassium: The mean intracellular potassium fell significantly from 139 to 107 mM/L in the soleus muscle (p<0.05). On refeeding, it rose to 124 mM/L (Pichard and Jeejeebhoy, 1989).

Membrane potential and intracellular potassium activity (aK_i) of the soleus muscle: The mean membrane potential of five hypocaloric rats fell significantly from the control value of -73.4 (17 rats) to -69.1 mV, and the aK_i fell significantly from 92.3 to 80.3 mM/L (p<0.05 for both) (Fong et al., 1987) (Table 2).

Phosphogens, free ADP, free Mg^{2+}, free energy change of ATP hydrolysis, and pH: Table 3 shows that there was no change in muscle ATP levels in the hypocaloric rats but there was a significant fall in muscle creatine phosphate (CrP) as well as a small but significant fall in pH (Pichard et al., 1988). Free muscle magnesium remained unchanged. The free

Table 2. Mean Intracellular Potassium Activity (aK_i) in Soleus Muscle

Nutritional Status	n	aK_i
		mM
Control	17	92.3
Hypocalorically fed	5	80.3

muscle ADP levels rose significantly and the free energy change of ATP hydrolysis fell significantly. Refeeding restored these values to normal.

DISCUSSION

The ratio of body potassium to nitrogen ratio in the stable individual is comparable to that observed by carcass analysis. The severely malnourished anorexic patient is relatively potassium-depleted. Furthermore, on refeeding there is a significant change in body potassium without a corresponding change in nitrogen. Therefore, the initial effect of refeeding is to cause a change in ion distribution rather than to increase protein synthesis. From total body measurements it is not clear where the extra potassium is situated, but studies of rat muscle suggest that, in part, there are changes in free ionic potassium (as indicated by the change in aK_i) as well as in total intracellular potassium. The question is, how do these changes come about?

Ion distribution is determined by either a change in the energetics of the Na^+-K^+ ATPase pump or in the permeability and selectivity of the cell membrane.

Energetics of the Na^+-K^+ ATPase pump: The muscle cell membrane maintains a gradient of high extracellular and low intracellular Na^+, and high intracellular and a low extracellular K^+. This gradient exerts an electrochemical force expressed as the free energy change of ion movement, which is counteracted by the free energy change of ATP hydrolysis. At ion equilibrium the two forces must also be in equilibrium. Since hypocaloric feeding of rats resulted in a fall in the the free energy change of ATP hydrolysis, on theoretical grounds there should be a change in ion gradient, with a fall in intracellular potassium and a rise in intracellular sodium.

The free energy change of an ion ΔG_{ion} = RT ln Ci/Ce + Z x F x MP (Kammermeier, 1987). Where R= universal gas constant, T= temperature in degrees Kelvin, Ci=intracellular concentration of an ion, Ce=extracellular concentration of the ion, Z=valency and F=Faraday constant, MP=Membrane potential (volt).

The effect of the change of free energy change of ATP hydrolysis on the change ratio of intra- to extra-cellular concentration of Na^+ can be calculated from:

$$\Delta G_{ATP}(J/mol) = \Delta G_{K+} = 3RT \ln Ci/Ce + (Z \times F \times MP)$$

because one mole of ATP is used to pump 3 moles of Na^+, then rearranging the equation,

Table 3. Muscle Energetics.

Groups	ATP	CrP	pH	Free Mg2+	Free ADP	ΔG_{ATP}
	mM/g wet wt.	mM/g wet wt.		μM	mol x10^5	kJ/mol
Control	8.0	30.1	7.15	532	39.2	-69.0
Hypocal	7.8	16.1	7.11	516	108.3	-65.9
Refed	7.7	26.2	7.15	628	41.9	-67.4

$$Ci/Ce = \exp((\Delta G_{ATP}\text{-}(Z \times F \times MP))/3RT)$$

$$Ci/Ce(control)/Ci/Ce(hypocaloric)=$$

$$\exp((\Delta G_{ATP}\text{-}(Z \times F \times MP))/3RT(control))/$$

$$\exp((\Delta G_{ATP}\text{-}(Z \times F \times MP))/3RT(hypocaloric))$$

Using the figures for ΔG_{ATP} and the membrane potential given above, the control/hypocaloric ratio of Ci/Ce is 0.70. Since Ce is unchanged, the hypocaloric muscle will have about 42% more sodium. Since the normal intracellular sodium is about 15 mM, it is likely that the hypocaloric muscle will have 15 x 1.42 = 21 mM of sodium. If Na/K exchange is in the ratio of 2/3 the outflow of potassium would be (21-15) x 3/2= 9 mM. This value is close to the figure of 12 mM obtained by direct measurements. Therefore, it is likely that the change of ΔG_{ATP} may contribute significantly to the fall in intracellular K$^+$.

<u>Permeability and selectivity</u>: Increased permeability to potassium or chloride will cause the membrane potential to become more negative (Darnell et., 1986) which is contrary to the observed change. However, if permeability to sodium increased, the membrane potential will become less negative. In this case we would have to postulate a selective conductance for Na$^+$.

In conclusion the effect of malnutrition and refeeding on body potassium is rapid and disproportionate to that of nitrogen. It is likely to be due to changes in cell energetics and/or selective permeability to Na$^+$ ions. Current evidence is in favour of the former mechanism.

REFERENCES

Darnell, J., Lodish, H., and Baltimore, D., 1986, Molecular cell biology, Scientific American Books, New York.

Fong, C.N., Jeejeebhoy, K.N., Charlton, M.P., 1987, Nutrition and muscle potassium: differential effect in rat slow and fast muscles, <u>Can. J. Physiol. Pharmacol.</u>, 65: 2188-2190.

Forbes, G.M., and Hursh, J.M., 1963, Age and sex trends in lean body mass calculated from ^{40}K measurements, <u>Ann. NY Acad. Sci.</u>, 110:225.

Jeejeebhoy, K.N., Baker, J.P., and Wolman, S.L., 1982, Critical evaluation of the role of clinical assessment and body composition studies in patients with malnutrition and after total parenteral nutrition, <u>Am. J. Clin. Nutr.</u>, 35(Suppl.):1117-1127.

Kammermeier, H., 1987, Interrelationship between the free energy change of
    ATP-hydrolysis, cystolic inorganic phosphate and cardiac performance
    during hypoxia and reoxygenation, _Biomed. Biochim. Acta_, 8:S 499.
Knight, G.S., Beddoe, A.H., Streat, S.J., et al., 1986, Body composition
    of two human cadavers by neutron activation and chemical analysis,
    Am. J. Physiol., 250:E179-185.
Mernagh, J., Harrison, J.E., and McNeill, K.G., 1977, In vivo
    determination of nitrogen using Pu-Be sources, _Phys. Med. Biol._,
    22:831.
Pichard, C., Vaughan, C., Struk, R., Armstrong, R.L., and Jeejeebhoy,
    K.N., 1988, The effect of dietary manipulations (fasting,
    hypocaloric feeding and subsequent refeeding) on rat muscle
    energetics as assessed by nuclear magnetic resonance spectroscopy, _J.
    Clin. Invest._, 82:895-901.
Pichard, C., and Jeejeebhoy, K.N, 1989, Intracellular Potassium (K+) and
    Membrane Potential in rat muscle during malnutrition and refeeding
    studied in vivo by K+-Sensitive microelectrode (KISE) and by atomic
    absorption spectrophometry (AAS), _Clin. Nutr._, 8: 51, 1989.
Report of the Task Group on Reference Man. International Commission on
    Radiological Protection Report No. 23, 1975, Pergamon Press, Oxford.
Shike, M., Russell, D.McR., Detsky, A.S., Harrison, J.E., McNeill, K.G.,
    Shepherd, F.A., Feld, R., Evans, W.K., and Jeejeebhoy, K.N., 1984,
    Changes in body composition in patients with small-cell lung cancer,
    Ann. Intern. Med., 101:(3), 303-309.

ASSESSMENT OF BODY COMPOSITION IN ANOREXIC PATIENTS

J. Hannan, S. Cowen, C. Freeman[*], A. Mackie,
and C.M. Shapiro[*]

University Departments of Medical Physics and Psychiatry[*]
Western General Hospital and Royal Edinburgh Hospital[*]
Edinburgh, Scotland

INTRODUCTION

The four compartment model, consisting of protein, minerals, water and
fat, provides the most complete delineation of body mass into its
component parts. This has been used to compare the body composition of
anorexic patients with that of control subjects, and to follow changes
during treatment. A simpler two compartment model, consisting of lean body
mass and fat, may provide sufficient information for routine patient
management. Methods based on measurements of total body potassium and
total body water have been evaluated for use with anorexics.

METHODS

Subjects

Measurements were made on 16 female control subjects and 20 female
anorexic patients. The anorexics were subdivided into two groups,
consisting of 14 patients with body mass index (BMI) less than 17 kg m^{-2}
and 6 patients with BMI greater than this value. A summary of the age,
weight and BMI distributions for each group is given in Table 1.

Total Body Protein

Prompt neutron activation analysis with ^{252}Cf was used to measure total
body nitrogen, using hydrogen as an internal standard. Construction of the
equipment has been described (Mackie et al., 1988). A 1 m length of the
subject, from the shoulders to a point around the knees, was scanned in
forty minutes. The mean width and thickness were used to calculate a
calibration factor (Q) which corrects for the difference in detection
efficiency between nitrogen and hydrogen. Q was determined as a function
of width and thickness using tissue-equivalent phantoms. Total body
protein (TBP) was calculated from the relationship:

$$TBP = \frac{6.25 \, [0.12 \, (TBWt - TBM) - 0.01 \, TBW]}{[H_c/(N_c \cdot Q)] + 0.3125} \qquad (1)$$

where TBWt, TBM and TBW represent the total body weight, minerals and
water, respectively. N_c and H_c are the counts from the nitrogen and

Table 1. Age, Weight and Body Mass Index (BMI) for the Female Control Subjects and the Female Anorexics

	Controls			Anorexics BMI < 17 kg m^{-2}			BMI > 17 kg m^{-2}		
	Mean	SD	Range	Mean	SD	Range	Mean	SD	Range
Age (y)	36.4	10.9	24-65	25.6	14.8	17-76	31.0	10.4	20-43
Weight (kg)	69.6	8.4	55-101	40.4	6.1	26-50	49.6	3.7	44-53
BMI (kg m^{-2})	23.9	3.6	20-34	14.5	1.5	11-16	18.3	0.8	17-19

hydrogen energy regions of the prompt gamma ray spectrum and 6.25 is the multiplicative factor relating TBP to total body nitrogen. The derivation of this relationship has been described by Mackie et al (in press).

Total Body Minerals

It has been shown that the total mineral compartment represents 6.22% by weight of the lean body mass in normal subjects (Beddoe et al., 1984). For our control subjects we therefore estimated TBM from the relationship:

$$TBM \ (kg) = 0.0622 \ \frac{TBW \ (1)}{0.73} \ k \qquad (2)$$

where 0.73 is the normal hydration coefficient, k = 1 for females less than 55 years, and k = 1 - 0.0067114 (AGE - 55) for females older than 55 years. This age correction is that given by Cohn et al. (1976) for the age dependence of bone mineral ash in females.

For the anorexic patients a normal hydration coefficient could not be assumed. Therefore, the mineral compartment was estimated from a regression equation generated from normal subjects, relating minerals to body habitus parameters which do not change in wasting disease (Beddoe et al. 1984).

Total Body Water

Subjects fasted from one hour before the start of the study until its completion three hours after the oral administration of 2 MBq tritiated water. Urine was collected during the three hour equilibration period and 10 ml blood was withdrawn at three hours. The activity of ^{3}H in an aliquot of urine was measured, together with the activity in 2.0 ml plasma. All samples were counted relative to a standard activity and were quench corrected. This correction was determined for the same scintillator quenched by plasma. The volume of total body water was obtained by dividing the activity retained at three hours by the activity per millilitre of plasma.

Total Body Potassium

Total body potassium (TBK) was measured from naturally occurring ^{40}K using a shadow-shield Whole Body Counter consisting of four 15 cm diameter by 10 cm NaI(Tl) crystals. The procedure consisted of 40 minute scan times for subject and background, and 20 minute scan time for a 70 kg Bush phantom containing 1 kg potassium. A body habitus correction was applied to correct for differences in detection efficiency between the subject and the phantom. These corrections had previously been determined from ^{42}K studies on 74 subjects.

Table 2. Body Composition as Percentage of Total Body Weight

	Controls		Anorexics BMI < 17 kg m^{-2}	
	Mean	SD	Mean	SD
Total body protein (TBP)	14.6	1.3	19.4	1.6
Total body minerals (TBM)	4.4	0.3	6.0	0.9
Total body water (TBW)	52.1	3.7	67.6	6.0
Total body fat (TBF)	28.9	5.2	7.0	6.5
Lean body mass (LBM)	71.1	5.2	93.0	6.5

RESULTS

Lean body mass (LBM) was calculated from the relationship:

$$LBM = TBP + TBM + TBW \tag{3}$$

Total body fat (TBF) was derived from the difference between TBWt and LBM.

The distribution of body composition is given in Table 2 for the control subjects and the anorexic patients (BMI < 17 kg m^{-2}). All compartments were significantly different (P < 0.001) between the two groups.

An example of how the four compartment technique may be used to follow changes in body composition during treatment is shown in Fig 1.

For a simple two compartment model LBM may be estimated from total body potassium (TBK). Many centres have calculated LBM from TBK using a linear relationship. When the values of LBM for the control subjects were

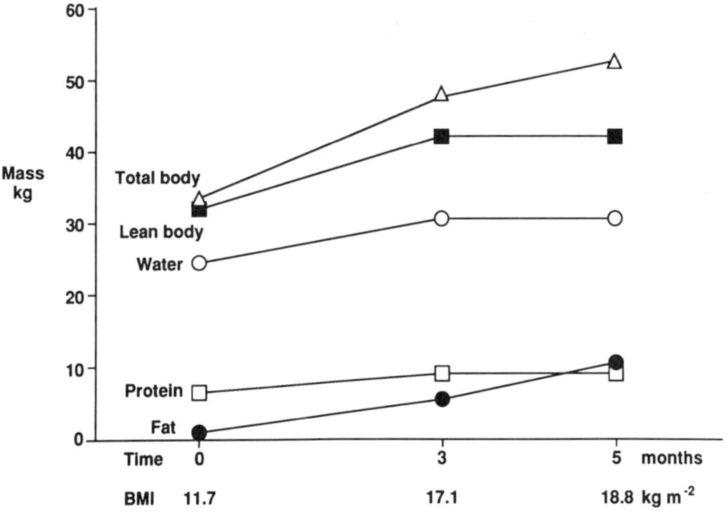

Fig. 1. Body composition in an anorexic patient during treatment.

regressed against TBK the following relationship was obtained:

$$\text{LBM (kg)} = \frac{\text{TBK (g)}}{2.414} \qquad (R^2 = 0.998, \; SER = 2.04 \; kg) \qquad (4)$$

The coefficient (2.414) is in good agreement with conversion factors published by other groups (Boddy et al., 1973).

However, when the results for LBM and TBK were compared for all groups it was apparent that there was a non-linear relationship. Fig. 2 shows a quadratic least-squares fit to the data. The standard error from the regression was 2.4 kg. If the normal linear relationship had been used, there would have been a large systematic underestimate of LBM for the anorexic patients.

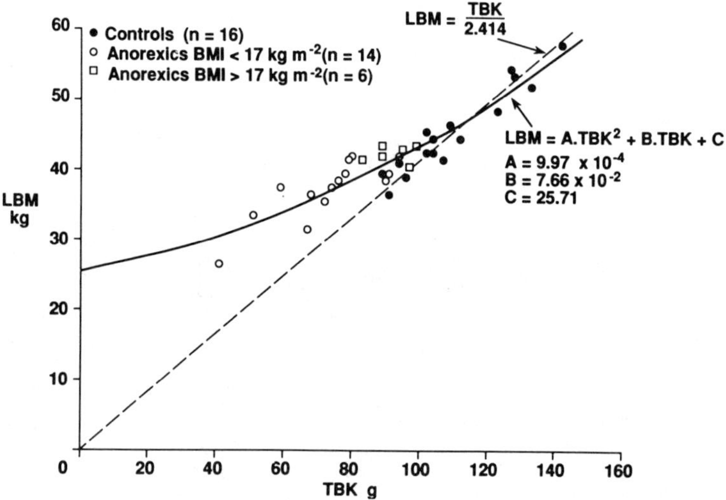

Fig 2. Relationship between TBK and LBM.

An alternative method of deriving the two compartment model is to calculate LBM from the TBW. The mean hydration coefficient for the control subjects was 0.733 ± 0.006 (SD). Therefore, LBM could be calculated from the relationship:

$$\text{LBM (kg)} = \frac{\text{TBW (1)}}{0.733} \qquad (5)$$

The mean hydration coefficient for the anorexic patients (BMI < 17 kg m^{-2}) was 0.729 ± 0.017 (SD). Although this is not significantly different from the control subjects, the hydration coefficients for some anorexic patients with very low BMI were greater than 0.75. The assumption of normal hydration in these patients results in an overestimate of LBM. Indeed in some patients the LBM calculated by this method is actually

greater than the total body weight. Fig 3 shows the relationship between
LBM calculated from equation (5) and that obtained from the sum of
protein, minerals and water. Where equation (5) gave a value of LBM
greater than TBWt it was set equal to TBWt. There was good agreement
between the values of LBM calculated by the two methods, with a standard
error from the regression of 0.63 kg.

DISCUSSION

The four compartment model provides the most complete description of
body composition. We have demonstrated significant differences in the
proportions of the various compartments between anorexics and control
subjects and have followed changes in body composition during treatment.
However, this involves the use of neutron activation analysis equipment.
This is only available in a few centres and is expensive to install. In
addition, the whole body dose equivalent from this technique was 0.17 mSv
(Mackie et al., in press). Therefore it may not be desirable to make
frequent use of the technique to follow changes during treatment. Simpler
techniques may be sufficient where it is only required to measure LBM and
TBF. A linear relationship has previously been used to calculate LBM from
TBK. However, this is not suitable for anorexic patients and may also be
inappropriate in other wasting diseases. This is presumably because the
loss of potassium in anorexic patients is preferentially associated with a
loss of muscle mass. The skeleton is relatively spared and so represents a
larger proportion of the lean body mass. We have established a non-linear
relationship between LBM and TBK which gives a standard error from the
regression of 2.4 kg. Although this may be satisfactory for group studies
it is unsuitable for use in individual patients. Calculation of LBM from
TBW requires the assumption of normal hydration and may overestimate LBM
in individual patients. However, in the present study, only patients with
BMI below 14 kg m^{-2} had a hydration coefficient which was significantly
different from the control group. For these patients the overestimate of
LBM was identified from the fact that it exceeded the total body weight.
These patients have such a low fat mass that for practical purposes LBM
may be considered to equal TBWt. The standard error from regression for
LBM was then only 0.63 kg; this was not significantly different for the
anorexic patients analysed separately. This method may therefore prove
sufficiently accurate to allow changes in LBM and TBF to be followed
during treatment. An assessment of protein malnutrition can, of course,
only be determined using the four compartment model.

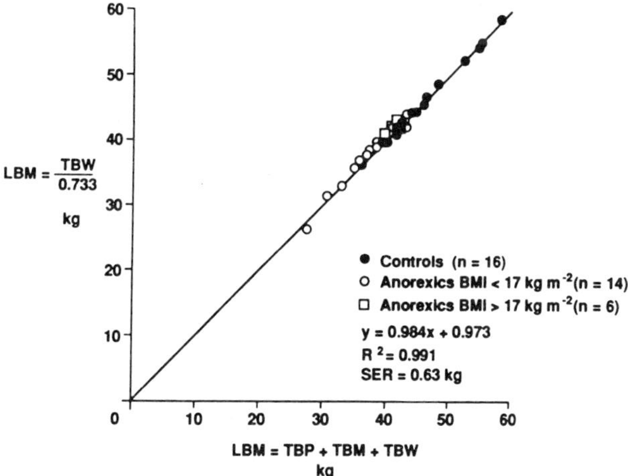

Fig. 3.   Comparison of LBM calculated from equation (5) with that derived
from equation (3).

REFERENCES

Beddoe, A.H., Streat, S.J., and Hill, G.L., 1984, Evaluation of an in-vivo
    prompt gamma neutron activation facility for body composition
    studies in critically ill intensive care patients: results on 41
    normal , Metabolism, 33:270.
Boddy, S.R., King, P.C., Womersley, J., and Durnin, J.V.G.A., 1973, Body
    potassium and fat free mass, Clin. Sci., 44:621.
Cohn, S.H., Vaswani, A., Zanzi, I., Aloia, J.F., Rodinsky, M.S., and
    Ellis, K.J., 1976, Changes in body chemical composition with age
    measured by total-body neutron activation, Metabolism, 25:85.
Mackie, A., Hannan, W.J., Smith, M.A., and Tothill, P., 1988, Apparatus
    for the measurement of total body nitrogen using prompt neutron
    activation analysis with Californium-252, J. Med. Eng. Technol.,
    12:152.
Mackie, A., Cowen, S., and Hannan, W.J., Calibration of a prompt neutron
    activation analysis facility for the measurement of total body
    protein, Phys. Med. Biol., (in press).

THE ROLE OF BODY PROTEIN STUDIES IN CLINICAL TRIALS

B. J. Allen[1], N. Blagojevic[1], I. Delaney[1], C. A. Pollock[2],
L. S. Ibels[2], M. A. Allman[3], D. J. Tiller[3], K. J. Gaskin[4],
L. A, Baur[4], D. L. Waters[4], C. Cowell[4], G. Ambler[4],
C. Quigley[4], and J. P. Fletcher[5]

[1]Australian Nuclear Science and Technology Organisation
PMB 1 Menai NSW 2234 Australia
[2]Royal North Shore Hospital St Leonards NSW 2065
[3]Royal Prince Alfred Hospital Camperdown NSW 2050
[4] Royal Alexandra Hospital for Children Camperdown NSW 2050
[5]Westmead Hospital Westmead NSW 2145

INTRODUCTION

Short-term protein accretion in malnourished patients does not
necessarily equate to a long-term gain in total body protein.
Consequently, classical cross-sectional studies are of limited value as
they examine the dynamics of protein metabolism at a single point of time.
Extrapolations to longitudinal changes in total body protein may therefore
be inappropriate.

In the past, body protein was estimated by indirect measures such as
mid-upper arm circumference for skeletal muscle (Blackburn et al., 1977)
and plasma proteins for non-muscle protein stores (McFarlane et al., 1962).
Other techniques to estimate body protein stores are total body potassium
(Cohn and Palmer, 1970) and the creatinine height index (Webster and
Garrow, 1985). However, total body potassium is not correlated with body
protein in renal disease (Schilling et al., 1985) and in cystic fibrosis
(Allen et al., 1987), and creatinine output exhibits large variations
because of diet, filtration, and secretion by the kidneys (Materson, 1971).

Total body protein can be readily measured by the prompt neutron
activation technique (Allen et al., 1987), particularly for monitoring
malnutrition. Malnutrition is associated with poor prognoses in most
diseases (Hill et al., 1977; Bistrian et al., 1975). In this paper,
malnourished patients with cystic fibrosis and end-stage renal failure are
identified by directly measuring the body protein compartment. In
addition, we examine the effect of aortic surgery and synthetic growth
hormone on body protein for normal subjects.

All body protein measurements were made with the ANSTO Body Protein
Monitor at the Lucas Heights Research Laboratories, now at Royal North
Shore Hospital.

The studies are approved by the ethics committees of the Royal
Alexandra Hospital for Children and the Royal Prince Alfred, Royal North
Shore and Westmead Hospitals. All subjects gave their written consent.

Advances in In Vivo Body Composition Studies
Edited by S. Yasumura *et al.,* Plenum Press, New York, 1990

METHODS

Total Body Nitrogen

The only body tissue which contains nitrogen is protein. The contribution of nitrogen containing metabolites such as urea and creatinine to the body pool is negligible, even in patients with renal failure. Therefore, total body protein can be obtained directly from the total body nitrogen, on the basis of 1 g nitrogen in 6.25 g protein.

With the neutron activation technique for the in vivo measurement of nitrogen (Biggin et al 1972; Harvey et al 1973), we can determine whole body protein directly by measurement of nitrogen relative to body hydrogen (Dabek et al., 1977; Vartsky et al., 1979).

The ANSTO technique used for the prompt measurement of neutron capture gamma rays in nitrogen was described previously (Allen et al., 1987). Patients lie on a moving table which carries them over a well-shielded and collimated beam of fast neutrons from a californium-252 neutron source. Capture gamma rays from nitrogen and hydrogen are measured by unmatched bilateral NaI detectors (diameter x depth of 20 x 15 and 10 x 10 cm, and later 20 x 15 and 20 x 10 cm).

We ensured that patients were lying centrally on the table to minimise uncertainties relating to different detector efficiencies. The administered radiation dose equivalent was typically 0.2 mSv, using the neutron quality factor Q=20. Nitrogen is determined relative to hydrogen, after corrections for background and gamma attenuation (Allen et al., 1987).

The coefficient of variation (CV) for the long-term reproducibility of the net nitrogen yield, after correction for the decay of the neutron source, is 2% CV, and is measured using a box phantom with known nitrogen concentration. A comparable CV was obtained from four measurements of one of the authors (BJA) over 18 months.

Anthropometry

Triceps, biceps, subscapular, and suprailiac fat skin-folds were measured with Holtain calipers. The sum of these four skin-folds was used to estimate the percentage body fat (Durnin and Womersly, 1974), which is subtracted from body weight to obtain the lean body mass. The mid-arm muscle circumference (MAMC mm) was calculated from (MAC- 3.142*TSF) where MAC is the mid-arm circumference and TSF is the triceps skinfold.

The results are expressed as a nitrogen index (NI), that is, the observed nitrogen/predicted nitrogen. The predicted nitrogen is calculated from the product of a sex-based coefficient (0.5 for male, 0.42 for female) multiplied by (height)**2.6) (ie the Toronto NI as defined by Harrison et al., 1984). Nitrogen per lean body mass (N/LBM g/kg) and nitrogen weight (N g) relative to sex-and-height matched controls are also calculated. Students t-test was used to test for significance in statistical analyses.

NORMAL CONTROLS

Over 72 measurements on 66 healthy control adults gave the normal body composition data (Table 1). The male to female ratio for body nitrogen is 1.5, but reduces to 1.05 for N/LBM and unity for the NI. Thus the Toronto NI satisfactorily removes the sex difference. The average NI is only 3% larger than the Toronto value of unity.

Table 1.   Normal Control Values

Adults age > 18 y	Male	Female	M/F
Number	31	35	
Age y	40 (11)	36.5 (8.4)	
Weight Kg	77.9 (12.8)	61.1 (13.2)	1.27
Height cm	176.9 (7.9)	161.3 (10.8)	1.10
BMI	24.7 (3.1)	23.4 (4.1)	1.06
Fat %	24 (6)	31 (6)	0.77
LBM g	59.0 (8.2)	41.2 (6.7)	1.43
N g	2290 (310)	1530 (250)	1.50
N/LBM g/Kg	38.9 (2.7)	37.0 (2.9)	1.05
NI	1.03 (0.10)	1.03 (0.14)	1.00
corr (N,Fat)	0.09	0.58	
corr (N,HT)	0.77	0.73	
corr (N,WT)	0.78	0.83	
corr (N,LBM)	0.89	0.90	

Children	Male		Female	
Age y	12-18	<12	12-18	<12
Number	4	4	12	5
Age y	14.2 (0.7)	10.0 (1.7)	14.7 (1.2)	8.3 (2.7)
BMI	19.8 (2.2)	17.1 (2.3)	19.6 (2.0)	17.9 (2.8)
Fat %	24.2 (6.9)	23.2 (6.8)	27.3 (6.4)	25.0 (7.4)
N g	1433 (310)	1063 (186)	1473 (224)	803 (244)
N/LBM g/Kg	39.6 (1.3)	39.2 (2.1)	40.6 (3.5)	35.8 (2.4)
NI	0.91 (0.05)	0.82 (0.04)	1.03 (0.11)	0.97 (0.05)

+   Serial values included for children.
    Standard deviation given in parenthesis.

With the exception of height, the smallest coefficient of variation is
for the N/LBM, being only 6.9% for males and 7.8% for females. This low
variance arises because of the high correlation between N and LBM (r=0.90).
Thus, comparisons of malnourished subjects with normal controls can best be
made by using N/LBM. However, N/LBM is subject to errors in the estimation
of skin folds and hence body-fat, which may be significant in renal disease
patients. Therefore, we use the NI more often to estimate nitrogen
depletion.

Correlation coefficients between body nitrogen and height and weight
are similar for males and females (r=0.7-0.8), but not for body fat where a
correlation occurs only for females (r=0.6).

The sample size is small for the children but the N/LBM data are
consistent with those of the adults, as is the NI for female children.
However, the NI for male children is substantially lower, as is the BMI for
all children.

END-STAGE RENAL FAILURE

In end-stage renal failure, Degoulet et al. (1980) showed a
significant association between malnutrition, morbidity and mortality.
Malnutrition occurs in patients with end-stage renal failure, either
managed by haemodialysis (HD) or by continuous ambulatory peritoneal
dialysis (CAPD) (Pollock et al., 1989). A proposed advantage of CAPD is

that anabolism is promoted, provided adequate dietary protein is ingested
(Blumenkrantz et al., 1982; Young et al., 1986; Stefanidis et al., 1983;
Guarnieri et al., 1983).

The role of prompt in vivo neutron activation in renal failure was
explored by Cohn et al. (1983) who found, for 15 haemodialysis patients,
that their body composition was not significantly different from that of
matched normal controls; Schilling et al. (1985) showed significant
protein depletion for 26 patients on peritoneal dialysis; Williams et al.
(1984), working with 75 patients with either renal failure or kidney
transplantation, found depletion of lean body mass and protein but not
potassium in dialysed patients, and an excess of fat in transplanted
patients. Higher nitrogen levels were observed for patients on long-term
dialysis for more than 5 years.

<u>Patients</u>

Thirty-six patients on long-term continuous ambulatory peritoneal
dialysis were assessed; twenty-two patients had used CAPD as the sole form
of therapy, and 7 patients had previous intermittent haemodialysis.
Twenty-nine patients on maintenance haemodialysis were measured for TBN,
seven on more than one occasion. Details of the patients' condition,
prescribed diet, and urea and dialysate analyses are given in Pollock et
al. (1989) for CAPD and in Allman et al. (unpublished results) for HD.
Patient parameters are summarised in Table 2.

<u>Protein Depletion</u>

Our results indicate that both haemodialysis and peritoneal dialysis
patients have similar average protein depletion levels relative to normal
controls (Table 3). The average nitrogen depletion levels are 14-17% for
CAPD and HD patients, respectively, and are similar to those for the N/LBM
(16-18%) and NI (12-14%). The most significant results are obtained for
N/LBM where the CV of 8-9% is much less than the 13-14% for N. The NI's of
HD and CAPD patients are plotted in Fig. 2a and 2b, which shows that most
results lie below the average control value of 1.03.

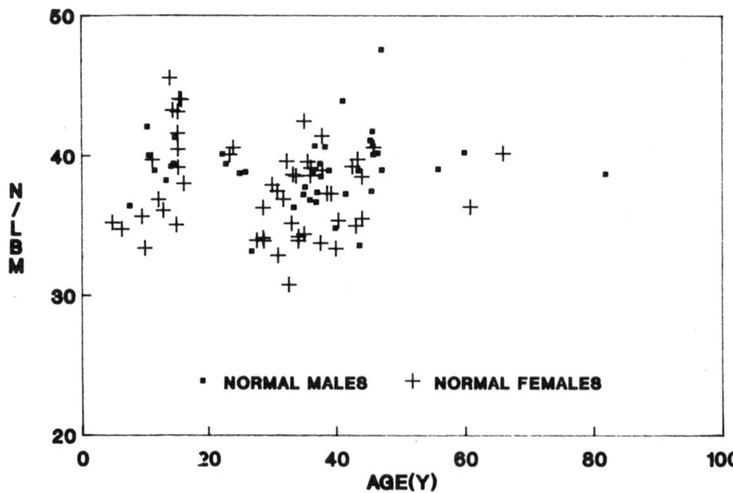

Fig. 1.   The age dependence of N/LBM is shown for normal adults and
          children.

158

Table 2. End-Stage Renal Failure Patient Parameters+

Sex	Haemodialysis		Peritoneal Dialysis	
	M	F	M	F
patient number	19	10	19	17
age	49(13)	38(9)	49(17)	51(12)
duration (month)	37(25)	33(22)	25(23)	30(17)
protein intake g/kg *	1.0-1.2		1.2-1.5	
height cm	173(7)	158(7)	173(9)	160(6)
P-value	NS	NS	0.1	NS
weight Kg	70.5(8.5)	57.9(10.3)	73.6(13)	65.8(13)
P-value	0.05	NS	NS	NS
Fat %	19(6)	26(7)	20(5)	35(8)
P-value	<0.01	0.05	<0.01	0.05
BMI	23.7(2.6)	23.1(4.2)	24.3(2.6)	25.9(5.3)
	NS	NS	NS	0.1

NS   Mean not significantly different to normal control value
*   Recommended intake in g per kg ideal body weight
+   End-stage renal failure for renal function <5% normal
    Standard deviation in parenthesis

In males and females, body nitrogen is correlated with weight and, to a lesser extent, height (r=0.8,0.5 resp). There is no correlation with body-fat, but, in females, body-fat is well correlated with weight (r=0.84). None of these parameters are correlated with the duration of either peritoneal dialysis or haemodialysis.

Body weight is reduced by 9% and 5% for male and female HD patients, while it is unchanged for male and increased by 8% for female CAPD patients; only in the first case is the loss significant. For male and female HD patients, significant fat depletions of 21% and 16% are observed, comparable to the 17% loss in male PD patients, but not the 13% gain in fat by female PD patients. The BMI is not significantly different from normals. Clearly, weight and fat changes provide no information on protein status.

The average nitrogen depletion of about 15% places half the dialysis patients within one standard deviation of the mean value of normals. However, 17% of dialysis patients have protein depletions exceeding 30% of the normal average. These patients could be at serious risk because of malnutrition.

<u>Longitudinal Variation</u>

With the exception of one patient on nasogastric and intensive oral feeding, the HD group showed relatively static body protein in a longitudinal study (Fig. 2a), although some patients were already substantially depleted (Allman et al., 1989). On the other hand, body protein increased significantly (Fig. 2b) in longitudinal studies of several CAPD patients (Pollock et al., 1989). The change in NI was the most significant longitudinal parameter for this group, with P<0.005 for the null hypothesis. The average change in NI (SD) is 0.080 ± 0.048 gpt 7 CAPD patients over 6.7 ± 1.0 months.

<u>Protein and Prognosis</u>

In a retrospective analysis of the peritoneal dialysis group, of the 8 patients with NI<85% of the average control value of NI=1.03,   6 were

Table 3. Nitrogen Depletion in End-Stage Renal Failure (SD)

| | Haemodialysis | | Peritoneal Dialysis | |
	M	F	M	F
N g	1820(260)*	1280(190)*	1940(410)*	1310(170)*
N %control	82	83	87	86
NI	0.88(.13)*	0.90(.13)*	0.90(.12)*	0.90(.13)*
NI % control	85	87	89	88
N/LBM	31.9(2.7)*	30.2(2.7)*	32.9(3.1)*	31.2(2.5)*
N/LBM % control	82	82	85	84

*P<0.01 for normal control values
 All HD data are not statistically different from PD data
 Repeat measurements included
 Standard deviation in parentheses

withdrawn from CAPD because of complications or death, the remaining two patients having metastatic melanoma or severe heart failure. Another patient who died initially had a normal NI, but on reassessment 8 months later, and 9 months before death, this had decreased by 10%. Defining these patients as a high-risk group, the average NI is significantly lower in the high-risk group than in the low-risk group, that is, NI=0.82(0.10) vs 0.96(0.13): P<0.01. Thus, for peritoneal dialysis, an association is observed between body nitrogen levels and prognosis.

A few studies have looked at the effects of nutritional status on morbidity and mortality in end-stage uraemia and patients undergoing dialysis (Degoulet et al., 1980). Where morbidity was assessed on the frequency of peritonitis, the inherent difference between short-lived episodes responsive to host defenses and antibiotics, and overwhelming infection requiring catheter removal, has not been emphasized. This latter condition appears to be associated with protein depletion in peritoneal dialysis (Pollock et al., 1989).

SYNTHETIC GROWTH HORMONE

Classical nitrogen balance studies (Brown et al., 1967; Prader et al., 1964; Richter et al., 1987; Rudman et al., 1979) have been used to investigate the short-term anabolic action of growth hormone via its mediator, insulin like growth factor (IGF1). It was long assumed that the increase in linear growth which occurs with growth hormone treatment reflects an increase in total body nitrogen.

Net protein accretion is diminished in disease states. By decreasing the rate of body protein catabolism, growth hormone may be able to promote a net protein accretion in states of either decreased protein synthesis or of increased protein breakdown.

Our aim was determine the effect of synthetic methionine-free growth hormone (Kabi Genotropin, Sweden) on total body nitrogen and to examine its relationship with changes in growth parameters and in the markers of the growth hormone effect (IGF1,IGF-binding protein). Children with idiopathic partial or complete growth hormone deficiency (GHD) and short, slowly growing (SSG) children with clinical, but not biochemical, GHD are included in the trial.

<u>Patients</u>

Three groups of 8 children are involved in the double blind trial:

-short, slowly growing (SSG) children with clinical but not biochemical GHD on a conventional dose regime (0.5 U/Kg/week),

-SSG children on high dose regime (1.0 U/Kg/week), and

-SSG children on a placebo regime.

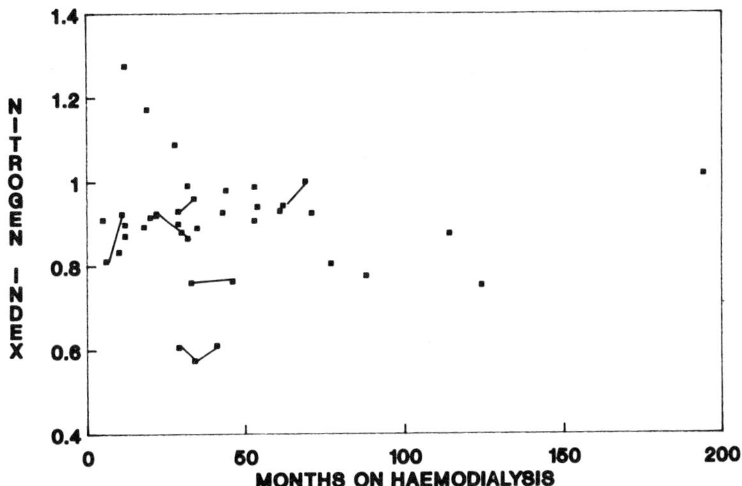

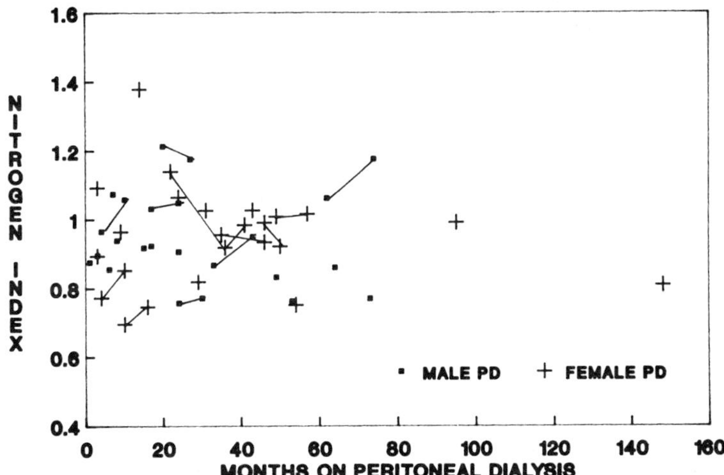

Fig. 2. Months on dialysis versus nitrogen index for end stage renal failure patients on haemodialysis (top) and peritoneal dialysis (bottom). Serial measurements of the same subject are connected by lines.

Table 4. Average Changes and Significance in SGH Trial

Parameter	Initial	Final	SD	T	P value(1)
N g	637	761	34	2.6	0.013
Fat %	18.8	13.4	3.4	6.0	<0.01
WT Kg	22.0	23.9	5.3	1.4	0.12
HT cm	116.2	121.2	11.8	1.6	0.16

Parameter	Difference	SD	SEM	T	P value(2)
N g	124.3	94.1	17.5	7.1	<0.001
Fat %	-5.3	2.3	0.44	12.3	<0.001
HT cm	5.0	4.2	0.8	6.3	<0.001
WT kg	2.0	2.0	0.4	5.5	<0.001

1)   Two sample t-test for N = 29.
2)   Null hypothesis

All children in the SSG trial are randomised and children on placebo growth hormone are not identified.

Children deficient in growth hormone are those with height < 3rd %ile, height velocity < 25th %ile for bone age, and growth hormone response of < 20 mIU/l on two pharmacological tests and one physiological test.

The TBN measurements are performed just before and after six months of the trial. Each child received a full auxological assessment.

Results

Table 4 shows the average values for 29 children after six months of the trial. At this stage, the trial remains blind and the results are considered as a single population. Significant gains in height are observed, varying from 2.2 cm to 6.8 cm, with corresponding changes in total body nitrogen from -32 to 256 g. All subjects lost body fat and all but one gained weight. However, one group of subjects thrived with improved gains in height of >4cm and in weight >2 kg, and increases in N of more than 100 g.

Body nitrogen is highly correlated with body weight (r=0.97) and height (r=0.92), as is expected for children in good health (Table 5). Average values for N/LBM and NI are consistent with those found for normal children (Table 1), but body fat fractions are considerably lower. However, body fat is not significantly correlated with any parameter. The changes in weight and height over the first six months (Table 5) show a moderate correlation with N (r=0.5-0.4). Body fat changes are not significantly correlated with those of N, body weight, or height.

The average initial and final values of subject parameters over the six months interval were tested for significance, using student's t-test (Table 4).   The most significant are those for body fat and body nitrogen; those for weight and height are not significant. However, the actual changes are all significantly different from the zero-change hypothesis.

AORTIC SURGERY

Major surgery initiates a stress response as a result of the body's neuroendocrine reaction to injury, which leads to the catabolism of muscle

Table 5.   Correlations at 6 Months in SGH Trial

	WT	HT	Fat
N	0.97	0.92	0.13
	dWT	dHT	dFat
dN	0.51	0.35	-0.10
	dWT/WT	dHT/HT	dFat/Fat
dN/N	0.43	0.40	-0.07

protein with increased urinary nitrogen excretion and a negative nitrogen balance. Skeletal muscle mass forms a major store of body protein (Daniel et al., 1977) and its depletion postoperatively may affect the patient's prognosis.

The severity of the stress response depends on the extent of injury, and stress categories have been defined on the basis of urinary nitrogen losses (Bistrian, 1979; Cerra et al., 1979). Aortic surgery for aneurysm or occlusive disease was shown to be associated with such severe negative nitrogen balance over several days postoperative that these patients were placed in the highest stress category (Fletcher and Little, 1986).

<u>Patients</u>

Six patients for aortic surgery were studied pre-operatively and post-operatively for a mean of 92 days (range 39 to 134 days). The subjects were normally nourished and none had a history of recent weight loss. All subjects had normal pre-operative serum albumin and transferrin.

The time course for change of nitrogen index (NI) is shown in Fig. 3a; Fig. 3b shows the relative change with respect to pre-operative body nitrogen. We saw similar falls in body nitrogen at 20-40 days post-operation, regardless of the initial value of patient's NI; that is, the duration and magnitude of the protein loss is independent of the initial protein status.

<u>Results</u>

Two patients had post-operative complications, patient 1 required aortic clamping at the renal artery origins and patient 2 developed a chylous fistula from a closed suction drainage tube site, requiring 15 days of total parenteral nutrition before spontaneous closure occurred. All other patients had uncomplicated recovery.

Maximum changes in body weight (-2.9%) and LBM (+1.0%) over the study period are small and not statistically significant. Average changes in N/LBM are in this range -10.7% to -8.2%, with P values of 0.14 and 0.2 respectively for zero change.

By expressing body nitrogen N relative to the pre-operative value, and interpolating the data between measured points, the mean fraction of retained body nitrogen is obtained with 95% confidence limits and P-values for zero change. These data are given in Table 6. The mean fractional nitrogen loss is significant out to 70 days (P < 0.05).

Most patients (5/6) lost from 8-18% body nitrogen over 20-40 days post-operation. Assuming that 30% of body nitrogen is bound up in stable connective tissue, these losses represent a 12-27% loss of metabolic protein. In one case (patient 5) there was no significant change from the pre- to the post-operative nitrogen level, and only patients 2 and 3 returned to pre-operative nitrogen levels after post-operative losses within the study period.   We have no explanation as to why patients 1 and 4 lost so much protein, while patient 5 lost none.   The data do not show a relationship between pre-operative NI, loss of protein and outcome of surgery.   In particular, there is nothing to suggest how one patient maintained protein status.

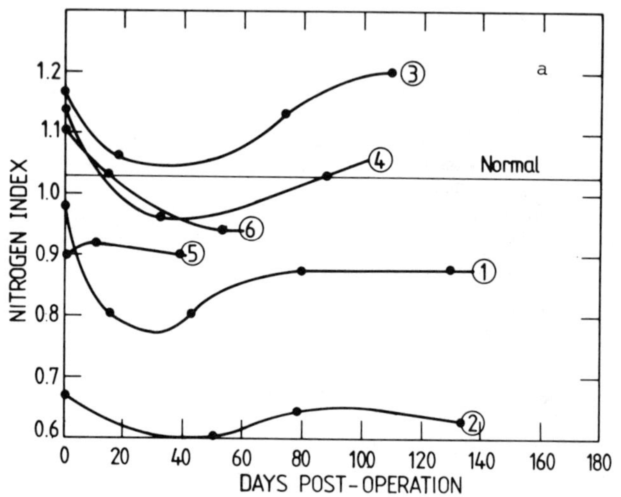

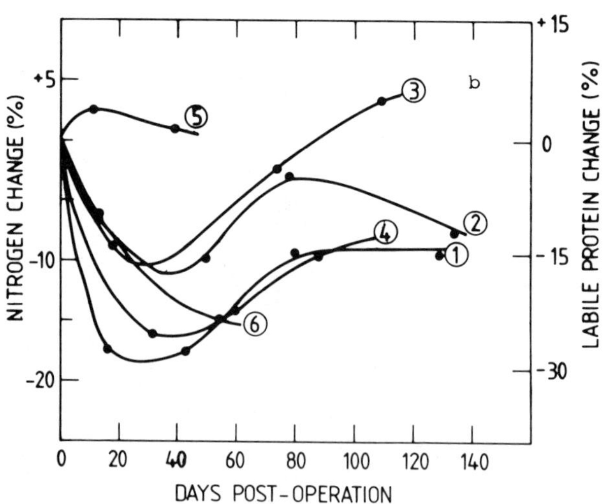

Fig. 3.   Change in body nitrogen after aortic surgery a) nitrogen index, and b) proportional changes in body nitrogen.

Table 6. Post-operative Changes After Aortic Surgery

Days Post-Op	Number of Subjects	Mean Relative Body Nitrogen	95% Confidence Limits	P Value
11	7	0.940	± 0.046	0.019
13	6	0.928	± 0.062	0.031
16	6	0.917	± 0.070	0.028
18	6	0.911	± 0.073	0.026
32	6	0.887	± 0.079	0.014
39	6	0.885	± 0.071	0.009
43	5	0.863	± 0.041	0.001
50	5	0.872	± 0.042	0.001
54	5	0.879	± 0.043	0.001
60	4	0.899	± 0.060	0.013
74	4	0.931	± 0.070	0.052
78	4	0.939	± 0.075	0.082
80	4	0.943	± 0.078	0.101
88	4	0.954	± 0.091	0.203
109	3	0.976	± 0.202	0.665

CYSTIC FIBROSIS

Our aim was to measure the rate of protein deposition in children with cystic fibrosis, and to investigate the relationship between protein deposition and growth parameters. In an earlier report (Allen et al., 1987) the nitrogen per lean body mass was found to be significantly lower for malnourished CF subjects than for normal adults. We now compare children with cystic fibrosis with a group of normal children.

Subjects

Twenty-one malnourished children with CF undergoing supplemental feeding were included, together with 21 normal children. Of the normal group, 12 were siblings of CF children and 9 had no family history of CF. All the children were either prepubertal or in early puberty. Measurements were made of height, weight, body fat (derived from the sum of four skinfolds using Brook (1971) and Durnin and Rahaman (1967)) and total body nitrogen (N).

Results

Table 7 compares the malnourished CF patients with normal controls for the first measurement in both groups. While age and sex are not significantly different, the CF subjects have significantly lower values for standardised height, standardised weight, lean body mass and total body nitrogen than the normal subjects (all with P<0.001). The body nitrogen is depleted to 67% of the control value, confirming that the CF group is malnourished.

TBN generally increased with increasing age for both the individual and the population. Only three subjects, not on enteral supplementation, had falls in TBN values. Changes in TBN are seen to be most rapid within six weeks of the start of enteral supplementation. Many subjects achieved relatively normal growth rates and maintained this status over the study.

Growth velocity (change in height per elapsed time period), weight velocity, and the rate of nitrogen deposition can be determined by comparing serial measurements of each parameter. Data are available for 39

Table 7. Comparison of CF Subjects and Controls (SD)

	CF subjects	Normal subjects
Number	21	21
Male:Female	12:9	6:15
Age y	12.3(3.1)	12.4(3.4)
Standardised height	-1.81(0.94)	-0.07(1.04)*
Standardised weight	-1.76(0.76)	-0.04(0.80)*
LBM kg	25.1(6.5)	31.5(7.7)*
N g	830(295)	1235(370)*

Standard deviation in parentheses
* P < 0.001
Standardised value = (actual value - median value)/SD

separate periods and are considered as a group. The nitrogen deposition
rate was found to bear a linear relationship with weight velocity, yielding
a correlation coefficient r=0.78, p<0.001, 95% confidence limits 0.61,
0.88. However, the relationship between nitrogen deposition rate and
growth velocity showed a wide variation. For subjects with an adequate
growth velocity (>4.0 cm/y) and weight velocity (>2 kg/y), the nitrogen
deposition rates are generally greater than 150 g/y. However, patients
with an adequate growth velocity but a weight velocity <2 kg/y had a low or
negative nitrogen deposition rate.

CONCLUSIONS

A comparison of average N/LBM and NI values for normal adults and
children reveals that the Toronto Nitrogen Index for adults is applicable
to both female teenagers and those under 12 y of age. For male children,
the female NI formula may be more appropriate. The N/LBM is only weakly
dependent on age.

Substantial and comparable body protein depletion is observed in both
haemodialysis and peritoneal dialysis patients. Dialysis patients are
largely static in their protein levels, regardless of the extent of
depletion or the duration of dialysis. Their protein depletion therefore
may occur before dialysis begins. Significant gains in protein were
observed for several dialysis patients during the study, and the body
protein of patients who continued with peritoneal dialysis was
significantly higher than that of patients who left the program because of
severe complications or death. Thus, there may be an association between
low body protein and a high risk of patient morbidity and/or mortality.

The data for the blind trial of synthetic growth hormone for short,
slowly growing children, are consistent with a range of different growth
rates, with changes in weight, height and body nitrogen being only
moderately correlated. Body fat is lost, and is uncorrelated with any other
parameter. Body nitrogen is the most sensitive indicator of growth.

Significant losses of body protein were observed in five of the six
patients following aortic surgery. The time course and magnitude of these
losses may have implications for deciding patient fitness for return to
full activity.

Serial body nitrogen measurements can be used to measure the rate of
protein deposition in growing children. In a group of clinically
malnourished children with cystic fibrosis, substantial nitrogen depletion

was observed relative to normal controls. Weight gain occurred when
nitrogen deposition occurred, but gains in height could occur with minimal
or even negative nitrogen deposition, although at the expense of weight
gain.

We conclude that the TBN method is an accurate and sensitive technique
to investigate the relationship between protein deposition and growth
parameters in children, and protein depletion and repletion in adults. The
results also suggest a significant association between protein status and
prognosis for end-stage renal failure patients on peritoneal dialysis.

REFERENCES

Allen, B. J., Blagojevic, N., McGregor, B. J., Parsons D.E., Gaskin K.,
    Soutter V., Waters D., Allman M., Stewart P., and Tiller D., 1987,
    In vivo determination of protein in malnourished patients, in: "In
    Vivo Body Composition Studies," K. J. Ellis, S. Yasumura, W. D.
    Morgan, eds., The Institute of Physical Sciences in Medicine,
    London. 10:77-82.
Allman, M. A., Allen, B. J., Stewart, P. M., Blagojevic, N. Tiller, D.J.,
    Gaskin, K., and Truswell, A.S., Body protein of patients undergoing
    haemodialysis, Eur. J. Clin. Nutr. Vol. 44, In press.
Biggin, H. C., Chen, N. S., Ettinger, K. V., Fremlin, J.H., Morgan, W.D.,
    Nowotny, R., and Chamberlain, M.T., 1972, Determination of protein
    in living plants. Nature, 236NB:187.
Bistrian, B. R., Blackburn, G. L., and Scrimshaw, N. S. et al., 1975,
    Cellular immunity in semi-starved states in hospitalised adults,
    Amer. J. Clin. Nutr., 28:1148.
Bistrian, B. R., 1979, A simple technique to estimate the severity of
    stress, Surg. Gynecol. Obstet., 148:675.
Blackburn, G. L., Bistrian, B. L., Maini, B. S., Schlamm, H. T., and Smith,
    M. F., 1977, Nutritional and metabolic assessment of the
    hospitalised patient, JPEN, 1:11.
Blumenkrantz, M. J., Koppler, J. D., Moran, J. K., and Coburn, J.W., 1982,
    Metabolic balance studies and dietary protein requirements in
    patients undergoing continuous ambulatory peritoneal dialysis,
    Kidney. Int., 21:849.
Brook, C. G. D., 1971, Determination of body composition of children from
    skinfold measurements, Arch. Dis. Child., 46:182.
Brown, G. A., Stimler, L., and Lines, J. G., 1967, Growth hormone induced
    nitrogen retention in children of short stature, Arch. Dis. Child.,
    42:239.
Cerra, F. B., Siegal, J. H., Border, J. R., Peters, D., and McMenamy, R.
    R., 1979, Correlations between metabolic and cardiopulmonary
    measurement in patients after trauma, general surgery sepsis, J.
    Trauma, 19:621.
Cohn, S. H., and Palmer, H. E., 1970, Recent advances in whole body
    counting, a review, J. Nuc. Biol. Med., 23: 875.
Cohn, S. H., Brennan, B. L., Yasumura, S., Vartsky, D., Vaswani, A.N., and
    Ellis, K.J., 1983, Evaluation of body composition and nitrogen
    content of renal patients on chronic dialysis by total body neutron
    activation, Amer. J. Clin. Nutr., 38:52.
Dabek, J. T., Vartsky, D., Dykes, P. W., Hardwick, J., Thomas, B.J.,
    Fremlin, J.H., James, H., 1977, Prompt gamma neutron activation
    analysis to measure whole body nitrogen absolutely: Its application
    to studies of in vivo changes in body composition in health and
    disease, J. Radioanal. Chem., 37:325.
Daniel, P. M., 1977, The metabolic homeostatic role of muscle and its
    function as a store for protein, Lancet 2:446.
Degoulet, P., Reach, I., Aime, F., Rioux, P., Jacobs, C., Legrain, M.,
    1980, Risk factors in chronic haemodialysis, Proc. E. D. T. A.,
    17:149.

Durnin, J. V., and Rahaman, M. M., 1967, The assessment of the amount of fat in the human body from measurements of skinfold thickness, <u>Br. J. Nutr.</u>, 21:681.

Durnin, J. V., and Womersley, J., 1974, Body fat assessed from total body density and its estimation from skinfold thickness: measurements on 481 men and women aged from 16 to 72 years, <u>Br. J. Nutr.</u>, 32:77.

Fletcher, J. P., and Little, J. M., 1986, The effect of intravenous fat and amino acid on the nitrogen balance following major surgery, <u>Southeast Asian J. Surg.</u>, 9:110.

Fletcher, J., Allen, B. J., and Blagojevic, N., 1989, Body protein loss in aortic surgery, <u>Aust. J. Surg.</u>, in press.

Guarnieri, G., Toigo, G., and Situlin, R., 1983, Muscle biopsy studies in chronically uraemic patients: evidence for malnutrition, <u>Kidney. Int.</u>, 24:s187.

Harrison, J. E., McNeill, K. G., and Strauss, A. L., 1984, A nitrogen index - total body protein normalised for body size - for diagnosis of protein status in health and disease, <u>Nutr. Res.</u>, 4:209.

Harvey, T. C., Dykes, R. W., Chen, N. S., Ettinger, K.V., Jain, S., James, H., Chettle, D.R., Fremlin, J.H., Thomas, B.J., 1973, Measurement of whole body nitrogen by neutron activation analysis, <u>Lancet</u>, August 25:395.

Hill, G. L., Blackett, R. L., Pickford, I., Burkinshaw, L., Young, G.A., Warren, J.V., Schorah, C.J., Morgan, D.B., 1977, Malnutrition in surgical patients: an unrecognised problem, <u>Lancet</u>, 1:689.

Materson, B. J., 1971, Measurement of glomerular filtration rate, <u>CRC Crit. Rev. Clin. Lab. Sci.</u>, 2:1.

McFarlane, H., Adcock, K. J., Cooke, A., Ogbeide, M. I., Adestna, H., Taylor, G. D., Reddy, S., Guiney, C. M., and Mordie, J. A., 1962, Biochemical assessment of protein calorie malnutrition, <u>Lancet</u>, 2:392.

Pollock, C. A., Allen, B.J., Warden, R.A., Caterson, R.J., Blagojevic, N., Cocksedge, B., Mahony, J.F., Waugh, D.A., and Ibels, L. S., 1989, a) Total body nitrogen as a prognostic indicator in CAPD, and b) Total body nitrogen by neutron activation in CAPD, unpublished results.

Prader, A., Illig, R., and Szeky, J., 1964, The effect of human growth hormone in hypopituitary dwarfism, <u>Arch. Dis. Child.</u>, 39:535.

Richter, I., Heine, W., Plath, C., Mix, M., Wutzke, K.D., and Towe, J., 1987, 15N tracer techniques for the differential diagnosis of dwarfism and prediction of growth hormone action in children, <u>J. Clin. Endocrinol Metab.</u>, July, 65(1):74.

Rudman, D., and Kutner, M. H., 1979, Normal variant short stature: subclassification based on responses to exogenous human growth hormone, <u>JCEM</u> 49:1,92.

Schilling, H., Wu, G., Pettit, J., Harrison, J., McNeill, K., Siccion, Z., and Oreopoulos, D.G., 1985, Nutritional status of patients on long term CAPD, <u>Peritoneal Dialysis Bulletin</u>, Jan-Mar:12.

Stefanidis, C. J., Hewitt, I. K., and Balfe, J. W., 1983, Growth in children receiving continuous ambulatory peritoneal dialysis, <u>J. Paediatr.</u>, 102:681.

Vartsky, D., Prestwich, W.V., Thomas, B.H., Dabek, J.T., Chettle, D.R., Fremlin, J.H., and Stammers, K., 1979, The use of body hydrogen as an internal standard in the measurement of nitrogen in vivo by prompt neutron capture gamma analysis, <u>J. Radioanalyt. Chem.</u>, 48:243.

Webster, J., and Garrow, J. S., 1985, Creatinine excretion over 24 hours as a measure of body composition or of completeness of urine collection, <u>Human Nutrition: Clinical Nutrition</u>, 39C:101.

Williams, E. D., Henderson, I. S., Boddy, K., Kennedy, A.C., Elliott, H.L., Haywood, J.K., and Harvey, I.R., 1984, Whole body composition in patients with renal failure and after transplantation studied using total body neutron activation analysis, <u>European J. Clin. Investigation</u>, 14:362.

Yasumura, S., Cohn, S. H., and Ellis, K. J., 1983, Measurement of
extracellular space by total body neutron activation, <u>Am. J. Phys.</u>,
244:36-40.
Young, G. A., Young, J. B., Hobson, S. M., Hildreth, B., Brownjohn, A.M.,
and Parsons, F.M., 1986, Nutrition and delayed hypersensitivity
during continuous ambulatory peritoneal dialysis in relation to
peritonitis, <u>Nephron</u>, 43:177.

TOTAL BODY NITROGEN AND POTASSIUM DETERMINATION IN PATIENTS

DURING CIS-PLATIN TREATMENT

J.Tölli, M.Alpsten, L.Larsson, K.Lundholm*,
S.Mattsson**, B.Unsgaard*** and A.Wallgren***

Departments of Radiation Physics, Surgery* and
Oncology***. University of Göteborg, Sahlgren
Hospital, S-413 45 Göteborg, Sweden
Department of Radiation Physics, Malmö**,
University of Lund, Malmö General Hospital
S-214 01 Malmö, Sweden

INTRODUCTION

Total body protein has often been estimated from the total body
potassium (TBK) content. The basic assumption is that there exists a
constant ratio between potassium and cellular nitrogen. In healthy persons
the intracellular nitrogen constitutes about 65 % of total body nitrogen
(TBN). In some diseases, such as cancer, the TBK and TBN may vary in
different ways. It has been shown (Jeejeebhoy et al., 1982) that TBK and
TBN do not maintain the same ratio in malnutrition nor during total
parenteral nutrition (TPN). Assuming that almost all of the body nitrogen
is bound to protein molecules, the best way to estimate total body protein
is by estimation of TBN content (Burkinshaw, 1987). The aim of this study
was to investigate changes in TBK and TBN in patients with testicular
carcinoma treated with cis-platin, with and without TPN.

MATERIALS AND METHODS

Total body nitrogen (TBN) and total body potassium (TBK) were
estimated in fourteen patients with testicular carcinoma during treatment
with cis-platin. Eight of these patients received total parenteral
nutrition (TPN) during the treatment. The TPN consisted of 30% fat and 70%
glucose. The patients also received 0.2 g of nitrogen per kg body weight
every day. The patients were given cis-platin once daily for 5 days per
week and 4 litres of infusion solution per treatment day (Einhorn and
Donohue, 1977). The treatment was repeated every third week up to a total
number of four courses. TBK was determined by counting the gamma radiation
from the naturally present radionuclide K-40, which is a constant fraction
of all natural potassium, in a high sensitivity 3π whole body counter
containing four plastic scintillators with a total volume of 0.7 m^3. The
total expected standard deviation in one determination is approximately $\pm$
100 mmol. For each patient the TBK was measured 12 times during three
months.

Total body nitrogen was measured by <u>in vivo</u> neutron activation
(Vartsky et al., 1979; Vartsky et al., 1984; Chettle and Fremlin, 1984;
McNeill et al., 1979; Morgan et al., 1984). The method is based on the

capture of low-energy (thermal) neutrons by N-14 nuclei. This reaction converts a certain amount of N-14 to an excited state of N-15, which de-excites promptly with the emission of 10.8 MeV gamma radiation. The 10.8 MeV gamma radiation is characteristic for nitrogen and was detected by two 15 cm (diam.) x 15 cm (height) NaI(Tl) detectors. Both detectors were mounted above the patient and were shielded with a boron/plastic shield to reduce neutron reactions inside the detectors (Larsson et al., 1987).

A Cf-252 source (5500 MBg) was used to produce the neutrons. The source was contained in a polyethylene block which forms a collimator, surrounded by a 140 cm (diam.) x 80 cm (height) water tank. The patient was irradiated from below by a 15 cm x 50 cm rectangular neutron field. By moving the stretcher it was possible to measure the nitrogen content in the whole body or parts of it. For each patient, TBN was measured at the start of the treatment and then between 3 and 7 times during the therapy, which lasted for 3 months.

The mean dose equivalent to the patient is less than 0.5 mSv per measurement. A quality factor Q=25 was used.

ANALYSIS OF DATA

A careful analysis of the measured pulse height distribution is necessary. We developed a deconvolution method based on Bayes' postulate (Tölli, 1989). Calibration factors have been evaluated by neutron irradiation of two different size human-shaped phantoms. The phantoms were filled with tap water for background evaluation and with aquaeous solution of Carbamide, $(NH_2)_2CO$, for the pulse height distribution from nitrogen. The weight of the phantoms were from 68.9 kg to 84.3 kg. After the deconvolution of the distributions of each pulse height, the net number of counts in the region of interest was calculated.

In phantom studies the nitrogen content can be measured with a precision of ± 2.0 % (1 S.D.). However, phantom studies do not take into consideration inhomogeneities or the nonuniform distribution of nitrogen in the body, which may give varying sensitivity for nitrogen measurements between patients. In our study, each patient was their own standard.

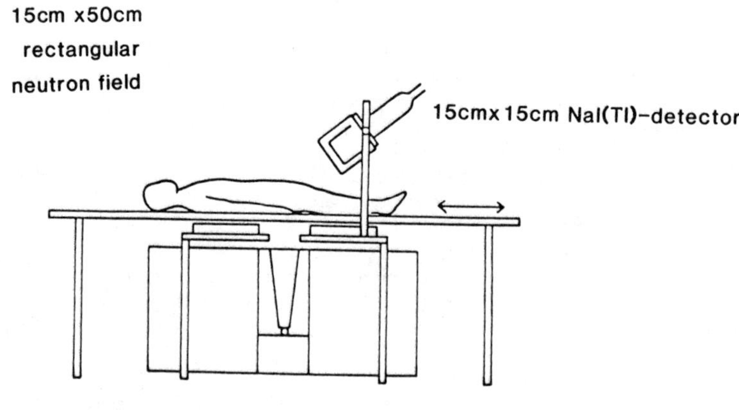

Fig. 1. The experimental arrangement for TBN measurements.

Table 1. Percentage Changes in TBK, TBN and Weight in all Patients

Patients receiving TPN			Patients without TPN		
TBK	TBN	weight	TBK	TBN	weight
-8.2	-10.5	+2.5	-5.6	-8.9	+2.7
-9.8	-9.8	+3.8	-10.2	-19.6	-2.8
-11.5	-19.9	+0.1	-4.0	-12.7	+5.1
-21.6	-29.4	-2.6	-14.9	-29.3	-8.7
-6.6	-20.0	+6.5	+1.5	-13.0	-5.4
+9.4	-15.9	+11.8	-4.3	-9.6	-14.8
-14.0	-26.9	-3.7			
-5.1	-11.2	+2.7			
Mean values of relative change %					
-8.4	-18.0	+2.6	-6.3	-15.5	-4.0

RESULTS AND DISCUSSION

One of the effects of malnutrition is the loss of body protein. Therefore one of the major objectives of nutritional support is either to minimize the loss of body protein or to maximize protein repletion.

It is important to assess the effects of nutritional support by changes in total body protein. As there is a fixed relation between TBN and body protein (1 g nitrogen = 6.25 g protein), changes in TBN reflects changes in the body protein content.

Fig. 2 shows the changes in TBK and TBN in two patient studies. Table 1 shows the results from both patient groups. These figures show that the

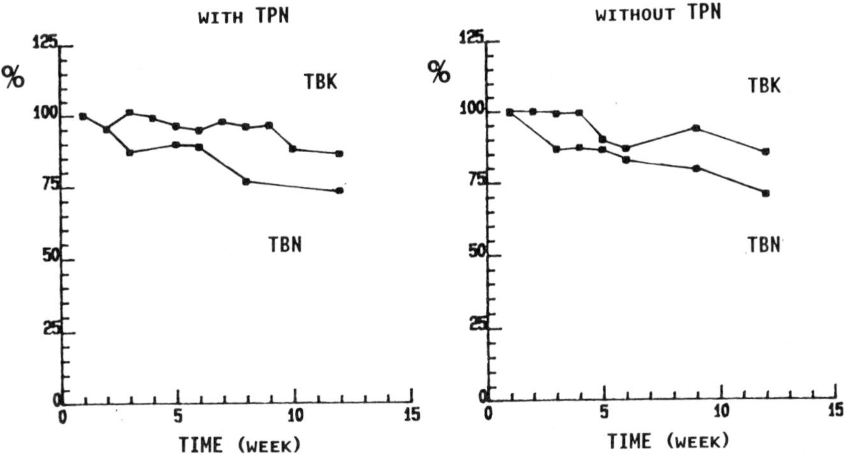

Fig. 2. Time variation of TBK and TBN in two patients during cis-platin treatment.

only significant difference between these two groups of patients is the changes in body weight. In both groups the loss of potassium was less than the loss of nitrogen. The ratio K/N increases during the treatment. There is a loss of nitrogen (protein) up to 30 %. The loss of nitrogen could be a direct toxic effect of cis-platin. These results indicate that the patients' body composition changes in a dramatic way during cis-platin treatment.

One possible explanation of these results is that there could be an increase in the extracellular water compartment and that when the cis-platinum is given to the patients the metal complex is bound to protein molecules (Sykes et al., 1986; Ollenschlaeger et al., 1989).

CONCLUSIONS

This study demonstrated a dramatic reduction in total body nitrogen (and potassium) during cis-platin therapy and shows the importance of continuous body composition studies in cancer patients.

In future work we will include new groups of patients treated with chemotherapy and hormone therapy. The effects of total parenteral nutrition will be evaluated and the studies will include the determination of exchangeable body water.

ACKNOWLEDGEMENTS

The authors wish to thank H. Andersson and I. Bosaeus at the Department of Clinical Nutrition, Sahlgren Hospital for valuable assistance and advice during this study.

REFERENCES

Burkinshaw, L., 1987, Models of the distribution of protein in the
        human body, in: "In Vivo Body Composition Studies," K. J. Ellis,
        S. Yasumura and W. D. Morgan, eds., The Institute of Physical
        Sciences in Medicine, London.
Chettle, D. R., and Fremlin, J. H., 1984, Techniques of in vivo neutron
        activation analysis, Phys. Med. Biol., 29:1011.
Einhorn, J., and Donohue, L., 1977, Cis-diamminedichloroplatinum,
        Vinblastine and Bleomycin combination chemotherapy in disseminated
        testicular cancer, Ann.Intern. Med., 87:293.
Jeejeebhoy, K. N., Baker, J.P., Wolman, S.L., et al., 1982, Critical
        evaluation of the role of clinical assessment and body composition
        studies in patients with malnutrition and after total parenteral
        nutrition, Am. J. Clin. Nutr., 35(suppl):1117.
Larsson, L., Alpsten, M., Mattsson, S., 1987, In vivo analysis for
        nitrogen using a ^{252}Cf source, J. Rad. Nucl. Chem., 114:181.
McNeill, K. G., Mernagh, J. R., Jeejeebhoy, K. N., Wolman, S. L., and
        Harrison, J. E., 1979, In vivo measurements of body protein based
        on the determination of nitrogen by prompt gamma analysis, Am. J.
        Clin. Nutr., 32:1955
Morgan, W.D., Vartsky, D., Ellis, K.J., and Cohn, S.H., 1984, A comparison
        of Cf-252 and Pu-238-Be neutron sources for partial body in vivo
        activation analysis, Phys. Med. Biol., 26:413.
Ollenschlaeger, G., Konkol, K.M., Wickramanayake, P.D., Schrappe-Baecher,
        M., and Mueller, J.M., 1989, Nutrient intake and nitrogen
        metabolism in patients during oncological chemotherapy, Am. J.
        Clin. Nutr., 50:454

Sykes, T. R., Noujaim, A. A., Stephens-Newsham, L. G., Ng, P. K.,
    Lentle B. C., and Turner, A. R., 1986, Effect of cis-platinum on
    the protein binding and pharmacokinetics of ^{67}Ga-Citrate, _Nucl._
    Med. Biol., 13:501.
Tölli,, J., 1989, A mathematical method for analysis of pulse height
    distributions obtained at in-vivo determinations of nitrogen by
    neutron activation, Internal Report, _Dep. Rad. Phys._, 89:05.
Vartsky, D., Ellis, K. J., and Cohn, S. H., 1979, In vivo measurement of
    body nitrogen by analysis of prompt gammas from neutron capture, _J._
    Nucl. Med. 20:1158.
Vartsky, D., Ellis, K. J., Vaswani, A. N., Yasumura, S., and Cohn, S. H.,
    1984, An improved calibration for the in vivo determination of body
    nitrogen, hydrogen and fat, _Phys. Med. Biol._, 29:209.

CLINICAL STUDIES OF TOTAL BODY NITROGEN IN AN AUSTRALIAN HOSPITAL

D.B.Stroud(1), D.J.Borovnicar(2), J.R.Lambert(3),
K.G.McNeill(4), S.J.Marks(1), R.P.Rassool(2),
H.C.Rayner(1), B.J.G.Strauss(3), E.H.Tai(1), M.N.Thompson(2),
M.L.Wahlqvist(3), B.A.Watson(3), C.M.Wright(2)

(1) Monash Medical Centre, Melbourne, Australia
(2) School of Physics, University of Melbourne, Australia
(3) Department of Medicine, Monash University, Australia
(4) Department of Physics, University of Toronto, Canada

INTRODUCTION

The first reports of the measurement of Total Body Nitrogen (TBN) by In
Vivo Neutron Activation Analysis (IVNAA) using a cyclotron date from the
early 1970's (Harvey et al., 1973).   By the late 1970's other laboratories
were using the prompt gamma IVNAA technique with isotopic neutron sources
(Mernagh et al., 1977; Vartsky et al., 1979a).   The prompt IVNAA method is a
relatively simple, non-invasive test which requires approximately 20 minutes
of the patient's time.   The radiation dose is minimal, and both capital and
running costs are small.   Clinically, Total Body Protein (TBPr) can be
reliably estimated from TBN (TBPr = 6.25 x TBN).   The many clinical
applications of the technique are described by Wahlqvist and Marks elsewhere
in this volume, and by McNeill (1988).   It is therefore surprising that so
few hospitals are currently using the technique.

A Body Composition Laboratory has been established recently at the
Monash Medical Centre in Melbourne, Australia, and a clinical facility for
the measurement of TBN has been available for patients since early 1988.
This paper reviews all of the results from the first 18 months of operation.

Costing of this facility shows that the measurement of TBN by IVNAA is
an inexpensive test.   The total cost of the equipment was approximately
A$40,000, which is less than the cost of refurbishing a suite of clinical
rooms to accommodate it. Staffing costs, which include a full-time physicist
and nurse, and a part-time physician, are the major cost.

EQUIPMENT

Fig. 1 is a cross-sectional view of the equipment.   The neutron source
is 10 micrograms of ^{252}Cf.   Preliminary design work for this system was
carried out at the University of Melbourne and the Toronto General Hospital
(McNeill et al., 1989) where the advantages of a relatively small ^{252}Cf
source were demonstrated. The source is positioned in a borated wax
collimator, and there is 7 cm of lead on top of the collimator to shield the
detectors.   The source is approximately 38 cm below bed level.   The
aperture is 24 cm wide and 43 cm long when projected to bed level.

Two 12.5 x 10 x 10 cm NaI detectors are located on either side of the bed;
they are surrounded by 3 cm of borated wax, and there is 50 cm of clear space
between them.   Standard electronics are used.

The patient is positioned on a sliding wooden stretcher mounted on a
long bench.   The section of the stretcher exposed to the neutron beam was
replaced by a thin panel of aluminium because the chipboard top was
contributing to the nitrogen background.

TECHNIQUE

The patient is placed in a supine position with the mid point between
the iliac crests and the umbilicus over the centre of the neutron aperture.
The nitrogen and hydrogen capture gamma rays are then counted for 1000 s
without moving the patient.   The data are analysed using the Toronto method
(Harrison et al., 1984) which uses H as an internal standard (Vartsky et al.,
1979).   The precision of the measurements is limited by the number of counts
in the nitrogen window and by the background in this window;   a precision of
4% is obtained with a 1000s patient count.

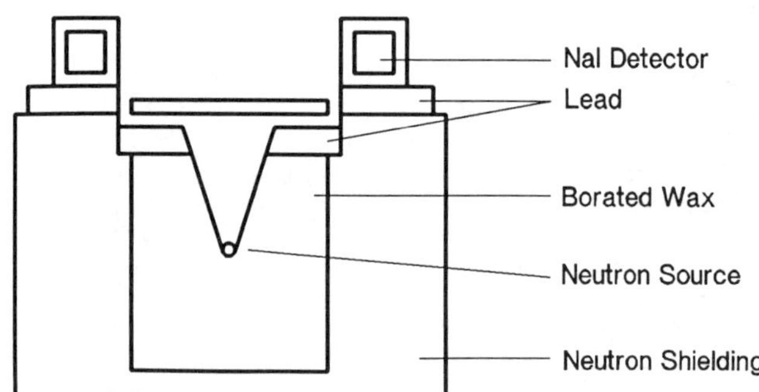

Fig. 1.   Cross sectional view of the total body nitrogen facility at the
Monash Medical Centre, Melbourne, Australia.

Techniques have not yet been finalised in this laboratory for applying
various corrections to the measured results.   Preliminary measurements with
phantoms indicate that chlorine in the body of the patient increases the
background in the nitrogen window by approximately 3%.   This background also
depends on the width of the patient, but this effect has been minimised by
using phantoms that are 29 cm wide, a width that is very close to the average
patient width.   Finally, in the present technique it is assumed that 10% of
the body weight is hydrogen;   Harrison et al (1984) have shown that this
assumption can lead to errors of $\pm$ 5% in extreme cases.

For obese patients, the magnitude of two of these corrections has been
determined by preliminary phantom measurements.   The N/H ratio increases by
10% when the width of the phantom is increased from 27 to 45 cm, because the
hydrogen gamma rays are absorbed more than the nitrogen gamma rays.

178

Table 1 - Measured Values of TBN/HAS$^{2.6}$ for Normals

	Melbourne			Toronto		
	No	Mean	Std Dev	No	Mean	Std Dev
Males	9	0.48	0.05	25	0.50	0.06
Females	6	0.43	0.03	63	0.42	0.04

The nitrogen background decreases by 14% when the width of a water-filled phantom is increased from 27 to 45 cm.

This supine-only, non-scanned technique was adopted only after measurements showed an invariant N/H ratio at several sites in a normal volunteer and in two pig carcasses. Further tests have been designed to discover those situations in which the assumption is invalid, and it has been demonstrated that the N/H ratio can vary significantly between the normal upper abdominal position and the thighs in obese patients.

PATIENT DOSE

The estimated patient dose is less than 0.15 mSv, based on the work of Allen et al (1985). The dose received by operators is approximately 0.004 mSv/h at 1 m from the patient; the equipment is located in a large room and the control console is positioned to minimise this dose.

VALIDATION

Fifteen healthy, non-obese volunteers (age range 21-59) have been measured, and the results compared with data published by Harrison et al (1984). The comparison technique is based on Harrison's definition of the Nitrogen Index, NI, viz.

$$NI = \frac{\text{Measured TBN}}{\text{Predicted TBN}}$$

where   Predicted TBN   =   Constant x HAS$^{2.6}$

HAS is the mean of height and armspan. By definition, the mean NI of the Toronto normals is 1.00. Harrison et al. (1984) determined the constant by calculating the mean value of (Measured TBN)/HAS$^{2.6}$ from the Toronto data for males and females separately; this procedure has been repeated for the Melbourne data. There is good agreement between the results from both laboratories as shown in Table 1. (See footnote)

Fig. 2 is a histogram of all of the actual NI values for the Melbourne patients. The other data in Fig. 2 are discussed below.

Two tests have been designed to study the long-term reproducibility of the TBN results. Firstly, two normal volunteers have been measured on three

NOTE ADDED IN PROOF - More recent measurements on a larger group of normals suggest that with the Melbourne method normal patients have a mean NI of 0.86 rather than 1.00. The interpretation of the patient results presented below has been altered accordingly. Measurements on normals are continuing.

Table 2. Measured Values of the Ratio* $\dfrac{(N/H)_{am}\ \text{urea phantom}}{(N/H)_{pm}\ \text{urea phantom}}$

Measured Mean	= 1.004
Measured Standard Deviation	= 0.03
Calculated Statistical Error	= 0.033
Calculated Standard Error	= 0.005 $(= 0.033/\sqrt{43})$

*Based on the results of 43 paired measurements of the same phantom.

separate occasions, once in August 1987, and twice in June 1989. The measured values of TBN for Volunteer 1 are 2.20, 2.13 and 2.15, mean = 2.16, Std Dev = 0.03. Values for Volunteer 2 are 1.93, 1.83, 1.81, mean = 1.86, Std Dev = 0.05. The volunteers had no significant change in weight over the period that the measurements were performed. For each volunteer, the observed standard deviation of the three measurements is smaller than the predicted counting error; this indicates good reproducibility.

The second stability test was conducted with phantoms. When measuring TBN in patients a phantom containing an aequeous solution of urea is measured for standardisation before and after the patient(s). There is generally at least 3 hours between the measurements. The N/H ratio calculated from the morning (am) measurement should equal the N/H ratio from the afternoon (pm) measurement to within the precision of the measurement. To test the system reproducibility, the ratio of these N/H ratios, as shown in Table 2, was calculated for each of 43 studies during a period of 10 months. The

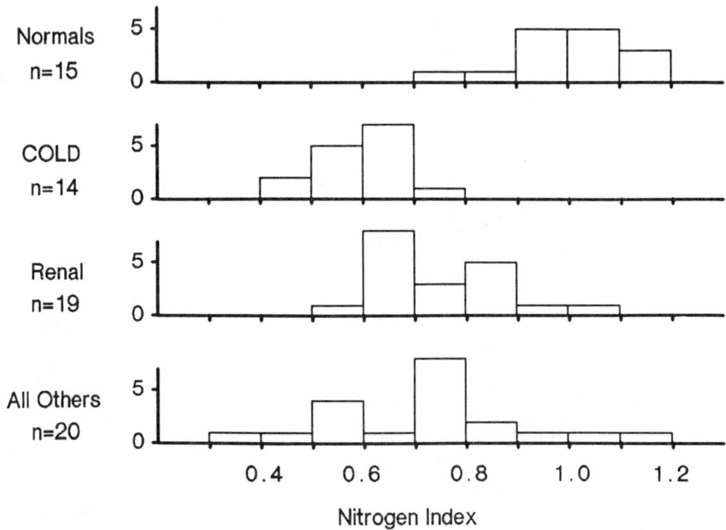

Fig 2 Histogram showing Nitrogen Index (NI = Measured TBN / Predicted TBN) for normals and for all patients.

180

measured mean should be unity.   The standard deviation should approximate
the calculated statistical error, and the calculated standard error should
approximate the measured mean minus unity.   The results indicate stability
to within the precision of the measurement.

CLINICAL STUDIES

Chronic Obstructive Lung Disease Patients

A select group of patients suffering from Chronic Obstructive Lung
Disease (COLD) has been studied as part of a randomised, placebo-controlled
trial of an oral nutrient supplement.   Subject selection criteria included
the presence of severe COLD, less than 90% of ideal body weight (Metropolitan
Life Insurance Table, New York 1978), and capability to perform respiratory
function and exercise tests.   Fourteen patients (Age Range 48-76) have been
studied so far.   Their mean Body Mass Index (BMI = Weight/Height^2) is
18.   Their mean NI is 0.58 with a standard deviation of 0.08; this mean is
approximately three standard deviations below normal.   All of their NI
values are shown in histogram form in Fig. 2;   the distribution of NI values
is rather narrow.   The study is continuing.

Short Bowel Patients

TBN has been measured in two short bowel syndrome patients about to be
transferred to home parenteral nutrition.   Both were referred for TBN
measurements immediately prior to discharge after many months in hospital and
were clinically well.   The first patient, age 37, had an NI = 0.72 which is
low, but normal.   The second patient, age 66, had NI = 0.66;   this is also
interpreted as normal since there is evidence of a decline in TBN in normals
over the age of 60 (Harrison et al., 1984).

Renal Patients

Monash Medical Centre contains the largest renal dialysis unit in
Australia.   Because of concerns about their nutritional status, TBN has
been measured in a slected group of patients consisting of 10 on
Haemodialysis, 4 on continuous ambulatory peritoneal dialysis, and 5
approaching end-stage chronic renal failure.   Dialysis patients were
studied immediately after dialysis.   The ages of these 19 patients range
from 22 to 69, their BMI's range from 16 to 33, and their NI's have a mean
value of 0.73 (SD = 0.16).   Their mean NI is approximately 13% below that
of normals. The histogram in Fig. 2 suggests a fairly broad distribution
of NI values.   A number of these patients display total body protein
depletion, despite having a BMI in the normal range (20-25) or higher.

Obese Patients

TBN measurements are important in obese patients because protein
malnutrition is a concern when dieting, and following surgery for Upper
Gastric Reduction (UGR).   However, it is difficult to measure TBN
accurately in obese patients because of the limited penetration of the
neutrons and because of the inhomogeneities caused by the thick outer
layers of nitrogen-deficient adipose tissue.   Siwek et al (1984) have
successfully measured obese patients by constructing a phantom based on CT
measurements for each individual patient.   Such an approach is
impractical except in a research environment.   Preliminary measurements
have been made on phantoms as described above, and patients in Melbourne
to develop a more practical approach.

In patients, TBN has been measured by applying the method described
above with no corrections. Measurements have been made at both the
normal abdominal position and in the thigh, since the outer layer of
adipose tissue may be thinner in the thigh. In six female patients with
weight exceeding 100 kg the mean NI observed in the abdominal area is 0.86
$\pm$ 0.12 (Std Dev); the thigh measurement was, on average, 20% higher than
the abdominal measurement. Further work is needed before it will be
possible to measure TBN accurately in obese patients.

CONCLUSIONS

Our experience with various groups of patients shows that prompt IVNAA
is a cheap, clinically useful and accurate method of assessing total body
protein.

ACKNOWLEDGEMENTS

We wish to acknowledge generous assistance from B.J. Allen of the
Australian Nuclear Science and Technology Organisation, and from
G.L. Hill's group in Auckland. Financial assistance was received from
Novo Laboratories Pty Ltd for the study of the COLD patients.

REFERENCES

Allen, B.J., Bailey, G.M., and McGregor, B.J., 1985, Dose equivalent
    distributions in the AAEC total body nitrogen facility, Fourth
    Australian Conference on Science and Engineering, Lucas Heights,
    6-8 Nov 1985.
Harrison, J.E., McNeill, K.G., Strauss, A.L., 1984, A nitrogen index -
    total body protein normalised for body size - for diagnosis of
    protein status in health and disease, Nutrition Research, 4:209.
Harvey, T.C., Dykes, P.W., Chen, N.S., Ettinger, K.V., Jain, S., James,
    H., Chettle, D.R., Fremlim, J.H., and Thomas, B.J., 1973,
    Measurement of whole-body nitrogen by neutron-activation analysis,
    Lancet, 2:395.
McNeill, K.G., 1988, Current techniques and applications of body
    composition analysis at Toronto General Hospital, Australasian
    Physical & Engineering Sciences in Medicine, 11:15.
McNeill, K.G., Borovnicar, D.J., Krishnan, S.S., Wang, H., Waana, C., and
    Harrison, J.E., 1989, Investigation of factors which lead to the
    background in the measurement of nitrogen by IVNAA, Phys. Med.
    Biol., 34:53.
Mernagh, J.R., Harrison, J.E., and McNeill, K.G., 1977, In vivo
    determination of nitrogen using Pu-Be sources, Phys. Med. Biol.,
    22:831.
Siwek, R.A., Burkinshaw, L., Oxby, C.B., and Robinson, P.A.J., 1984,
    Multi-element analysis of the obese subject by in vivo neutron
    activation analysis, Phys. Med. Biol., 29:687.
Vartsky, D., Ellis, K.J., Cohn, S.H., 1979a, In vivo measurement of body
    nitrogen by analysis of prompt gammas from neutron capture, J.
    Nucl. Med., 20:1158.
Vartsky, D., Prestwich, W.V., Thomas, B.J., Dabek. J.T., Chettle, D.R.,
    Fremlin, J.H., and Stammers, K., 1979b, The use of body nitrogen
    as an internal standard in the measurement of nitrogen in vivo by
    prompt neutron capture gamma-ray analysis. J. Radioanal. Chem.,
    48:243.

BODY COMPOSITION FOR THE INVESTIGATION OF OBESITY

John Garrow

Rank Department of Human Nutrition
St Bartholomew's Hospital Medical College
Charterhouse Square, London EC1 6BQ, UK

DIAGNOSIS AND CLASSIFICATION OF OBESITY

It is generally agreed that obesity is a condition in which the fat stores of the body are too large. However, the limitations of this simple definition soon become apparent: a young woman of normal body build may wish to be thinner, so in her opinion her fat store is too large - is she therefore obese? In a sense she is, but with such a subjective definition of obesity it is not possible to write sensibly about the measurements of body composition which are required for the investigation of obesity.

The simplest objective evidence that fat stores are too large is obtained when increasing fatness is associated with increasing mortality or morbidity. However, life insurance companies (who have the largest data bases concerning predictors of mortality) do not routinely measure body fat; they rely instead on measurements of weight and height, and publish empirical weight-height ranges within which life expectancy is maximum. Quetelet (1869) observed that among normal adults of different stature weight is proportional to the square of height, so the ratio weight divided by height squared may be called Quetelet's Index (QI). It was rediscovered a century later by Keys et al (1972) who named it body mass index (BMI). Fortunately, the lower limit of the insurance companies' desirable range of weight for height corresponds quite closely to a QI of 20 (where W is measured in kg, and H in m), and the upper limit of the desirable range is about QI = 25. It may be objected that the subjects on whom the insurance societies base their ranges are a self-selected and probably atypical section of the general population, but other more representative samples of the population show a similar relation of QI to mortality (Garrow, 1979).

I have explained elsewhere the practical reasons for defining obesity as QI > 25, and for classifying it according to severity, so QI=25-29.9 is Grade I, 30-40 is Grade II, and >40 is Grade III (Garrow, 1988a). Others disagree and claim that more harm than good is done by considering Grade I (as defined above) to be in any way abnormal (Abraham and Mira, 1988). The threshold for making the diagnosis depends on the authors' estimate of the relative risks of setting the threshold too low or too high. On one hand, we do not want to cause unnecessary concern about trivial overweight; on the other hand, we do not want mildly overweight people to be lulled into complacency so they become significantly overweight before they begin to take the problem seriously. A good case can be made for

using the lower threshold for young people, and for those who have
diabetes, hypertension, or a family history of these diseases (NIH
Consensus Conference 1985), and a higher threshold for older people with
no predisposition to the complications of obesity. Since in the matter of
health education through the media it is not possible to know whom you are
addressing it seems wise to say to people in Grade I that, although their
obesity carries little health risk at present, they should be particularly
vigilant to prevent further weight gain.

REQUIREMENTS FOR TREATMENT OF OBESE PATIENTS

The obese patient and his/her doctor need answers to two questions:

(a). Is it likely that loss of weight (fat) would confer an objective
health benefit on the patient?

(b). Is the prescribed treatment achieving the desired effect?

If we adopt as a definition of obesity that increasing fatness brings
increasing mortality and morbidity, and chose QI > 25 as a threshold at
which this effect is apparent, it follows by definition that having a QI >
25 is bad for health. It does not necessarily follow that weight loss
improves health, but there is good evidence that this is so which is
reviewed in detail elsewhere (Garrow, 1988a). The main pieces of evidence
are that overweight people who are refused life insurance at normal rates
solely because they are overweight, and who subsequently reduce weight,
are found thereafter to have normal life expectancy (Dublin, 1953), and
that longitudinal studies of normal populations have shown that increases
in weight are associated with increases in risk factors for heart disease
(blood pressure, plasma cholesterol, blood glucose) and that weight loss
is associated with a decrease in these risk factors (Borkan et al., 1986).

I conclude that to answer the questions above the clinician needs
only to measure the weight and height of the patient: if the initial value
for QI is greater than 25 then the answer to question (a) is yes, and if
on subsequent visits the weight of the patient is decreasing the answer to
question (b) is also yes (but the question of rate of weight loss also
needs to be considered, but not here).

I realise that this simple approach will bring down on my head the
scorn of professional anthropometrists, who deride QI (or BMI) as a
measure of obesity (Garn et al., 1986, McLaren, 1987, Micozzi and Albanes,
1987, Ross et al., 1988). Of course it is not a perfect measure of
fatness (but nor is any other test), and of course it tells us nothing
about the distribution of body fat. The pioneering work of Behnke et al
(1942) was undertaken because very fit muscular young men were
"overweight" by normal standards, and old people may have a normal weight-
for-height yet have large fat stores and a small fat-free mass (Lesser et
al., 1971). Despite these limitations I think QI is the most useful
measure for those who treat obese patients, since the alternatives are
worse.

Ross et al (1988) report rather poor correlations between QI and the
sum of 5 skinfold thicknesses in a survey of 12,282 men and 6,593 women in
Canada, and conclude that this "further indicts (the use of QI) for the
purpose of assessing adiposity status or monitoring change in
individuals." The implication is that if a clinician wants to assess
adiposity status, or monitor change in an individual, he should measure
five skinfolds rather than weight and height. With severely obese
subjects this advice is impractical: it is technically impossible to make

valid measurements of skinfold thickness, and even if the measurements
were made there are no tables which would permit the calculation of body
fat from these data in severely obese subjects. With moderately obese
people it is possible to measure skinfolds, but how are these data used to
answer question (a) above? To my knowledge there are no studies relating
skinfold thickness and mortality or morbidity, nor studies relating change
in skinfold thickness to change in mortality or morbidity. Certainly
skinfold measurements are very unsatisfactory for following change in
adiposity, as the following example will show. Suppose a patient who
weighs 100 kg loses 5 kg between visits to the doctor, this will involve a
reduction in body fat from about 40% to 38% of body weight, which will be
reflected in a change of about 10% in skinfold thickness. However, an
apparent change of 10% in skinfold reading is within the measurement error
variance, even for normal-weight subjects (Mueller and Malina, 1987), so
in an obese subject a 10% decrease in skinfolds may or may not indicate
loss in body fat. For this reason skinfolds are much less satisfactory
than weight for answering question (b).

<u>Quetelet's Index as a Measure of Fatness</u>

The three primary methods for estimating total body fat and fat-free
mass (FFM) in living adults are based on density, water or potassium measurements respectively. The assumptions are that the density of fat is 0.90
g/ml and that of FFM 1.10 g/ml; or that FFM is 73% water and fat is anhyd-
rous; or that FFM contains 60 mmol K/kg in women, or 66 mmol K/kg in men,
and that fat contains no K. All other methods (anthropometry, skinfolds,
conductivity, etc) have themselves been calibrated against one of the
primary methods. Imaging techniques can be made to yield estimates of fat
and FFM by summing serial slices, and neutron activation can measure the
body content of elements which are fairly constantly related to FFM.

To answer the question - how accurately does QI measure body fat? it
is necessary to agree a gold-standard measurement of body fat with which
it can be compared. Garrow and Webster (1985) analysed data on 104 women
aged 32 (10) years - mean (sd) - with a QI of 34.7 (7.4), and 24 men aged
36 (13) years with a QI of 27.2 (7.8). Each individual had a
determination of body fat by density, water and potassium, which yielded,
in women, values of 35.5 (15.3) kg fat by density, 40.6 kg (14.7) fat by
water, and 42.8 (15.3) kg fat by potassium. Among the men the values were
16.9 (15.9), 23.6 (18.8) and 24.4 (17.8) kg fat by the same three methods.
The three methods gave answers which were significantly different from
each other (p<0.001). In order to obtain a gold-standard estimate the
mean values for the three methods were used, which yielded estimates of
41.6 (9.4) kg fat for the women and 23.6 (13.9) kg fat for the men.

The estimates of body fat in each individual by each method, and on
the basis of QI, were then compared with the gold-standard estimate for
that individual. The standard deviation of the estimates based on
individual measurements when compared with the gold standard was 3.4 kg,
3.3 kg, 3.5 kg and 4.2 kg of fat for density, water, potassium and QI in
women, and 5.8 kg, 3.8 kg, 4.1 kg and 5.8 kg in men. These deviations are
too large to be explained by measurement errors: they reflect the fact
that the assumptions underlying each of the methods about the density,
water and potassium content of FFM are not quite true, and deviate to
different extents in different people. We therefore conclude that, at
least among our clinic population of obese patients, QI is almost as valid
a measure of body fat as density, water or potassium measurements.

One criticism of this conclusion which can be anticipated is that QI
correlates less well than, for example, skinfold thickness, with
percentage fat estimated by density. Womersley and Durnin (1977) performed

a linear regression of percentage fat on QI in a series of 324 women and
derived the equation:

$$\%fat = 1.371 \,(QI) - 3.467$$

However, it must be obvious that %fat, which is fat/fat+FFM, cannot be
linearly related to QI, which is fat+FFM/H*H. Indeed, the Womersley and
Durnin formula predicts that anyone with a QI >75.47 should be >100% fat,
which is absurd. It should be remembered, therefore, that QI predicts the
weight of fat in the body, not % fat. Also, at a given QI, body fat
differs between men and women. One of the reasons why Ross et al (1988)
obtained a correlation coefficient of only 0.50 between QI and skinfolds
was that they pooled the data on both sexes before analysis.

<u>The composition of excess weight in obesity</u>

If, as has been suggested, QI indicates fatness in obese adults, and
weight change is a valid indicator of fat change, it is necessary to show
that change in weight is proportional to change in fat. This is difficult
to demonstrate directly, in view of the errors in the methods for
measuring fat change, which will be discussed later. However, Fig. 1
shows the data of Webster et al (1984) in which the "gold-standard"

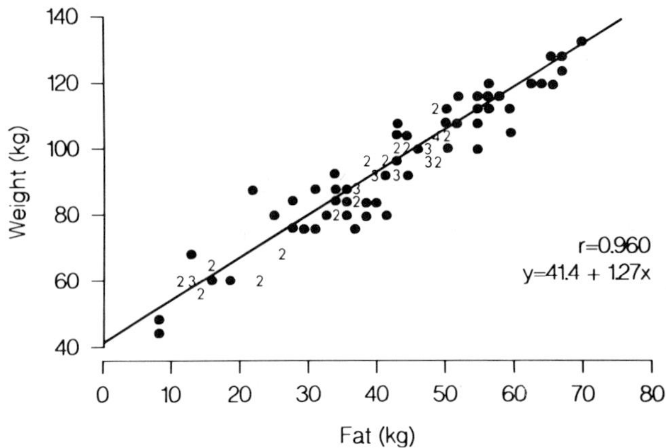

Fig. 1. Body fat vs body weight in a series of 104 women. (Data of
Webster et al 1984) 2,3,4 indicate superimposed values.

estimate of fat is plotted against body weight for 104 women who range
from slim (<10 kg fat) to very fat (>70 kg fat). The correlation
coefficient of 0.960 indicates that 92% of the very large variation in
body weight (40 - 130 kg) is explained by variation in body fat, and there
is no evidence of departure from linearity over this range. (Probably
there would have been if we had studied very thin subjects, but this paper
is considering body composition in obesity, not cachexia.) The slope of
the line is 1.27, indicating that a difference of 1.0 kg in fat content is
associated with a difference of 1.27 kg in body weight, and neither this
slope, or the correlation coefficient, is materially altered if the
estimate of fat is based on density, or water, or potassium alone, or if
both axes are divided by height squared, so the y axis becomes Quetelet's
Index (Webster et al., 1984).

These data indicate that the excess weight in obese people consists of about 75% fat and 25% FFM, and this conclusion seems to apply as strongly to the difference between a person who weighs 70 kg compared with one of 60 kg, as it does to one of 120 kg compared with one of 110 kg.

Fat Distribution

There is good evidence that, for a given degree of adiposity, the risk of ischaemic heart disease, stroke and total mortality among men aged 36-68 years is increased if the fat is in the abdominal region, as indicated by a high waist-hip ratio (Larsson, 1984), and a similar association is also found among women (Lapidus & Bengtsson, 1984). It may be argued, therefore, that the distribution of fat is one of the pieces of information required for the treatment of obese patients. However, this assumes that there is some difference in the type or intensity of treatment which is appropriate for the android type of obese patient compared with the gynoid obese patient. I will not consider further the distribution of body fat because this is the subject of another paper, and also because I do not think that fat distribution is altered by dieting (Garrow, 1988b). The objective of treatment for the obese patient is therefore to remove excess fat, however that fat may be distributed in the body.

EVALUATION OF TREATMENTS FOR OBESITY

The obvious way to rank treatments for obesity is to compare the rates of weight loss which they produce: usually obese patients would consider that the treatment which caused the most rapid weight loss was best. However, this is obviously unsound, because a single oral dose of a diuretic (60 mg frusemide) will cause the loss of about 2 kg in 3 hours in normal subjects, but will not affect fat stores at all.

Quite subtle differences in diet programmes may cause differences in the rate of weight loss which does not reflect accurately the loss of fat. For example, patients who were given a diet supplying 800 kcal and 20 g protein as one meal a day for a week lost on average 271 g/d, but when crossed over to a diet supplying 800 kcal and 30 g protein a day in five meals per day they lost only 193 g/d. However, N balance studies showed that on the single-meal low-protein diet average daily N loss was 2.66 gN/d, whereas on the higher-protein five-meal diet it was only 0.62 gN/d (Garrow et al., 1981). The extra 2 gN/day implies the loss of an extra 60 g fat-free mass, so the difference in the rates of weight loss between the two diets is almost entirely explained by the extra loss of lean tissue.

Measuring the Composition of Weight Loss

There are five ways in which the proportion of fat to FFM in weight loss can be estimated; each method depends on assumptions which are not entirely true.

First, it may be assumed that weight lost is either fat, with an energy value of 9000 kcal/kg, or FFM with an energy value of 1000 kcal/kg. If the weight lost and the energy deficit during the weight loss is known, the proportion of fat and FFM in the weight loss can be calculated. For example if a person lost 2.0 kg over a period when their energy deficit was 10,000 kcal, this indicates a loss of 1.0 kg fat and 1.0 kg FFM. No other combination would explain the weight loss and energy deficit given the assumptions mentioned above.

Second, it can be assumed that FFM contains 3 gN/kg, and of course

fat contains no N. If the N balance is known the contribution of FFM to
weight loss is calculated and, by difference, the remainder is fat.

Third, the composition of the body can be estimated from density,
before and after weight loss, and the difference will show the loss of FFM
and fat.

Fourth, the same calculation can be done having estimated the
composition of the body before and after weight loss from measurements of
total body water.

Fifth, the same calculation can be done using total body potassium.

We have reported a study in which all five of these methods were used
to estimate the composition of the 5.43 kg of weight lost by 19 obese
women who were in a metabolic ward (Garrow et al., 1979). The estimated
fat loss was by energy balance 2.77+0.71 kg, by N balance 2.69+1.23 kg, by
change in density 2.83+2.32 kg, by change in water 2.37+2.38 kg and by
change in potassium 2.90+3.54 kg. The mean estimated fat loss by the five
methods agrees well, but the SD increases from 0.71 kg by energy balance
to 3.54 kg by change in potassium. Some of this variation is true
variation between individuals: there is no reason to believe that all 19
women in fact lost exactly the same amount of fat, but the SD also
contains a component due to the error of the estimate.

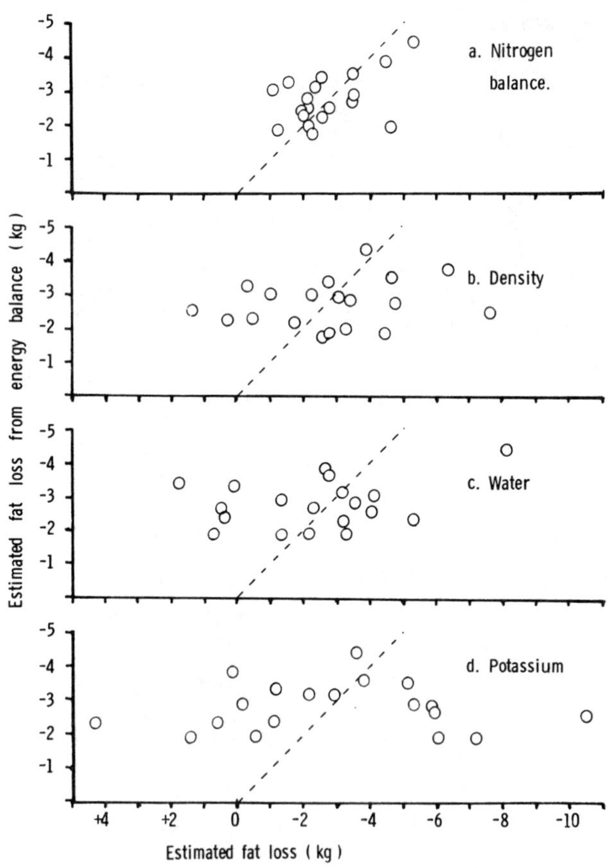

Fig 2. Estimated fat loss in 19 women, based on energy balance,
nitrogen balance, and change in body density, water or potassium.

We cannot distinguish true variation from error in any one set of
measurements, but we can assume that the energy balance method, having the
smallest SD, had the smallest error component. If we assume that the
error component in the energy balance data is zero we can calculate, for
each woman, the error in the other estimates. This calculation is shown
graphically in Fig 2. If any pair of methods had yielded identical
estimates of fat loss in this group of women all the points would fall on
the broken line, which is the line of identity. It is evident that none
of the methods agree perfectly, nor is there any predictable pattern of
disagreement. For example, the outlying values for potassium are one
woman who by potassium lost 10.53 kg of fat, who by energy balance, N
balance, density and water showed changes of -2.54, -2.89, -7.67 and +0.44
kg, respectively, while another woman who by potassium gained 4.22 kg fat
by the other methods showed changes of -2.32, -2.09, -0.57, and -5.38 kg
fat, respectively.

CONCLUSION

     Weight-height indices, despite their theoretical limitations, serve
well to distinguish those who are obese from those who are not, and to
follow the progress of weight loss. The composition of weight lost on a
given diet can be measured by tedious balance studies in a "closed"
metabolic ward, but estimates of body composition by density, water or
potassium involve errors of about 2.3 kg to 3.5 kg of fat when they are
used to estimate the composition of weight lost. Thus, they can only be
useful when the weight loss is large (say >20 kg).

REFERENCES

Abraham, S., and Mira, M., 1988, Hazards of attempted weight loss,
     Med.J.Austr., 148:324.
Behnke, A.R., Feen, B.G., and Wenham, W.C., 1942, The specific gravity of
     healthy men; body weight and volume as an index of obesity,
     J.Am.Med.Assn., 118:495
Borkan, G.A., Sparrow, D., Wisniewski, C., Vokonas, P.S., 1986, Body
     weight and coronary heart disease risk: patterns of risk factor
     change associated with long-term weight change, Am.J.Epidemiol.,
     124:410.
Dublin, L.I., 1953, Relation of obesity to longevity, New.Eng.J.Med.,
     248:971.
Garn, S.M., Leonard, W.R., and Hawthorne, V.M., 1986, Three limitations of
     the body mass index, Am.J.Clin.Nutr., 44:996.
Garrow, J.S., 1979, Weight penalties, Br.Med.J, 2:1171.
Garrow, J.S., 1988a, "Obesity and related diseases," Churchill
     Livingstone, London.
Garrow, J.S., 1988b, Is body fat distribution changed by dieting?,
     Acta.Med.Scand, Suppl, 723:199.
Garrow, J.S., Durrant, M.L., Blaza, S., Wilkins, D., Royston, P., and
     Sunkin, S., 1981, The effect of meal frequency and protein
     concentration on the composition of the weight lost by obese
     subjects, Br.J.Nutr., 45:5.
Garrow, J.S., Stalley, S., Diethelm, R., Pittet, Ph., Hesp, R., and
     Halliday, D., 1979, A new method for measuring the body density of
     obese adults, Br.J.Nutr., 42:173.
Garrow, J.S., and Webster. J., 1985, Quetelet's index as a measure of
     fatness, Int.J.Obes., 9:147.
Keys, A., Fidanza, F., Karvonen, M.J., Kimura, N., Taylor, H.L., 1972,
     Indices of relative weight and obesity, J.Chron.Dis., 25:329.

Lapidus, L., Bengtsson, C., Larsson, B., Pennert, K., Rybo, E., and
    Sjostrom, L., 1984, Distribution of adipose tissue and risk of
    cardiovascular disease and death: a 12 year follow-up of
    participants in the population study of women in Gothenburg,
    Sweden, Br.Med.J, 289:1257.
Larsson, B., Svardsudd, K., Welin, L., Wilhelmsen, L., Bjorntorp, P., and
    Tibblin, G., 1984, Abdominal adipose tissue distribution, obesity,
    and risk of cardiovascular disease and death: 13 year follow up of
    participants in the study of men born in 1913, Br.Med J., 288:1401.
Lesser, G.T., Deutsch, S., and Markofsky, J., 1971, Use of independent
    measurement of body fat to evaluate overweight and underweight,
    Metabolism, 20:792.
McLaren, D.S., 1987, Three limitations of body mass index, Am.J.Clin.Nutr.
    46:121.
Micozzi, M.S., Albanes, D., 1987, Three limitations of body mass index,
    Am.J.Clin.Nutr., 46:376.
Mueller, W.H., and Malina, R.M., 1987, Relative reliability of circum-
    ferences and skinfolds as measures of body fat distribution,
    Am.J.Phys.Anthropol., 72:437.
National Institutes of Health Consensus Development Panel on the
    Health Implications of Obesity, 1985, Health Implications of
    Obesity, Ann.Int.Med., 103:1073.
Quetelet, L.A.J., 1869, Physique Sociale. Brussels, C.Muquardt,
    2:92.
Ross, W.D., Crawford, S.M., Kerr, D.A., Ward, R., Bailey, D.A., and
    Mirwald, R.M., 1988, Relationship ofthe body mass index with
    skinfolds, girths, and bone breadths in Canadian men and women aged
    20-70 years, Am.J.Phys.Anthropol., 77:169.
Webster, J.D., Hesp, R., and Garrow, J.S., 1984, The composition of excess
    weight in obese women estimated by body density, total body
    water and total body potassium, Hum.Nutr:Clin.Nutr, 38C:299.
Womersley, J., Durnin, J.V.G.A., 1977, A comparison of the skinfold
    method with extent of 'overweight' and various weight-height
    relationships in the assessment of obesity. Br.J.Nutr., 38:271.

DUAL PHOTON ABSORPTIOMETRY IN OBESITY:

EFFECTS OF MASSIVE WEIGHT LOSS

Elisabeth W McKeon, Jack Wang, Richard N Pierson Jr.,
and John G. Kral[*]

Body Composition Unit, St. Luke's-Roosevelt Hospital,
Columbia University, New York, N.Y.
[*]Department of Surgery, SUNY, HSC Brooklyn, N.Y.

INTRODUCTION

Obesity is associated with increases in fat-free mass (FFM) and organ
sizes, in addition to body fat (BF). Several studies show increased bone
mass in extreme obesity, but data are conflicting (Steiniche et al., 1986).

In treating morbid obesity (excess body weight >45 kg) it is
desirable to reduce fat as much as possible without a disproportionate loss
of lean tissue or bone mineral. After malabsorptive operations for morbid
obesity (intestinal or gastric bypass), decreases of varying degree in FFM
and bone mineral content have been described (Compston et al., 1984;
Crowley et al., 1984, Halverson et al., 1979). Results have been equivocal
after gastric restrictive procedures.

This study establishes baseline parameters of body composition
studied by dual photon absorptiometry (DPA) validated against underwater
weighing (UWW) in morbidly obese women before and after weight loss by
anti-obesity surgery.

METHODS

Patients

Twenty-eight morbidly obese white and Hispanic premenopausal women
with a mean weight of 124 kg, height 164 cm, body mass index 46.2 kg/m2,
and eighty age- and height-matched women of normal weight (58 kg, BMI 21.5)
were studied by DPA to determine total body bone mineral (TBBM), bone
density (BD), and body fat (Table 1). Conventional hydrodensitometry was
performed by UWW.

Twelve morbidly obese women were also studied a mean of 26 months
after gastroplasty, after loss of 64% of excess weight (Revised
Metropolitan Life Insurance standards by Knapp, 1983), at a weight of 81
kg. Another ten obese women were studied after a loss of 88% of excess
weight, at a mean weight of 71 kg, a mean of 40 months after intestinal
bypass. All patients received supplemental calcium and vitamin D.

Table 1. Morbidly Obese Patients Before and After Surgical Weight Reduction
         Compared to Controls (mean ± sd)

	Preop. Obese	Gastroplasty	Intestinal bypass	Controls
Subjects(n)	28	12	10	80
Age	33.1 ± 7.6	36.8 ± 6.3	39.5 ± 8.0	33.2 ± 6.6
Weight,kg	123.7 ± 18.2	81.3 ± 13.9	70.5 ± 8.4	58.3 ± 5.7
Height,cm	164.0 ± 6.0	163.0 ± 8.0	166.0 ± 3.0	165.0 ± 6.0
BMI(kg/m2)	46.2 ± 7.7	30.8 ± 5.7	25.6 ± 3.0	21.5 ± 1.9

Bone Density and Body Fat Measurements

DPA (Lunar DP4) with ^{153}Gd with photon emissions of 44 and 100 keV was
used for a whole-body rectilinear scan to estimate total body bone mineral
(TBBM), average bone density (BD), and body fat (Mazess et al., 1984;
Gotfredsen et al., 1986). The DPA system was calibrated weekly with 7 bags
of ground beef in which the fat content ranged between 3-83% based on
chemical analysis (Wang et al., 1989). The reproducibility of TBBM, BD, and
fat% by DPA ranged between 0.97 and 0.99 in human subjects measured five
times, and was 0.99 in phantom studies. The radiation dose is small; less
than 3 millirem, or 0.03 milliSievert. A single study showed high
correlation between calcium and TBBM by DPA (Mazess et al., 1981), and
total body calcium by in-vivo neutron activation (Cohn, 1981). The
comparison was accomplished by converting TBBM (from DPA) to calcium by
using the constant fraction of 0.37 (Peppler and Mazess, 1981). Bone
density is bone mineral content expressed in gm/cm2, which is calculated
from TBBM divided by the total area of scanned bone.

Underwater-weighing was performed by total submersion of the subject
sitting on a weighing platform in a tank of warm water. Residual lung
volume was measured by a standard spirometry system before submersion, and
fat estimated from density according to Siri (1956). Reproducibility of
fat% by UWW was ± 1.3% in five subjects studied five times. Statistical
significances of differences between groups were tested by Student's
t-test; Pearson's correlation coefficients were calculated.

RESULT

Body Fat

Body fat by DPA was 68 kg, or 55% of body weight, at a level of 124
kg body weight in the morbidly obese, compared to 16 kg, or 26% of body
weight in the control women weighing 58 kg (p<0.001, Table 2a). The
correlation of body fat mass between UWW and DPA was 0.88 in the morbidly
obese (n=21) and 0.85 in the control subjects (n=80), and SEE was 0.09 kg
and 0.06 kg, respectively (Fig. 1 and 2). For women having gastroplasty,
body fat was 35 kg or 42% of a body weight of 81 kg. For women who had
intestinal bypasses, body fat was 25 kg or 34% of body weight of 71 kg
(Table 2a). Comparison of body fat measured by UWW and DPA shows that there
were significantly greater levels of fat by DPA than by UWW (Table 2b).
Based on the fat by DPA, the 65 kg difference in weight between the
morbidly obese and the control women consisted of 53 kg of fat, or 81% of
excess weight.

Table 2a. Body Fat, Bone Density, and Bone Mineral by DPA

	Preop. obese	Gastroplasty	Intestinal bypass	Controls
Subjects (n)	28	12	10	80
Weight,kg	123.7 ± 18.2	81.3 ± 13.9	70.5 ± 8.4	58.3 ± 5.7
Body fat:				
Kg	68.4 ± 14.3	34.6 ± 11.5	24.5 ± 7.9	15.5 ± 4.3
%	55.0 ± 5.4	41.7 ± 8.4	34.1 ± 8.4	26.4 ± 6.1
Lean body mass:				
Kg	55.3 ± 9.0	46.7 ± 6.9	46.0 ± 4.1	42.5 ± 5.0
%	45.0 ± 5.4	58.3 ± 8.4	65.9 ± 8.4	73.6 ± 6.1
TBBM,gm	3849 ± 522	3104 ± 380	2915 ± 285	2401 ± 263
BD (g/cm2)	1.34 ± .10	1.24 ± .07	1.17 ± .06	1.12 ± .07

Table 2b. Comparison of Body Fat between UWW and DPA

	Obese	Controls
Subjects (n)	21*	80
Body fat by UWW:		
Kg	64.9 ± 13.1	14.2 ± 4.2
%	51.7 ± 5.5	24.1 ± 5.8
Body fat by DPA:		
Kg	68.9 ± 15.1	15.3 ± 4.3
%	54.9 ± 5.7	26.4 ± 6.1

* Subgroup of the 28 patients in Table 2a
Significance of difference between UWW and DPA, p<0.001

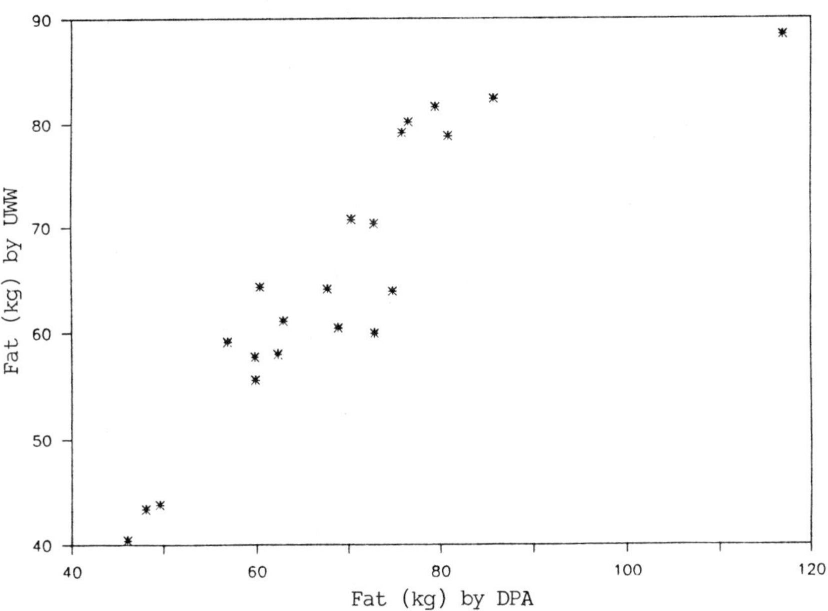

Fig. 1. Relationship between underwater-weighing (UWW) and dual photon
absorptiometry (DPA) for body fat determination in 21 morbidly
obese women. Fat(UWW) = 0.77 x Fat(DPA) + 11.9, r = 0.88, SEE =
0.09 kg, p<0.001.

Six women were studied by DPA before and after gastroplasty. The composition of the 31 kg of weight loss in these six patients was 25 kg of fat, or 81% of the total weight loss. In two patients with intestinal bypass, the composition of 75 kg of weight loss was 67 kg of fat, or 90% of the total weight loss. Total bone mineral content (3849 vs 2401 gm) and bone density (1.34 vs 1.12 gm/cm2) were both significantly greater in the obese than the lean subjects (p<0.001).

There were statistically significant positive correlations between body weight and TBBM in both the obese and the lean groups (r=0.62 vs 0.70, p<0.001). Body fat correlated with TBBM (0.68) and BD (0.50) in the obese (p<0.05). After significant losses in bone mineral content associated with weight loss in both groups of women with surgical procedures, the TBBM was similar in both groups (2915 vs 3104, ns), and correlated with body weight (0.57 vs 0.72 for intestinal bypass and gastroplasty, respectively.) The total bone density was significantly different for the two groups with 1.174 $\pm$ 0.056 gm/cm2 for the patients with malabsorptive procedures compared with 1.235 $\pm$ 0.067 gm/cm2 in patients with gastroplasty (p<0.05), yet remained significantly greater than controls.

Figure 3 demonstrates the correlation between TBBM and body weight in all subjects (r=0.91, p<0.001). The regression equations for body weight against TBBM were similar for the two groups who lost weight, (TBBM = 19.5*BW + 1538) after intestinal bypass, and (TBBM = 19.7*BW + 1500) after gastroplasty. However, the slope and the intercept were significantly different between, on the one hand, the obese pre- and post-operative groups, and on the other, the controls.

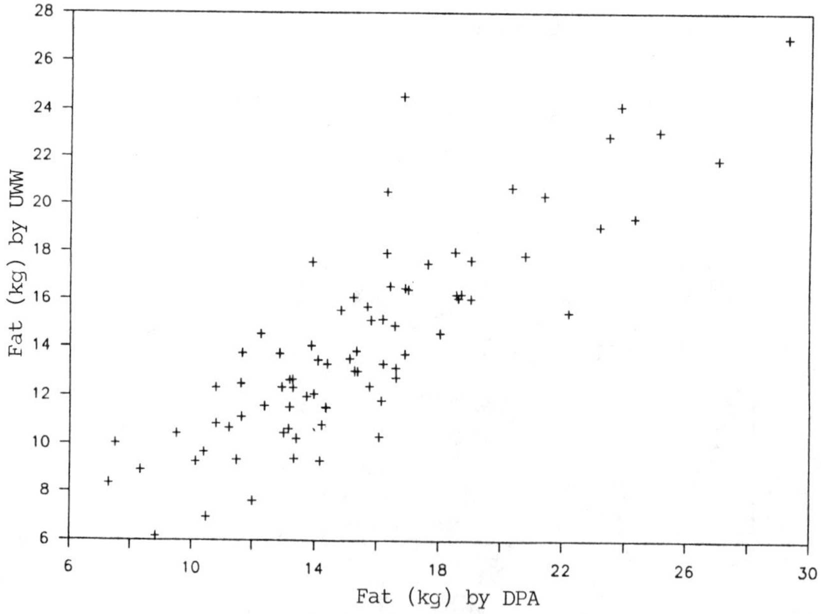

Fig. 2. Relationship between underwater-weighing (UWW) and dual photon absorptiometry (DPA) for body fat determination in 80 normal women of normal weight. Fat(UWW) = 0.83 x Fat(DPA) + 1.28, r = 0.85, SEE = 0.06 kg, p<0.001.

DISCUSSION

    This study revealed absolute increases in body fat, fat-free mass,
bone mineral content, and bone density in morbidly obese premenopausal
women using dual photon absorptiometry. These findings support the concept
of a "large frame" in morbid obesity (Naeye and Roode, 1970; Van Itallie,
1985). The results are in agreement with several other studies using
different methods for different components of body composition.

    DPA determines TBBM and percentage of fat from measurements of
differential attenuation of two different energy photons from decay of
153Gd. Attenuation is a function of mass and density, and is influenced by
the thickness of tissue. It is likely that the extreme thickness of soft
tissue in morbidly obese so decreases the yield of photons surviving
attenuation, and that a large amount of fat might be interpreted as a
smaller amount of bone, leading to an over-estimation of the TBBM. The
finding that percentage of fat was higher by DPA than by UWW can be
explained by a recognized overestimation of fat-free mass by UWW in the
morbidly obese (Wang, 1989).

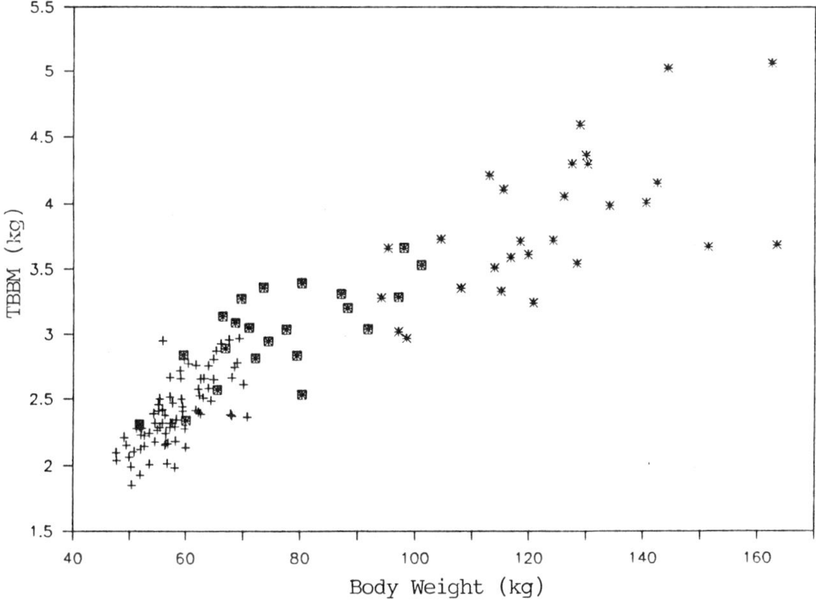

Fig. 3. Total body bone mineral (TBBM) vs body weight in controls (+),
        obese (*) and post-operative obese women with weight loss (▩).

    The composition of excess weight in the morbidly obese and the
composition of the excess weight loss in the post-operative patients are
remarkably similar in this study (82 and 82.6%, respectively). In obese
women with a mean weight of 92 kg, Webster et al. (1984) estimated the
contribution of the FFM to excess weight as being 22% (and thus fat = 78%)
using densitometry, body water, and body potassium. This estimate is in
good agreement with the present results in heavier patients. Even though we
find an absolute decrease in fat-free mass with the loss of 63-88% of
excess weight, and an absolute decrease in bone mineral (Table 2), the
results demonstrate that post-surgical obese patients, nevertheless,
maintain elevated levels of lean tissue and bone mineral for at least 2 to
4 years, compared to control subjects.

        This study supports the concept of a "large frame" in morbid obesity.
Loss of 63 to 88% of excess weight over 26 to 40 months after surgical
treatment of morbid obesity is associated with maintenance of a "large
frame". Longer periods of observation will be required to verify the
apparent lack of detrimental effect of extreme weight loss on body
composition.

REFERENCES

Cohn, S. P., 1981, In vivo neutron activation analysis: State of the art
        and future propects, Med. Phys., 8:145.
Compston, J. E., Vedi, S., Gianetta, E., Watson, G., Civalleri, D.,
        and Scopinaro, N., 1984, Bone histomorphometry and vitamin
        D status after biliopancreatic bypass for obesity, Gastro. 87:350.
Crowley, L. V.,Seay, J., and Mullin, G., 1984, Late effects of gastric
        bypass for obesity, Am. J. Gastro., 79:850.
Gotfredsen, A., Jensen, J., Borg, J., and Christiansen, C., 1986,
        Measurement of lean body mass and total fat using dual
        photon absorptiometry, Metabol., 35:88.
Halverson, J. D., Teitelbaum, S. L., Haddad, J. G., and Murphy,
        W. A., 1979, Skeletal abnormalities after jejunoileal
        bypass, Ann. Surg., 189:785.
Knapp, T. R., 1983, A methodological critique of the "ideal weight"
        concept, JAMA, 250:506.
Mazess, R.B., Peppler, W. W., Chestnut, C. H. 3rd, Nelp, W. B., Cohn, S.H.,
        1981, Total body bone mineral and lean body mass by dual-photon
        absorptiometry, 2. Comparison with total body calcium by neutron
        activation analysis, Calcif. Tissue Intern., 33:361.
Mazess, R. B., Peppler, W. W., and Gibbons, M., 1984, Total body
        composition by dual photon (153Gd) absorptiometry, Am. J.
        Clin. Nutr., 40:834.
Naeye, R. L., and Roode, P., 1970, The sizes and numbers of cells in
        visceral organs in human obesity, Am. J. Clin. Nutr., 54:251
Peppler, W. W., and Mazess, R. B., 1981, Total body bone mineral and lean
        body mass by dual-photon absorptiometry I, Calcif. Tissue Int.,
        33:353.
Siri, W. E., 1956, The gross composition of the body, Adv. Biol.
        Med. Phys. 4:239.
Steiniche, T., Vesterby, A., Eriksen, E. F., Mosekilde, L., and
        Melsen, F., 1986, A histomorphometric determination of iliac
        bone structure and remodeling in obese subjects, Bone,
        7:77.
Van Itallie, T. B., 1985, When the frame is part of the picture,
        AJPH, 75:1054.
Wang, J., Heymsfield, S. B., Aulet, M., Thornton, J. C., and
        Pierson, R. N. Jr., 1989, Body fat from body density:
        Underwater weighing vs dual photon absorptiometry, Am. J.
        Physiol., 256:E829.
Webster, J. D., Hesp,, R. and Garrow, J. S., 1984, The composition of
        excess weight in obese women estimated by body density, total body
        water and total body potassium, Human Nutrition: Clin. Nutr.,
        38c:299.

BODY FAT AND ADIPOSE TISSUE DETERMINATIONS BY COMPUTED TOMOGRAPHY AND BY

MEASUREMENTS OF TOTAL BODY POTASSIUM

H. Kvist[§], L. Sjöström[+], B. Chowdhury[+], M. Alpsten[#],
B. Arvidsson[#], L. Larsson[#] and Å. Cederblad[#]

Depts. of Radiology[§], Medicine I[+] and Radiation Physics[#],
Sahlgren's Hospital, University of Göteborg, Göteborg, Sweden

Correspondence to Lars Sjöström, Dept. of Medicine I, Sahlgren's
Hospital, 413 45 Göteborg, Sweden

INTRODUCTION

Recent investigations have demonstrated associations between hypertension, hyperinsulinemia, diabetes, hypertriglyceridemia, cardiovascular
disorders, stroke and death on one hand, and obesity and adipose tissue
(AT) distribution on the other (Krotkiewski et al., 1983; Evans et al.,
1983; Larsson et al.,1984; Lapidus et al., 1984; Seidell et al., 1988;
Sjöström et al., 1989). One important prerequisite for an increased
knowledge in this field is improved body composition techniques.

The majority of all human body fat determinations have been performed
with indirect techniques, which use different assumptions about the
chemical and/or physical properties of fat free mass (FFM) or lean body
mass (LBM). Usually two-compartment models have been used according to:

$$BW = BF + FFM \qquad\qquad \text{eq 1)}$$

$$BW = AT + LBM \qquad\qquad \text{eq 2)}$$

In eq 1 body weight (BW) consists of FFM and body fat (BF). FFM is thus
the non-lipid part of the body, including extracellular water (ECW) and
stromal vascular cells of adipose tissue (AT), as well as cell membranes,
cytoplasm and organells of the adipocytes. In eq 2 BW is composed of LBM
and AT. LBM contains all tissues of the body except AT. In these models FFM
or LBM are estimated by determining total body potassium (TBK) or total
body water (TBW) and by using different assumptions about the potassium
content of FFM or LBM (Sjöström, 1989). Alternatively, the body density is
determined by under water weighing (Behnke et al., 1942) or by means of
pletysmographic techniques (Garrow et al., 1979). By assuming a density of
1.1 for FFM and a density of 0.9 for BF the percentage of BF can then be
calculated (Siri, 1956). Modified density assumptions have also been used
(Brozek et al., 1963).

TBK and TBW determinations have also been used in a 4-compartment model
according to (Bruce et al., 1980):

$$BW = BF + FFECS + BCM + ECW \qquad eq\ 3)$$

In this model, it is assumed that the respirating (intracellular) body cell mass (BCM) contains 120 mmol K/kg. Thus, BCM = TBK/120. Furthermore, it is assumed that the water content of BCM is 75%. Thus, the extracellular water (ECW) is equal to TBW - 0.75 · BCM. Finally, fat free extracellular solids (FFECS) are assumed to be 12% of ideal weight in relation to height. BF can then be solved in eq. 3.

Relationships between various body compartments are illustrated in Fig 1.

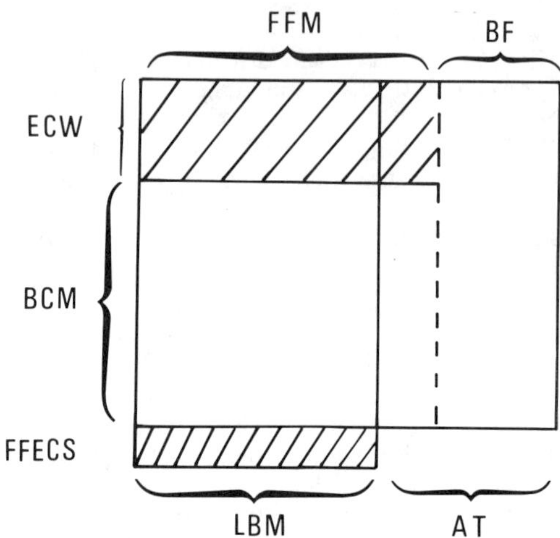

Fig. 1. Relationships between various body compartments.
FFM, fat free mass; BF, body fat; LBM, lean body mass; AT, adipose tissue; ECW, extracellular water; BCM, body cell mass; and FFECS, fat free extracellular solids.

The conventional body composition techniques mentioned above have errors in the range of 3 to 15 per cent (Sjöström, 1988b) and are not able to describe the AT distribution. Computed tomography (CT) and magnetic resonance imaging (MRI) are the only methods available for cross sectional characterization of soft tissue distribution in vivo. Area measurements by MRI can give unreliable results (Zhu et al., 1986). CT is probably the most accurate and reproducible AT technique available today. From several CT scans and the distances between these scans total, as well as regional AT volumes can be determined with high precision (Kvist et al., 1986; Kvist, 1988; Kvist et al., 1988 a,b; Kvist et al., unpublished data; Sjöström et al., 1983; Sjöström, 1985; Sjöström et al., 1985; Sjöström et al., 1986a,b; Sjöström, 1987; Sjöström and Kvist, 1988; Sjöström, 1988; 1989).

Although not suitable for large series (due to the expense and time constraints), the CT technique is ideal as a standard when calibrating simpler techniques. In this report, we describe the procedures of our CT-based technique for measurements of total and regional AT volumes in man. Furthermore, the assumptions of the total body potassium technique are "calibrated" by means of CT. The ^{40}K technique itself is also briefly described and a correction procedure for ^{134}Cs-contaminations is discussed in some detail. Finally, CT calibrated anthropometric methods for the determination of total and visceral AT are described.

METHODS

Scanning Parameters

Scanning was performed with a Philips Tomoscan 310 CT scanner at 120
kVp, using a tube filtration of 4.0 mm Al and a shaped "bow tie" Al filter.
Slice thickness was 12 mm. Exposure and scan times were 1.2 and 4.8 s,
respectively.

The CT Number Interval of Adipose Tissue and Determinations of AT Areas

The upper limit of the CT number interval characteristic of AT was
determined as the mean value of circles containing 50% muscle and 50% AT as
well as by determining the intersections between the CT number distributions of AT and muscle (Fig. 2) (Kvist et al., 1986; Kvist et al.,
unpublished data, Sjöström et al., 1986a).

The lower limit of the CT number interval of AT could not be exactly
determined. The limit was deduced from in vitro and in vivo measurements of
AT, in which considerations were taken to the partial volume and beam
hardening effects, as well as to the CT number distribution of pure AT
around its mean value. Direct area determinations of the CT number interval
of AT could then be made (Kvist et al., 1986; Sjöström et al., 1986a).

Volume Determinations

Distances between scans were obtained from frontal scanograms to the
nearest mm. From the distances between scans and the AT areas of scans, AT
volumes were calculated (Kvist et al. 1986, Sjöström et al. 1986a)
according to three different formulae:

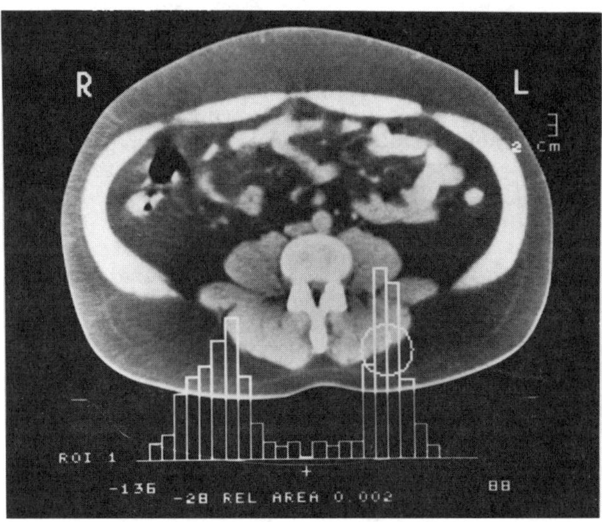

Fig. 2. The upper attenuation limit of adipose tissue.
The mean attenuation of the circumscribed area
was -30 HU and the intersection between the
muscle and AT distributions was -28 HU.

Reproduced with permission from Int. J. Obesity,
1988, 12, 249-266

a)

$$V = \sum_{1}^{i=23} a_i \frac{(b_i + c_i)}{2} \qquad \text{eq 4)}$$

where a_i is the distance between scans and b_i, c_i are the AT areas of adjacent scans.

The volume of the subcutaneous AT between adjacent scans was also obtained as the difference between truncated cones. The visceral AT between scans was calculated as truncated cones. The total AT volume was then determined as:

b)

$$V = \sum_{1}^{i=23} \frac{a_i}{3} \left[(y_{1_i} + y_{2_i} + \sqrt{y_{1_i} \cdot y_{2_i}}) - [(y_{1_i} - y_{3_i}) + (y_{2_i} - y_{4_i}) + \sqrt{(y_{1_i} - y_{3_i})(y_{2_i} - y_{4_i})}] + (x_{1_i} + x_{2_i} + \sqrt{x_{1_i} \cdot x_{2_i}}) \right] \qquad \text{eq 5)}$$

where y_1, y_2 are the total area of adjacent scans and y_3, y_4 are the subcutaneous areas of the scans. x_1, x_2 are the visceral areas and a_i is the distance between the scans.

Finally, the subcutaneous AT between scans was determined as the volume of a truncated cone and not as the difference between cones, in the same way as the calculations of visceral AT in equation 2:

c)

$$V_{subcut} = \sum_{1}^{i=23} \frac{a_i}{3} (y_{3_i} + y_{4_i} + \sqrt{y_{3_i} \cdot y_{4_i}}) \qquad \text{eq 6)}$$

Regional Distribution of Adipose Tissue

The visceral area was determined by encircling the abdominal and thoracic cavities with a cursor half way through the muscle wall. The vertebrae were always considered to be subcutaneous parts of the body. No attempts were made to separate retro- from intraperitoneal AT.

The legs were defined as the volume up to the caudal edge of the symphysis. The visceral volume was calculated from the proximal edge of the symphysis to the caudal edge of the first rib. The arms, head and neck volumes were measured from the level where the soft parts of the trunk could be separated into neck and arms, approximately the C4-5 level (Kvist et al., 1986; Kvist et al., 1988a,b; Sjöström et al., 1986a).

Artifactual Influence of Bone

In order to test the artifactual influence of bone on the CT numbers of soft tissues, a Radiation Measurements Inc. (RMI) phantom was used (Kvist et al., unpublished data). The energy dependence of the artifactual influence of structures of bone on adipose tissue (AT) was tested by additional filtration of 0.2 mm Cu and 0.2 mm Cu + 0.3 mm brass. In these tests, the scanner was calibrated with Plexiglas© according to the regulations of the manufacturer to ensure proper water correction (Joseph, 1981) for beam hardening and scattered radiation. An additional calibration to ensure correct water scaling was also made with a regular water phantom.

<u>Measurements of Total Body Potassium</u>

The most reliable method to measure TBK takes advantage of the fact that a small fraction ($1.18 \cdot 10^{-4}$) of natural potassium is radioactive and emits gamma radiation. Approximately 4000 of these nuclear transformations occur every second in the body of an average man. Every tenth of these gives rise to a photon which may be detected outside the body. There are two prerequisites for successful measurements of these photons: a massive shield to reduce the background radiation and an efficient detector setup.

Our whole body counter (WBC) and its detector arrangement has been described previously (Sköldborn et al., 1972).

The counting efficiency for photons from K-40 in a source with no self-attenuation is more than 50 per cent if the full pulse height distribution is included. Self-attenuation in the human body, as well as considerations such as the background pulse height distribution, and the occurrence of internal radioactive contaminants cause, however, a reduction of this figure to about 20 per cent.

Moreover, the counting efficiency decreases with increasing body weight according to the relationship of Fig. 3. The basis for this graph is an experimental study in which volunteers of various BMI had ingested K-42, which emits gamma radiation similar to that of K-40 (Arvidsson et al., 1972).

The background radiation corresponding to the efficiency figure given above causes a count rate of about 100 counts per second (cps). From the information given, the signal count rate for a man of 70 kg BW and average BMI can be calculated to 80 cps. A count duration of one hundred seconds for the patient count and one thousand seconds for the background count will make an uncertainty due to the counting statistics of 1.7 per cent (1 SE).

Other sources of error such as individual variations from the efficiency calibration, internal radon contamination, and slow variations of the background count rate increase the uncertainty to about 2.5 per cent (1 SE).

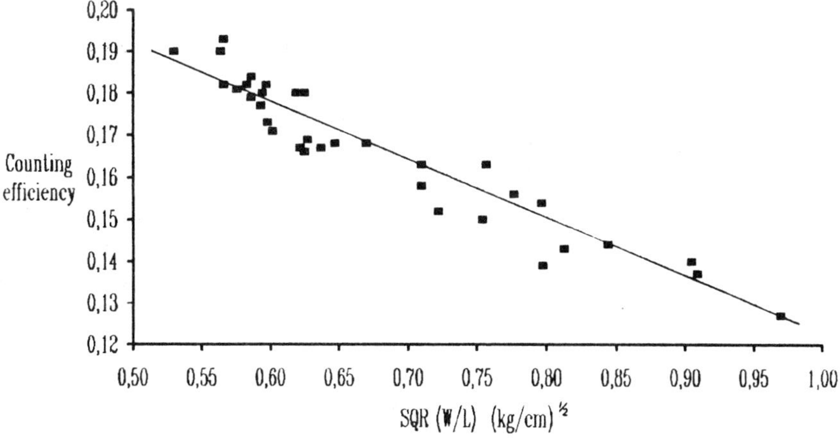

Fig. 3. The relationship between body stature and the sensitivity of the whole body counter as determined by ^{42}K-administration.

The Chernobyl disaster in April 1986, however, drastically changed these conditions since virtually all Swedes were contaminated with radionuclides from the reactor. As a consequence elevated false K-40 values appeared. The most important isotope was Cs-134, which gives signals in the same pulse height region as K-40. Therefore, the two nuclides could not be separated by the whole-body counter using normal measuring procedures.

However, a number of cascades occur in the decay of Cs-134. The same type of cascade events do not occur in the K-40 decay. A coincidence measurement procedure could therefore be used to detect the presence of Cs-134. To make a correct assessment, it is necessary that no other radionuclide with similar decay properties is present in significant amounts. After the Chernobyl accident this condition was approximately fulfilled as far as contamination of patients was concerned. A simple method of correction for Cs-134 contribution in K-40 measurements was developed in the following way:

In our whole-body counter two detectors were situated above the patient and two below. Pulse height windows were set on the two pairs of detectors. By applying a coincidence condition for pulses appearing in different energy regions, gamma cascades in the Cs-134 decay were detected in the presence of other events. Pulses due to photons in the 750-1400 keV region had to be coincident with pulses corresponding to 604 keV photons to be accepted as Cs-134. The logic pulses from the single channel analyzers setting the energy conditions were 0.5 µs long. The time window for the pulses to be accepted as coincident was 2 µs. In spite of this relatively long time for acceptance, randomly coincident pulses appeared to be relatively few (0.1 per second).

Two measurements were needed to obtain a corrected K-40 value:

1. The pulse rate in the K-40 energy window in the single pulse height distribution (R_a).

2. The pulse rate of coincidence pulses in the Cs-134 windows (R_b).

The following efficiencies were measured as discussed below –

e_1 and e_2, the efficiencies for detection of Cs-134 and K-40, respectively, in the coincidence measurement, and

e_3 and e_4, the corresponding efficiencies for Cs-134 and K-40 in the K40 pulse height window.

If A_k is the amount of K-40 and A_{cs} the activity of Cs-134,

$$R_a = A_k {}^* e_4 + A_{cs} {}^* e_3$$

$$R_b = A_k {}^* e_2 + A_{cs} {}^* e_1$$

From these equations both Cs-134 and K-40 activities were calculated.

The efficiency figures e_1, e_2, e_3 and e_4 were determined by measurements on three healthy male volunteers. One thousand Bq Cs-134 were administered per os in a water solution. Whole body measurements were carried out both by scanning and by the coincidence procedure in the whole body counter.

Urine samples were collected in standard bottles and the Cs-134 content of these was measured by a shielded NaI(Tl)-crystal. Corrections were made for urine volume.

In order to determine the influence of body stature on the efficiency
figures, body-like phantoms with different dimensions were built of plastic
bottles filled with water.

Conversions Between Body Fat and Adipose Tissue Weights and Volumes

To convert BF mass to volume the density of human triglycerides at 37°C,
0.900 g/m^3, was used. It was assumed that AT was composed of 83% lipid, 15%
water and 2% protein by weight with the densities 0.900, 1.012 and 1.4
g/cm^3, respectively. This implies that 1 kg of AT has a volume of 1.084 l
and a density of 0.923 g/cm^3 (Kvist et al., 1988a).

Deduction of the Potassium Content of Fat Free Mass and Lean Body Mass

Adipose tissue volumes determined by CT were regressed versus BF_K
(body fat as estimated from total body potassium, litres) calculated for
different assumptions regarding the potassium content of FFM (Sjöström et
al., 1986a). The AT volumes were also regressed against AT_K (adipose tissue
as estimated from total body potassium, litres) for different assumptions
of the potassium content of LBM. The regression analyses converged the
lines of equations through the origin (Ryan et al., 1976). The potassium
content of LBM was assumed to be correct for a regression coefficient of
1.0 between $AT_{(CT)}$ and $AT_{(K)}$. Since it was assumed that AT contained 85%
fat by volume, the potassium content of FFM was assumed to be correct for a
regression coefficient of 1/0.85 = 1.176 for $AT_{(CT)}$ versus $BF_{(K)}$ (Sjöström
et al., 1986a; Kvist et al., 1988a).

Anthropometric Measurements

Weight was obtained in underwear to the nearest 0.1 kg on a calibrated
scale. Height was obtained in standing position without shoes to the
nearest 0.01 m. The other anthropometric measurements were obtained from CT
scans, but chosen in such a way that true anthropometric correlates easily
could have been obtained.

Dosimetry

Thermoluminiscence dosimetry with LiF detectors revealed a dose to skin
surface in each slice of 25-50 mSv. The effective dose equivalent for 22
scans was calculated to be 2-4 mSv.

Materials and Procedures

Nineteen female and twenty-four male volunteers (age 24-64) with average
weights of 72.6 kg (range 46-124 kg) and 101.4 kg (range 58-145 kg), respectively (Table 1) were used for the CT examination. Scans were examined
at the levels indicated in Table 2.

During the examination, the arms were stretched over the head to
minimize artifacts. To obtain predictive anthropometric equations of total
and visceral AT volumes, the results of 17 men and 10 women were used as
primary data. From these groups predictive anthropometric equations were
constructed. The equations were then tested on two cross-validation groups
consisting of 7 males and 9 females (Table 1) (Kvist et al., 1988b).

The effects of negative energy balance were tested on 9 male volunteers
(age 26-65) with an average weight of 114 kg who were examined immediately
before and after 7 days of semistarvation (365 kcal/d, not shown in
tables). In the Cs-134 experiments three healthy volunteers with the body
weights 82.5, 64.6 and 62.7 kg were used to determine detection sensitivities.

Table 1. Age, Weight, Height, Body Mass Index, Total and Visceral Adipose Tissue Volumes of Examined Materials

Materials		Age	Weight	Height	BMI	Total AT volume	Visceral AT volume
		y	kg	m	kg/m²	l	l
FOR PREDICTIVE EQUATIONS							
Males n = 17	x ± SD	38±9	102.2±23.3	1.81±0.07	31.5±7.3	34.9±18.1	7.3±4.5
Females n = 10	x ± SD	39±8	70.3±19.9	1.64±0.06	25.8±5.8	30.4±17.5	2.5±1.9
FOR CROSS-VALIDATION							
Males n = 7	x ± SD	44±10	100.6±28.6	1.78±0.08	31.6±7.1	37.7±19.5	8.2±4.8
Females n = 9	x ± SD	43±11	75.0±19.7	1.67±0.05	26.8±5.8	31.5±15.1	2.7±2.2

Pooled material from Kvist et al., 1986; 1988a; 1988b and Sjöström et al., 1986a.

Table 2. Levels of examined scans in the complete 22-scan model

---

1	foot, mid metatarsal level
2	ankle
3[+]	calf lower 1/3 (ankle-knee joint distance x 1/3)
4 [*]	calf upper 2/3 (ankle-knee joint distance x 2/3)
5[+]	kneejoint
6[+]	thigh lower 1/4 (knee joint-iliac crest distance x 1/4)
7[+]	thigh mid (knee joint-iliac crest distance x 1/2)
8 [*]	symphysis caudal edge
9[+][*]	symphysis proximal edge
10 [*]	sacro-iliac joint caudal edge
11[+][*]	L4-L5
12	L3-L4
13 [*]	L2-L3
14	Th12-L1
15[+][*]	Th8-Th9
16[+][*]	sternoclavicular joint caudal edge
17 [*]	where soft parts of the trunk are separated into neck and arms. Approx C4-C5.
18	mid-humerus
19[+][*]	elbow
20	mid-forearm
21	wrist
22	hand, mid-metacarpal level

Bold underlined numbers indicate scans used in the 15-scan model (Kvist et al., 1988b)
[+] indicate scans used in the female 9-scan model (Kvist et al., 1986; Sjöström et al., 1986a)
[*] indicate scans used in the male 10-scan model (Kvist et al., 1988a)

All volunteers were informed in detail about the procedures and gave their oral consent. All examinations were approved by the Ethical Committée and the Isotope Committée of the Medical Faculty, University of Göteborg.

RESULTS AND DISCUSSION

Artifactual Influence of Structures of Bone

The bone rod substitutes of the RMI phantom caused a lucent streak artifact. The artifact was strongly reduced by the extra filtration of 0.2 mm Cu and 0.2 mm Cu + 0.3 mm brass, respectively (Kvist et al., unpublished data.

The scatter corrections assigned to different technique factor sets, in CT examinations, subtract undesirable scattered and off focal radiation as well as a contribution of beam hardening from the received detector signal. Different technique factor sets and different CT scanners may therefore yield different artifactual influences on soft tissues situated between structures of bone.

Since photoabsorption in bone removes low energy photons the percentage of scattered high energy photons reaching the detectors increases in the presence of beam hardening. Scattered radiation may therefore have a

greater influence on the imaging process in the presence of bone than
without (Glover, 1982).

Different authors have used different CT number intervals in their
attempts to characterize adipose tissue distribution. The different
intervals can, to a certain extent, be a consequence of the use of
different scanners with different photon energy distributions and different
reconstruction algorithms. However, reducing the artifacts between structures of bone by additional filtration is not easy. Many manufacturers do
not give the user the opportunity of independent selection of scanning
parameters. Arithmetic area corrections of artifacts are therefore
necessary (see below).

The Upper Limit of the CT Number Interval of Adipose Tissue

The average upper limit of the AT interval was determined to -30 HU
(range -35 to -26 HU) for both women and men (Fig. 2). The standard error
of the two methods for determination of the upper attenuation limit was on
the order of 9 to 15%, as determined from double determinations made by one
or two independent investigators (Kvist et al., 1988a).

The Lower Limit of the CT Number Interval of Adipose Tissue

In the presence of air and between structures of bone, the CT numbers of
AT was lowered beyond -250 HU. This was especially true for the upper arm
regions where the humeri and the vertebral column strongly affected the CT
numbers of soft tissues in between (Kvist et al., 1988a). In vitro phantom
examinations of human omental AT with dispersed air revealed that about
1.5% of the CT numbers obtained values lower than -190 HU (Kvist et al.,
1986). Similar partial volume phenomena may occur in vivo due to interaction between visceral AT and intestinal gas. However, extreme effects of
partial volume and beam hardening cannot be considered as typical for the
rest of the body. Therefore, the lower limit of the AT interval was set to
-190 HU. Keeping the upper limit of the interval constant at -30 HU and
raising the lower limit to -180 and -170 HU, resulted in decreases of AT
areas in men with 0.2 and 0.5% and in women with 0.3 and 0.7%, respectively
(Kvist, 1988a).

Area Determinations

The area of the AT interval was visualized by the interval detection
function and the computer automatically gave an uncorrected value of the
"adipose tissue" (Fig. 4b). However, it was evident both from in vitro and
in vivo examinations (Kvist et al. 1988a; Kvist et al., unpublished data)
that some areas of muscle obtained CT numbers below -30 HU due to the beam
hardening and scattering effects of bone (Fig. 4b, c). These muscle areas
were thus erroneously counted as adipose tissue. AT areas, which obtained
CT values below -190 HU, were on the other hand not accounted for (Fig.
4c). These inconsistencies in the area determinations had to be corrected
in femoral and humural scans. Areas within muscles in the CT interval from
-190 to -30 HU were subtracted from the uncorrected AT areas mentioned
above. AT areas with CT numbers below -190 HU were determined independently
by calculating AT areas with CT numbers between -1000 and -190 HU. These
areas were added to the uncorrected AT areas (Kvist et al., 1988a).

The error of these combined correction procedures, as determined by
double determinations in four men, was calculated to be less than 0.5%,
while the corrections as such could constitute up to 20% of the uncorrected
"adipose tissue".

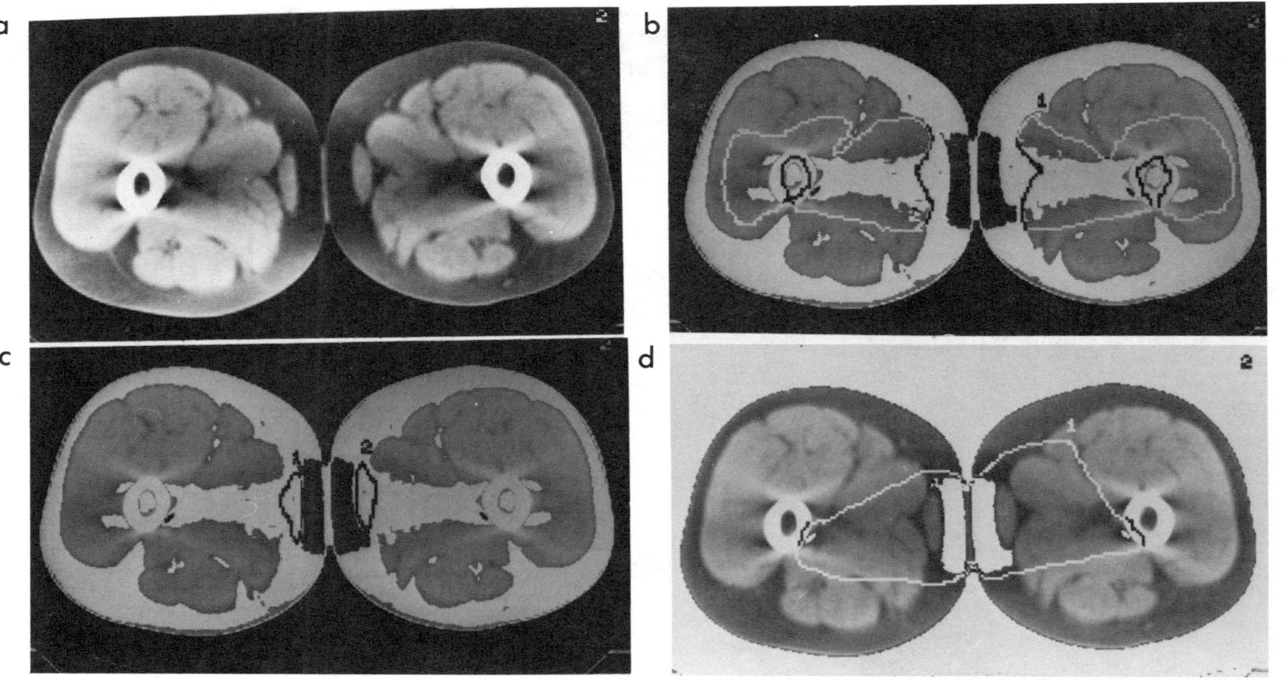

Fig. 4. Manual corrections of muscle erroneously accounted for as adipose tissue and adipose tissue not accounted for because of beam hardening. a) An ordinary mid-thigh scan without interval detection. b) and c) All area elements (pixels) within the interval from -190 HU to -30 HU are indicated in white. The uncorrected area of the interval is 253.0 cm². The circumscribed areas of muscle in b) and c) show that 59.5 cm² are erroneously counted as adipose tissue. The black areas medially within the adipose tissue are not accounted for. d) The uncounted adipose tissue is circumscribed and determined to be 22.4 cm² by using an interval from -1000 HU to -190 HU. Thus, the correct AT area of this scan is 253.0-59.5+22.4=215.9 cm².
Reproduced with permission from   Int. J. Obesity, 1988, 12:249-266.

Table 3. Percentage of AT in Different Regions of the Body

	Men (n=17)	Women (n=8)
Subcutaneous		
legs	29.0±7.3	31.8±5.6
trunk	41.4±7.4	48.6±5.1
arms	6.8±1.0	7.6±1.2
head and neck	1.9±1.0	1.8±0.2
Visceral	20.9±7.0	10.3±1.8

Reproducibility of Determinations

Double determinations with 9 scans in four females with weights ranging from 80.0 to 124.3 kg revealed a standard error of the method of 0.6% (Sjöström et al., 1986a). The reproducibility of the CT technique is thus very high, and the corresponding error for the TBK method was at least three times as high.

In order to achieve a precision of this order it is crucial to use skeletal landmarks for the positioning of the CT scans. By using internal landmarks it is also easier to follow changes in AT distribution with gain or loss of weight.

Number of Indispensable Scans

From a practical point of view it is very important to establish the minimum number of scans which give acceptable standard errors against the 22-scan model. These evaluations were performed by using eq. 4 (method a) with a reduced number of scans and by regressing AT areas of single scans against the total AT volume calculated from 22 scans (Kvist et al., 1986; 1988a; 1988b; Sjöström et al., 1986a). Both in men and women the total AT volume showed high correlations and small standard errors versus the AT areas at the level of the L4-5 scan (r = 0.967 and 0.991, SE = 9.0 and 5.2%, respectively), the sternoclavicular joint (r = 0.957 and 0.985) and at the level of the elbow (r = 0.973 and 0.921). By using eq. 4 it was found that at least nine scans in women and ten scans in men had to be included in order to retain a high accuracy. Fewer scans resulted in correlations and standard errors of the estimate in the same order as those obtained by regressing the AT area of single scans versus the total AT volume. In women, results from 22 scans versus results from 4 and 9 scans resulted in standard errors of 4.6 and 1.1%, respectively. In men the corresponding standard errors for 5 and 10 scans were 5.4 and 3.2%, respectively.

Volume Determinations and Regional Adipose Tissue Distribution

In 24 males with body weights ranging from 58 to 145 kg the total AT volume ranged from 5 to 71 litres (Kvist et al., 1988b) (cf. Table 1). The corresponding figures from females were 46-119 kg and 12-62 litres, respectively. The regional distribution of AT in eight females (Kvist et al., 1986) and 17 males (Kvist et al., 1988a) is given in Table 3. As compared to men, women tended to have higher fractions of subcutaneous AT in all regions expect the head and neck region. The relative amount of visceral AT was twice as high in males as in females.

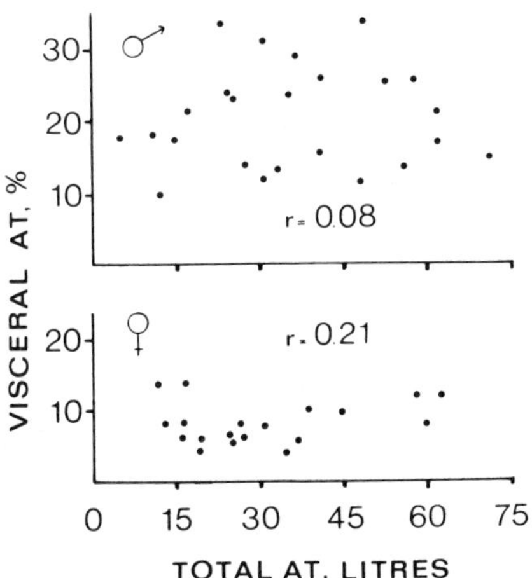

Fig. 5. The relative amount of visceral AT in
relationship to total AT volume. Males,
n=24; females, n=19. The correlations
were non-significant in both sexes.
Visceral and total AT determined with CT.
Reproduced with permission from Am. J.
Clin. Nutr., 1988, 48:1351-61.

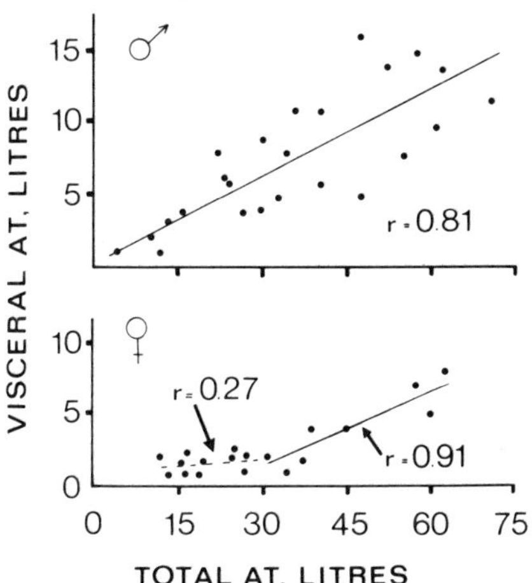

Fig. 6. The absolute amount of visceral AT in
relation to total AT. Conditions as
in Fig. 5.
Reproduced with permission from Am. J.
Clin. Nutr., 1988, 48:1351-61.

Neither in men nor in women were there any relationships between the
relative amount of visceral AT and total AT volumes (Fig. 5). The range of
% visceral AT was much larger in men (Fig. 5) (Kvist et al., 1988b).
Expressed in litres, however, there was a positive relationship between the
visceral and total AT volumes over the whole AT range in men (Fig. 6).
Women, on the other hand, seemed to be protected from visceral fat accumulation up to a total AT volume of about 30 litres. Above this amount of
total AT the visceral AT volume in females increased with a slope comparable to that of men (Fig. 6). It is tempting to suggest that these sex
patterns might be related to earlier findings, which showed that women can
accumulate some 20-30 kg body fat more than men until they reach comparable
degrees of metabolic aberrations (Krotkiewski et al. 1983).

Total Body Potassium (TBK) Determinations

TBK determinations were performed before the Chernobyl disaster without
Cs-134 corrections and after Chernobyl with corrections according to the
coincidence technique.

K-40 values with and without correction for Cs-134 for one of the
volunteers are found in Fig. 7.

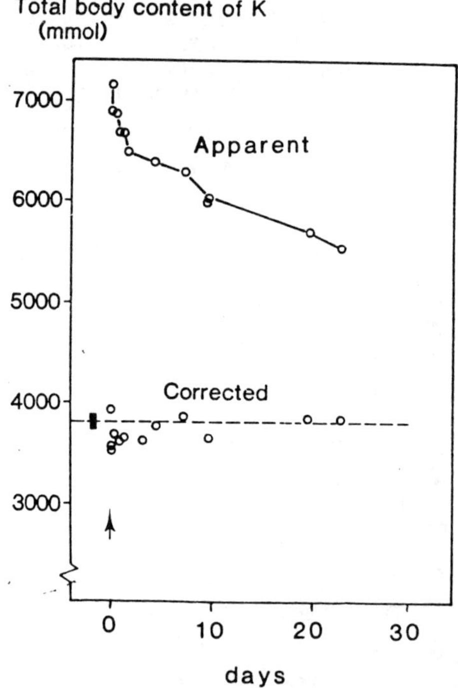

Fig. 7. Apparent total body content of potassium
evaluated with and without correction for
Cs-134 by means of the coincidence technique.
Filled bar indicates SD of mean value ±1 SD
of measured body content of potassium in
one subject during three months before
administration of 1000 Bq of Cs-134 (arrow).

The presence of 1000 Bq of Cs-134 in a person subjected to K-40 measurement corresponded to a false contribution of about 3000 millimoles of potassium.

When this error is corrected, the potassium value obtained obviously will be less reliable than would have been the case without Cs-134 contamination. This is most pronounced immediately after the administration of Cs-134. The deterioration is, however, relatively modest (Fig. 7).

Normally a potassium determination of an uncontaminated person can be made with an uncertainty of about 100 millimoles (1 SD). If the Cs-134 contamination is 1000 Bq, the uncertainty rises to 120 millimoles if the initial period with rapid redistribution is excluded. When the Cs-134 has been ingested continuously, no such period exists. The contamination level among people in the Göteborg region is in fact far less, usually below 100 Bq Cs-134.

This means that the potassium values determined with Cs-134 correction are almost as reliable as they were before Chernobyl, but to the cost of doubling the measuring time. However, uncertainties may increase if persons with statures very different from those of the volunteers are measured.

<u>Deduction of the Potassium Content of Fat Free Mass and Lean Body Mass by Means of the CT Method</u>

Womersley et al., (1976) have observed that the potassium content of FFM is likely to differ by age, gender, muscularity and degree of obesity. By comparing TBK and density measurements, they were able to suggest a range of 62-69 mmol/kg FFM for men and 55-63 mmol/kg FFM for women.

Our materials covered wide weight ranges. The appearance of the regression for AT_{CT} versus AT_K or BF_K did not indicate that obesity would notably affect the potassium concentrations of LBM or FFM (Sjöström et al., 1986a, 1989; Sjöström, 1987; Kvist et al., 1988a; Sjöström, 1989). Our iterative regression technique for deduction of the potassium content of FFM and LBM is illustrated in Fig. 8. The potassium content was 62.0 mmol/kg FFM in women and 64.7 mmol/kg FFM in men. These CT calibrated figures for potassium content per kg FFM are lower than the original suggestions by Forbes (1961), but in approximative agreement with the findings of Womersley et al (1976).

The potassium concentration of LBM must be higher than the concentration in FFM due to the low potassium content of AT. Since the LBM/FFM ratio is smaller in women than in men, the sex difference with respect to potassium content per kg of LBM can be expected to be smaller or even reversed as compared to the sex differences regarding the potassium content per kg FFM. We observed 72-74 and 71 mmol potassium per kg LBM in females and males, respectively (Kvist et al., 1988a; Sjöström et al., 1986a; Sjöström, 1989). The potassium content of LBM has not previously been measured, but a value of 69 mmol/kg has been assumed (Garrow, 1974). Larger materials covering several age groups and both sexes need to be examined in order to specify confidence limits for potassium content per kg LBM.

The investigated groups of men and women were deliberately heterogenous regarding age and body weight. Our data are too limited to permit division into different age or weight groups. Both density and potassium content of FFM may change with age and body weight. Despite this, high correlations and low standard errors have been obtained between CT- and TBK-based measurements of AT. This might indicate that age and degree of obesity influence the potassium content of FFM and LBM less than previously believed. On the other hand, two- or four-compartment models based on total body water determinations by means of tritium dilution give results in poor agreement with the CT technique (Sjöström et al., 1986a; Sjöström 1989). The main reason for this is probably that the tritium and deuterium spaces overestimate total body water (Sheng and Huggins, 1979). In our experiments this phenomenon was particularly disturbing at lower body weights (Sjöström et al., 1986a).

Anthropometric Predictions of Total and Visceral Adipose Tissue Volumes

Standard errors of 1.5–2.0% were observed when comparing true anthropometric measurements of the sagittal diameters with those obtained from the CT scans. The same was true for waist and hip circumferences

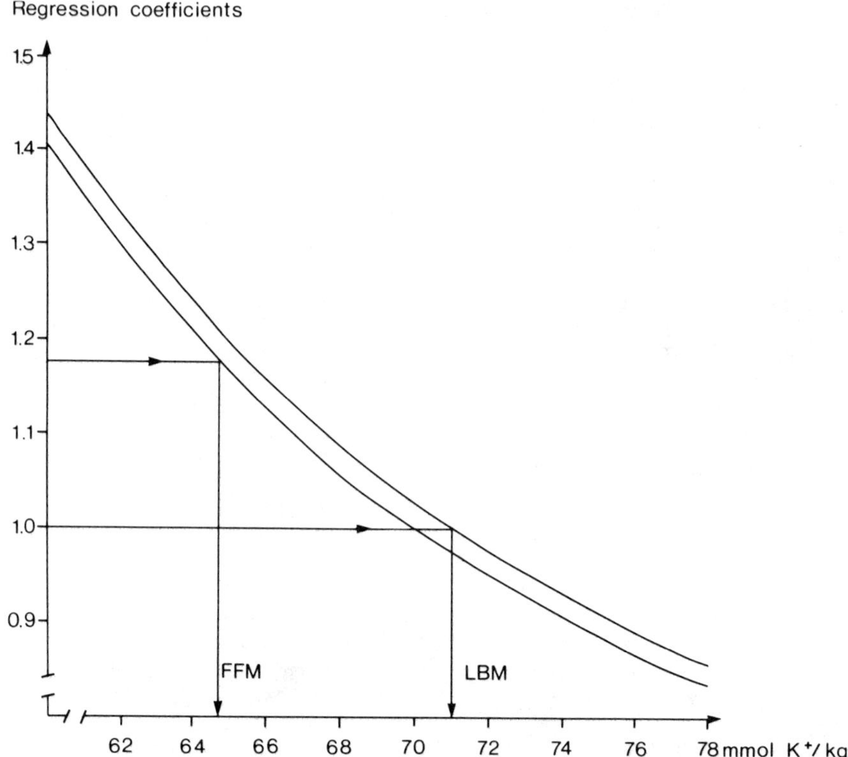

Fig. 8. Deduction of potassium contents in FFM and LBM in males. The y-axis gives the regression coefficients for $AT_{CT(1)}$ vs $AT_{K(1)}$ or $BF_{K(1)}$. A regression program is used that forces the lines of equations through origin. The regression coefficients are plotted versus different assumed values of the potassium content of LBM and FFM. With 85 per cent fat by volume in AT, a slope of $1/0.85 = 1.176$ is expected when $AT_{CT(1)}$ is regressed vs $BF_{K(1)}$, while a slope of 1.0 is expected for $AT_{CT(1)}$ vs $AT_{K(1)}$. In this way the potassium content of FFM and LBM was estimated to 64.7 and 71.0 mmol/kg, respectively.
Reproduced with permission from Int. J. Obesity, 1988, 12, 249–266.

(Table 4). The anthropometric circumference at the L4-5 level was measured
just above the iliac crest and that at the L3-4 level three cm above
crista. In standing position these circumferences increased with 4.1 and
1.3 cm, respectively.

Table 4. Differences in Anthropometric Versus CT Determined Measurements on
Nine Males Weighing 79-97 kg. (Additional material not included in other
studies).

Level	Sagittal diameter		
	Anthrop.	CT	SE
L3-4	22.2±2.3	22.3±2.3	1.8%
L4-5	21.7±2.5	21.6±2.6	1.7%

Level	Circumference			
	Anthrop. standing	Anthrop. supine	CT supine	SE supine
L3-4	93.3±8.9	92.0±8.6	90.8±8.4	1.4%
L4-5	95.8±8.9	91.7±8.2	90.1±7.8	1.6%

No single diameter, circumference or AT thickness showed higher
correlation versus total AT volume than the weight/height ratio (r = 0.967
for men and 0.984 for women). As judged from absolute residuals, weight/
height^2 resulted in significantly lower correlations vs total AT. In both
sexes predictive equations based on weight/height resulted in total AT
volumes with standard errors of the estimate in the order of ±10%, both in
the predictive and in the cross-validation groups. The explained variances
increased only marginally and insignificantly when adding other predictors
to weight/height (Table 5) (Kvist et al., 1988b; Sjöström et al., 1985;
1986b; 1988; Sjöström, 1987; 1988a; 1988b).

In both sexes the sagittal diameters at the L3-4 and L4-5 levels showed
the highest correlations versus the visceral AT volume; r = 0.9 for both
genders. Predictive equations were obtained with standard errors of
residuals of less than 20% for both men and women in primary and cross-
validation materials (Table 5). The absolute errors were of the same order
in both sexes, ±0.5 litre.

Table 5. Anthropometric predictive equations of total and visceral AT volumes (l) in 17 men and 10 women and errors obtained from cross-validation in 7 men and 9 women. Equations based on weight/height (W/H) transverse (d_i), sagittal (d_2) diameters and circumferences of CT scans.

	Level of measurements	Predictive material		Cross-validation material
		$R^2 \cdot 100$	Standard error of residuals	Standard error of differences
		%	%	%
MALES				
Total AT = 1.36 W/H*** −42.0		93.2	9.0	11.4
Total AT = 1.10 W/H*** +0.659 d_2 − 44.2	L4–L5	93.5	8.5	11.7
FEMALES				
Total AT = 1.61 W/H*** −38.3		96.5	6.8	8.5
Total AT = 1.5 W/H** +0.301 circumf. −44.4	L3–L4	97.2	5.8	7.7
MALES				
Visceral AT = 0.731 d_2*** −11.5	L4–L5	81.0	17.9	11.5
Visceral AT = 0.934 d_2*** −0.096 W/H −11.3	L4–L5	81.1	17.3	11.5
FEMALES				
Visceral AT = 0.370 d_2*** −4.85	L4–L5	79.6	21.7	16.7
Visceral AT = 0.554* d_2 −0.0549 circum −3.50	L4–L5	79.2	20.4	18.4

R^2 adjusted for degrees of freedom. Weight (W) given in kg, height (H) in m, diameters and circumferences in cm.

Significances of predictors: *, p<0.05; **, p<0.01; ***, p<0.001; non-significance is not listed. The significance of R^2 and of whole equations was in all cases highly significant (p<0.001).

* Errors calculated on differences in results between volume determinations by CT and predictions according to equations. Formula according to Dahlberg (1948). Data derived and reproduced with permission from Am. J. Clin. Nutr., 1988, 48:1351–61.

Table 6. Correlations between relative and absolute amounts of visceral AT
and selected variables. Pooled primary and cross-validation
materials (24 men, 19 women).

	Relative amount of visceral AT percent of total		Absolute amount of visceral AT(1)	
	males r	females r	males r	females r
Determinations obtained at the L3-L4 level:				
Visceral AT area	0.72***	0.61**	0.95***	0.98***
Visceral/total AT area ratio	0.86***	0.73***	0.41ns	0.25ns
Visceral/subcut AT area ratio	0.87**	0.60***	0.40ns	0.14ns
Sagittal diameter	0.44ns	0.33ns	0.92***	0.94***
Waist circumference	0.36ns	0.19ns	0.88***	0.85***
Waist/hip circumference ratio	0.59**	0.48**	0.77***	0.83***

Significance level: ns, non significant; *, $p < 0.05$; **, $p < 0.01$;
***, $p < 0.001$
Reproduced with permission from Am. J. Clin. Nutr., 1988; 48:1351-61

Predictions of Total and Visceral Adipose Tissue Volumes From Adipose Tissue Areas of Single Scans

Predictive equations based on the total AT area of the L4-5 scan
resulted in errors of 9 and 7% in the primary groups of men and women,
respectively. The errors were of the same order for the cross-validation
groups. The precision increased with about 2% in all groups by adding
weight/height as a second predictor (Kvist et al., 1988b).

Predictions of visceral AT volume based on the visceral AT area of the
L3-4 or L4-5 scan resulted in errors in the order of 10% in both sexes
(Kvist et al., 1988b).

Relative and Absolute Amount of Visceral Adipose Tissue

The sagittal diameter and the waist circumference measured at the L3-4
and L4-5 levels were closely related to the absolute but not to the relative amount of visceral AT. The waist/hip ratio was the only anthropometric
measurement that showed significant correlations against both the absolute
and relative amounts of visceral AT (Table 6). It has previously been
reported that the waist/hip circumference ratio predicts cardiovascular
disease better than the waist circumference in both men and women (Lapidus
et al., 1984; Larsson et al., 1984).

It has also been reported that the visceral/total AT area ratio of
abdominal CT scans is positively correlated to metabolic aberrations
related to cardiovascular disease (Fujioka et al., 1987). The present study
shows that the visceral/subcutaneous and the visceral/total AT area ratios
of abdominal scans correlate poorly with the visceral AT volume. However,
these ratios correlate significantly with the relative amount of visceral

AT in both sexes (Table 6). It may therefore be the relative rather than
the absolute amount of visceral AT that is statistically related to
metabolic aberrations and cardiovascular disease.

<u>Early Regional Changes in AT Depots During Negative Energy Balance</u>

Prior to their week of starvation nine obese men had 21.4, 45.7, 25.2,
6.4 and 1.3 per cent of their AT located in the visceral region, subcutaneous trunk, legs, arms and head plus neck, respectively. In relative
terms the visceral AT region was reduced more (9.6%) than other regions
(2-7%). A change in the fat pattern was also demonstrated by significant
changes of the volume ratios visceral AT/subcutaneous trunk AT and visceral
AT/total AT ratios (p<0.023 and p<0.025). By using the anthropometric
equations of Table 5, the loss of total and visceral AT were estimated to
3.3 l and 1.2 l as compared to the CT obtained changes of 2.54 l and 0.86
l, respectively. The volume changes obtained with CT and with anthropometry
were significantly correlated (delta total AT: r = 0.76, p = 0.01; visceral
AT: r = 0.85, p<0.01).

CONCLUSIONS

-   The interval of CT numbers characteristic for adipose tissue was
    determined to be -190 HU to -30 HU for the Tomoscan 310 CT scanner at
    120 kVp using a filtration of 4.0 mmAL

-   Correction procedures have been developed which diminish artifactual
    influence from bone structures caused by beam hardening and scattered
    radiation. Between structures of thick bones, adequate corrections are
    necessary in order to obtain correct adipose tissue area determina-
    tions.

-   Three different mathematical calculation procedures have been used
    which give negligible differences in the assessments of adipose tissue
    volumes.

-   Total as well as regional adipose tissue volumes have been determined
    with high precision in both males and females. Women seem to be
    protected against visceral fat accumulation up to a total adipose
    tissue volume of approximately 30 l.

-   A coincidence technique to correct TBK determinations for Cs-134
    contaminations has been developed.

-   The CT technique has been used to estimate the potassium content of fat
    free mass (FFM) and lean body mass (LBM) in healthy living subjects.

-   Equations have been constructed which predict total as well as visceral
    adipose tissue volumes from anthropometric variables. The total adipose
    tissue volume is best predicted by weight over height and the visceral
    adipose tissue volume is best predicted by the sagittal diameter
    measured at the L4-5 level in the supine position.

-   The fat patterning of males is changed by a short period of negative
    energy balance. In relative terms the visceral AT is reduced more than
    other depots.

REFERENCES

Arvidsson, B., Sköldborn, H. and Isaksson, B., 1972, Clinical application
    of a high-sensitivity whole-body counter. 3rd Int. Conf. Medical
    Physics, including Medical Engineering, Göteborg, 41:2.

Behnke, A.R., Feen, B.G. and Welhamn, W.C., 1942, The specific gravity of healthy men: body weight and volume as an index of obesity, JAMA 118:495.

Brozek, J., Grande, FR., Andersson, J.T. and Keys, A., 1963, Densitometric analysis of body composition: revision of some quantitative assumptions, Ann NY Acad Sci 110:113.

Bruce, Å., Andersson, M., Arvidsson, B. and Isaksson, B., 1980, Body composition. Prediction of normal body potassium, body water and body fat in adults on the basis of body height, body weight and age, Scand. J. Clin. Lab. Invest. 40:461.

Dahlberg, G., 1948, "Statistical Methods for Medical and Biological Students", Allen and Unwin, London.

Evans, D.J., Hoffman, R.G., Kalkhoff, R.K. and Kissebah, A.H., 1983, Relationship of androgenic activity to body fat topography, fat cell morphology, and metabolic aberrations in premenopausal women, J. Clin. Endocrinol. Metab. 57:304.

Forbes, G.B., Gallup, J. and Hursh, J.B., 1961, Estimation of total body fat from potassium-40 content, Science 133:101.

Fujioka, S., Matsuzawa, Y., Tokunaga, K. and Tarui, S., 1987, Contribution of intra abdominal fat accumulation to the impairment of glucose and lipid metabolism in human obesity, Metabolism 36:54.

Garrow, J.S., 1974, "Energy Balance and Obesity in Man," Elsevier, New York.

Garrow, J.S., Stalley, S., Diethelm, R., Pittet, Ph., Hesp, R. and Halliday, D., 1979, A new method for measuring the body density of obese adults, Br. J. Nutr. 42:173.

Glover, G.H., 1982, Compton scatter effects in CT reconstruction, Med. Phys. 9:860.

Joseph, P.M., 1981, Artifacts in computed tomography. in: "Radiology of the Skull and Brain," Vol. 5, T.H. Newton and D.G. Potts, eds., The C.V. Mosby Co., St. Louis.

Krotkiewski, M., Björntorp, P., Sjöström, L. and Smith, U., 1983, Impact of obesity on metabolism in men and women - importance of regional adipose tissue distribution, J. Clin. Invest. 72:1150.

Kvist, H., Sjöström, L. and Tylén, U., 1986, Adipose tissue volume determinations in women by computed tomography: technical considerations, Int. J. Obesity 10:53.

Kvist, H., 1988, Adipose tissue volume determinations by computed tomography. Dissertation. Medical faculty, University of Göteborg, Sweden. Vasaplatsens Tryckeri, Göteborg.

Kvist, H., Chowdhury, B., Sjöström, L., Tylén, U. and Cederblad, Å., 1988a, Adipose tissue volume determination in males by computed tomography and ^{40}K, Int. J. Obesity 12:249.

Kvist, H., Chowdhury, B., Grangård, U., Tylén, U. and Sjöström. L., 1988b, Total and visceral adipose tissue volumes derived from measurements with computed tomography in adult men and women: predictive equations, Am. J. Clin. Nutr. 48:1351-61.

Kvist, H., Starck, G., Alpsten, M., Ytterberg, C. and Tylén, U. The influence of beam hardening and scattered radiation on adipose tissue determination by computed tomography. Unpublished data.

Lapidus, L., Bengtsson, C., Larsson, B., Pennert, K., Rybo, E. and Sjöström, L., 1984, Distribution of adipose tissue and risk of cardiovascular disease and death: a 12 year follow up of participants in the population study of women in Gothenburg, Sweden, Brit. Med. J. 289:1261.

Larsson, B., Svärdsudd, K., Welin, L., Wilhelmsen, L., Björntorp, P. and Tibblin, G., 1984, Abdominal adipose tissue distribution, obesity and risk of cardiovascular disease and death: a 13 year follow up of participants in the study of men born in 1913, Br. Med. J. 288:1401.

Ryan, T.A., Joiner, B.L. and Ryan, B.F., 1976, Minitab student handbook. Duxbury Press, Boston.

Seidell, J.C., Bjorntorp, P., Sjostrom, L., Krotkiewski, M., and
    Sannerstedt, R. 1988, Abdominal obesity and metabolism in men-
    possible role of behavioural characteristics, in: "Obesity in Europe,"
    P. Bjornstorp and S. Rossner, eds., John Libbey, London.
Sheng, H.P. and Huggins, R.A., 1979, A review of body composition studies
    with emphasis on total body water and fat, Am. J. Clin. Nutr.
    32:630.
Siri, W.E., 1956, Body composition from fluid spaces and density: analysis
    of methods. Univ. Calif. Radiation Lab. Rept. 3349. Donner
    Laboratory of Biophysics and Medical Physics, University of
    California.
Sjöström, L., Tylén, U., Kvist, H., Hallgren, P. and William-Olsson, T.,
    1983, Determination of adipose tissue volume and distribution by a
    new computed tomography technique. The 4th International Congress on
    Obesity, New York. (abstract)
Sjöström, L., 1985, A review of weight maintenance and weight changes in
    relation to energy metabolism and body composition, in: "Recent
    Advances in Obesity Research IV," T. Van Itallie and J. Hirsch,
    eds., J. Libbey, London.
Sjöström, L., Kvist, H. and Tylén, U., 1985, Methodological aspects of
    measurements of adipose tissue distribution, in: "The Metabolic
    Complication of Human Obesities," J. Vague, P. Björntorp, B.
    Guy-Grand, M. Rebuffé and P. Vague, eds., Elsevier.
Sjöström, L., Kvist, H., Cederblad, Å. and Tylén, U., 1986a, Determination
    of total adipose tissue and body fat in women by computed tomo-
    graphy, ^{40}K, and tritium, Am. J. Physiol. 250:E736.
Sjöström, L., Kvist, H., Lapidus, L., Bengtsson, C. and Tylén, U., 1986b,
    Weight-height indices and adipose tissue distribution - measurements
    and health consequences, in: "Human body composition and fat distri-
    bution, N.G. Norgan, ed., Euro-Nut Workshop London.
Sjöström, L., 1987, New aspects of weight-for-height indices and adipose
    tissue distribution in relation to cardiovascular risk and total
    adipose tissue. volume, in: "Recent Advances in Obesity Research V,"
    E.M. Berry, S.H. Blondheim, H.E. Eliahou and E. Shafrir, eds.,
    Libbey, London.
Sjöström, L. and Kvist, H., 1988, Regional body fat measurements with
    CAT-scan and evaluation of anthropometric prediction. Acta Med.
    Scand., suppl. 723:169.
Sjöström, L., 1988, Measurement of fat distribution, in: "Fat Distribution
    During Growth And Later Health Outcomes," C. Bouchard. and F.E.
    Johnston, eds., Alan. R. Liss, New York.
Sjöström, L., 1989, Recent methods in the study of body composition, in:
    "Perspectives in the science of growth and development," J.M.
    Tanner, ed., Smith-Gordon and Co Ltd, London.
Sjöström, L., Backman, L., Bouchard, C., Bengtsson, C., Dahlgren, S.,
    Jonsson, E., Larsson, B., Lindstedt, S., Näslund, I., Olbe, L.,
    Sjöberg, L., Sullivan, M. and Wedel, H., 1989, Swedish Obese
    Subjects, SOS.   An intervention study of obesity. Int. J. Obesity,
    abstr. no 48.
Sköldborn, H., Arvidsson, B. and Andersson, M., 1972, A new whole-body
    monitoring laboratory. Acta Radiol. Diagn. suppl. 233:313.
Womersley, J., Durnin, J.V.G.A., Boddy, K. and Mahaffy, M., 1976, Influence
    of muscular development, obesity and age on the fat free mass of
    adults. J. Appl. Physiol. 41:223.
Zhu, X.P., Checkley, D.S., Hickey, D.S. and Isherwood, I., 1986, Accuracy
    of area measurements made from MR images compared with computed
    tomography. J. Comput. Assist. Tomogr. 10:96.

STUDIES OF SOFT TISSUE BODY COMPOSITION USING SINGLE AND DUAL PHOTON ABSORPTIOMETRY

Christian Hassager, Jan Pødenphant, Elsebeth
Iversen and Claus Christiansen

Department of Clinical Chemistry, Glostrup
Hospital, University of Copenhagen, Glostrup
Denmark

INTRODUCTION

Both single and dual photon absorptiometry were originally designed
to measure bone mineral content, but we have recently shown that these
methods may also be used to estimate the body composition of soft tissue
(Gotfredsen et al., 1986; Hassager et al., 1989 a,c,d). In the present
report we shall summarize some of our studies of soft tissue body
composition with photon absorptiometry under four headings: 1) The
methods and their validation; 2) The effect of partial immobilization
due to hemiplegia on local body composition; 3) The effect of estrogen/
progestogen substitutional therapy on the body fat of postmenopausal
women; and 4) The effect of the anabolic steroid nandrolone decanoate on
the body composition of postmenopausal women.

The Methods and Their Validation

Single photon (125-I) absorptiometry (SPA) is widely used for
measurements of bone mineral content (BMC) in the distal forearm. In our
set-up, using a two dimensional scanning equipment, the forearm is
immersed in water to give a constant thickness of soft tissue. The mass
attenuation coefficients and densities for water and soft lean tissue
are almost equal and higher than the values for fat. Comparison of the
count rates in pure water with the count rates in soft tissue pixels
beside the forearm bones may therefore give a direct measurement of the
subcutaneous fat content in the distal forearm (FATarm). The method is
described in detail elsewhere (Hassager et al., 1989a). The obtained
values (BMC and FATarm) are the mean of 12 scans, six on each forearm.
Measurements of both forearms take less than 10 min. The precision error
(CV%) is 1% for a BMC measurement and from 2% (obese subject) to 9%
(normal subject) for a FATarm measurement (Hassager et al. 1989a).

Dual photon (153-Gd) absorptiometry (DPA) may be used for
measurements of total body bone mineral, total body fat mass (FM), and
soft tissue lean body mass (LBM). Details and technical specifications
are given elsewhere (Gotfredsen et al., 1984a,b, and 1986). About 4000
body pixels are measured during a 90-minute rectilinear transmission
scan of the entire body, 2000 of which contain only soft tissue. The
ratio of attenuation at the two energies (44 and 100 KeV), measured in
pure soft tissue pixels, can be used as an indicator of the soft tissue
composition (Gotfredsen et al., 1986). As DPA gives a direct measurement

of both the lean and the fatty components in any pixel of the body, both
local and total body composition can be measured. The long-term
precision errors (CV%) of total body LBM and FM measurements are 2.1%
and 7.6%, respectively (Hassager et al., 1989d).

To assess the accuracy of these two direct measuring methods for
fat (FATarm by SPA and FM by DPA) we recruited 12 healthy men (aged
21-37 years, height 180-196 cm, weight 70-95 kg) and 17 healthy women
(aged 20-36 years, height 162-184 cm, weight 51-89 kg). All were
measured once by SPA and once by DPA. In addition, all had their body
density (BD) measured by underwater weighing and their total body
potassium (TBK) measured in a whole body scintillation counter. From
each of these measurements (DPA,BD, and TBK) we calculated the
percentage of body fat as the fat mass in per cent of body weight
(FAT%(DPA), FAT%(BD), and FAT%(TBK), respectively). Table 1 shows the
results from correlations and linear regressions between these methods.
Agreement was found between the DPA method and the BD method, whereas
the comparison between DPA and TBK indicated a slight problem of
accuracy with at least one of them, as the slope was significantly
different from 1. This may partly be due to interindividual variation in
the concentration of potassium in the lean body mass, which is assumed
to be constant for the TBK method (we used 68.1 mmol/kg, but lower
values gave similar results) (Hassager et al., 1989d). The ability of
the FATarm measurement to predict the total body fat percentage was
approximately equal to that of other easy, indirect measurements.
However, the SEEs from the FATarm regressions were so large that the
FATarm method may mainly be valuable for research purposes where groups
of subjects are compared. It is, however, noteworthy that the SEEs from
the regressions between the reference methods (DPA, BD and TBK) were in
the same range.

The Effect of Partial Immobilization Due to Hemiplegia on Local Body Composition

It is a well-known clinical observation that immobilization of a
limb leads to muscular atrophy and demineralization of the bone of that
limb. With DPA and SPA, we tried to quantify the changes in soft tissue
and bone mineral content occurring in the paretic extremities of
hemiplegic patients after a stroke. Fifteen patients (7F,8M), who had
had a stroke with unilateral involvement of both upper and lower
extremity 6 to 9 months earlier (mean 29 weeks) were studied (Iversen et
al., 1989). All were still hemiplegic 4 weeks after the stroke and none
earlier had had any diseases with known asymmetrical involvement of the
limbs. All had both forearms measured once by SPA and all had their
local content of bone mineral, lean tissue, and fat tissue in both arms
and in both legs measured once by DPA 6 to 9 months after the stroke. By
comparing the paretic limb with the non-paretic limb it is possible to
eliminate a large interindividual variance. Fig. 1 shows the median
differences between the paretic and the non-paretic limbs. Assuming
equality between the right and left sides before the stroke, then Fig. 1
illustrates the changes that have occurred in the paretic limbs 7 months
later. The figure shows that partial immobilization, due to paresis
after a stroke, results in a significant decrease in bone mineral and
lean tissue and an increase in fat tissue in the affected limb.

The Effect of Estrogen/Progestogen Substitutional Therapy on the Body Fat of Postmenopausal Women

Bone mass decreases and body fatness increases with advancing age
(Hassager et al. 1986; Gotfredsen et al. 1987). Postmenopausal hormone

Table 1. Relation between Body Fat Percentage Measured by DPA, BD, and
          TBK and FAT$_{arm}$ Measured by SPA

Regression equation		r	SEE
FAT% (BD)	= 1.0 x FAT% (DPA) + 2.3	0.86	4.0
FAT% (TBK)	= 1.4 x FAT% (DPA) - 5.2	0.90	4.2
FAT% (TBK)	= 1.1 x FAT% (BD) - 2.3	0.83	6.0
FAT% (DPA)	= 5.9 x FAT$_{arm}$ + 13	0.88	3.4
FAT% (BD)	= 6.6 x FAT$_{arm}$ + 14	0.86	4.1
FAT% (TBK)	= 8.9 x FAT$_{arm}$ + 13	0.85	5.8

Data from Hassager et al. 1989 c, d.

replacement therapy has proved to be effective in maintaining bone mass
(Christiansen et al., 1981), but its effect on the soft tissue body
composition is largely unknown. Using SPA and DPA, we studied the effect
of postmenopausal hormone replacement therapy on body fatness in two
double-blind 2-year placebo-controlled trials. The women were allocated
blindly to four treatment groups: 1) oral cyclical combination of
estradiol valerate and cyproterone acetate (n=32); 2) oral placebo
(n=33); 3) percutaneous 17β-estradiol, supplemented by oral progesterone
during the second year (n=20); or 4) percutaneous placebo cream (n=25),
(n=number who completed 2 years of treatment). For details see Hassager
and Christiansen (1989e). All women were measured by SPA every three
months and by DPA once a year. Both hormone therapies prevented the
postmenopausal bone loss (Riis et al., 1987a,b).

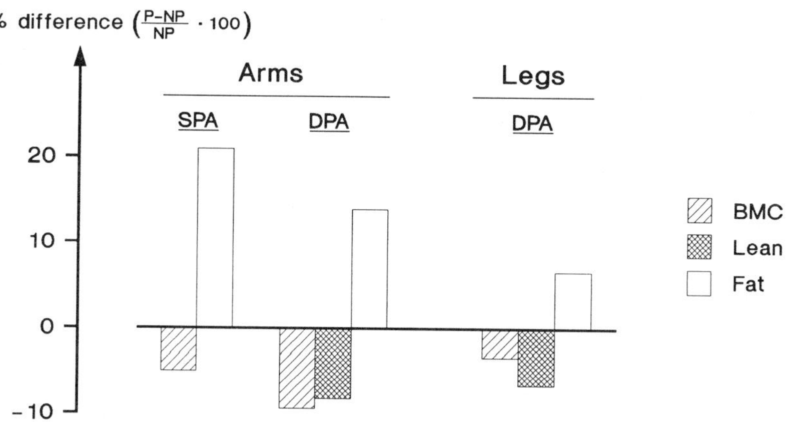

Fig. 1. Relative differences 100x(P-NP)/NP) in bone mineral content
         (BMC), lean tissue, and fat between the paretic (P) and the
         non-paretic (NP) extremity. Bars indicate median values, all of
         which were significantly different from zero (p<0.05). Data
         from Iversen et al., 1989.

221

Fig. 2 shows the changes in FATarm and FM in the four groups during the study. Both oral and percutaneous estradiol therapy prevented the increase in fatty tissue observed in the placebo groups, although a statistically significant difference between the changes in the hormone and that in the placebo groups was only found for FATarm in the percutaneous study. The oral hormone seemed to have a less pronounced effect on FATarm. This may reflect a dose-response relation, as serum estradiol, the most potent natural estrogen, was considerably higher during the percutaneous than during the oral hormone therapy (Hassager et. al, 1987).

These results are consistent with those of another double-blind placebo-controlled study (Hassager et al. 1989a), where postmenopausal women aged 55-75 years (mean age 65) were given either an oral combination of 17β-estradiol and nor-ethisterone acetate or placebo tablets for one year. Compared with placebo, the hormone therapy resulted in a significant decrease in FATarm without affecting body weight (Hassager et al. 1989a).

<u>The Effect of the Anabolic Steroid Nandrolone Decanoate on the Body Composition of Postmenopausal Women</u>

In the last double-blind study under review we investigated the changes in the soft tissue body composition that occur in postmenopausal women, treated experimentally with metabolic steroids for established osteoporosis. Thirty-nine osteoporotic (prior spine or Colles' fracture), but otherwise healthy postmenopausal women were allocated to receive blindly either 50 mg nandrolone decanoate (Deca-Durabolin R) (n=20) as an intramuscular depot or placebo (n=19) injections every three weeks for one year (Hassager et al, 1989b; Johansen et al., 1989).

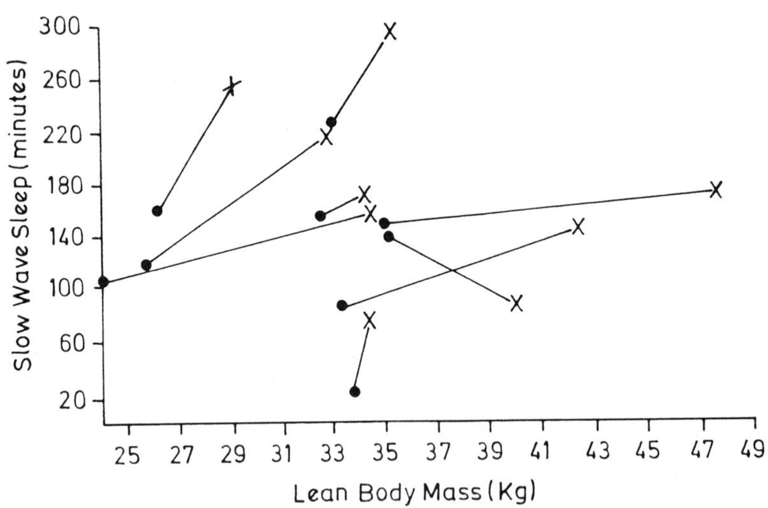

Fig. 2. Changes in forearm fat content (FATarm) measured by SPA and total body fat mass (FM) measured by DPA during 2 years of estrogen/progestogen replacement therapy (▼) or placebo (•). Values are given as the difference from the initial values as mean+SEM. Used with permission, Hassager and Christiansen, 1989e.

Thirty-six women (92%) completed the study. One year of nandrolone
decanoate therapy resulted in an increase in forearm bone mineral
content of 3% (Johansen et al., 1989). Fig. 3 shows the changes in soft
tissue body composition that had occurred in the two groups after one
year. Compared with placebo, nandrolone decanoate therapy produced a 4.0
kg increase (p<0.001) in LBM and a corresponding decrease (3.0 kg,
p<0.01) in FM, both measured by DPA. The FATarm, measured by SPA,
decreased 40% percent in the nandrolone decanoate group. This change
corresponded to a decrease in total body fat of 5 kg (Hassager et al.,
1989a), i.e of the same magnitude as that found by DPA. As LBM consists
of both muscle and non-muscle lean tissue, the observed 4.0 kg increase
could be caused by a change in either of these compartments. The 24-hour
urinary creatinine excretion did, however, increase approximately 20%
(from 8.5±1.6 mmol/24 h initially to 10.2±1.9 mmol/24 h at one year,
p<0.001, (mean±SD)), which suggests an increase in muscle mass. Assuming
equivalence between 1 g creatinine/24 h and 18 kg muscle, as proposed by
Heymsfield et al (1983), the increase in the 24-hour urinary creatinine
excretion observed in the nandrolone decanoate group corresponds to an
increase in muscle mass of 3.4 kg.

The change in soft tissue composition towards a leaner and more
muscular body found in the present study agrees with the findings of
Aloia et al. (1981), who reported a 10% increase in total body potassium
in osteoporotic women during treatment with methandrostenolone.

Finally, we have recently investigated the effect of withdrawal of
therapy in an open follow-up of the present study (Hassager et al.
1989f). We found that withdrawal of nandrolone decanoate leads to bone
loss and the soft tissue body composition returns to its pretreatment
condition (Hassager et al. 1989f).

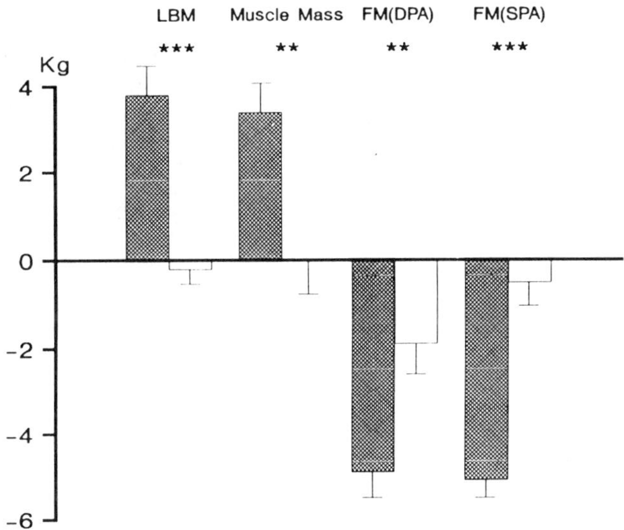

Fig. 3. Changes in body composition after 1 year of nandrolone
         decanoate (□) or placebo (□) therapy. Total body fat mass (FM)
         and soft tissue lean body mass (LBM) was measured by DPA. FM
         (SPA) is the total body fat mass estimated from forearm fat,
         according to Hassager et al. 1989a. Muscle mass was estimated
         on the 24-hour urinary creatinine excretion, according to
         Heymsfield et al. (1983). Values are given as the difference
         from initial values as mean±SEM.
         The difference between the changes: **=p<0.01; ***=p<0.001.
         Data from Hassager et al., 1989b.

CONCLUSION

From these four studies we may draw the following four conclusions:

1) Both single and dual photon absorptiometry may be used to measure the body composition of soft tissue for research purposes.

2) Partial immobilization, due to paresis after a stroke, brings about a decrease in bone mineral and lean tissue and an increase in fat tissue in the affected limb.

3) Estrogen/progestogen replacement therapy in postmenopausal women seems to prevent the age-related increase in body fat.

4) Nandrolone decanoate therapy changes body composition in postmenopausal women towards a leaner and more muscular body with a decreased fat content.

REFERENCES

Aloia, J.F., Kapoor, A., Vaswani, A., and Cohn, S.H., 1981, Changes in body composition following therapy of osteoporosis with methandrostenolone, _Metabolism_, 30:1076.

Christiansen, C., Christensen, M.S., and Transbøl, I., 1981, Bone mass in postmenopausal women after withdrawal of oestrogen/gestagen replacement therapy, _Lancet_, i: 459.

Gotfredsen, A., Borg, J., Christiansen, C., and Mazess, R.B., 1984a, Total body bone mineral _in vivo_ by dual photon absorptiometry. I. Measurement procedures, _Clin Physiol_, 4:343.

Gotfredsen, A., Borg, J., Christiansen, C., and Mazess, R.B., 1984b, Total body bone mineral _in vivo_ by dual photon absorptiometry. II. Accuracy, _Clin Physiol_, 4:357.

Gotfredsen, A., Jensen, J., Borg, J., and Christiansen, C., 1986, Measurement of lean body mass and total body fat using dual photon absorptiometry, _Metabolism_, 35:88.

Gotfredsen, A., Hadberg, A., Nilas, L., and Christiansen, C., 1987, Total body bone mineral in healthy adults, _J Lab Clin Med_, 110:362.

Hassager, C., Gotfredsen, A., Jensen, J., and Christiansen, C., 1986, Prediction of body composition by age, height, weight, and skinfold thickness in normal adults, _Metabolism_, 35:1081.

Hassager, C., Riis, B.J., Strøm, V., Guyene, T.T., and Christiansen, C., 1987, The long-term effect of oral and percutaneous estradiol on plasma renin substrate and blood pressure, _Circulation_, 76:753.

Hassager, C., Borg, J., and Christiansen, C., 1989a, Measurement of the subcutaneous fat in the distal forearm by single photon absorptiometry, _Metabolism_, 38:159.

Hassager, C., Pødenphant, J., Riis, B.J., Johansen, J.S., Jensen, J., and Christiansen, C., 1989b, Changes in soft tissue body composition and plasma lipid metabolism during nandrolone decanoate therapy in postmenopausal women, _Metabolism_, 38:238.

Hassager, C., Nielsen, B., and Christiansen, C., 1989c, Estimation of total body composition from single photon absorptiometry measurement of forearm fat content, _Scand J Clin Lab Invest_, 49:233.

Hassager, C., Sørensen, S.S., Nielsen, B., and Christiansen, C., 1989d, Body composition measurement by dual photon absorptiometry: Comparison with body density and total body potassium measurements, _Clin Physiol_, 9:353.

Hassager, C., and Christiansen, C., 1989e, Estrogen/gestagen therapy changes soft tissue body composition in postmenopausal women, _Metabolism_, 38:662.

Hassager, C., Riis, B.J., Pødenphant, J., and Christiansen, C., 1989f,
     2 years treatment of postmenopausal osteoporosis with nandrolone
     decanoate and the effect of withdrawal of therapy, _Maturitas_, in
     press.
Heymsfield, S.B., Arteaga, C., McManus, C., Smith, J., and Moffitt, S.,
     1983, Measurement of muscle mass in humans: validity of the
     24-hour urinary creatinine method, _Am J Clin Nutr_, 37:478.
Iversen, E., Hassager, C., and Christiansen, C., 1989, The effect of
     hemiplegia on bone mass and soft tissue body composition, _Acta
     Neurol Scand_, 79:155.
Johansen, J.S., Hassager, C., Pødenphant, J., Riis, B.J., Hartwell, D.,
     Thomsen, K., and Christiansen, C., 1989, Treatment of
     postmenopausal osteoporosis: is the anabolic steroid nandrolone
     decanoate a candidate, _Bone and Mineral_, 6:77.
Riis, B.J., Thomsen, K., Strøm, V., and Christiansen, C., 1987a, The
     effect of percutaneous estradiol and natural progesterone on
     postmenopausal bone loss, _Am J Obstet Gynecol_, 156:61.
Riis, B.J., Jensen, J., and Christiansen, C., 1987b, Cyproterone
     acetate, an alternative gestagen in postmenopausal oestrogen/
     gestagen therapy, _Clin Endocrinol_, 26:327.

COMPARISON OF CONDUCTIVITY, IMPEDANCE AND DENSITY METHODS FOR BODY

COMPOSITION ASSESSMENT OF OBESE WOMEN

Nancy L. Keim, Teresa F. Barbieri and Marta Van Loan

Western Human Nutrition Research Center, USDA/ARS
P.O.Box 29997
Presidio of San Francisco, CA   94129

INTRODUCTION

Body composition assessment should be used routinely to monitor progress
of obese individuals participating in weight control programs. The advent of
newer techniques based on bioelectrical properties has simplified body
composition assessment, but accuracy of these techniques is still uncertain
when applied to the obese. We report here our experience with 3 methods: 1)
total body conductivity (TBC), 2) bioelectrical impedance (BIA) and 3) body
density (DEN) used for body composition assessment of obese women.

METHODS

We estimated body fat-free mass (FFM) and body fat mass (FAT) by DEN, TBC
and BIA. DEN was determined by underwater weighing (Akers and Buskirk, 1969)
with the simultaneous measurement of residual lung volume, and percent body
fat was calculated by the Siri equation (1956). TBC was measured with the
HA-2 body composition analyzer (EM-SCAN, Inc., Springfield, IL). FFM was
predicted from TBC using the Van Loan equation (1987). BIA measurements were
made using a four terminal impedance plethysmograph (RJL Systems, Model 101,
Detroit, MI). FFM was predicted from the manufacturer's equation.

Body composition data were obtained from 47 women participating in
various energy metabolism studies. All studies shared a common objective of
describing FFM and FAT changes during supervised programs of diet and/or
exercise. Relative body weight ranged from 116-181% of desirable weight for
height, and body mass index (wt/ht^2) ranged from 25.5 to 40.8.

RESULTS

Absolute Values for FFM and FAT

At the initial test, all methods yielded FFM values that were highly
correlated; FAT values were also highly correlated (Table 1). Despite good
correlations between methods, FFM$_{TBC}$ and FFM$_{BIA}$ values were higher than FFM$_{DEN}$,
and conversely, FAT$_{TBC}$ and FAT$_{BIA}$ values were lower than FAT$_{DEN}$. The plot of
FFM$_{TBC}$ vs. FFM$_{DEN}$ (Fig. 1) illustrates that FFM$_{TBC}$ > FFM$_{DEN}$, since the majority
of data points lie above the line of identity. The discrepancy between

Table 1. Correlation Coefficients for FFM and FAT Values
Determined by Different Methods.

	FFM	FAT
DEN vs. TBC	0.826	0.920
DEN vs. BIA	0.757	0.911
TBC vs. BIA	0.864	0.869

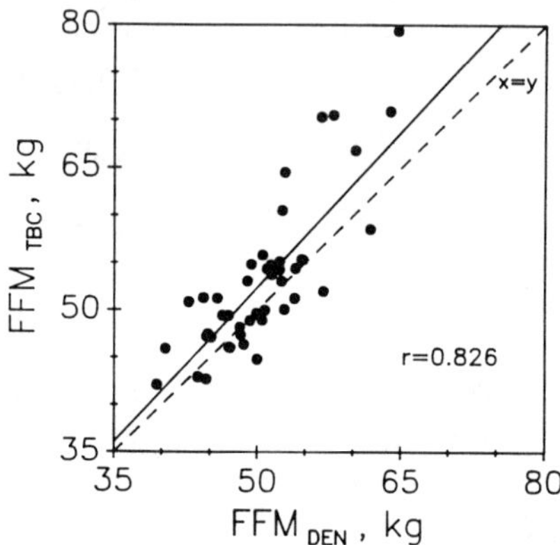

Fig. 1. Relationship between FFM Determined by DEN and
by TBC. Each point represents values obtained
during the first body composition test day, n=47.

FFM_{TBC} and FFM_{DEN} was particularly apparent in larger women with
FFM_{TBC} > 60 kg (Fig. 1).

<u>Changes in FFM and FAT</u>

Since the three methods produced different FFM and FAT values, we also
compared the methods for body composition changes occurring with weight
loss. In study 1, four obese women, 147-179% of desirable weight, lost 14.2
± 2.0 kg (mean ± SEM) during 12 wk of moderate caloric restriction. Initial
mean (± SEM) FFM_{TBC} and FFM_{BIA} were greater (p<0.01) than FFM_{DEN}, 57.2 ± 3.2,
57.5 ± 3.4, and 52.6 ± 2.6 kg, respectively. Initial mean FAT_{TBC} and FAT_{BIA}
were lower (p<0.01) than FAT_{DEN}, 41.7 ± 3.5, 41.4 ± 2.7, and 46.3 ± 3.9 kg,
respectively. After 12 wk of energy restriction FFM loss was 4.1 ± 2.4,
4.3 ± 2.8, and 3.7 ± 0.7 kg for TBC, BIA and DEN methods, and respective FAT
loss was 10.1 ± 0.9, 9.9 ± 1.5, and 10.5 ± 1.7 kg. Small discrepancies in
FFM and FAT changes among methods were not significantly different, p=0.93.

In study 2, ten mildly obese women, 119-141% of desirable weight, were
assigned to either an exercise program (n=5) or a diet restriction plus
exercise program (n=5). After 12 wk, the diet + exercise group lost
12.6 ± 0.6 kg, and the exercise only group lost 5.1 ± 0.7 kg. As in study 1,
initial mean FFM_{TBC} and FFM_{BIA} were greater than FFM_{DEN}. The loss of
FFM and loss of FAT did not differ by method in either group (Table 2)
(p values ranged from 0.22 to 0.32). However, for the diet + exercise group,
the calculated fat-to-lean ratio of weight loss tissue varied by twofold: DEN
= 3:1; TBC = 1.5:1; BIA = 2:1. For the exercise only group, the ratio varied
by more than threefold: DEN = 7:1; TBC = 3:1; BIA = 2:1.

Table 2. Mean ($\pm$ SEM) Loss of FFM and FAT by Different Methods.

	FFM kg	FAT kg
Diet + Exercise:		
DEN	-3.1 $\pm$ 0.3	-9.5 $\pm$ 0.5
TBC	-5.0 $\pm$ 1.3	-7.6 $\pm$ 0.8
BIA	-4.3 $\pm$ 0.9	-8.3 $\pm$ 0.5
Exercise Only:		
DEN	-0.7 $\pm$ 0.4	-4.4 $\pm$ 0.5
TBC	-1.3 $\pm$ 1.0	-3.8 $\pm$ 1.1
BIA	-1.9 $\pm$ 0.5	-3.2 $\pm$ 0.5

CONCLUSIONS

In the obese, the DEN method yields lower FFM and higher FAT than the
values obtained by TBC and BIA methods. Increased body buoyancy leading to
greater difficulty in performing the underwater weighing procedure could
contribute error to DEN values; also, the use of the Siri equation produces
a higher percent body fat in the obese (Lukaski, 1987). Factors that could
contribute error to TBC or BIA values include hydration of adipose tissue
leading to increased body water in the obese (Garrow, 1982) which, in turn,
would affect body bioelectrical properties, and an effect of body geometry
that produces increased conductivity in large individuals (Harker, 1973).
FFM and FAT changes measured over time are similar by method, but when weight
lost is small, the fat-to-lean ratio of weight loss tissue varies by method.
In energy balance studies it is important to know composition of weight loss
tissue accurately. Because basic assumptions underlying each method may not
hold true in the obese, no method can be singled out as the most accurate.
Comparison of these methods to absorptiometry, neutron activation analysis or
magnetic resonance imaging should provide insight into the relative accuracy
of DEN, TBC and BIA for application to obesity.

REFERENCES

Akers, R., and Buskirk, E.R., 1969, An underwater weighing system utilizing
    force cube transducers, J. Appl. Physiol. 26:649.
Garrow, J.S., 1982, New approaches to body composition, Am. J. Clin. Nutr.
    35:1152.
Harker, W.H. (Inventor), and EMME (Assignee), 1973, Method and apparatus for
    measuring fat content in animal tissue either in vivo or in slaughtered
    and prepared form, US Patent 3,735,247.
Lukaski, H.C., 1987, Methods for the assessment of human body composition:
    traditional and new, Am. J. Clin. Nutr. 46:537.
Siri, W.E., 1956, The gross composition of the body, in: "Advances in
    Biological and Medical Physics," Vol 4, Tobias, C.A., and Lawrence, J.H.,
    eds., Academic Press, New York.
Van Loan, M., and Mayclin, P., 1987, A new TOBEC instrument and procedure for
    the assessment of body composition: use of Fourier coefficients to predict
    lean body mass and total body water, Am. J. Clin. Nutr. 45:131.

SLEEP AND BODY COMPOSITION

C.M. Shapiro[*] H. Driver[+] K. Cheshire[*] A. Carver[*]
and J. Hannan[x]

[*]Department of Psychiatry and [x]Medical Physics,
Edinburgh University, Morningside Park, Edinburgh, Scotland,
EH10 5HF and [+]Department of Physiology University of
Witwatersrand

INTRODUCTION

An early statement concerning sleep function suggested that sleep
was a time "related to the Sisyphean labor of anabolically building up
what each day's catabolic activity has torn down" (Johnson and Lubin
1966). This restorative theory of sleep is widely presumed to be
correct and the scientific evidence in support has been marshalled by
Oswald (1980). One aspect of this evidence is the relation of metabolic
activity to sleep composition (Shapiro et al., 1984). A range of
hypermetabolic states (e.g. thyrotoxicosis and extreme exercise)
increase sleep duration and depth (both quantitatively and
qualitatively) and conversely situations of lowered metabolic activity
(e.g., quadraplegic and hypothyroid patients) have less slow wave sleep
(SWS) and shorter sleep duration.

Several studies have considered the relationship between body mass
and sleep. Adam (1977; 1987) showed a relationship between mass and
rapid eye movement (REM) sleep. Lacey et al. (1975) showed an increase
in sleep continuity, sleep length and REM sleep with weight gain in
anorexic patients. Surgical weight reduction in obesity has been shown
to decrease wakefulness and to increase both REM and SWS (Charuzi et
al., 1985). Other studies on the effect of weight loss on sleep have
yielded conflicting results (Crisp et al., 1973; Ogilvie and Broughton,
1977; Ho et al., 1978).

If the critical interlinking factor between body mass and sleep is
metabolic rate, then the body composition may explain the discrepancy
in reported studies concerning mass and sleep. Halladay et al. (1979)
have shown that resting metabolic rate is most highly correlated with
lean body mass (LBM) as measured by total body potassium ($r = 0.84$,
$p < 0.001$). Using different techniques to assess body composition, we
(Paxton et al., 1984; Shapiro et al., 1987) have obtained conflicting
results regarding the influence of body composition on sleep. In these
two studies, the subjects were fit and unfit, healthy young men, or a
group of unfit men undergoing a training programme during which body

composition did alter. In this study we describe for the first time the
impact of body composition on SWS in three groups of female subjects:
a group of anorexics gaining weight; a group of obese subjects losing
weight; and a group of athletes undergoing a training programme during
which fitness increased but no consistent change in body composition
(on average) was recorded (Meintjes et al., 1989).

METHODS

In each of the three groups studied, sleep and body composition
measures were carried out on two occasions. Sleep was recorded in a
standard way and sleep records were subsequently scored blind (by HD)
according to standardised criteria (Rechtshaffen and Kales, 1968). At
each time of sleep measurement, subjects had an adaptation night prior to
the recorded night. The technique of measurement of lean body mass was
that of total body potassium (Boddy et al., 1972) with **appropriate**
correction factors applied in using this technique in anorexic women
(Hannan et al., this volume).

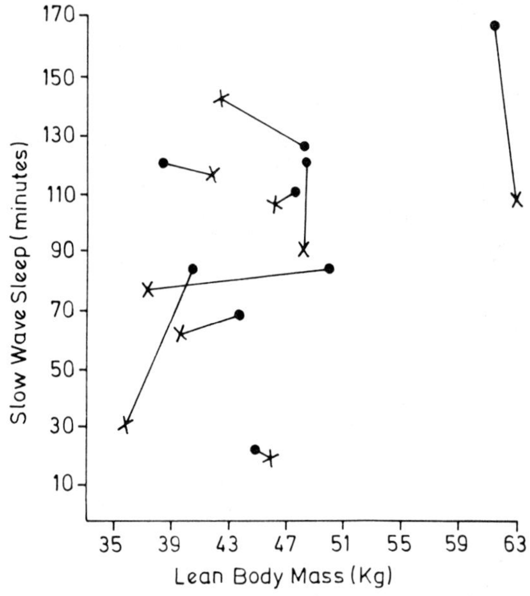

Figure 1 Slow wave sleep and LBM in a group losing weight.
● = initial measure   x = final measure

In the group of training females eight of nine starters completed the
12 week training programme and were recorded a second time. Eleven
overweight women entered the study, nine succeeded in losing 6 kg in
weight and were recorded for a second time (on average 12 weeks later;
range 7 to 16 weeks). For the anorexic patients (n = 9), the initial
recording was carried out several weeks after admission to an inpatient
unit with the subsequent recording carried out on average 17 weeks later
(range 9 to 32 weeks).

Table 1. Weight (kg) and lean body mass (kg) (mean $\pm$ standard deviation) in two conditions for three female groups

	Weight	LBM	Weight	LBM
Anorexic group (n = 9) Pre- and post-weight gain	37.8 $\pm$ 1.6	31.1 $\pm$ 1.6	47.5 $\pm$ 3.2	36.6 $\pm$ 2.0
Weight loss group (n = 10) Pre- and post-weight loss	74.5 $\pm$ 4.8	46.0 $\pm$ 2.2	68.6 $\pm$ 4.5	43.5 $\pm$ 2.6
Training group (n = 8) Pre- and post-training	59.4 $\pm$ 1.3	46.8 $\pm$ 1.1	60.4 $\pm$ 1.3	47.6 $\pm$ 1.2

RESULTS

Table 1 shows the weights and LBM measures at the two points for each of the three groups. The relationship between slow-wave sleep (SWS) and body composition in the three groups is shown in Figs. 1, 2 and 3.

DISCUSSION

The LBM does not alter parri passu with body mass. One anorexic patient increased her LBM from 26 to 29 kg while retaining a total mass of 34 kg. One of the obese subjects lost 8.9 kg while increasing LBM by 1.6 kg. These examples imply that there are alterations in metabolism in complex ways during weight change. Metabolic rate is related to body composition but the impact of the load the body is carrying and the likelihood of activity changing with weight change are factors which have not been considered in this study.

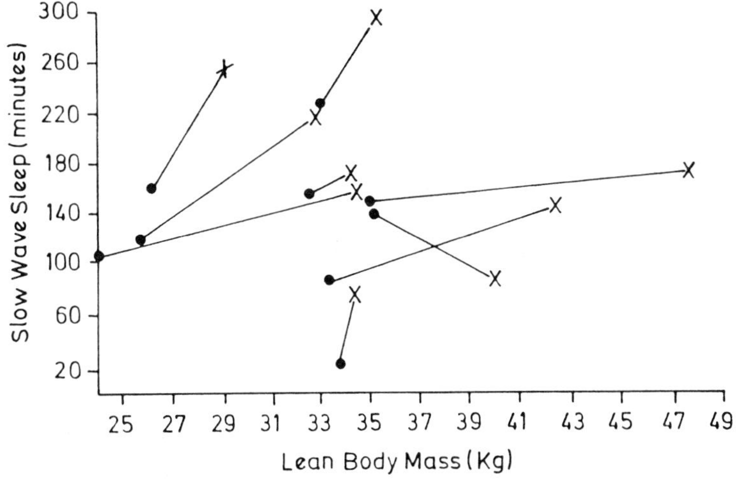

Figure 2 Slow wave sleep and LBM in a group of anorexic patients.

In the training group there is a significant positive relation
between LBM and SWS which is not evident in the weight loss group. All
the anorexic group show a positive relationship between LBM increase
and SWS other than for the only known binger and vomiter in this group.

There are many circumstances in which an increase in metabolic rate
leads to an increase in slow wave sleep. These include performance of
exercise, increased levels of fitness, hyperthyroidism and pregnancy.
Conversely, situations of lower activity and metabolism are associated
with decreased levels of slow wave sleep, for example hypothyroidism
and Sheehan's syndrome. Low levels of slow wave sleep have been
recorded in blind and paraplegic patients. The increase in slow wave
sleep in eight of the nine anorexic patients who would have increased
metabolic rate as a result of increased lean body mass and as a result
of the greater work associated with increased total body weight is

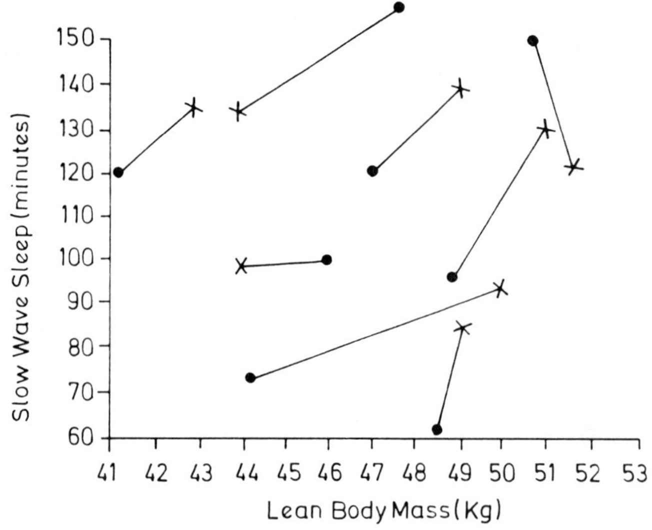

Figure 3 Slow wave sleep and LBM in a training programme group

consistent with previous observations relating metabolism and sleep.
In the group of women undergoing a training programme, lean body mass
rather than trained or untrained state appears to be the major
determinant of slow wave sleep (see Meintjes et al., 1989). Here
again metabolic rate determined by lean body mass has the major
influence. In the case of the group losing weight, there is no clear
association between lean body mass and slow wave sleep. It is likely
that other factors, including activity level, may have a more profound
influence on total metabolism and hence mask any relationship between
slow wave sleep and lean body mass in this group. In this study it was
not possible to establish whether in the course of weight loss the
women became more active and as a result increased metabolic rate.
There is some suggestion of this in that some of the weight loss group
increased their lean body mass despite an overall fall in weight.

It would appear from these three studies that LBM is a major
determinant of SWS but other facets of metabolism and metabolic rate

influence SWS. Therefore no simple relationship between SWS and body
composition can be deduced.

ACKNOWLEDGEMENTS

We thank M. Dodd for typing the manuscript. Funding was made
available by the Edinburgh Sleep Research Trust (REST).

REFERENCES

Adam, K., 1977, Body weight correlates with REM sleep, Br. Med. J.,
    1: 813.
Adam, K., 1987, Total and percentage REM sleep correlate with body weight
    in 36 middle aged people, Sleep, 19: 69.
Boddy, K., King, P.C., Hume, R., and Weyers, E., 1972, The relation of
    total body potassium to height, weight, and age in normal adults,
    J. Clin. Path., 25: 512.
Charuzi, I., Ovnat, A., Peiser, J., Saltz, H., Weitzman, S., and Lavie, P.,
    1985, The effect of surgical weight reduction on sleep quality in
    obesity-related sleep apnea syndrome, Surgery, 97: 535.
Crisp, A.H., Stonehill, E., Fenton, G.W., and Fenwick, P.B.C., 1973, Sleep
    patterns in obese patients during weight reduction, Psychother.
    Psychosom., 22: 159.
Halliday, D., Hesp, R., Stalley, S.F., Warwick, P., Altman, D.G., and
    Garrow, J.S., 1979, Resting metabolic rate, weight, surface area and
    body composition in obese women, Int. J. Obesity, 3: 1.
Hannan, J., Cowen, S., Freeman, C., Mackie, A., and Shapiro, C.M., 1989,
    Assessment of body composition in anorexic patients, International
    Symposium of in vivo body composition studies, Toronto (this volume).
Ho, A., Fekete-Mackintosh, A., Resnikoff, A.M., and Grinker, J., 1978,
    Sleep timing changes after weight loss in the severely obese,
    Sleep Res., 7: 69.
Johnson, L.C. and Lubin, A., 1966, Spontaneous electrodermal activity
    during waking and sleeping, Psychophysiol., 3: 8.
Lacey, J.H., Crisp, A.H., Kalucy, R.S., Hartmann, M.K., and Chen, C.N.,
    1975, Weight gain and sleeping EEG: a study of 10 patients with
    anorexia nervosa, Br. Med. J., 4: 556.
Meintjes, A.F, Driver, H.S. and Shapiro, C.M., 1989, Improved physical
    fitness failed to affect the sleep patterns of young women, Eur.
    J. App. Physiol., 39: 123.
Ogilvie, R.D., and Broughton, R.J., 1977, Sleep recordings from obese
    people before and after weight loss, Sleep Res., 6: 194.
Oswald, I., 1980, Sleep as a restorative process: human clues, Progr.
    Brain Res., 50: 279.
Paxton, S.J., Trinder, J., Montgomery, I., Oswald, I., Adam, K., and
    Shapiro, C., 1984, Body composition and human sleep, Aust. J.
    Psychol., 36: 181.
Rechtschaffen, A. and Kales, A. Eds., 1968, "A Manual of Standardized
    Terminology, Techniques and Scoring System for Sleep Stages of
    Human Subjects", Government Printing Office, Washington, D.C.
Shapiro, C.M., Goll, C.C., Cohen, G. and Oswald, I., 1984, Heat
    production during sleep, J. App. Physiol., 56: 671.
Shapiro, C.M., Catterall, J., Warren, P., Trinder, J., Paxton, S.,
    East, B., and Oswald, I., 1987, Lean body mass correlates with non-
    REM sleep, Br. Med. J., 294: 22.

ISSUES OF LEAD TOXICITY

Doris M. Nicholls[1] and Donald R.C. McLachlan[2]

[1]Dept. of Biology, York University, North York, Ontario
[2]Dept. of Physiology, University of Toronto, Toronto,
Ontario, Canada

INTRODUCTION

The purpose of this article is to review the current status of lead
neurotoxicity. Evidence that chronic, long-term exposure to lead may result
in altered cognitive function in both the developing and mature brain is
now compelling. Recommendations for the further investigation of the
consequences of lead toxicity upon human health are discussed.

It is widely accepted that exposure of infant animals and children to
lead (Pb) at high dosages leads to neuronal damage and behavioral
abnormalities. Children and young rats with acute Pb intoxication have
blood Pb of 100-800 ug/dl and brain Pb of 2-12 ug/g wet weight compared to
control values of 11 $\pm$ 6 ug/dl and 0.2 $\pm$ 0.14 ug/g, respectively (Chisholm,
1980; Mykkanen et al., 1979). In addition, studies suggest that low levels
of exposure to Pb in children resulting in blood Pb of 40-60 ug/dl, may
cause developmental defects, learning disabilities and hyperactivity. Davis
and Svendsgaard (1987) conclude that even blood levels as low as 10-15
ug/dl (compared to control values of <6 ug/dl) are linked with undesirable
neurobehavioral effects and lowered birth weight. The pioneering work of
Needleman et al. (1979; 1988) (humans) and Rice (1984; 1985; 1988)
(monkeys) studying cognitive development following prenatal and postnatal
exposure to Pb is widely recognized.

McMichael et al. (1988) studied pregnant mothers and their children,
up to the age of 4 years, who were living near a Pb smelter in Australia.
They concluded that postnatal blood Pb concentration was inversely related
to cognitive development in children. Gilbert and Rice (1987) reported
their elegant studies of adult monkeys dosed orally throughout 10 years to
0, 50 or 100 ug/Pb/kg/day. This treatment resulted by 7 months of age in
blood Pb levels of 3, 15 and 25 ug/dl, respectively. The blood Pb
concentration decreased over the next 5 months to a steady-state
concentration of 3, 11 and 13 ug/dl. The monkeys at 10 years of age were
tested in a series of "spatial discrimination reversal" problems in which
the higher Pb dose group exhibited greater impairment than the lower Pb
dose group, which was impaired relative to the controls. Dentine Pb levels
of deciduous teeth provide a biological averaging mechanism for Pb
exposure, largely circumventing the problems associated with measuring
highly variable blood Pb levels (Cohen, 1980; Needleman et al., 1972;
1979). The neuropsychologic performance of 58 children with high and 100
with low dentine Pb levels was compared. Children with high Pb levels (>20
ug/g dentine) had significantly lower performance on several different
tests of intelligence than those with low Pb levels (< 10 ug/g dentine).

FACTORS INVOLVED IN THE ENTRY OF Pb INTO TISSUES

Lead has been administered to experimental animals by several routes: in the diet, by intraperitoneal, subcutaneous, or intravenous injections, gastrointestinal gavage, or inhalation of suspended particles. The uptake and retention of Pb depend greatly upon the route of administration. Intravenous, subcutaneous, and intraperitoneal injections of Pb acetate in solution result in the rapid deposition of insoluble Pb compounds, mostly carbonate, from which Pb is slowly leached.

The form of Pb compound, eg. Pb citrate compared to Pb acetate, and route of entry affect the rate of appearance of Pb in blood (reviewed by Jaworski, 1978). In guinea pigs, the intratracheal route of Pb entry provided higher absorption of ^{210}Pb than the i.p. route (32% compared to 24%) while the oral route was much lower (5%). The increase in blood Pb was greater with relatively coarser dust particles (1.6 um) than fine dust (0.8 um), given a constant gravimetric Pb concentration in air of 2000 ug/m^3. Of significance was the comparison of discontinuous (occupational) exposure with continuous (environmental) exposure to airborne Pb (150 ug/m^3 Pb, 7.5 h/day, 5 days/week, compared to 10.9 ug/m^3 continuously). Regression analysis showed that blood Pb increased nearly 7 times more rapidly for the continuous exposure group (Jaworski, 1978). Skin absorption of organic alkyl Pb compounds also can be a significant entry route.

A major factor in the uptake of Pb in young rats exposed through food or water appears to be the facilitation of intestinal absorption by lactose (Bushnell and DeLuca, 1981).

Cerebral growth is impaired in infant rats exposed via the Pentschew and Garro model, (i.e. via the milk from mother rats receiving 4% Pb carbonate in their diet compared to pair-fed controls) (Michaelson, 1973; Krigman et al., 1974; Kennedy et al., 1983). Retardation of neuronal growth and maturation leads to a reduction in the amount of cerebral grey and white matter. Hypomyelination is observed and a decreased number of synapses per neuron. The synapses themselves appear to be normal, and their formation is not delayed. However, there are fewer synapses because of the limited development of neural dendritic fields. Similarly, the hypomyelination is primarily related to retarded neuronal growth and maturation (Krigman and Hogan, 1974).

Many recent experiments have used Pb^{2+} in drinking water as an approach to chronic low-dose toxicity studies and have measured blood and tissue Pb concentrations in young and adult animals (Aungst et al., 1981; Michaelson and Bradbury, 1982; Rader et al., 1983). In Michaelson and Bradbury's protocol (1982), lactating mother rats receive drinking water containing 0.1% Pb acetate (i.e. 0.05% Pb or 546 ppm Pb) acidified to prevent the precipitation of insoluble Pb salts. The pups after weaning drink this same water. The Pb-intoxicated pups display normal growth characteristics but elevated levels of blood Pb comparable to those seen in asymptomatic Pb-exposed children (<60 ug/dl). The blood Pb values that Michaelson and Bradbury reported in pups are 15 ± 2 ug/dl at 2 days of age, 35-40 ug/dl between 13 and 23 days of age, and 55 ± 9 ug/dl at 55 days of age compared to control values below the lower limits of detectability (< 5 ug/dl).

Similar exposure of mice to Pb in the drinking water (0.04% Pb acetate) for 6 months resulted in a rise of blood Pb from 5.0 ± 0.1 ug/dl in controls to 61.6 ± 4.7 ug/dl in 30 days with no further increment. Kidney, liver, and bone Pb continued to rise over 6 months, with by far the largest amount in bone, i.e. 50 ug/g wet weight compared to control levels of approximately 1 ug/g wet weight. The concentration in the brain was not significantly elevated until the fourth month of treatment, when it reached a level of 0.45 ug/g wet weight compared to a control level of 0.25 ug/g

wet weight. Blood delta-aminolevulinic acid dehydratase (ALAD) activity
decreased during the first month but did not decrease further during the
study (Torra et al., 1989).

There are species differences in Pb tolerance and the rat is known to
tolerate far higher levels of Pb than does the human (Mahaffey, 1983).
Mahaffey (1980; 1983) reviewed the influence of nutrients on Pb toxicity.
The data show that reduced overall food intake and mild to severe
deficiencies of calcium, iron, zinc, and phosphorus increase the percentage
of Pb that is absorbed from the gastrointestinal tract. Mahaffey (1983)
reports that the higher intake of Pb by children than adults, on a body
weight basis, results from the higher requirement for calories, fluid, and
air. Adult humans absorb 5 - 10% of ingested Pb while infants and children
absorb 42% and retain 32% of intake when Pb exposure exceeds 5 ug/kg/day.

Pb INTERACTIONS WITH MOLECULES AND ENZYME REACTIONS

Jaworski (1978) reviewed evidence for a number of Pb interactions.
Many biological ligands interact so strongly with Pb that they are used to
dissolve Pb minerals (e.g. glycine dissolves Pb oxide) and this may
contribute to neurotoxicity. Pb catalyzes a non-enzymatic hydrolysis of
nucleoside triphosphate, particularly ATP, so that energy deficits can
occur. From the competition of Pb^{2+} and H^+ it was concluded that Pb binds
to the carboxyl groups of serum albumin. Pb^{2+} binds to vacant iron binding
sites on transferrin but does not displace bound iron. Transferrin is
believed to play an important part in the tissue distribution of Pb. Other
proteins binding Pb are ferritin (400 atoms of Pb per molecule of
ferritin), collagen, metallothionein, possibly haemoglobin (although the
evidence is controversial), and the acidic nuclear proteins of kidney and
brain, forming intranuclear inclusion bodies.

Pb affects enzyme activity by binding either to the enzyme or to the
reactants, and Pb induced membrane damage may confound the effect _in vivo_.
The inhibition of ALAD is a sensitive and early indicator of short-term
changes in blood Pb level (at 10-20 ug/dl) but is not a good indicator of
neurotoxicity (see Torra et al., 1989). The urinary excretion of ALA is not
a reliable indicator of Pb toxicity.

In vitro studies (reviewed by Vallee and Ulmer, 1972) have shown that
Pb at a concentration of 10^{-4}M inhibits most enzymes bearing a single
functional sulphydryl group, but less readily than either mercury or
cadmium. Much lower concentrations of Pb inhibit a small select group of
enzymes including certain ATPases, lipoamide dehydrogenase, acetylcholine
esterase, and aminolevulinic acid dehydratase. ($Na^+ K^+$) ATPase inhibition
by Pb has been proposed to account for Pb toxicity and may depend on concomitant cellular influx of Cu^{2+}, a potent inhibitor of this enzyme
(Tiffany-Castiglioni et al., 1987).

Decreases were observed in the activity of brush border membrane
enzymes involved in amino acid transport, but not in the activity of (Na^+
K^+) ATPase of the basolateral border following acute or chronic oral
exposure to Pb. The decreases were not well correlated with the tissue Pb
or membrane Pb levels (approximately 10^{-7} - 10^{-9} M) and may be secondary to
selective membrane damage (Nicholls et al., 1983; Teichert-Kuliszewska
and Nicholls, 1985). Stimulation of the alkaline phosphatase activity in
the brush border may result from an interference by Pb with normal
phosphate metabolism.

Markovac and Goldstein (1988a, b) demonstrated that picomolar
concentrations of Pb are equivalent to micromolar calcium in stimulating
protein kinase C activity of isolated, immature microvessels. Nanomolar
concentrations of Pb stimulate the Ca-dependent ATPase of red cell
membranes (MacDonald et al., 1986).

BLOOD AND TISSUE Pb CONCENTRATION

Jaworski (1978) summarized studies on the distribution of Pb in soft tissues and noted that the only tissues reflecting an increasing body burden are brain, liver, and the aorta, though muscle accumulates Pb in cases of poisoning. Pb crosses the placenta and is secreted with milk of mammals. "It is not known whether it continues to accumulate in soft tissues with age or body burden" (Jaworski, 1978) but it does increase with age in bone. Pb quickly accumulates in cancellous bone with lower amounts in bone marrow. It can be mobilized from bone along with Ca^{2+}, by infections, fever, and other debilitating diseases: thus, disease may create late neurotoxicity problems (Cohen, 1980).

Jaworski (1978) reviewed the data on Pb binding and distribution. In one study, the body burden in soft tissues in children was 28% compared to 5% in adults, probably due to high bone turnover in children. The lower proportion of red cells in blood of children, compared to adults, yields lower blood Pb levels (since 90% of blood Pb is bound to red cells). Pentschew and Garro (1966) showed that doses of Pb which caused brain damage in suckling mice and rats were tolerated by the nursing dams. Further, the retention of radioactive tracer Pb in the brain of infants remained unchanged over time while it decreased slowly in adults.

In humans, a blood Pb level of 27 ug/dl whole blood is equivalent to 78,000 atoms of Pb bound to the red blood cell, possibly the phospholipids and lipoproteins of the membrane. At higher blood Pb concentrations the Pb is bound to plasma constituents. _In vitro_ it was found that 90% of ^{203}Pb was removed from plasma to red cells in 15 min, while _in vivo_, in human volunteers, it took 100 h for ^{203}Pb bound to red cells to reach equilibrium with body tissues and fluids (Chamberlain et al., 1975). The half-life of Pb in humans and animals is approximately 2 weeks.

Several experiments confirmed that blood Pb is not always an indicator of neurotoxicity. For example, in the study of Lefauconnier et al. (1983) Pb encephalophy was produced in infant rats by the i.p. administration of 60 mg Pb acetate/kg (i.e. approximately 30 mg Pb/kg) body weight daily from day 5 following birth. The cerebellum had increased Pb after 2 injections before any pathology was detected (i.e. 1.09 $\pm$ 0.15 ug/g wet weight compared to controls of < 0.05 ug/g, the detection limit). After 4-6 days hemorrhagic Pb encephalopathy developed. From 11-14 days 25% of the animals improved, while 75% progressed to abnormal signs and death. Surprisingly, high blood Pb levels were sometimes found in animals that improved.

No effects of age (adult vs. weanling) on levels of Pb in blood or brain were observed when dosing (via the drinking water or weekly p.o.) was equivalent on the basis of body weight (Rader et al., 1983). However, weanling rats retained significantly higher fractions of ingested Pb in brain and femur than did adults under both schedules of Pb administration. The brain Pb concentration in young rats that exhibit signs of Pb neurotoxicity is 4.07 $\pm$ 0.81 ug/g wet weight compared to control values of 0.20 $\pm$ 0.14 (Mykkanen et al., 1979) and, in other studies (Michaelson, 1973), up to 12 ug/g wet weight.

THE BLOOD-BRAIN BARRIER

Vascular pathology in the brain following exposure to high levels of Pb in humans and animals is well documented, but the effects of low levels of Pb on blood-brain permeability remains controversial. Animal studies in rats, which withstand Pb toxicity better than do humans, clearly demonstrate changes after high Pb doses in the endothelial cells of infant rat brain 24 h after the first administration of 600 mg/kg Pb acetate by gastric gavage (Press, 1985). The effect of chronic low level Pb exposure in infant rats exposed via milk and resulting in 20 - 80 ug/dl blood was described by Moorehouse et al. (1988). Although regional cerebrovascular permeability to nutrient tracers representing seven blood-brain barrier transport classes was not impaired by Pb, nevertheless, the permeability for lysine, histidine, and thiamine was greater in some regions, especially the frontal cortex.

Bradbury and Deane (1986; 1988) reviewed evidence for the influence of Pb on transport of neurotransmitter precursors or of Ca^{2+} and Mg^{2+} *in vivo* and concluded there was very little effect. However, an unexpected high permeability of the blood-brain barrier to Pb itself (> 1,000 times greater than to Ca^{2+}), particularly at low plasma (Pb^{2+}) levels, was found. They suggested that the high affinity of Pb for certain membrane glycosoamino-glycans may be a mechanism by which low concentrations of Pb could influence neuron development.

LEAD AND NERVE FUNCTION

Fjerdingstad et al. (1974) found that the hippocampal region of normal (unexposed) rat brain concentrated Pb to a level 10 times higher than the rest of the brain. Thus, it contained 50% of the total brain Pb (1.65 ug/g compared to 0.27 ug/g wet weight). Since this region is also a site of Al accumulation, Pb may have additive or synergistic effects on Al neurotoxicity. An association of Pb and Al neurological effects is suggested by Marlowe et al. (1985). The neurofibril changes reported in human and rabbit brain after inorganic and organic Pb poisoning (Niklowitz, 1974; 1975) may be related to changes in Alzheimer's disease and in Al neurotoxicity. The organic metabolites, triethyl Pb and trimethyl Pb are further metabolized to inorganic Pb (Jaworski, 1978).

Pb-induced pathology in the hippocampus, including reduced width of cell layers and reduced synaptic profiles, has been attributed to the late maturation of the hippocampus. Burdette and Goldstein (1986), however, observed a depressive effect of Pb on 6 - 7 Hz frequencies independent of the developmental stage.

In early studies of Pb, where high exposure was used, catecholaminergic function was found inhibited. However, results of lower exposure to Pb demonstrate that there is a stimulation of this function (reviewed by Winder, 1982). Receptor changes appear to be involved, since foetal or neonatal Pb exposure approximately doubled the apparent density of forebrain α_1 adrenoreceptors, β-cortical adrenoreceptors, striatal D-2 dopamine receptors, and hippocampal indoleamine receptors, while decreasing the striatal muscarinic receptors (Rossouw et al., 1987). In these infant rats blood Pb rose to 62 ug/dl while brain Pb rose to 0.84 ± 0.06 ug/g (compared to controls of 0.21 ± 0.03 ug/dl and 0.24 ± 0.08 ug/g, respectively).

Although whole brain catecholamines are generally increased in Pb-exposed animals (Shih and Hanin, 1978), new data on discrete areas of brain shows quite different Pb effects. Using acute and chronic low Pb exposures, Meredith et al. (1988) found reductions in catecholamine neurotransmitters and in the activity of the rate-limiting enzyme, tyrosine hydroxylase. These changes were localized in the median eminence, periventricular nucleus, and anterior hypothalamus but were not seen in the paraventricular nucleus, posterior hypothalamus, caudate putamen, or globus pallidus.

By transmission electron microscopy, Holtzman et al. (1987) observed that neurons are more Pb sensitive than astrocytes studied *in vivo* or *in vitro* in primary cultures. Pb was located only in lysosomes of neurons, but it was in the nuclear and cytoplasmic inclusions of astrocytes. By atomic absorption analysis these cells took up Pb from the medium and concentrated it at least 55 times that of the extracellular concentration, supporting the hypothesis that these cells act as Pb "sinks" in brain (Tiffany-Castiglioni et al., 1986).

In rats, a Pb resistant species, acute exposure to Pb results in its binding to certain cytosolic proteins and nuclear proteins in the kidney, brain and red cells but not liver (Choie and Richter, 1980; Shelton and Egle, 1982; Oskarsson et al., 1982; Raghavany et al., 1981). According to Goering et al. (1986), brain Pb binding protein is 12,000 d compared to kidney Pb-binding protein which is 9,000 d but others report that brain protein is 23,000 d (Duval and Fowler, 1989). More work is required to establish whether the Zn binding protein of brain (thought to be a form of metallothionein), is the same as the Pb-binding protein. Goering et al. (1986) inferred from their results that there were no major regional differences in the distribution of the brain Pb-binding protein.

GENE EXPRESSION

Inorganic Pb is a mitogen in kidney and liver of rats and mice but its mechanism of action is not clear. Recent studies of sperm chromatin of Pb-exposed mice reported an increase in the stabilization of chromatin such that there was an interference in normal chromatin decondensation and fertility (Johansson and Pellicciari, 1988). A single i.v. injection of Pb nitrate induces DNA synthesis and hyperplasia in liver and kidney (Ledda-Columbano et al., 1987). One injection i.p. of 5 mg Pb^{2+}/kg body weight (i.e. 26 umole Pb^{2+}/kg body weight) caused an increase in mRNA translation of several proteins in the liver of rats, especially serum acid proteins thought to be acute phase reactants (Nicholls et al., 1984). While no change in metallothionein mRNA was found in the kidney of these rats, there was an increase in the synthesis and activity of kidney plasminogen activator (Kuliszewski and Nicholls, 1983; Nicholls and Kuliszewski, 1984).

Pb^{2+} catalyzes tRNA depolymerization by cleaving phosphodiester bonds (Brown et al., 1983), and also decreases amino acid acceptance and ribosomal binding at 10^{-4} - 10^{-3} M. Pb^{2+} depolymerizes mRNA at 10^{-6} M and can inhibit mRNA release from the nucleus at 10^{-5} - 10^{-4} M (Farkas, 1968). The above effects, combined with aminoacyl synthetase enzyme effects, contribute to the reported overall inhibition of brain protein synthesis. In the cell-free protein synthesizing systems of infant rat brains, we observed a decreased protein synthesis, following Pb^{2+} exposure via milk of the dam. The results demonstrated that this could be attributed to decreases in the activity of phenylalanyl-tRNA and leucyl- tRNA synthetase enzymes, as well as in mRNA activity (Kennedy et al., 1983).

CONCLUSIONS AND RECOMMENDATIONS

1. Research into bone and brain Pb levels has been neglected, yet such data could give insight into Pb effects on neurodegenerative diseases.

2. The targets of Pb toxicity in brain, kidney, and bone may be multiple, e.g. membrane barriers, polynucleotides and other phosphate containing molecules, and additional acidic molecules.

3. Brain and bone Pb levels should be determined in patients dying from various types of encephalopathy and compared to other disease and accident victims. Correlations with Al levels in various brain regions should be measured.

4. Animal species sensitive to Pb should be sought and studied in comparison with Pb-resistant species. These should be animals with a short life span in comparison with monkeys, where the cost and time for aging studies are very high. A combination of Al and Pb neurotoxicity studies in aging humans and animals are needed.

5. Long-term prospective studies should be initiated on high- and low-Pb
    exposure groups of children, with the goal of assessing neurological
    performance at each decade.

6. In an aging population it may be feasible to obtain tooth dentine Pb
    levels and set up a data base of Age/Pb levels/Neurological Function
    Tests.

ACKNOWLEDGMENTS

    Supported by Natural Sciences and Engineering Research Council of
Canada and Medical Research Council of Canada.

REFERENCES

Aungst, B.J., Dolce, J.A., and Fung, H.L., 1981, The effect of dose in the
    disposition of lead in rats after intravenous and oral
    administration, Toxic. Appl. Pharm., 61:48.
Bradbury, M.W.B., and Deane, R., 1986, Rate of uptake of lead - 203 into
    brain and other soft tissues of the rat at constant radiotracer
    levels in plasma, Ann. N.Y. Acad. Sci., 481:12.
Bradbury, M.W.B., and Deane, R., 1988, Brain endothelium and interstitium
    as sites for effects of lead, Ann. N.Y. Acad. Sci., 529:1.
Brown, R.S., Hingerty, B.E., Dewan, J.C., and Klug, A., 1983, Pb (II)-
    catalysed cleavage of the sugar-phosphate backbone of yeast
    tRNA^{Phe}-implications for lead toxicity and self-splicing RNA,
    Nature, 303:543.
Burdette, L.J., and Goldstein, R., 1986, Long-term behavioral and electro-
    physiological changes associated with lead exposure at different
    stages of brain development in the rat, Devel. Brain Res., 29:101.
Bushnell, P.J., and DeLuca, H.F., 1981, Lactose facilitates the intestinal
    absorption of lead in weanling rats, Science, 211:61.
Chamberlain, A.C., Clough, W.S., Heard, M.J., Newton, D., Scott, A.N.B.,
    and Walls, A.C., 1975, Uptake of inhaled lead from motor exhaust,
    Postgrad. Med. J., 51:790.
Chisholm, J.J., 1980, Lead and other metals: a hypothesis of interaction.
    in: "Lead Toxicity", R.L. Singhal and J.A. Thomas, eds., Urban and
    Schwarzenberg, Baltimore.
Choie, D.D., and Richter, G.W., 1980, Effects of lead toxicity on the
    kidney, in: "Lead Toxicity", R.L. Singhal and J.A. Thomas, eds.,
    Urban and Schwartzenberg, Baltimore.
Cohen, M.M., 1980, Neurotoxic effects of heavy metals and metalloids. in:
    "Biochemistry of Brain", S. Kumar, ed., Pergamon, New York.
Davis, J.M., and Svendsgaard, D.J., 1987, Lead and child development,
    Nature, 329:297.
Duval, G., and Fowler, B.A., 1989, Preliminary purification and
    characterization studies of a low molecular weight, high affinity
    cytosolic lead-binding protein in rat brain, Biochem. Biophys. Res.
    Commun., 159:177.
Farkas, W.R., 1968, Depolymerization of ribonucleic acid by plumbous iron,
    Biochim. Biophys. Acta, 155:401.
Fjerdingstad, E.J., Danscher, G., and Fjerdingstad, E., 1974, Hippocampus:
    selective concentration of lead in the normal rat brain, Brain Res.,
    80:350.
Gilbert, S.G., and Rice, D.C., 1987, Low-level lifetime lead exposure
    produces behavioral toxicity (spatial discrimination reversal) in
    adult monkeys, Toxic. Appl. Pharm., 91:484.
Goering, P.L., Mistry, P., and Fowler, B.A., 1986, A low molecular weight
    lead-binding protein in brain attenuates lead inhibition of
    -aminolevulinic acid dehydratase: comparison with a renal lead-
    binding protein, J. Pharm. Exp. Therap., 237:220.

Holtzman, D., Olson, J.E., DeVries, C., and Bensch, K., 1987, Lead
    toxicity in primary cultured cerebral astrocytes and cerebellar
    granular neurons, <u>Toxic. Appl. Pharm.</u>, 89:211.
Jaworski, J.J., 1978, Effects of lead in the environment. Quantitative
    aspects, National Research Council, Canada. NRCC No. 16736. ISSN
    0316-0114.
Johansson, L., and Pellicciari, C.E., 1988, Lead-induced changes in the
    stabilization of the mouse sperm chromatin, <u>Toxicol.</u>, 51:11.
Kennedy, J.L., Girgis, G.R., Rakhra, G.S., and Nicholls, D.M., 1983,
    Protein synthesis in rat brain following neonatal exposure to lead,
    <u>J. Neurol. Sci</u>., 59:57.
Krigman, M.R., and Hogan, E.L., 1974, Effect of lead intoxication on the
    postnatal growth of the rat nervous system, <u>Envir. Health Perspec.</u>,
    7:187.
Krigman, M.R., Druse, M.J., Traylor, T.D., Wilson, M.H., Newell, L.R., and
    Hogan, E.L., 1974, Lead encephalopathy in the developing rat:
    Effect on cortical ontogenesis, <u>J. Neuropath. Exp. Neurol.</u>, 33:671.
Kuliszewski, M.J., and Nicholls, D.M., 1983, Translation of mRNA from rat
    kidney following acute exposure to lead, <u>Int. J. Biochem.</u>, 15:657.
Ledda-Columbano, G.M., Columbano, A., Dessi, S., Coni, P., Chiodino, C.,
    Faa, G., and Pani, P., 1987, Hexose monophosphate shunt and
    cholesterogenesis in lead-induced kidney hyperplasia, <u>Chem. Biol</u>.
    <u>Interact.</u>, 62:209.
Lefauconnier, J.M., Hauw, J.J., and Bernard, G., 1983, Regressive or
    lethal lead encephalopathy in the suckling rat, <u>J. Neuropath. Exp.</u>
    <u>Neurol.</u>, 42:177.
MacDonald, E., Hellevuo, K., and Komulainen, H., 1986, Lead does not
    affect calmodulin-induced activation of calcium-dependent adenosine
    triphosphatase in human red blood cell membranes, <u>Arch. Toxic</u>.
    <u>Suppl.</u>, 9:397.
Mahaffey, K.R., 1980, Nutrient-lead interactions, <u>in:</u> "Lead Toxicity",
    R.L. Singhal and J.A. Thomas, eds, Urban and Schwarzenberg,
    Baltimore.
Mahaffey, K.R., 1983, Biotoxicity of lead: influence of various factors,
    <u>Fed. Proc.</u>, 42:1730.
Markovac, J., and Goldstein, G.W., 1988a, Picomolar concentrations of lead
    stimulate brain protein kinase C, <u>Nature</u>, 334:71.
Markovac, J., and Goldstein, G.W., 1988b, Lead activates protein kinase C
    in immature brain microvessels, <u>Toxic. Appl. Pharm.</u>, 96:14.
Marlowe, M., Stellern, J., Errera, J., and Moon, C., 1985, Main and
    interaction effects of metal pollutants on visual-motor performance,
    <u>Arch. Env. Health</u>, 40:221.
McIntosh, M.J., Meredith, P.A., Petty, M.A., and Reid, J.L., 1988,
    Influence of lead exposure on catecholamine metabolism in discrete
    rat brain nuclei, <u>Comp. Biochem. Physiol</u>., 89C:211.
McMichael, A.J., Baghurst, P.A., Wigg, N.R., Vimpani, O-.V., Robertson,
    E.F., and Roberts, R.J., 1988, Port Pirie cohort study: Environ-
    mental exposure to lead and children's abilities at the age of four
    years, <u>New Engl. J. Med.</u>, 319:468.
Meredith, P.A., McIntosh, M.J., Petty, M.A., and Reid, J.L., 1988, Effects
    of lead exposure on rat brain catecholaminergic neurochemistry.
    <u>Comp. Biochem. Physiol</u>., 89C:215.
Michaelson, I.A., 1973, Effects of inorganic lead on RNA, DNA and protein
    content in the developing neonatal rat brain, <u>Toxic. Appl. Pharm.</u>,
    26:539.
Michaelson, I.A., 1980, An Appraisal of rodent studies on the behavioral
    toxicity of lead, <u>in:</u> "Lead Toxicity", R.L. Singhal and J.A. Thomas,
    eds., Urban and Schwarzenberg, Baltimore.
Michaelson, I.A., and Bradbury, M., 1982, Effect of early inorganic lead
    exposure on rat blood-brain barrier permeability to tyrosine or
    choline, <u>Biochem. Pharm.</u>, 31:1881.

Moorehouse, S.R., Carden, S., Drewitt, P.N., Eley, B.P., Hargreaves, R.J., and Pelling, D., 1988, The effect of chronic low level lead exposure on blood-brain barrier function in the developing rat, Biochem. Pharmac., 37:4539.

Mykkanen, H.M., Dickerson, J.W.T., and Lancaster, M.C., 1979, Effect of age on the tissue distribution of lead in the rat, Toxic. Appl. Pharm., 51:447.

Needleman, H.L., Tuncay, O.C., and Shapiro, L.M., 1972, Lead levels in deciduous teeth of urban and suburban American children, Nature, 235:111.

Needleman, H.L., and Bellinger, D., 1988, Recent developments, Envir. Res., 46:190.

Needleman, H.L., Gunnoe, C., Leviton, A., Reed, R., Peresic, H., Mahler, C., and Barret, P., 1979, Deficits in psychological and classroom performance of children with elevated dentine lead levels, New Engl. J. Med., 300:689.

Nicholls, D.M., and Kuliszewski, M.J., 1984, Kidney urokinase activity following acute exposure to lead, Biochem. Pharmac., 33:181.

Nicholls, D.M., Teichert-Kuliszewska, K., and Kuliszewski, M.J., 1983, The activity of membrane enzymes in homogenate fractions of rat kidney after administration of lead, Toxic. Appl. Pharm., 67:193.

Nicholls, D.M., Wassenaar, M.L., Girgis, G.R., and Kuliszewski, M.J., 1984, Does lead exposure influence liver protein synthesis in rats? Comp. Biochem. Physiol., 78C:403.

Niklowitz, W.J., 1974, Ultrastructural effects of acute tetraethyl lead poisoning on nerve cells of the rabbit brain, Envir. Res., 8:17.

Niklowitz, W.J., 1975, Neurofibrillary changes after acute experimental lead poisoning, Neurology, 25:927.

Oskarsson, A., Squibb, K.S., and Fowler, B.A., 1982, Intracellular binding of lead in the kidney: The partial isolation and characterization of postmitochondrial binding components, Biochem. Biophys. Res. Commun., 104:290.

Pentschew, A., and Garro, F., 1966, Lead encephalopathy of the suckling rat and its implications on the porphyrinopathic nervous diseases, Acta. Neuropath. (Berl.) 6:266.

Press, M.F., 1985, Lead-induced permeability changes in immature vessels of the developing cerebellar microcirculation, Acta. Neuropath. (Berl)., 67:86.

Rader, J.I., Celesk, E.M., Peeler, J.T., and Mahaffey, K.R., 1983, Retention of lead acetate in weanling and adult rats, Toxic. Appl. Pharm., 67:100.

Raghavan, S.R.V., Culver, B.D., and Gonick, H.C., 1981, Erythrocyte lead-binding protein after occupational exposure, J. Toxic. Env. Health, 7:561.

Rice, D.C., 1984, Behavioral deficit (delayed matching to sample) in monkeys exposed from birth to low levels of lead, Toxic. Appl. Pharm., 75:337.

Rice, D.C., 1985, Chronic low-lead exposure from birth produces deficits in discrimination reversal in monkeys, Toxic. Appl. Pharm., 77:201.

Rice, D.C., 1988, Schedule-controlled behavior in infant and juvenile monkeys exposed to lead from birth, Neurotox., 9:75.

Rossouw, J., Offermeier, J., and van Rooyen, J.M., 1987, Apparent central neurotransmitter receptor changes induced by low-level lead exposure during different developmental phases in the rat, Toxic. Appl. Pharm., 91:132.

Shelton, K.R., and Egle, P.M., 1982, The proteins of lead-induced intra-nuclear inclusion bodies, J. Biol. Chem., 257:11802.

Shih, T.M., and Hanin, I., 1978, Chronic lead exposure in immature animals: neurochemical correlates, Life Sciences, 23:877.

Teichert-Kuliszewska, K., and Nicholls, D.M., 1985, Rat kidney brush border enzyme activity following subchronic oral lead exposure, Toxic. Appl. Pharm., 77:211.

Tiffany-Castiglioni, E., Zmudzki, J., and Bratton, G.R., 1986, Cellular targets of lead neurotoxicity: in vitro models, Toxicol., 42:303.

Tiffany-Castiglioni, E., Zmudzki, J., Wu, J.N., and Bratton, G.R., 1987, Effects of lead treatment on intracellular iron and copper concentrations in cultured astroglia, <u>Metab. Brain Dis</u>., 2:61.

Torra, M., Rodamilans, M., To-Figueras, J., Hornos, I., and Corbella, J., 1989, Delta aminolevulivic acid dehydratase and ferrochelatase activities during chronic lead exposure in mice, <u>Bull. Env. Contam. Toxicol</u>., 42:476.

Vallee, B.L., and Ulmer, D.D., 1972, Biochemical effects of mercury, cadmium, and lead, <u>Ann. Rev. Biochem</u>., 41:91.

Winder, C., 1982, The interaction between lead and catecholaminergic function, <u>Biochem. Pharmac</u>., 31:3717.

MEASUREMENTS OF TRACE ELEMENTS <u>IN VIVO</u>

D.R. Chettle, R. Armstrong, A.C. Todd, D.M. Franklin,
M.C. Scott and L.J. Somervaille

Medical Physics Group, School of Physics and Space
Research, University of Birmingham, P.O. Box 363
Birmingham B15 2TT England

INTRODUCTION

When measurements of potentially toxic trace elements are to be made
<u>in vivo</u>, the usual constraints of keeping any radiation dose as low as
possible and the non-standard, extended shape of humans, are compounded by
the fact that the target element is present, by definition, only in small
quantities. Lower limits of detection, are often, therefore, vital
parameters with which to characterise measurement system peformance.

There is a considerable amount of literature which has been reviewed,
exhaustively at the time by Cohn (1980), from a technical standpoint by
Chettle and Fremlin (1984) and concentrating specifically on trace and
minor elements by Scott and Chettle (1986) and Chettle et al (1987a).
This paper will, therefore, concentrate on developments since 1986, which
have focussed on lead (Pb) and cadmium (Cd) with further work on platinum
(Pt) and a few other trace elements.

LEAD

Pb normally accumulates in bone and is therefore an archetypically
high atomic number (z) element embedded in a much lower z matrix, lending
itself to X-ray fluorescence. Three basic types of systems have been used
for <u>in vivo</u> human studies, one based on measurement of Pb L X-rays, the
other two using K X-rays, but with different γ-ray sources exciting Pb
X-rays and different source-sample-detector geometries.

L X-ray Fluorescence

The L X-ray system has evolved from the use of radioisotopic sources
to excite Pb L X-rays (Wielopolski et al., 1983) to the latest system,
which employs an X-ray tube and polarises its output (Wielopolski et al.,
1989). The performance quoted for the present instrument is of a
detection limit 3 σ(background) of 6.4 μg Pb (g wet bone)$^{-1}$ in a child's
tibia with 4 mm of overlying tissue for a local skin dose of 10 mGy and an
effective dose equivalent of 2.5 μSv. The performance depends quite
strongly on the thickness of overlying tissue, so that the detection limit
rises from 4.0 to 10.6 μg g^{-1} as tissue thickness goes from 3 to 5 mm.
The calibration procedure is based on measurements of 10 amputated legs

measured intact, and as bare bones, samples of which were then analysed by atomic absorption spectrometry. However, four were discarded from the calibration line. One was excluded because it was clearly ulcerated, but three were observed as distinct outliers. It was later deduced that two of these had been subject to positioning error whilst, for the third, an incorrect determination of overlying tissue thickness was surmised.

<u>K X-ray Fluorescence</u>

The two K X-ray systems for Pb measurement use either the original [57]Co excitation source method with 90° scattering angle (Ahlgren and Mattsson, 1979) or the subsequent [109]Cd based technique in an approximate backscatter geometry (Somervaille et al, 1985). Either system can be simply and reliably calibrated using appropriate phantoms. Recently, the two sets of apparatus were used in a collaborative survey of a group of Pb-exposed workers (Somervaille et al., 1989). Finger bone lead was measured by the [57]Co-based system, whilst tibia and calcaneus were each measured using [109]Cd. The performance of the systems was compared by calculating the root mean square standard deviation of the human results (ie. $((\Sigma\sigma_i^2)/n)^{\frac{1}{2}}$, where σ_i is the measurement error on one of n individuals. The results are summarised in Table 1. A practical detection limit was taken as twice the root mean square standard deviation, in this case the figure was quoted as μg Pb (g bone mineral)$^{-1}$. A better performance figure for tibia using [109]Cd has since been reported (Chettle et al., 1989). It appears that a major reason for the greater precision of the tibia measurements was the larger volume of bone sampled, as reflected by the higher effective dose equivalent (EDE).

<u>Choice of Bone Pb X-ray Fluorescence System</u>

There are several factors affecting the choice of a bone lead X-ray fluorescence measurement system (Hu et al., in press). From the foregoing it can be seen that precision favours either an L X-ray system or a [109]Cd based K X-ray system. The approximately equivalent precisions are obtained for very similar EDEs, although the L X-ray method has a substantially higher entrance dose. A contrast between the techniques lies in the volume of bone sampled. The L X-rays have the lower energy and can escape only from superficial bone (<0.5 mm deep), whereas the higher energy K X-rays effectively sample up to 20 mm of bone. Some data show marked variation of lead concentration in tibia on a scale of 0.01 - 0.1 mm (Jones et al., this volume), but other studies show no such variation on the 1-10 mm scale (Wittmers et al., 1988). Therefore [109]Cd K X-ray measurement can more easily be interpretted as representative of average bone Pb content. Further to this, the [109]Cd based K X-ray fluorescence system permits accurate robust quantitation by normalising the amplitudes of Pb K X-rays to that of the elastically scattered incident γ-rays (Somervaille et al., 1985); this contrasts with the difficulties noted for the calibration of the L X-ray system. On both these grounds the [109]Cd K X-ray system seems preferable.

<u>Applications of In Vivo Bone Pb Measurements</u>

A number of studies have been undertaken in which bone Pb has been measured by X-ray fluorescence as a marker of body burden or cumulative Pb dose. These have included studies of bone Pb metabolism (Schütz et al, 1987a), where vertebral biopsies were taken and their Pb content compared with finger bone Pb in active and retired workers, from which the authors deduced that there was a significantly shorter residence time for Pb in the vertebral material than in the finger. The same group have also examined the relationship between chelatable Pb and that in either vertebra or finger (Schütz et al., 1987b). They observed an association between chelatable Pb and vertebral Pb, but not between chelatable Pb and finger Pb. In both cases the mean ratio of chelatable Pb to bone Pb was higher in active than in retired workers. Taken together these observations suggest that chelatable Pb reflects current exposure at least as much, if not more, than body burden. The polarised L X-ray Pb measurements have also been compared with chelatable Pb, in this instance in the assessment of children exposed to excessive quantities of Pb (Rosen et al., 1989). It was found that L X-ray fluorescence could be a useful indicator of Pb intoxication; there was also a significant correlation between tibia Pb, measured by L X-ray fluorescence, and chelatable Pb; however, the association between chelatable Pb and blood Pb was stronger. Other work has demonstrated the use of _in vivo_ bone Pb measurement, in this case tibia Pb measured using K X-rays excited by [109]Cd, in monitoring chelation treatment in an individual suffering from chronic Pb poisoning (Batuman et al., 1989). This person showed a very significant clearance of Pb as a result of treatment assessed by _in vivo_ X-ray fluorescence, iliac crest biopsy and chelation test. Bone Pb assessed by K X-ray fluorescence has elsewhere been shown to reflect cumulative exposure or Pb dose as determined by an integrated blood Pb index (Christoffersson et al 1984; Somervaille et al., 1988). It is suggested that _in vivo_ measurement of tibia Pb can provide a simple measurement of body burden, a contention which receives support from the work of Wittmers et al., (1988) who measured Pb at 5 bone sites at autopsy and concluded that the single measurement which was likely best to reflect body burden was tibia Pb. A [109]Cd K X-ray system has also been used to measure tibia lead in a group of 74 non-occupationally exposed urban dewellers, in whom a clear positive correlation of tibia lead with age was observed (Morgan et al., in press).

CADMIUM

Measurement Techniques

      Cd is measured _in vivo_ by neutron activation, in which prompt γ-rays are monitored, or by polarised X-ray fluorescence. The techniques have been largely covered in a recent review (Scott and Chettle, 1986). Recent work on technical development has included the use of a Si(Li) rather than a Ge detector in the polarised X-ray fluorescence system (Nilsson et al., this volume), which resulted in a factor of two reduction in detection limit. A new [252]Cf neutron activation system has been designed to be used in a range of applications, including organ Cd measurements (Ryde et al., 1987).

      At Birmingham the neutron activation system has also been redesigned, both to reduce the ambient dose by improving the neutron shielding and to investigate the effect of modifying the neutron beam energy characteristics. This new system (Fig. 1) uses iron and high density polythene to thermalise fast neutrons and is encased in boron-doped resin to absorb low energy neutrons. The ambient neutron dose around the new system was a factor of 7 lower than for its predecessor of the same size, which was based on boric oxide filled iron tanks. Beryllium (Be) was chosen to modify the output spectrum of [238]Pu/Be neutron sources. Cylinders of Be of thickness 20 mm, 40mm and together, 60 mm were mounted in the beam tube in front of the sources (Fig. 1). The effects are presented in Table 2, which shows the dose, the Cd signal, from 60 mg of Cd in a kidney phantom and the resulting lower limit of detection (LLD), defined as twice the error on the net peak area. Also shown is a performance index, taken as LLD x dose ½, so that a lower value indicates a better performance.

Table 1. Performance Characteristics of K X-ray Pb Measurement Systems.

Source	Bone sampled	Detection limit		Skindose	EDE
		μg Pb (g bone mineral)$^{-1}$	μg Pb (g wet bone)$^{-1}$	mGy	μSv
^{57}Co	finger	50.0	29.1	3.0	0.1
^{109}Cd*	tibia	14.8	8.3	1.2	4.5
^{109}Cd*	calcaneus	33.2	-	1.5	2.7
^{109}Cd	tibia	10.0	5.6	0.5	2.1

*Electronics design fault led to considerable loss in counts for these measurements.

The new system was then applied to patient measurements, using 20 mm of Be, since this gave the biggest improvement in performance.

The results could then be compared with the previous system and other neutron activation kidney Cd systems, taking into account different efficiencies of detectors (Table 3). The figure of merit, F, is (skin dose x detection efficiency) ½ x LLD; again, a low value of F indicates a good performance.

Thus, in this new system, not only has the ambient dose been significantly reduced but, with the use of 60 mm of Be, the signal to dose ratio has been increased by approximately a factor of 1.4. This is comparable with the results obtained by Morgan et al. (1981), when comparing ^{252}Cf with ^{238}Pu/Be and it can therefore be seen that using Be with ^{238}Pu/Be sources has all the advantages of ^{252}Cf without the disadvantage of its short half life (2.65y) and eventual higher cost (at least in the U.K.).

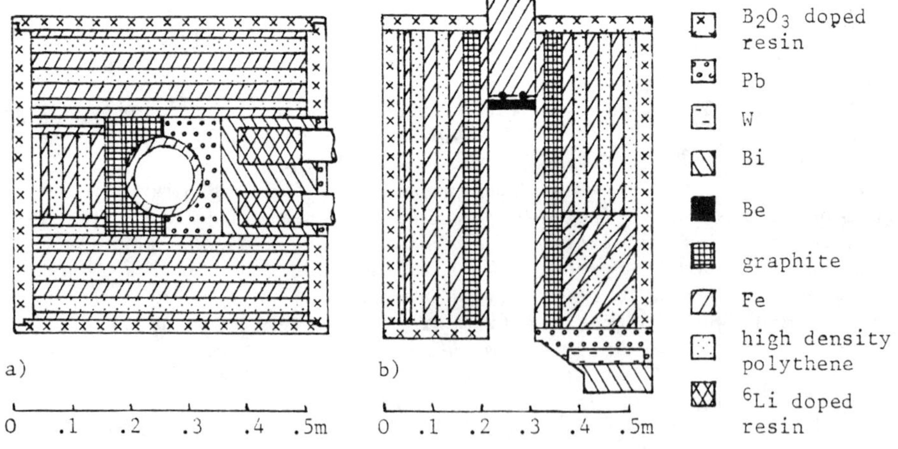

Fig. 1. The new transportable kidney cadmium measurement system
a) Front view with boric oxide doped resin slab removed.
b) Plan view in cross section through centre.

Table 2 Effect of Be on System Performance

Thickness of Be cylinder (mm)	Skin dose (mSv)	Cd signal (counts)	LLD (mg)	Performance index
0	4.2	8964	3.34	6.82
20	3.6	9324	3.25	6.18
40	2.9	7841	3.58	6.14
60	2.2	6834	3.96	5.91

Measurements of Exposed Workers

Results of a cross sectional survey of workers exposed to Cd in the production of copper-cadmium alloys have been analysed to explore in detail the known impact of Cd exposure on kidney function (Mason et al., 1988). In addition to measurements of Cd in liver and kidney a Cd exposure index was derived, based on the factory's air monitoring records. Thresholds for the onset of kidney dysfunction, as assessed by a variety of biochemical indicators, ranged from 20-55 μg Cd g^{-1} in liver. Urinary total protein, β_2-globulin, retinol binding protein and albumin all showed the onset of raised excretion at cumulative Cd in air levels of about 1100 y.μg m^{-3}. Further data from the same survey (Davison et al., 1988) showed clear evidence of a loss of lung function, consistent with emphysema, and some indication of progressive pulmonary decrement, both with increasing liver Cd and with cumulative Cd in air. In an entirely separate study, a group of workers at a Cd smelter has been re-examined after a 7 year interval (Ellis, in press). In this case the accumulation of Cd in kidney was measured as a function of previously determined Cd in air levels and the duration of exposure. In people with normal kidney function the increase was 24-26 μg y^{-1} per μg Cd m^{-3} in air. From previously estimated critical levels of Cd in kidney cortex of 200-250 μg g^{-1}, and allowing for background Cd levels, it was then estimated that the cumulative exposure required to reach the critical level was 890-1140y.μg m^{-3} with corresponding uptake into the liver of 23-35 μg Cd g^{-1}. Ellis was also able to estimate a half-life of 7 to 11 years for Cd in liver. The pattern of changes in organ Cd burdens in this survey were complex, but the overall appearance is similar to that reported by Guthrie et al., (1987) on a much smaller data base.

Measurements of Referent Groups

In vivo measurements of liver and kidney Cd in smokers and non-smokers were reported some time ago (Ellis et al., 1979), and the accumulation of Cd as a result of smoking was clearly shown. Since then two further non-occupationally exposed data sets have become available (Morgan et al., in press; Franklin et al., in press) so it is possible to make comparisons, as shown in Table 4. Morgan et al. (in press) also reported slight revisions to an earlier data set, dated 1980 in Table 4, in which liver Cd was not measured. Liver Cd burden has been calculated from measured concentrations on the assumption of a mean liver mass of 1.8 kg. Body burden has then been estimated from the report of Nordberg et al (1985) to be:

Table 3 Performance Characteristics of Kidney Cd Measurement Systems

Laboratory	Neutron source	Skin dose (mSv)	Detection efficiency (%)	LLD (mg)	F	Reference
Swansea	^{252}Cf	3.0	40	3.4	37.2	Ryde et al., 1987
Brookhaven	^{252}Cf	2.0	50	3.1	31.1	Ellis et al., 1983
Birmingham	^{238}Pu/Be	0.9	28	6.4	32.1	Chettle et al., 1987b
Birmingham (new system)	^{238}Pu/Be +20mm Be	1.0	15	5.2	20.2	

$$\frac{\text{liver (mg)} + 2 \times \text{kidney (mg)}}{.16 + .53}$$

or, in the case of Morgan's 1980 data:

$$\frac{2 \times \text{kidney (mg)}}{.53}$$

Minimal uncertainties have been estimated from the reported detection
limits for the different systems. There are differences in the absolute
Cd levels reported, which could in part arise from the different
populations sampled. More striking is the agreement in the estimates of
the impact of smoking.

OTHER ELEMENTS

There are earlier reports of measurements of a wide range of trace
elements including beryllium, silicon, iron, lithium, silver, copper and
strontium, briefly summarised in Chettle et al., (1987). There is also
work reported on aluminium (Ellis and Kelleher, 1987). However, recent
work summarised here will concern platinum, mercury (Hg) and gold (Au).

Platinum

In vivo measurements of Pt are designed to assist in the study of the
pharmacokinetics of Pt containing drugs, used in the treatment of various
cancers. As the side effects of these drugs include nephrotoxicity,
measurements are made in the kidney as well as at the tumour site. A
detailed technical description has recently been published of an
instrument which uses the filtered and polarised output of an X-ray
generator and is able to achieve a minimum detectable concentration of 8
μg g^{-1} (3 x background ½) for an organ depth of 40 mm (20 mm overlying
tissue) (Jonson et al., 1988). The skin dose was 4 mGy and the EDE 2
μSv. This system has been used to follow the uptake of Pt into the kidney
and into brain tumours (Jonson, 1988).

Table 4. Cadmium Levels in Non-occupationally Exposed Subjects - Smokers
          and Non-Smokers.

	Ellis et al., 1979 (n)	Morgan et al., 1980 (n)	Morgan et al., 1989 (n)	Franklin et al., 1989) (n)
Liver Cd (mg) Non-smokers	4.14 (8)	-	1.8 (20)	0.0 (19)
Smokers	7.38 (12)	-	2.88 (26)	1.26 (82)
Kidney Cd (mg) Non-smokers	3.1 (8)	3.0 (26)	1.9 (20)	0.6 (19)
Smokers	5.8 (12)	4.7 (19)	3.9 (27)	3.2 (82)
Smoking index (pack years)	38.7	29.9	18.9	26.3
Increase in Cd body burden from smoking (μg pack year^{-1})	324±69	215±86	390±100	368±140

Unexpected findings of very large Pb signals, assumed to emanate from
the kidney, have been reported (El-Sharkawi et al., 1986). These
observations were made during the earlier phases of a study of
cis-platinum pharmacokinetics, using X-ray fluourescence. It was
noteworthy that particularly high levels were observed in one person with
a long history of occupational exposure to Pb.

In an attempt to confirm this observation an X-ray fluorescence
system for simultaneous analysis of kidney Pb and Pt was developed at
Birmingham. Design parameters included the need to maximise the X-ray
signal, with respect to both the dose delivered and the uncertainty in the
underlying background. A Monte Carlo photon transport program was written
to aid the design by simulating the operation of different systems under
consideration. This code can readily be extended to a wide range of
analogous problems, but, in this case, it was used to model the effect of
three source/geometry combinations, namely ^{109}Cd/180°, - ^{57}C°/90°, and
^{99m}Tc/180°; where the angle is 180° minus the angle between the source,
subject and detector. In each case the program computed the ratio of
X-ray signal to dose (both to skin and to kidney) as a function of the
amount of soft tissue overlying the kidney. For Pb measurement, the ^{109}Cd
system was predicted to be best of these three for kidneys having 50 mm or
less overlying tissue; but Pt could not be measured with such a system as
the main Compton scatter peak obscures the Pt Kα X-rays. The ^{57}Co and
^{99m}Tc systems appeared similar to each other in their performance. The
^{99m}Tc system was chosen both because a backscatter geometry is simpler to
arrange than 90° and because a sufficient supply of ^{99m}Tc was available
from used isotope generators. An annular collimator was made from
tungsten alloy; this incorporated silver linings to filter out tungsten K
x-rays and contained a 2 ml annular source cavity, for the ^{99m}Tc. The
collimator/source holder was mounted co-axially with, and in front of, a
hyperpure germanium detector. This basic system was used in two ways;
initially, with a fixed source to kidney distance, with the aim of

ensuring that the entire kidney was always, in view of the detector
crystal. This approach resulted in an LLD of about 44 μg g^{-1} for either
Pb or Pt in a kidney with 20 mm of overlying tissue and a kidney dose of
1.7 mSv. The dosimetry for this system was studied and the comparison of
thermoluminescent dosimetry with the results of the Monte Carlo programme
agreed well, the mean difference between the two being 4.1%. Six patients
were studied, both before infusion with cis-platinum, and at times after
the start of infusion which varied from 9 hours to 22 days. None of these
patients showed observable accumulations of either Pb or Pt. However,
this arrangement suffered from the disadvantage that the shape of the
observed spectrum of scattered photons varied quite markedly as the source
to skin distance varied.

In the second approach, the source to skin distance was kept
constant; LLDs were improved to 20 μg Pt g^{-1} and 24 μg Pb g^{-1}, in each
case for 20 mm of tissue overlying the kidney and the same kidney dose as
previously. With the improved system each patient was followed more
closely, although fewer people were studied. Of the three subjects in
this group, two showed no significant Pb or Pt burdens. However, the
third showed no significant X-ray peaks prior to his first cis-platinum
infusion, but 17 hours post infusion the Pb level in kidney was 85±35 μg
g^{-1}. Prior to his second infusion, twenty days later, 52±34 μg Pb g^{-1} was
observed; 23 hours after this infusion started the X-ray fluorescence
measurement showed 97±37 μg Pb g^{-1} in his kidney. A further 20 days later
a third cis-platinum infusion was administered; on this occasion Pb was
not observed, either before or after infusion. On the occasion when the
highest Pb concentration was seen, Pt was also observable. These findings
cannot, on their own, resolve the question of possible mobilisation of Pb
body stores by cis-platinum, but they do suggest that any such process is
less common and of a smaller magnitude than previously suggested,
particularly in patients without a history of heavy exposure to Pb.

Mercury

Either X-ray fluorescence or neutron activation can be used to
measure Hg. X-ray fluorescence measurements using a ^{57}Co source have been
reported (Bloch and Shapiro, 1986), but the detection limit (2 x
σ(background)$^{1/2}$) (20 μg g^{-1}) remains rather high. Jonson (1988) reports
preliminary measurements using polarised X-rays with significantly lower
detection limits (6-15 μg g^{-1}), but this is still under development. The
use of ^{109}Cd to excite Hg K X-rays has also been studied at Birmingham.
Earlier studies, in which analysis was based on Hg K$_\alpha$ X-rays, yielded an
LLD (2 x net peak error) of 49 μg g^{-1} for a skin dose of 1.4 mGy and 30 mm
of tissue overlying the kidney. More recently this work has been repeated
and Hg K X-rays have also been analysed with a resulting improvement in
LLD to 31 μg g^{-1}. Neutron activation developments have featured the use
of a mobile teaching reactor. Early feasibility studies (Chang et al.,
1987) reported a detection limit for kidney Hg of 58 μg g^{-1} for a skin
dose 23.9 mSv. This system has since been improved, chiefly by
thermalising the neutron beam, to produce a detection limit of 16 μg g^{-1}
for a skin dose of 4.7 mSv (Chung et al., 1988). In both cases the
detection limit is taken as 2 σ(background).

Gold

Au salts are used in the treatment of rheumatoid arthritis, but they
can produce nephrotoxic side effects and little is known about their
distribution in the body following administration. A feasibility study
for an in vivo X-ray fluorescence measurement system has recently been
reported (Scott and Lillicrap, 1988), which examined the use of ^{133}Xe in a
backscatter geometry for this purpose.

DISCUSSION AND CONCLUSIONS

For _in vivo_ measurements of bone Pb it seems it is clear that a K
X-ray measurement of tibia Pb using ^{109}Cd is now the method of choice to
meet the commonly identified need in toxicological research for an
indicator of cumulative internal Pb dose (Landrigan, 1989). Additionally,
however, other bone sites, such as calcaneus may be used in trying to
elucidate Pb toxicokinetics. and it should be noted that the ^{109}Cd K X-ray
system has not yet been applied to the study of childhood Pb intoxication,
whereas the L X-ray method has been usefully applied to this problem.

In the case of Cd measurements, the most striking feature of recent
work is the way independent systems, studying different populations, have
arrived at remarkably similar conclusions. The estimates of both liver Cd
and Cd in air levels required to produce renal dysfunction emerging from
the work of Ellis (1989) and Mason et al., (1988) are extraordinarily
close. The agreement between these sets of measurements and the sets of
data from which the impact of smoking was examined underlines the
reliability of these _in vivo_ measurements. As with Pb, _in vivo_
measurements of organ Cd clearly provide both the best available indicator
of internal cumulative dose and a valuable tool in the study of the
toxicokinetics of these elements.

If measurements of Pb and Cd have reached a kind of maturity, then
those of Pt are clearly developing in that direction. Further progress
can be expected towards a practicable _in vivo_ Hg measurement. Work on Au
and other elements will doubtless continue, but its outcome is less
predictable.

ACKNOWLEDGEMENTS

Work on _in vivo_ measurements of toxic heavy metals has received
extensive support from the Health and Safety Executive. The Pt studies
were made possible by grants from the Colt Foundation (who supported
A.C.T.) and the Wellcome Trust (who funded L.J.S.) and the collaboration
in the clinical studies of R.P. Beaney and E.J. Buxton of the Queen
Elizabeth Hospital, Birmingham and R.A. Braithwaite of the Regional
Toxicology Laboratory, Dudley Road Hospital, Birmingham. This support and
collaboration is gratefully acknowledged.

REFERENCES

Ahlgren, L., and Mattsson, S., 1979, An X-ray fluorescence technique for
    in vivo determination of lead concentration in a bone matrix,
    Phys. Med. Biol., 24: 136.
Batuman, V., Wedeen, R.P., Bogden, J.D., Balestra, D.J., Jones, K., and
    Schidlovsky, G., 1989, Reducing bone lead content by chelation
    treatment in chronic lead poisoning: An _in vivo_ X-ray fluorescence
    and bone biopsy study, Env. Res., 48: 70.
Bloch, P., and Shapiro, I.M., 1986, An X-ray fluorescence technique to
    measure in situ the heavy metal burdens of person exposed to these
    elements in the workplace, J. Occup. Med., 28: 609.
Chang, P.S., Ho, Y.H., Chung, C., Yuan, L.J., and Weng, P.S., 1987, In
    vivo measurement of organ mercury by prompt gamma activation
    anlaysis using a mobile nuclear reactor, Nuclear Technology, 76:
    241.
Chettle, D.R., and Fremlin, J.H., 1984, Techniques of _in vivo_ neutron
    activation analysis, Phys. Med. Biol., 29: 1011.

Chettle, D.R., Scott, M.C., Ellis, K.J., and Morgan, W.D., 1987a, In vivo monitoring of trace elements in medicine and research, in: "In vivo Body Composition Studies", K.J. Ellis, S. Yasumura and W.D. Morgan, eds., IPSM, London.

Chettle, D.R., Franklin, D.M., Guthrie, C.J.G., Scott, M.C., and Somervaille, L.J., 1987b, In vivo and in vivo measurements of lead and cadmium, Biological Trace Element Research, 13: 191.

Chettle, D.R., Scott, M.C., and Somervaille, L.J., 1989, Improvements in the precision of in vivo bone lead measurements, Phys. Med. Biol., 34: 1295.

Christoffersson, J.O., Schütz, A., Ahlgren, L., Haeger-Aronsen, B., and Mattsson, S., 1984, Lead in finger-bone analysed in vivo in active and retired lead workers, Am. J. Industr. Med., 6: 447.

Chung, C., 1988, In vivo partial body activation analysis using filtered neutron beam, Appl. Rad. Isot., 39: 93.

Cohn, S.H., 1980, The present state of in vivo neutron activation analysis in clinical diagnosis and therapy, Atom. Ener. Rev., 18:599.

Davison, A.G., Fayers, P.M., Newman Taylor, A.J., Venables, K.M., Darbyshire, J., Pickering, C.A.C., Chettle, D.R., Franklin, D.M., Guthrie, C.J.G., Scott, M.C., O'Malley, D., Holden, H., Mason, H.J., Wright, A.L., and Gompertz, D., 1988, Cadmium fume inhalation and emphysema, Lancet, ii, 663.

Ellis, K.J., Vartsky, D., Zanzi, I., Cohn, S.H., and Yasumara, S., 1979, Cadmium: in vivo measurement in smokers and non-smokers, Science, 205: 323.

Ellis, K.J., Vartsky, D., and Cohn, S.H., 1983, In vivo monitoring of heavy metals in man: cadmium and mercury, Neurotoxicology, 4:164.

Ellis, K.J., and Kelleher, S.P., 1987 In vivo bone aluminium measurements in patients with renal disease, in: "In vivo Body Composition Studies", K.J. Ellis, S. Yasumura, and W.D. Morgan, eds., IPSM, London.

Ellis, K.J, In vivo measurements of a cadmium smelter population, Proceedings 6th International Cadmium Conference (in press).

El-Sharkawi, A.M., Morgan, W.D., Cobbold, S., Jaib, M.B.M., Evans, C.J., Somervaille, L.J., Chettle, D.R., and Scott, M.C., 1986, Unexpected mobilisation of lead during cisplatin chemotherapy, Lancet, ii: 249.

Franklin, D.M., Guthrie, C.J.G., Chettle, D.R., Scott, M.C., Mason, H.J., and Newman Taylor, A.J., In vivo neutron activation analysis of organ cadmium burdens - referent levels in liver and kidney and the impact of smoking, Nuclear Analytical Methods in the Life Sciences (in press).

Guthrie, C.J.G., Franklin, D.M., Scott, M.C., Chettle, D.R., Mason, H.J., Smith, N.J., Wright, A.L., and Blindt, M., 1987, A longitudinal survey of exposure to cadmium fume preliminary findings from in vivo body burden measurements in: "In vivo Body Composition Studies", K.J. Ellis, S. Yasumura, and W.D. Morgan, eds., IPSM London.

Hu, H., Milder, F.L., and Burger, D.E., X-ray fluorescence: issues surrounding the application of a new tool for measuring burden of lead, Env. Res., (in press).

Jones, K.W., Schidlovsky, G., Burger, D.E., and Milder, F.L., this volume.

Jonson, R., Mattsson, S., and Unsgaard, B., 1988, A method for in vivo analysis of platinum after chemotherapy with cisplatin, Phys. Med. Biol., 33: 847.

Jonson R., 1988, "Radioanalytical methods for the determination of bone mineral and heavy metals in vivo", Thesis, University of Göteborg, Sweden.

Landrigan, P.J., 1989, The toxicity of lead at low dose, editorial. Br. J. Indust. Med., 46: 593

Mason, H.J., Davison, A.G., Wright, A.L., Guthrie, C.J.G., Fayers, P.M., Venables, K.M., Smith, N.J., Chettle, D.R., Franklin, D.M., Scott, M.C., Holden, H., Gompertz, D., and Newman Taylor, A.J., 1988, Relations between liver cadmium cumulative exposure, and renal function in cadmium alloy workers, <u>Br. J. Indust. Med.</u>, 45: 793.

Morgan, W.D., Vartsky, D., Ellis, K.J., and Cohn, S.H., 1981, A Comparison of ^{252}Cf and ^{238}Pu, Be neutron sources for partial body in vivo activation analysis, <u>Phys. Med. Biol.</u>, 26:413.

Morgan, W.D., Ryde, S.J.S., Jones, S.J., Wyatt, R.M., Hainsworth, I.R., Cobbold, S.S., Evans, C.J., and Braithwaite, R.A., In vivo measurements of cadmium and lead in occupationally exposed workers and in an urban population, Nuclear Analytical Methods in the Life Sciences. (in press).

Nilsson, U., Ahlgren, L., Christoffersson, J.O., and Mattsson S., this volume.

Nordberg, G.F., Kjellstro m, T, and Nordberg, M., 1985, Kinetics and metabolism, <u>in</u>: "Cadmium and Health", L. Friberg, C.G. Elinder, T. Kjellström, and G.F. Nordberg, eds., CRC Press, Boca Raton, Florida, vol I.

Rosen, J.F., Markowitz, M.E., Bijur, P.E., Jenks, S.T., Wielopolski, L., Kalef-Ezra, J.A., and Slatkin, D.N., 1989, L-line X-ray fluorescence of cortical bone lead compared with the CaNa EDTA test in lead toxic children: Public Health implications, <u>Proc. Nat. Acad. Sci. USA</u>, 86: 685.

Ryde, S.J.S., Morgan, W.D., Sivyer, A., Evans, C.J., and Dutton, J., 1987, A clinical instrument for multi-element in vivo analysis by prompt, delayed and cyclic neutron activation using ^{252}Cf, <u>Phys. Med. Biol.</u>, 32: 1257.

Schütz, A., Skerving, S., Christoffersson, J.O., Ahlgren, L., and Mattsson, S., 1987a, Lead in vertebral biopsies form active and retired workers, <u>Arch. Environ. Health.</u>, 42: 340.

Schütz, A., Skerfving, S., Christoffersson, J-O., and Tell, I., 1987b, Chelatable lead vs lead in human trabecular and compact bone, <u>Sci. Tot. Environ.</u>, 61: 201.

Scott, J., and Lillicrap, S., 1988, ^{133}Xe for the X-ray fluorescence assessment of gold in vivo, <u>Phys. Med. Biol.</u>, 33: 859.

Scott, M.C., and Chettle, D.R., 1986, In vivo elemental analysis in occupational medicine, <u>Scand. J. Work. Environ. Health</u>, 12: 81.

Somervaille, L.J., Chettle, D.R., and Scott, M.C., 1985, In vivo measurements of lead in bone using X-ray fluorescence, <u>Phys. Med. Biol.</u>, 30: 929.

Somervaille, L.J., Chettle, D.R., Scott, M.C., Tennant, D.R., McKiernan, M.J., Skilbeck, A., and Trethowan, W.N., 1988, In vivo tibia lead measurements as an index of cumulative exposure in occupationally exposed subjects, <u>Br. J. Indust. Med.</u>, 45: 174.

Somervaille, L.J., Nilsson, U., Chettle, D.R., Tell, I., Scott, M.C., Schütz, A., Mattsson, S., and Skerfving, S., 1989, <u>In vivo</u> measurements of bone lead - a comparison of two X-ray fluorescence techniques used at three different bone sites, <u>Phys. Med. Biol.</u>, 34: 1833.

Wielopolski, L., Rosen, J.F., Slatkin, D.N., Vartsky, D., Ellis, K.J., and Cohn, S.H., 1983, Feasibility of noninvasive analysis of lead in the human tibia by soft X-ray fluorescence, <u>Med. Phys.</u>, 10: 248.

Wielopolski, L., Rosen, J.F., Slatkin, D.N., Zhang, R., Kalef-Ezra, J.A., Rothman J.C., Maryanski, M., and Jenks, S.T., 1989, In vivo measurement of cortical bone lead using polarized X-rays, <u>Med. Phys.</u>, 16: 521.

Wittmers, L.E., Aufderheide, A.C., Wallgren, J., Rapp, G., and Alich, A., 1988, Lead in bone IV. Distribution of lead in the human skeleton, <u>Arch. Environ. Health.</u>, 43: 381.

THE MEASUREMENT OF BONE LEAD CONTENT IN PATIENTS WITH END STAGE
RENAL FAILURE

Sarah J. Jones, A.J. Williams*, Hana Kudlac*,
I.R. Hainsworth+ and W.D. Morgan

Swansea In Vivo Analysis Research Group
Departments of Medical Physics and +Chemical
Pathology, Singleton Hospital, and *Department of
Renal Medicine, Morriston Hospital, Swansea SA6 6NL

INTRODUCTION

Acute renal failure due to lead exposure was a common
problem in 19th century industrial society, but is now rarely
seen. The role of lead in the development of chronic renal
failure is, however, less certain. Although epidemiological
studies of populations exposed to lead have suggested an increased incidence of renal insufficiency in these groups, only
about 1% of all cases of end-stage renal disease requiring
renal replacement therapy are reported to be caused by toxic
nephropathy. This may merely be due to underdiagnosis. For
example, in a recent study from Belgium (Van de Vyver et al.,
1988), 5% of the dialysis population were found to have bone
lead concentrations that approximated levels found in active
lead workers.

Various techniques have been used to assess cumulative
lead exposure. Measurement of the urinary excretion of lead
after EDTA chelation has for many years been a favoured method,
but is not of value in dialysis patients. Iliac crest bone
biopsy allows accurate measurement of bone lead, and it has
been shown that there are no significant differences between
the lead : calcium ratio from iliac crest, transiliac or
tibial specimens (Van de Vyver et al, 1988). Bone biopsy is,
however, inappropriate for screening purposes, but the development of an X-ray fluorescence (XRF) method for measuring lead
in vivo provides the opportunity to make a 'non-invasive biopsy' (Wedeen et al, 1988).

The purpose of this study was to investigate the relationship between in vivo (tibia XRF) and in vitro (transiliac bone
biopsy, AAS) measures of bone lead in dialysis patients, and
subsequently to compare the in vivo measurements with those
obtained in a control group, selected at random.

METHODS

The Swansea Group has adopted the Birmingham University technique for analysing tibial lead which uses a Cd-109 X-ray source and a planar germanium detector in a 180° scattering geometry (Somervaille et al., 1985). The method has been validated by cross comparison of bone samples measured using AAS (Morgan et al., 1989). The minium detectable concentration (2 SD of net peak counts) is approximately 11 μg Pb g^{-1} wet bone (corresponding to ~19 μg g^{-1} bone mineral) for a maximum skin dose of 0.6 mGy. Spectra have been acquired using pile-up rejection but only the Kα peaks have been analysed so far whereas the inclusion of Kβ peaks (Chettle et al., 1989) will improve the precision in the future.

Transverse iliac biopsy specimens were dried, digested in nitric acid, and analysed for lead and calcium using electro-thermal and flame atomic absorption spectrometry, respectively.

STUDY POPULATION

The renal group consists of 30 patients with end stage renal failure requiring chronic renal replacement therapy. Each subject had already consented to transiliac bone biopsy as part of another study, but each provided additional informed written consent to this project in accordance with a protocol approved by West Glamorgan District Ethical Committee.

So far, 27 of the group have received a tibia XRF measurement, and 20 bone biopsies have been analysed of which 17 pairs of data can be compared.

The control group consists of 74 adult volunteers recruited as part of another study (Morgan et al., 1989), of whom 59 had in vivo lead measurements of the tibia.

RESULTS

The mean ($\pm$SD) blood lead was 8.1$\pm$3.5 μg dl^{-1} (n=26) compared with 10.0$\pm$4.2 μg dl^{-1} (n=59) in the controls.

There was no significant relationship between blood lead and tibial lead in either patients or controls, but in both groups the tibial values were significantly dependent upon age (Fig. 1).

Linear regression analysis of the in vitro lead measurements of the ilium and the in vivo measurements of the tibia yielded the relationship

tibia Pb = 0.91 x (ilium Pb) + 0.06   (r=0.42)

where all values are expressed as μg g^{-1} bone mineral.

DISCUSSION AND CONCLUSIONS

Although the relative concentrations of lead from different bone sites vary with age (Wittmers et al., 1988), the

ratio of lead : calcium is less sensitive to such change. Using
this ratio, Wedeen et al (1988) demonstrated that the relationship between tibial XRF lead and the EDTA mobilisation test in
one group of subjects could be superimposed on the 95%
confidence limits obtained from an earlier EDTA and iliac bone
biopsy study. Our data provide direct confirmation of this
relationship in the same subjects.

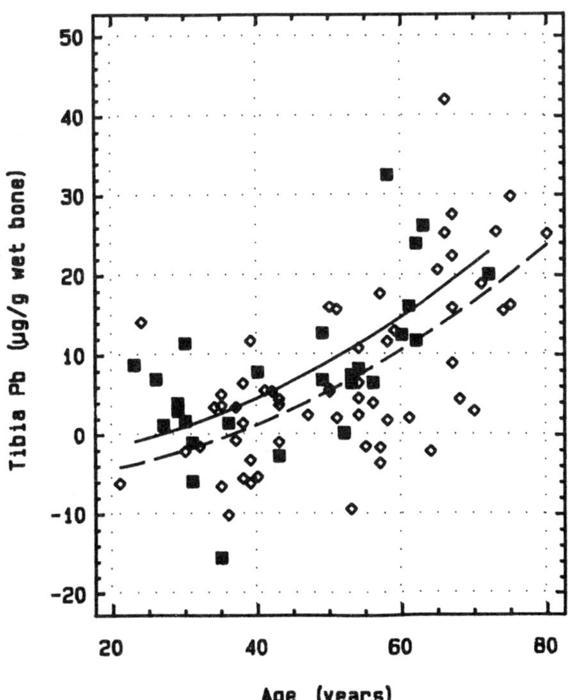

Fig. 1. Distributions of tibia lead with age for the
renal (■ , -) and control (◇ ,--) groups.
Fitted functions are
Pb=0.0051 x age^2-3.55   (r=0.66) and
Pb=0.0047 x age^2-6.23   (r=0.65), respectively.

Figure 1 shows that generally higher values of lead were
measured in the renal failure subjects than in the controls.
Although the difference between the two sets of data is not
statistically significant, it is of interest to note that 30% of
the control group are above the renal regression line; and
similarly only 22% of the renal patients have tibia values below
the regression line for the controls. These data confirm the
relatively high prevalence of elevated body lead burden in
patients with chronic renal failure (Koster et al, 1989) but
futher work is needed to establish the extent to which lead is
a cause of renal disease.

ACKNOWLEDGEMENTS

The equipment used in this study was funded by the Medical Research Council and the Welsh Office. One of the authors (SJJ) received support from the Morgan-Williams Trust and the Renal Research Fund.

REFERENCES

Chettle, D.R., Scott, M.C., and Somervaille, L.J., 1989, Improvements in the precision of bone lead measurements, Phys. Med. Biol., in press.

Koster, J., Erhardt, A., Stoeppler, M., Mohl, C., and Ritz, E., 1989, Mobilizable lead in patients with chronic renal failure, Eur. J. Clin. Invest., 19 : 228.

Morgan, W.D., Ryde, S.J.S., Jones, S.J., Wyatt, R.M., Hainsworth, I.R., Cobbold, S., Evans, C.J., and Braithwaite, R.A., 1989, In-vivo measurements of cadmium and lead in occupationally-exposed workers and in an urban population, J. Biol. Trace Elem. Res., in press.

Somervaille, L.J., Chettle, D.R., and Scott, M.C., 1985, In vivo measurement of lead in bone using x-ray fluorescence, Phys. Med. Biol., 30 : 929.

Van de Vyver, F.L., D'Haese, P.C., Visser, W.J., Elseviers, M.M., Knippenberg, L.J., Lamberts, L.V., Wedeen, R.P., and DeBroe, M.E., 1988, Bone lead in dialysis patients, Kidney Int., 33 : 601.

Wedeen, R.P., Van de Vyver, F.L., D'Haese, P.C., Visser, W.J., Elseviers, M.M., Knippenberg, L.J., Lamberts, L.V., DeBroe, M.E., Batuman, V., Schidlovsky, G., and Jones, K., 1988, Bone lead and the diagnosis of lead nephropathy, Contr. Nephrol., 64 : 102.

Wittmers, L.E., Wallgren, J., Alich, A., Aufderheide, A.C., and Rapp, G., 1988, Lead in bone. IV. Distribution of lead in the human skeleton, Arch. Environ. Health, 43 : 381.

IN VIVO MEASUREMENTS OF LEAD IN BONE

Jari Erkkilä, Vesa Riihimäki, Jukka Starck, Anne Paakkari,
Boris Kock*

Institute of Occupational Health, Laajaniityntie 1
SF-01620 Vantaa, Finland
*South-Saimaa Central Hospital, SF-53130 Lappeenranta
Finland

INTRODUCTION

Occupational exposure to lead is commonly monitored by means of
repeated blood lead determinations. Blood lead reflects mainly current
lead exposure and dose response studies indicate that it is an important
predictor of health effects. Lead is a highly cumulating metal, however,
and more than 95 % of the total body burden resides in the skeleton
(Barry, 1975). It is therefore important to determine bone lead levels
for the evaluation of the cumulative long-term exposure.

MATERIALS AND METHODS

A battery factory where the lead workers were exposed to the dust
and fumes of metallic lead and lead oxide was chosen for the study. The
subjects consisted of 54 current lead workers, 9 retired lead workers, 21
office workers from the same factory and 22 control subjects from the
Institute of Occupational Health.

The method for measuring bone lead concentration in vivo was based
on X-ray fluorescence. The method was developed originally at the
University of Birmingham (Sommervaille et al., 1985). The characteristics
of this method are: Cd-109 radiation source, back scattering geometry,
possibility for normalization, low detection limit and a reasonable
radiation dose.

Lead concentrations in two bones, the tibia (middle part of the leg)
and the radius/ulna (distal end of the forearm) were measured. The
measurement time was 25 minutes, and the radiation dose was about 0.1 mSv
for both sites. The current concentrations of lead in blood and urine
were analyzed by conventional AAS methods. For current and retired lead
workers the history of lead exposure was estimated using time-weighted
average concentrations and integrated concentrations over time (area
under the curve) of lead exposure.

RESULTS AND DISCUSSION

Table 1 summarizes the data on lead exposure and the measured lead
concentrations in blood, urine and bone. Tibial concentrations of lead
showed distinct differences in lead accumulation between the current lead
workers, retired workers, office personnel (working in a slightly

Table 1. Concentration of Lead in the Tibia and Radius/Ulna as well
as in the Blood and Urine of Subjects Variously Exposed.
The Lead Concentrations in Bones are Expressed in Terms of
Bone Minerals.

Subject category	Current lead workers		Retired lead workers		Office workers		Controls	
Parameter	Mean	+ SD	Mean	+ SD	Mean	+ SD	Mean	+ SD
Age (years)	40.0	9.0	58.8	6.2	42.1	7.9	35.5	8.9
Years at exposure	10.9	8.3	13.0	6.1	13.0	9.6		
Tibia-Pb (μg/g)	18.1	16.5	38.3	41.9	11.2	10.0	4.6	11.3
Radius/Ulna-Pb (μg/g)	63.1	49.9	32.9	35.8	32.8	22.9	33.1	31.7
B-Pb[1] (μmol/l)	1.42	0.44	0.67	0.33	0.33	0.36	0.18	0.08
U-Pb[2] (μmol/l)	0.20	0.10	0.07	0.03	0.05	0.07	0.03	0.01
B-Pb-ave[3] (μmol/l)	1.57	0.42	1.83	0.75				
B-Pb-int[4] (μmol*yr/l)	15.4	12.2	31.3	17.1				

Abbreviations:
[1]B-Pb: lead concentration in blood
[2]U-Pb: lead concentration in urine
[3]B-Pb-ave: time weighted average blood lead concentration
[4]B-Pb-int: integrated blood lead concentration over time

contaminated environment) and the controls. The retired workers exhibited
by far the highest tibial lead levels which is readily explained by their
longer and more intensive past exposure history.

The current concentrations of lead in the blood and urine are not
reliable indicators of the skeletal lead. This was clearly demonstrated by
the ratios of blood, urine and tibial lead between the differently exposed
groups (Table 2). By contrast, two exposure parameters, the time-weighted
average concentration of blood lead and, especially, the integrated blood
lead over time, predicted quite well the corresponding differences in
tibial lead among the current and retired lead workers. Tibial lead,
therefore, appeared to indicate the cumulative dose and body burden of
lead. Sommervaille et al. (1988) have earlier come to the same conclusion.

The lead concentrations in the radius/ulna, which are the first ones
reported _in vivo_, unfortunately did not make a distinction between the
different groups of subjects. There may be many reasons for this. The
measurement location was chosen near the distal end of the forearm because
of the overlying soft tissues and muscles elsewhere. At this site the
amount of bone minerals was small and the bone type was mainly trabecular.
Secondly, in spite of wiping the skin with an alcoholic solution, slight
skin contamination by lead cannot be excluded.

Table 2. Ratios of Different Parameters of Lead Exposure Between Differently Exposed Groups of Subjects. For Explanation of Abbreviations See Table 1.

| | Subject category ratios | | |
Parameter	Current/retired	Current/office	Office/controls
B-Pb	2.1	4.3	
U-Pb	2.9	6.1	2.0
Tibia-Pb	0.5	1.6	2.4
B-Pb-ave	0.9		
B-Pb-int	0.5		

These results constitute part of the collaborative study between the Institute of Occupational Health in Helsinki and the Medical Physics Group at the University of Birmingham.

The sensitivity of the system was moderate. The detection limit (2 times standard deviation of the net peak area), in terms of bone mineral, was 17.5 μg Pb (g bone mineral)$^{-1}$. The results also suggest sex dependence; for males the detection limit was 15.5 μg/g and for females 20 μg/g. This difference was probably due to the differences in bone size and overlying soft tissue. The lowest detection limit published in the literature, was about 10 μg/g (Chettle et al., 1989). The applied definition for a detection limit is practical. It gives the lead concentration above which the measurement values can be used for individual diagnosis. Lead concentrations under the detection limit can be used, however, for the calculation of group estimates. The present results confirmed the validity of the method for an estimation of occupational lead exposure.

REFERENCES

Barry, P. S. I., 1975, Comparison of concentrations of lead in human tissues, Br J Ind Med, 32: 119.
Chettle, D. R., Scott, M. C., and Sommervaille, L. J., 1989, Improvements in the precision of in vivo bone lead measurements, Phys Med Biol, 34:1295.
Sommervaille, L. J., Chettle, D. R., and Scott, M. C., 1985, In vivo measurement of lead in bone using X-ray fluorescence, Phys Med Biol, 30:929.
Sommervaille, L. J., Chettle, D. R., Scott, M. C., Tennant, D. R., McKiernan, M. J., Skilbeck, A., and Trethowan, W. N., 1988, In vivo tibia measurements as an index of cumulative exposure in occupational exposed subjects, Br J Ind Med, 45:174.

DISTRIBUTION OF LEAD IN HUMAN BONE:

I. ATOMIC ABSORPTION MEASUREMENTS*

H. Hu and T. Tosteson
Channing Laboratory
Department of Medicine, Brigham and Women's Hospital
Harvard Medical School
Boston, Massachusetts 02115

A. C. Aufderheide and L. Wittmers
University of Minnesota-Duluth School of Medicine
Duluth, Minnesota 55812

D. E. Burger and F. L. Milder
ABIOMED, Inc.
Danvers, Massachusetts 01923

G. Schidlovsky and K. W. Jones
Brookhaven National Laboratory
Upton, New York 11973

INTRODUCTION

X-ray fluorescence (XRF) provides a convenient and accurate means of
estimating lead content in bone. However, before one can accurately
interpret XRF results to obtain a measurement of lead exposure in either
clinical medicine or epidemiologic studies, several issues regarding the
distribution of lead in the human skeleton and the histomorphometry of
bone need to be addressed.

LEAD IN BONE: DISTRIBUTION IN TRABECULAR VS. CORTICAL BONE IN CADAVERS

Several studies suggest that skeletal lead is unevenly distributed
between trabecular bone and cortical bone. Age-stratified analyses of
data on lead in bone that is primarily dense (cortical) show a linear
increase in concentration with advancing age (Barry and Mossman, 1970;
Holtzman et al., 1970; Barry, 1975), whereas similar analyses of data on
lead in bone that is primarily spongy (trabecular) show a linear increase
in lead concentration with age until the fifth decade, when concentrations show a levelling off or decrease (Schroeder and Tipton, 1968;

*Research supported by NIEHS ST 32ES07069 (HH), NIH HL 07427 (TT),
and NIH/SBIR 2R44ES03918-02 (DEB,FLM). Development of the synchrotron x-
ray microscope is supported by the US DOE Division of Chemical Sciences,
Office of Basic Energy Sciences, under Contract No. DE-AC02-76CH00016,
and its use for biomedical measurements by NIH Biotechnology Research
Resource Grant No. P41RR01838 (KWJ,GS).

Schroeder and Balassa, 1961; Nusbaum et al., 1965; Cherry et al., 1974).
In studies of lead-exposed workers in which lead concentrations in mostly
cortical bone were compared with lead concentrations in mostly trabecular
bone among active and retired workers, markedly different ratios were
found, suggesting kinetic differences in the handling of lead (Skerfving
et al., 1983; Christoffersson et al., 1987; Somervaille et al., 1987).
General metabolic turnover rates between the two bone types, as measured
by isotope studies, have long been known to differ (Rivera, 1965;
Rabinowitz et al., 1976). This difference may be a function of the ratio
between bone surface exposed to remodeling activity and bone volume, this
ratio being 2 to 40 times greater in trabecular bone than in cortical
bone (Frost, 1973).

The two-compartmental model of bone lead kinetics suggested by XRF
data implies that a single summary estimate of skeletal lead content may
be less informative than separate estimates of trabecular and cortical
bone. Lead content in each compartment may have distinct implications
for toxicity.

No studies, however, have specifically measured lead content in
cortical and trabecular bone that have been separated from each other at
multiple bone sites, thereby testing the assumption that within each bone
compartment (trabecular versus cortical) bone lead concentration is
fairly homogenous. Results from such experiments will provide a rationale for selecting the proper bone location and meaningful interpretation
of XRF measurements.

METHODS

We have preliminary results on 27 cadavers: 15 males and 12
females, with an age at death ranging from 50 to 91 years. The cadavers
came from a variety of general community hospital sources through Harvard
Medical School's Organ Donation Program; specific environmental/
occupational histories relevant to lead exposure are unknown.

From each of the cadavers, wedge specimens were extracted at eight
bone sites: the skull (SKU), rib (RIB), vertebra (VER), iliac crest
(ILI), proximal tibial metaphysis (T-M), mid-shaft tibial diaphysis
(T-D), patella, and calcaneus. At the time of this analysis, lead
analyses were not complete on calcaneus and patella specimens; thus, they
are excluded from the present analysis. In addition, whole 2-centimeter
thick sections of bone (simulting the target volume of an XRF measurement) were taken from the mid-shaft tibial diaphysis (T-D-M) and proximal
tibial metaphysis (T-M-M). Trabecular (T) bone was separated from cortical (C) bone for each sampling site. Both portions of bone were then
ashed, weighed, and analyzed for lead content using flameless atomic
absorption spectroscopy (AAS) (Wittmers et al., 1981). All lead concentrations reported in this paper are expressed in units of μg lead/gm
ashed bone.

STATISTICAL METHODS

As a preliminary descriptive analysis, bone site- and compartment-
specific mean lead contents were calculated, and bone site-specific cortical lead values were compared with trabecular values. Correlations
between site- and compartment-specific bone lead concentrations were
calculated. A factor analysis was performed to identify possible
groupings of sites.

To evaluate the relative contribution of distinguishing trabecular from cortical bone to the overall variance of bone lead content a general linear modelling program was used to perform analysis of covariance. Site- and compartment-specific bone lead concentration formed the dependent variable in one set of models, with age, sex, site, and compartment as the independent variables.

Three additional variables were created, MEAN-C and MEAN-T, and MEAN-ALL, which represent, respectively, the lead concentration averaged over all cortical (MEAN-C) and all trabecular (MEAN-T) bone tissue measured, and all (MEAN-ALL) bone tissue measured (excluding thick segment bones), for each individual.

Since the mid-shaft of the tibial diaphysis is comprised of 95% cortical bone (by weight) and is a primary target for XRF estimates of cortical bone lead content (Hu et al., 1989), the correlations of thick bone segment T-D-M to MEAN-C and other variables were explored.

Similarly, since the proximal end of the tibial metaphysis is comprised of 90% trabecular bone (by weight), and thus is a primary target for an XRF estimate of trabecular bone lead content (Hu et al., 1989), we explored the relation of thick bone segment T-M-M to MEAN-T.

Finally, additional linear models were developed to assess the contribution of T-D-M and T-M-M towards predicting the total mean lead content in each individual (MEAN-ALL). T-D-M and T-M-M were plotted against MEAN-ALL.

RESULTS

Some samples were not of adequate size (<25 mg) to provide reliable separate estimates of trabecular and cortical bone lead and were, therefore, omitted from this analysis (Wittmers et al., 1981). Lead concentrations in all remaining bones ranged from a low of 1.9 μg/gm in VER-C to 111 μg/gm in a T-M-T specimen. Mean values at individual bone sites ranged from 19.7 μg/gm in RIB-C to 43.3 μg/gm in T-M-T (Table 1). Cortical bone lead values differed significantly ($p<0.05$) from trabecular bone lead only at the vertebra, and possibly the rib ($p=0.05$, Table 1); moreover, mean cortical lead levels were <u>lower</u> than trabecular bone at one site (rib), but <u>higher</u> in the other site (vertebra).

The correlation matrix (Table 2) demonstrated coefficients ranging from a high of 0.95 (T-M-T vs. T-M-C) to a low of 0.17 (ILI-T vs. T-D-C). The factor analysis on the covariance matrix indicated that a single factor model was appropriate, and that this factor, a weighted combination of all the site- and compartment-specific concentrations, explained 72% of the sum of the variances of the original variables. The largest weight was 0.95 for T-M-C and the smallest weight was 0.62 for ILI-T. T-M-T had a weight of 0.94.

Statistical models with site- and compartment-specific lead concentration as the dependent variable showed that sex and site were statistically significant ($p<0.01$) predictors of lead concentration, whereas age and compartment were not (Table 3). An interaction term of sex with site did not significantly contribute to the model (Table 3).

The association of T-D-M with MEAN-C had a correlation coefficient of 0.55 ($p=.017$, Fig. 1). The association of T-M-M with MEAN-T had a correlation coefficient of 0.84 ($p=.0001$, Fig. 2).

Table 1. Mean Lead Content by Bone Site and Compartment (Type) in 27 Human Cadavers

Site[2] n	Type	Lead[1] (s.d.)	Type	Lead[1] (s.d.)	Difference (cortical - trabecular)[1]	Paired t-test p-value
ILI 27	-C	24.7 (8.4)	-T	22.7 (8.7)	2.0	0.13
RIB 16	-C	19.7 (11.5)	-T	24.2 (15.6)	-4.5	0.05
SKU 24	-C	26.8 (13.7)	-T	26.2 (15.1)	0.6	0.77
T-D 18	-C	33.8 (15.0)	-T	33.5 (13.5)	0.3	0.91
T-M 18	-C	40.1 (19.9)	-T	43.3 (25.2)	-3.2	0.16
VER 24	-C	25.1 (18.8)	-T	16.9 (11.7)	8.2	>0.01
MEAN	-C	26.7 (9.7)	-T	26.4 (11.1)		
MEAN	-ALL	26.4 (10.0)				

[1]mean value expressed as μg of lead/gram bone ash
[2]Wedge specimens of: ILI = iliac crest; RIB = rib; SKU = skull; T-D = mid-shaft tibial diaphysis; T-M = mid-shaft tibial metaphysis; VER = vertebra

Table 2. Pearson Correlation Coefficients

Site[1]	SKU-C	SKU-T	RIB-C	RIB-T	ILI-C	ILI-T	T-D-C	T-D-T	T-M-C	T-M-T
SKU-C										
SKU-T	0.76									
RIB-C	0.58	0.75								
RIB-T	0.65	0.65	0.86							
ILI-C	0.50	0.48	0.30	0.57						
ILI-T	0.39	0.50	0.43	0.68	0.70					
T-D-C	0.54	0.55	0.70	0.64	0.39	0.17				
T-D-T	0.57	0.53	0.79	0.69	0.37	0.25	0.67			
T-M-C	0.71	0.68	0.64	0.71	0.64	0.59	0.54	0.68		
T-M-T	0.58	0.62	0.67	0.69	0.63	0.60	0.57	0.70	0.95	

[1]The suffix -C specifies the cortical component of corresponding bone. The suffix -T specifies the trabecular component of corresponding bone. All other abbreviations are as in Table 1.

Plots of T-D-M and T-M-M versus total mean lead content (MEAN-ALL) demonstrate a recognizable relationship between the two large bone segment lead values and MEAN-ALL (Fig. 3).

Linear modelling of T-D-M and T-M-M on MEAN-ALL demonstrated that T-M-M had a more significant relationship than T-D-M to MEAN-ALL (Table 4a and 4b). When included in the same model, T-M-M clearly had a much closer relationship to MEAN-ALL than T-D-M (Table 4c), and it would not appear that T-D-M contributes significantly to the prediction if T-M-M is already known.

Table 3. General Linear Models Procedure. Dependent variable: lead
          concentration

Independent Variables	Freedom	Sum of Squares	F-Value	Prob>F
age	1	109.7	0.52	0.13
sex	1	4131.3	19.66	0.0001
site	5	11367.4	10.82	0.0001
compartment	1	5.4	0.03	0.873

Total R-Square: 0.23

age	1	127.7	0.61	0.43
sex	1	4008.7	19.21	0.0001
site	5	10407.5	9.97	0.0001
sex*site	5	1183.3	1.13	0.34

Total R-Square: 0.25

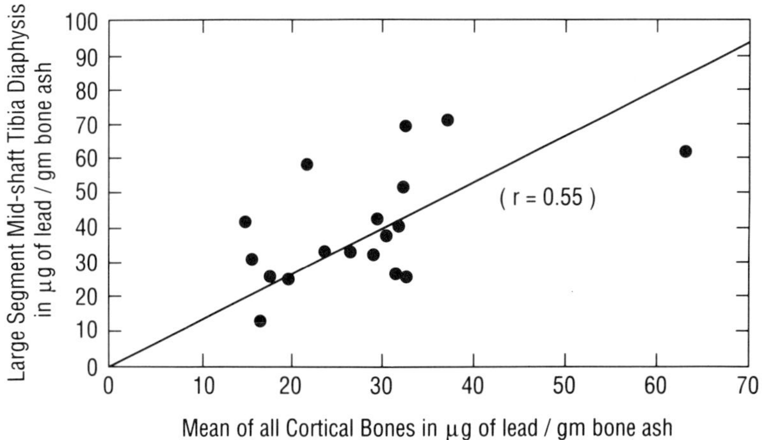

Fig. 1. Lead concentrations in the mid-shaft tibia diaphysis are shown
         as a function of the mean lead concentration in all cortical
         bones.

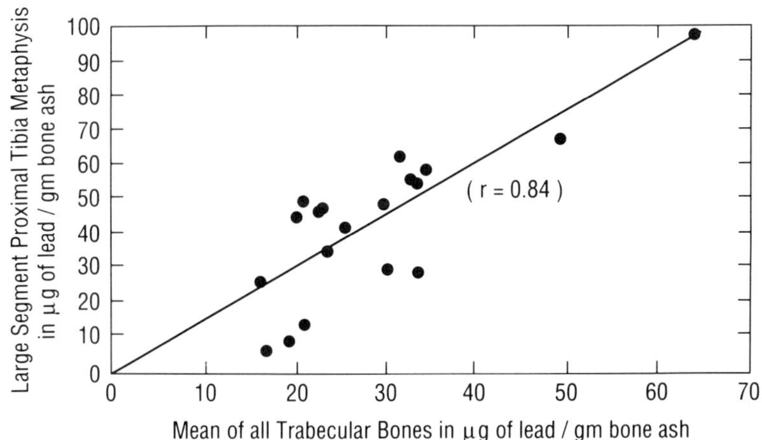

Fig. 2. Lead concentrations in the large segment proximal tibial meta-
         physis are shown as a function of the mean lead concentration
         in all trabecular bones.

Table 4. General Linear Models Procedure. Dependent variable: mean
lead concentration of all bones

Independent Variables	Freedom	Sum of Squares	F-Value	Prob>F
(a) age	1	179.6	6.25	0.026
sex	1	0.2	0.01	0.93
T-M-M	1	1414.8	49.25	0.0001

Total R-square = 0.81

(b) age	1	0.72	0.01	0.93
sex	1	57.3	0.57	0.46
T-D-M	1	405.8	4.03	0.065

Total R-square = 0.35

(c) age	1	151.74	4.99	0.044
sex	1	0.09	0.00	0.958
T-D-M	1	7.1	0.23	0.637
T-M-M	1	1016.2	33.44	0.0001

Total R-square = 0.82

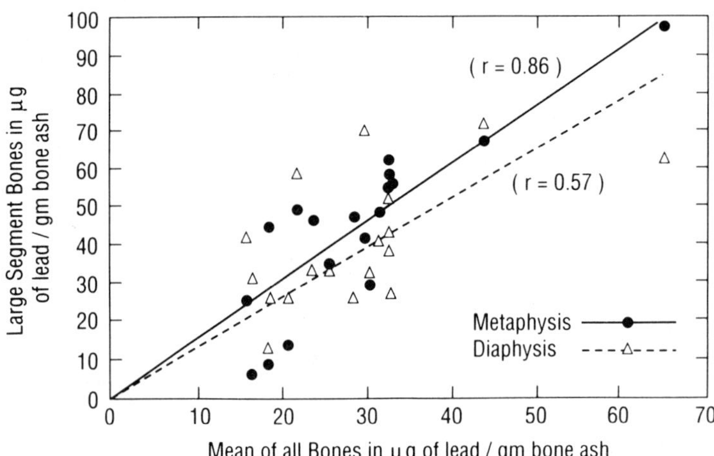

Fig. 3. Lead concentrations in large segment bones are shown as a
function of the mean lead concentration in all bones.

DISCUSSION

These results suggest that in this population of cadavers, trabecular and cortical bone collected lead at similar rates (or equilibrated
over time) at four of the six bone sites measured. The significance of
possible differences between trabecular and cortical bone in the rib and
vertebra is mitigated by the opposite nature of their directionality.
The concentrations were greatly dissimilar between bone sites, however.
The factor analysis suggests that a single bone measurement, particularly
at the tibia metaphysis, is highly representative of this group of site-
and compartment-specific lead concentrations. The implication is that
when selecting a bone site for XRF measurement in this population, focussing on a single bone site, such as the tibial metaphysis, may be adequate for characterizing an individual's lead burden.

272

It is likely that the relative lack of difference in lead content between cortical and trabecular bone probably reflects the rather narrow age range of this cadaver population, and the low-level lead exposures these individuals probably sustained during life. Site-specific mean bone lead levels (Table 1) are comparable to those seen by Wittmers et al. (1988) among the oldest subjects in a random sample of autopsies from a northern Minnesota community hospital. In addition, the tibia diaphysis and ilium lead levels are similar to those seen by Van de Vyver et al. (1988) after correcting to the same units of concentration using the factors given by Wittmers et al. (1988).

The significant sex differences in bone lead content seen in this data, with females having lower lead levels, is consistent with previous studies. In our data, these differences did not appear to be site-specific.

The values for T-M-M represent what an XRF measurement would derive from an examination of the proximal metaphysis. Similarly, the values for T-D-M represent what an XRF measurement would derive from an examination of the tibial mid-shaft. The plots suggest that if a compartment-specific bone lead measurement were desirable, T-M-M would approximate quite closely the mean of all trabecular bones; T-D-M would be somewhat less reliable.

Similar to the results of factor analysis, the general linear models of T-M-M and T-D-M on MEAN-ALL suggest that if the average of all bone lead levels were taken as the most representative indicator of lead burden in each cadaver, a single measurement of the proximal metaphysis with XRF would be more useful in predicting this parameter than would a single measurement of the tibial mid-shaft; making both measurements would not contribute additional information.

It is problematic to assume that MEAN-C, MEAN-T, and MEAN-ALL are the parameters with the most toxicologic significance. Elevations of lead in particular bones might have more significance. However, no data are available to assess this possibility. Testing related hypotheses would require difficult epidemiologic studies.

We emphasize that these observations were made on a population that probably did not have lead exposures significantly above community background exposures. We still believe that individuals with recent and/or heavier lead exposure may benefit from multiple XRF measurements that distinguish trabecular from cortical bone sites (Hu, et al., 1989). The present findings are most useful in guiding the approach of epidemiologic studies, using XRF on community-exposed individuals with relatively low body lead burdens and "normal" blood lead levels. Such studies will likely occur in the future, particularly when considering issues such as the relationship between lead exposure and blood pressure (Victery et al., 1988).

REFERENCES

Barry, P. S. I., and Mossman, D. B., 1970, Lead concentrations in human tissues, Br. J. Ind. Med., 27:339.
Barry, P. S. I., 1975, A comparison of concentrations of lead in human tissues, Br. J. Ind. Med., 32:119.
Cherry, W. H., Esterby, S. R., Finch, A., and Forbes, W. F., 1974, Studies on trace metal levels in human tissues. II. The investigation of lead levels in rib samples of 100 Canadian residents,

in: "Proc. of the 9th Conf. on Trace Substances in Environmental Health," D. D. Hemphill, ed., University of Missouri Press, Columbia, Missouri.

Christoffersson, J.-O., Schütz, A., Skerfving, S., Ahlgren, L., and Mattsson, S., 1987, A model describing the kinetics of lead in occupationally exposed workers, in: "In Vivo Body Composition Studies, Proceedings of an International Symposium," K. J. Ellis, S. Yasumura, and W. D. Morgan, eds., Bocardo Press Limited, Oxford.

Frost, H. M., 1973, "Bone Remodeling and Its Relationship to Metabolic Bone Diseases," Charles C. Thomas, Springfield, Illinois. Holtzman, R. B., Lucas, H. F., Jr., and Ileewics, F. H., 1970, The concentration of lead in human bone, in: "Argonne National Laboratory--Radiological Physics Division Annual Report," ANL-7489, pp. 43-49.

Hu, H., Milder, F., and Burger, D., 1989, X-ray fluorescence: issues surrounding the application of a new tool for measuring lead burden, Environ. Res., in press.

Nusbaum, R. E., Butt, E. M., Gilmour, T. C., and Didio, S. L., 1965, Relation of air pollutants to trace metals in bone, Arch. Environ. Health, 10:227.

Rabinowitz, M. B., Wetherill, G. W., and Kopple, J. D., 1976, Kinetic analysis of lead metabolism in healthy humans, J. Clin. Invest., 58:260.

Rivera, J., 1965, Human bone metabolism inferred from all-out investigations, Nature 207:1330.

Schroeder, H. A., and Balassa, J. J., 1961, Abnormal trace metal in man: lead, J. Chronic Dis., 14:408.

Schroeder, H. A., and Tipton, I. H., 1968, The human body burden of lead, Arch. Environ. Health 17:965.

Skerfving, S., Ahlgren, L., Christoffersson, J.-O., Haeger-Aronsen, B., Mattsson, S., and Schütz, A., 1983, Metabolism of inorganic lead in occupationally exposed humans, Arh. Hig. Rada. Toksikol., 34:341.

Somervaille, L. J., Chettle, D. R., Scott, M. C., Krishnan, G., Browne, C. J., Aufderheide, A. C., Wittmers, E., and Wallgren, J., 1987, XRF of lead in vivo: simultaneous measurement of a cortical and trabecular bone in a pilot study, in: "In Vivo Body Composition Studies, Proceedings of an International Symposium," K. J. Ellis, S. Yasumura, and W. D. Morgan, eds., Bocardo Press Limited, Oxford.

Van de Vyver, F. L., D'Haese, P. C., Visser, W. J., Elseviers, M. M., Knippenberg, L. J., Lamberts, L. V., Wedeen, R. P., and DeBroe, M. E., 1988, Bone lead in dialysis patients, Kidney International, 33:601.

Victery, W., Tyroler, H. A., Volpe, R., and Grant, L. D., 1988, Summary of discussion sessions: symposium on lead-blood pressure relationships, Env. Health Persp., 78:139.

Wittmers, L. E., Jr., Alich, A., and Aufderheide, A. C., 1981, Lead in bone I. Direct analysis for lead in milligram quantities of bone ash by graphite furnace atomic absorption spectroscopy, Am. J. Clin. Path., 75:80.

Wittmers, L. E., Jr., Aufderheide, A. C., Wallgren, J., Rapp, G., and Alich, A., 1988, Lead in bone IV. Distribution of lead in the human skeleton, Arch. Env. Health, 43:381.

DISTRIBUTION OF LEAD IN HUMAN BONE:

II. PROTON MICROPROBE MEASUREMENTS*

G. Schidlovsky and K. W. Jones
Brookhaven National Laboratory
Upton, New York 11973

D. E. Burger and F. L. Milder
ABIOMED, Inc.
Danvers, Massachusetts 01923

H. Hu
Channing Laboratory
Department of Medicine, Brigham and Women's Hospital
Harvard Medical School
Boston, Massachusetts 02115

INTRODUCTION

Little is known about the distribution of lead in the human tibia on a microscopic scale. The radial distribution of lead in a 2-mm thick section of the human femur has been measured by Lindh et al. (1978), who observed that the concentration peaked in a region close to the periosteal and endosteal surfaces. The distribution in the interior of the bone was relatively uniform with the exception of a peak located about 1.8 mm from the periphery along a straight radial scan from periosteum to endosteum. Lindh (1980; 1981) also mapped the distribution of lead over a single osteon and showed that the concentration was highest at the edges.

We have investigated the radial distribution of lead in the human tibia. Our motivation was to obtain data that can be used to understand the biological mechanisms for deposition of lead in bone and for use in the interpretation of in-vivo bone x-ray fluorescence (XRF) measurements of lead concentration.

This paper presents the results of proton microprobe line scans of several tibial sections from periosteum to endosteum with spatial resolutions from less than 100 micrometers to about 1000 micrometers. The results are complementary to those reported in the companion paper by Jones et al. (this volume).

---

*Research supported by NIH Biotechnology Research Resource Grant No. P41RR01838 (GS,KWJ); NIH/SBIR Grant No. 2R44ES03918-02 (DEB,FLM), and NIEHS ST Grant No. 32ES07069 (HH).

ANALYTICAL TECHNIQUE

The lead concentrations were determined using the technique of proton-induced x-ray emission (PIXE). Measurements were made with the proton beams produced by the Brookhaven Research Electrostatic Accelerator. Proton beams with an energy of 3.5 MeV were momentum analyzed, collimated, and focussed using an electrostatic quadrupole triplet. The beam size on the target was typically about 100 micrometers in diameter and the beam currents averaged 50 nA. Low resolution work was done with beams of about 1000 micrometers by turning off the quadrupole focussing lens. A mechanical stage with micrometer adjustment was used to position the sample in the beam. Scans were made by manually adjusting the micrometer drive on the stage.

The x-rays produced by the proton bombardment were detected with a 30 mm^2 Si(Li) detector placed at 135 degrees to the incident beam at a distance of 3 cm. Polyimide filters with thicknesses of about 900 μm were used to reduce the intensity of the calcium K x-rays relative to the lead L x-rays.

A typical pulse height spectrum produced by proton bombardment of a bone section is shown in Fig. 1. The detection limit for determination of lead was about 3 ppm (parts per million) wet weight for a 30-minute run at 50 nA beam current based on measurement of an International Atomic Energy Authority H-5 bone mineral standard. Our results are reported in units proportional to the ratio of the number of lead atoms to the number of calcium atoms. The conversion to lead values in absolute units of ppm requires a knowledge of the local density of the specimen which is not readily available. Hence, values of the lead/calcium atomic ratios are reported in arbitrary units. The use of the lead to calcium ratio is also useful for looking at the concentrations at the bone surface because ratio measurements with the beam partially on the bone are not affected by the portion of the beam which does not hit the bone. Note that the data shown in Figs. 2-4 were obtained at different times using different experimental conditions. Further, the data in Figs. 2 and 3b were obtained using the Pb Lß x-ray yield, while those in Fig. 4 were found from the Pb Lα x-rays.

SPECIMEN PREPARATION

Portions of tibia from three humans were investigated. The specimens were prepared in several different ways. First, thick sections around five millimeters were cut. The surface to be investigated was then polished with an abrasive and the wet specimen placed in the vacuum chamber. The resulting vacuum-dried specimens were then investigated with the microprobe. Second, a section 100 micrometers in thickness was prepared. The bone was defatted in a 1:1 mixture of ether-ethanol and then dried in an oven. The final thickness was produced using an abrasive. Third, an ashed section was prepared by cutting a one-millimeter thick section and then carrying out the ashing process at a temperature of 425°C for 18 hours.

The investigations were made on specimens from patients who were more than fifty years of age and suffered from a variety of health problems. None of them were occupationally exposed or otherwise unusually exposed to lead.

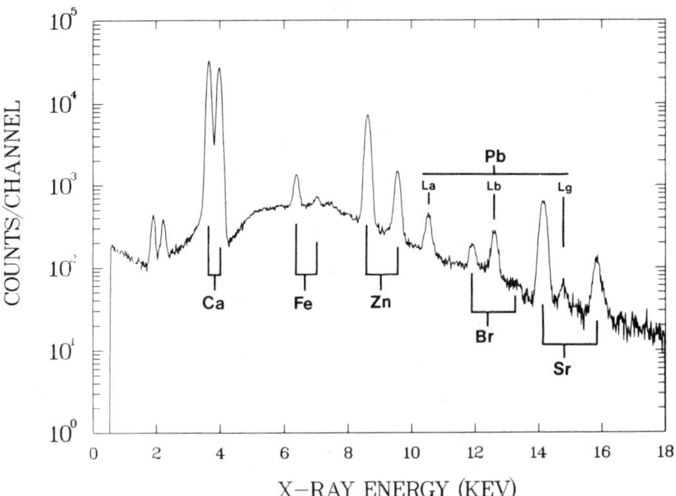

Fig. 1. Typical x-ray pulse height spectrum produced by the bombardment of thick section of human tibia #1 by 2.5-MeV protons.
The x-ray detector was a 30 mm^2 Si(Li) detector.

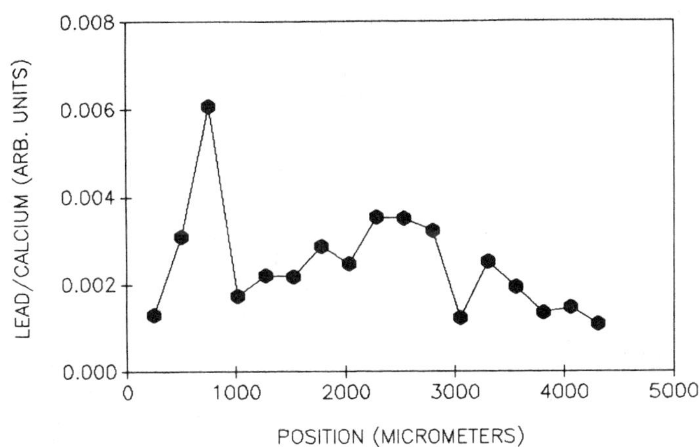

Fig. 2. The lead-to-calcium atomic ratio obtained in a scan of a section of human tibia #2 using a proton beam diameter of 250
micrometers is shown for the region from the periosteum (at
the left of the figure) to the endosteum (at the right of the
figure). A strong peaking of the ratio at the periosteal
surface can be seen.

RESULTS

Several scans were made with a resolution and step size of 250 micrometers to investigate the degree of uniformity of lead deposition over small spatial regions in the tibia. The results of one such scan from the periosteum to the endosteum are shown in Fig. 2. Peaking at the periosteal surface is clearly demonstrated. A scan with a resolution and step size of 100 micrometers is shown in Fig. 3. In this case, the ratio is a very strong function of the actual position on the sample.

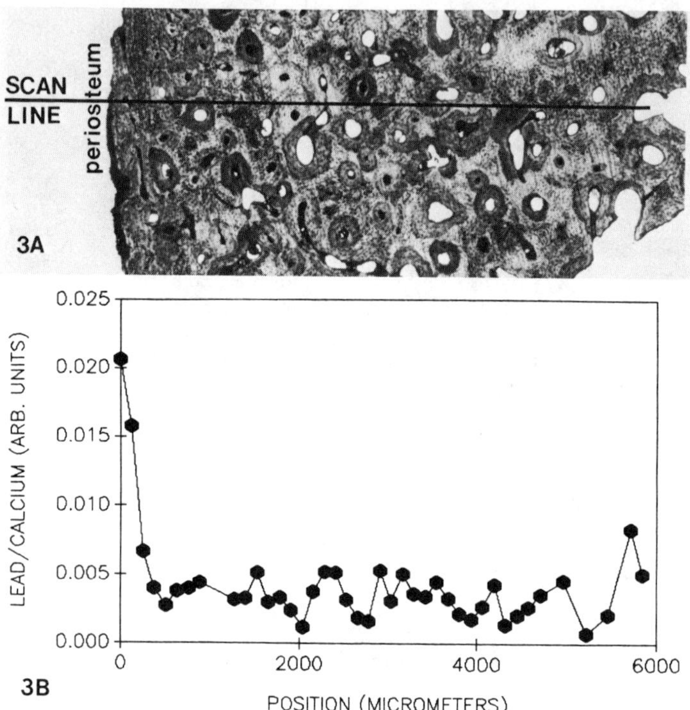

Fig. 3.   A photomicrograph of a section of human tibia #3 is shown in Fig. 3A. The heterogeneity of the bone is clearly visible. The line across the bone is the path taken to obtain the data shown in Fig. 3B. In Fig. 3B the lead-to-calcium atomic ratio obtained in a scan with a beam of 100 micrometers diameter is shown for the region from the periosteum (at the left of the figure) to the endosteum (at the right of the figure). A strong peaking of the lead/calcium atomic ratio is seen at the periosteum and, to a lesser extent, at the endosteum. The fluctuations of the ratio observed in the interior portion of the bone in Fig. 3B, reflect the morphological variations observed along the scan line in Fig. 3A.

A scan obtained using a 1-mm beam diameter and step size over a wet bone is shown in Fig. 4. It was found during the course of making this scan that the value of the ratio changed as a function of bombardment time. This effect can be attributed to the volatilization of the lead caused by beam-target interactions. A second scan was made on the

same region of the bone following ashing. The values are also shown in
Fig. 4 and are substantially less than the measurements on the wet sample, again indicating a loss of lead relative to calcium at high sample
temperatures. At one sample point, on the periosteal surface, the lead/
calcium ratio increased after ashing; this was a deviation from the
general trend and may indicate additional sample/beam interactions yet
not understood.

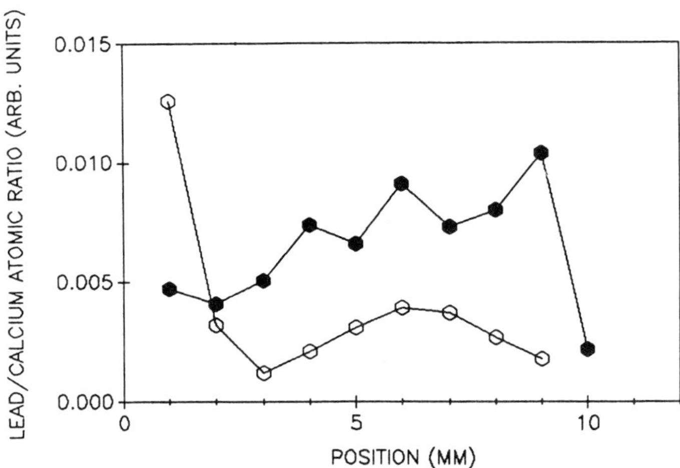

Fig. 4. The lead-to-calcium atomic ratio for scans of a section of
human tibia #1 which start from the periosteum (at the left of
the figure) and go to the endosteum (at the right of the
figure). A beam diameter of 1 mm was used to minimize the
effects of the bone heterogeneity. Scans were made over the
diaphysis of the wet bone sample (filled symbols) and ashed
sample (open symbols) over approximately the same path. The
large size of the beam reduces the sensitivity of the measurement to the surface effects shown in Figs. 2 and 3. The
ashing procedure produces a marked reduction in the lead/
calcium ratio which implies that lead is lost during ashing.
The periosteal surface-peaking may indicate slightly different
beam positions or an actual difference in the effects of the
ashing on different types of bone.

DISCUSSION AND CONCLUSIONS

The data obtained are in agreement with the results of Lindh
(1981) on the human femur in that there is peaking of the lead concentration at the periosteal, and sometimes to a lesser degree, the
endosteal surfaces. Our results do, however, show much more variability
in the interior regions than does the scan shown by Lindh. A strong
spatial dependence is to be expected though, based on the map made by
Lindh et al. (1978) over an individual osteon. A similar variation is

also reported by Jones et al. (this volume) using synchrotron x-ray microscopy.

The in-vivo measurements made using K x-ray detection give an average value of the lead concentration for the entire bone. Measurements with a larger size proton beam are useful for considering the variation of the lead for more direct comparison with the in-vivo work. Scans made with beams of about 1 mm determine the average variation over the bone. These scans are not sensitive to the local variation at the surface since that takes place over a region that is small compared to the beam size.

Low-resolution scans (of which Fig. 4 is typical) show that the lead concentration is fairly uniform over the entire region of the tibia. Values of the lead concentration derived from the in-vivo measurements using K x-rays will thus be representative of the entire bone lead value.

The high-resolution scans show that lead concentrations vary strongly at the surface of the bone. This is the region that is sampled in an in-vivo L x-ray measurement of the tibial lead concentration. In one sense, a surface enhancement will help to increase the sensitivity of the determinations made using L x-rays. The variability of lead concentration throughout this region, however, raises two important issues with regard to the significance of L x-ray fluorescence measurements: 1) the L x-ray fluorescence may not offer a precise measurement that is comparable from individual to individual, and 2) the L x-ray fluorescence measurement may not be comparable to the integrated value provided by K x-ray fluorescence or representative of the average lead concentration.

The work reported here when combined with the complementary investigation of Jones et al. (this volume) demonstrates the overall complexity of relating the actual bone lead in-vivo measurements to the biological questions raised by the microscopic variations that are observed. Further measurements on a suite of samples covering a wider range of ages and exposure conditions are needed. Measurements on specimens from children are particularly relevant for understanding the mechanisms of the deposition process and for interpretation of the results obtained using L x-rays.

The apparent loss of lead during the ashing process suggests that a fraction of the lead may be associated with the cellular component of the bone which is removed during the ashing process. The synchrotron x-ray microscopy measurements described by Jones et al. (this volume) show that the lead and zinc distributions in the tibia are similar. This finding is in agreement with the present experiment because zinc is known to be associated with the cellular component of bone.

REFERENCES

Lindh, U., Brune, D., and Nordberg, G., 1978, Microprobe analysis of lead in human femur by proton induced x-ray emission (PIXE), Sci. Tot. Envir., 10:31.
Lindh, U., 1980, A nuclear microprobe investigation of heavy-metal distribution in individual osteons of human femur. Int. J. App. Rad. and Isotopes, 31:737.
Lindh, U., 1981, The nuclear microprobe applied to bioenvironmental studies, Nucl. Instrum. and Meth., 181:171.

DISTRIBUTION OF LEAD IN HUMAN BONE:

III. SYNCHROTRON X-RAY MICROSCOPE MEASUREMENTS*

K. W. Jones and G. Schidlovsky
Brookhaven National Laboratory
Upton, New York 11973

D. E. Burger and F. L. Milder
ABIOMED, Inc.
Danvers, Massachusetts 01923

H. Hu
Channing Laboratory
Department of Medicine, Brigham and Women's Hospital
Harvard Medical School
Boston, Massachusetts 02115

INTRODUCTION

Measurement of the microdistribution of lead and other elements in bone is of interest for two reasons. First, the correlation of the lead concentrations with histological features in the bone may shed light on the biological mechanisms involved in the absorption and redistribution of lead in bone structures and help identify the compartments in which the lead is stored. Second, a knowledge of the distribution is required in order to provide a firm basis for the interpretation of in vivo x-ray measurements of bone lead concentrations based on the detection of characteristic K or L x-rays.

The measurement of the distributions with detection sensitivities on the order of parts-per-million (ppm) is not feasible with conventional analytical methods. In the present paper, we present results obtained with a synchrotron x-ray microscope (XRM) using the technique of synchrotron radiation induced x-ray emission (SRIXE). Results obtained using a proton microprobe based on proton-induced x-ray emission (PIXE) are given by Schidlovsky et al. (this volume). SRIXE offers complementary results with much finer spatial resolution and higher sensitivity as compared to PIXE.

Present methods for the in vivo determination of lead in bone are based on the production of vacancies in the lead L or K shell and the subsequent detection of characteristic lead L or K x-rays. In the case of L x-rays, only a thickness of a few hundred micrometers at the

*Research supported by NIH Biotechnology Research Resource Grant No. P41RR01838 (KWJ,GS); NIH/SBIR Grant No. 2R44ES03918-02 (DEB,FLM), and NIEHS ST Grant No. 32ES07069 (HH).

surface of the bone is sampled. In the second case, an average value of
the lead concentration over the entire tibia is obtained relative to the
calcium concentration. In both cases, knowledge of the microdistribution is necessary for interpretation of the results of the in vivo
determinations.

ANALYTICAL TECHNIQUE

The distribution of lead, calcium, and other elements in sections
of human tibia were determined using the Brookhaven X-Ray Microscope
located on the X-26 beam line of the National Synchrotron Light Source
(NSLS). The general principles of the XRM have been outlined in a
review of the field by Jones and Gordon (1989). In brief, the continuous energy x-ray beam from the NSLS was collimated to a size of about 50
μm x 50 μm. The position of the beam was determined by viewing its
position on a scintillator. The sample was then scanned, in known increments, past the beam using computer-controlled stages. The x-rays
were observed with an energy-dispersive Si(Li) x-ray detector, positioned at 90 degrees to the incident beam. An optical microscope was
used to view the sample and to position specific regions of interest
under the beam.

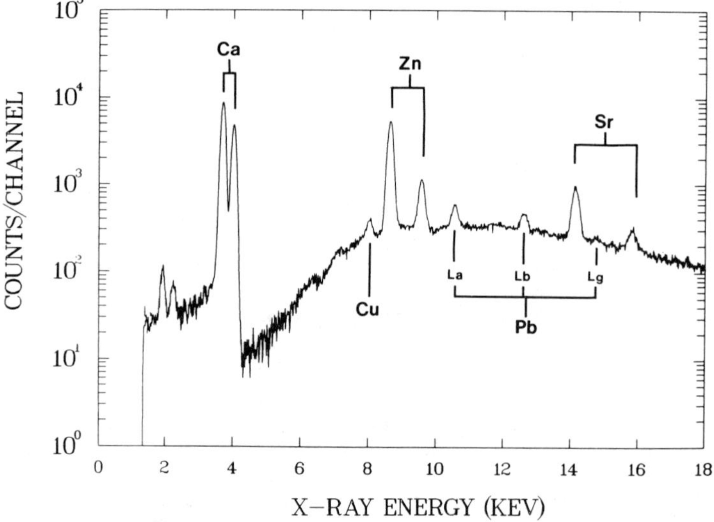

Fig. 1.   A typical spectrum produced using the SRIXE technique is shown
          for one of the thick sections of human tibia. The peaks pro-
          duced by x-rays from lead and calcium and from other elements
          are identified.

A typical spectrum obtained from the irradiation of a thick section of human tibia is shown in Fig. 1. The peak corresponding to the L
x-rays of lead are marked as are the peaks corresponding to the presence
of other elements. The sensitivity of the method was established by use
of standard bone samples such as the International Atomic Energy
Authority H-5 bone mineral standard. The minimum detection limit for
the measurement is estimated to be approximately 1 ppm for an observation time of 300 s, the beam size mentioned above, and a stored electron
current in the NSLS x-ray ring of 100-200 mA.

The results are given in terms of the number of lead L x-ray
counts divided by the number of calcium K x-ray counts. This ratio is
proportional to the actual atomic ratio. The observed values are modified by absorption in the sample and by absorbers used in front of the
detector and in the path of the beam. Values for the absolute concentrations of each element are not derived since the composition of the
bone specimen from point to point is not necessarily constant.

SAMPLE PREPARATION

Portions of tibial diaphysis investigated in this work were obtained from an amputation specimen taken from a male patient 54 years
old. Thick (approx. 5 mm) transverse sections were cut with a saw,
defatted in several changes of a 1:1 mixture of ether-ethanol, and dried
at 60°C overnight. For some specimens, the thickness of these sections
was further reduced to approximately 100 micrometers by grinding the
surfaces on ultrafine abrasive carborundum paper. Complete removal of
the abrasive was achieved by a 5-minute sonication of the thin section
in a 1:1 ether-ethanol bath and drying at room temperature. The use of
a thin section is advantageous, since it helps to ensure that the x-rays
sample a relatively uniform thickness in the direction normal to the
sample face.

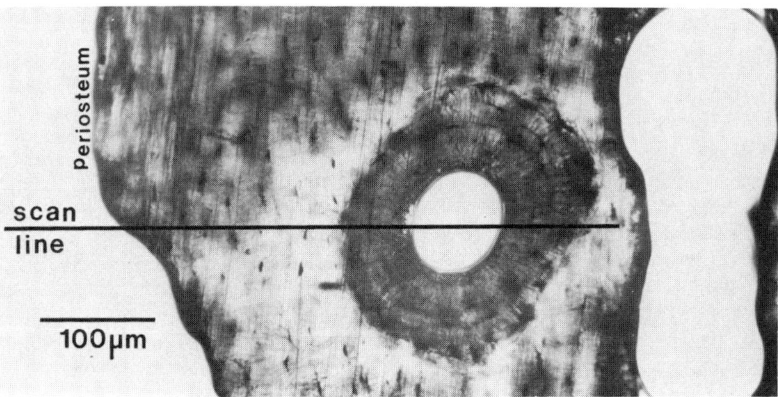

Fig. 2.   Photomicrograph of a typical thin section of bone investigated
           using the SRIXE technique. The periosteal surface is at the
           left of the figure. The lead/calcium atomic ratio shown in
           Fig. 3 was obtained along the line shown in the figure.

MEASUREMENTS OF ELEMENTAL DISTRIBUTIONS

A microphotograph of one of the specimens investigated is shown in
Fig. 2, which emphasizes the heterogeneity of the material and shows the
need for detailed elemental measurement with microscopic resolution
(<100 μm beam size). A line scan, which started about 100 μm outside
the periosteum and included an osteon with its central, circular, Haversian canal, was made for a distance of about 0.5 mm (40 μm x 40 μm beam
spot, 50-μm steps).

The ratio of the Pb Lα/Ca Kα intensities in arbitrary units along
the line scan are shown in Fig. 3. The results are proportional to the
atomic ratio of the two elements. The notable features are the strong
peaking in the ratio observed in the periosteal region and the additional structure observed in the scan across the osteon.

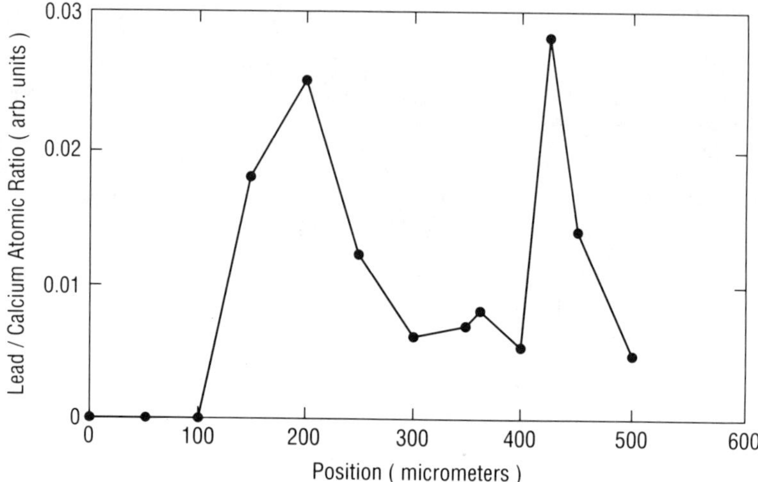

Fig. 3.   The lead/calcium atomic ratio is shown along the scan line
           marked in Fig. 2. The scan shows a strong enhancement of the
           ratio at the periosteal surface and in the interior, in the
           vicinity of the osteon and Haversian canal.

The second step in our investigation was to make a map of the elemental distributions over a 1 mm x 1 mm area that included the periosteal surface. In making the map, net peak areas were found by making a
linear interpolation of the background under the peak, using channels
below and above the peak as reference points. The elements considered
were calcium, zinc, strontium, and lead. The map was made using a beam
size of 50 μm x 50 μm and steps of 100 micrometers between adjacent
points in an 11 x 11 matrix.

A photomicrograph of the region of bone scanned is shown in Fig.
4A. The size of the bone investigated was about 1 mm x 1 mm. The
region mapped is indicated. The elemental maps for calcium, strontium,
zinc, and lead are shown in pseudocolor display in Figs. 4B, 4C, 4D, and
4E, respectively. The color scale has the lowest intensity corresponding to black and the highest intensity to white. The ratio of Pb Lα/Ca
Kα is shown in Fig. 4F.

DISCUSSION AND CONCLUSIONS

Examination of the line scan and the elemental maps shows that the
lead and calcium distributions change as a function of the bone struc-

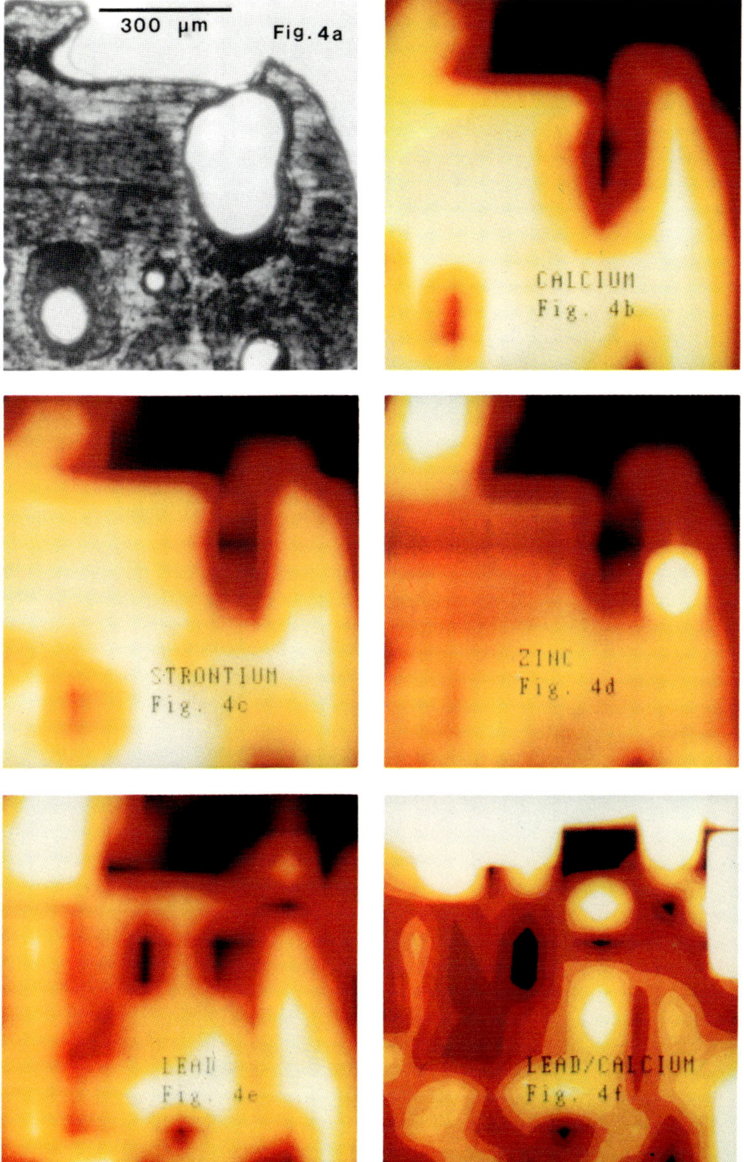

Fig. 4.   This composite figure shows the results of a map of the elemental concentrations for several elements taken over a section of bone about 1 mm x 1 mm in area located at the periosteal surface. A photomicrograph of the surface with the region of scan indicated is shown in Fig. 4A. The elemental concentrations for calcium, strontium, zinc, and lead are shown in Figs. 4B, 4C, 4D, and 4E, respectively. The lead/calcium atomic ratio is shown in Fig. 4F. These maps show that the spatial distribution of a major fraction of the lead concentration is similar to the zinc concentration. The calcium concentration decreases at the surface of the bone; but although the relative value of the lead/calcium atomic ratio does increase at the surface, it also fluctuates throughout the area of the scan as seen in Fig. 4F.

ture examined. The lead-to-calcium ratio determined in the line scan shows that there is a surface peaking at the periosteum. The maps of the two distributions individually show that both the calcium and the lead concentrations decrease at the periosteal surface. However, the map of the lead/calcium ratio shows that the lead concentration decreases less rapidly than does the calcium concentration in the first few hundred micrometers, as shown by the narrow bands of white and yellow at the surface in Fig. 4F.

The results show that the concentration of individual elements and the lead/calcium ratio are non-uniform in the bulk material of the bone. The line scan shows that there is a marked variation across the surface of the large osteon. Measurements were also made on two osteons which had different appearances in an optical photomicrograph. Some variations in the elemental distributions were also observed for these two osteons.

It can be noted that the distributions of calcium and strontium shown in the elemental maps are very similar, as are the distributions of the lead and zinc. Previous studies of the fate of ^{89}Sr and ^{90}Sr present in radioactive fallout have also observed that the behavior of strontium in bone parallels that of calcium (Vaughn, 1981). High levels of dietary zinc have been shown to be followed by a decrease in lead content in rat tibias (Petering, 1978) and zinc has an antagonistic effect on the effect of lead on δ-aminolevulinic acid dehydratase activity in vitro (Abdulla et al., 1979), suggesting a parallel metabolic relationship between the two metals.

Our observations showing the non-uniformity of lead concentration at the periosteum and osteon are in agreement with work by Lindh (Lindh et al., 1978; Lindh, 1980; Lindh, 1981) on the femur and an individual osteon, and with the results of Schidlovsky et al. (this volume), which were also obtained on the tibia and calcaneus.

The total set of observations to date supports two remarks about the interpretation of in-vivo measurements of tibial lead concentration using x-ray fluorescence. First, the assignment of absolute values of average lead concentrations from the L x-rays is hazardous because of the strong attenuation of the low energy exciting x-rays and the emitted x-rays by the bone. Accurate determination of the fluence of these x-rays is subject to assumptions about the structure of the bone. To obtain a correct assessment, the lead distribution as a function of depth in the tibia would have to be included in the analysis of the final lead L x-ray yield. However, an accurate calculation of the lead concentration is almost impossible, since even the gross spatial distribution is not known a priori. Second, looking at K x-rays gives an accurate representation of the atomic ratio of lead to calcium averaged over the entire bone volume (Jones et al., 1987; Somervaille, et al., 1985). The results of K x-ray measurements include contributions from blood contained in the Haversian canals and other fluids in the bone.

Acquisition of data with better spatial resolution on selected bone samples covering a wider range of ages and exposure conditions is necessary for several reasons. First, this will help to elucidate the mechanisms by which the lead is deposited in the bone, the compartments in which it is stored, and the lifetimes for storage. Second, further data is urgently needed for the understanding of results obtained using

L x-ray detection in in-vivo measurements (Rosen, 1989) and should also
be of help in refining the interpretation of the results of the K x-ray
work.

REFERENCES

Abdulla, M., Svensson, B. S., and Haeger-Aronson, B., 1979, Antagonistic
     effects of zinc and aluminum on lead inhibition of δ-
     aminolevulinic acid dehydratase, <u>Arch. Environ. Health</u>, 34:463.
Jones, K. W., Schidlovsky, G., Williams, F. H., Jr., Wedeen, R. P., and
     Batuman, V., 1987, In-vivo determination of tibial lead by K x-
     ray fluorescence with a ^{109}Cd source, <u>in</u>: "In Vivo Body
     Composition Studies, Proceedings of an International Symposium,"
     K. J. Ellis, S. Yasumura, and W. D. Morgan, eds., Institute of
     Physical Sciences in Medicine, London.
Jones, K. W., and Gordon, B. M., 1989, Trace element determinations with
     synchrotron-induced x-ray emission, <u>Anal. Chem.</u>, 61:341A.
Lindh, U., Brune, D., and Nordberg, G., 1978, Microprobe analysis of
     lead in human femur by proton induced x-ray emission (PIXE),
     <u>Sci. Tot. Envir.</u>, 10:31.
Lindh, U., 1980, A nuclear microprobe investigation of heavy-metal
     distribution in individual osteons of human femur. <u>Int. J. App.
     Rad. and Isotopes</u>, 31:737.
Lindh, U., 1981, The nuclear microprobe applied to bioenvironmental
     studies, <u>Nucl. Instrum. and Meth.</u>, 181:171.
Petering, H. G., 1978, Some observations on the interaction of zinc,
     copper and iron metabolism on lead and calcium toxicity,
     <u>Environ. Health Perspec.</u>, 25:141.
Rosen, J. F., Markowitz, M. E., Bijur, P. E., Jenks, S. T.,
     Wielopolski, L., Kalef-Ezra, J. A., and Slatkin, D. N., 1989, L-
     line x-ray fluorescence of cortical bone lead compared with the
     CaNa$_2$EDTA test in lead-toxic chlidren: public health
     implications, <u>Proc. Natl. Acad. Sci. USA</u>, 86:685.
Somervaille, L. J., Chettle, D. R., and Scott, M. C., 1985, In-vivo
     measurement of lead in bone using x-ray fluoresence, <u>Phys. Med.
     Biol.</u>, 30:929.
Vaughn, J., 1981, "The Physiology of Bone," Clarendon Press, Oxford.

AUTOMATED BONE LEAD ANALYSIS BY K-X-RAY FLUORESCENCE

FOR THE CLINICAL ENVIRONMENT

D.E. Burger, F.L. Milder, P.R. Morsillo, B.B. Adams and H. Hu[*]

ABIOMED, Inc., Danvers, MA and [*]Channing Laboratory
Harvard Medical School, Boston, MA

INTRODUCTION

The use of K-x-ray fluorescence (K-XRF) for measuring the content of
lead in bones began over a decade ago (Ahlgren et al., 1976). In K-XRF,
lead atoms are made to emit K-x-rays by exciting them with gamma radiation
from a radioisotope. It was realized that extracting the signal (the
fluorescence photons) from the large background (Compton scattered photons)
was the chief impediment to be overcome. The use of carefully chosen
geometries coupled with radioisotopes with gamma emissions of a single or
few energies has succeeded in this regard (Ahlgren et al., 1976; Price et
al., 1984; Somervaille et al., 1985; Jones et al., 1987). For radiation
safety reasons, as well as ease of access to bone material with little
overlying tissue, measurement sites are bones in the appendicular skeleton
(Ahlgren et al., 1980; Somervaille et al., 1987; Somervaille et al.,1985;
Chettle et al., 1989).

In this paper we describe a system which incorporates all of the
important features of previous researchers and adds several significant
improvements. The development of our system was supported by the NIH SBIR
grant program, the function of which is to foster transfer of research
technology from the laboratory into the general clinical environment. To
this end, our device had to: (1) be useable in a clinical setting by persons with no previous knowledge of bone lead measurements or any aspect of
x-ray measurements, (2) be self-calibrating, (3) produce real-time results
and (4) be usable on any site on either lower limb.

MATERIALS

Fig. 1 shows a block diagram of the organization of the system. Fig.
2 shows a picture of the system, positioned to acquire data at the mid-
tibia diaphysis. The source consists of 200 mCi of Cd-109 (Du Pont, Bed-
ford, MA) in a tungsten source holder which also serves as the collimator.
Within the collimator is a computer controlled tungsten shutter. Data is
collected in a 164 degree backscatter geometry. The patient chair is
mounted on a platform which is mounted on lockable casters. The chair and
platform are lead free and of low-Z composition to reduce Compton scattering. The source to skin distance is nominally 2.5 cm.

The detector (Canberra Industries, Meriden, CT) is a 50 mm diameter

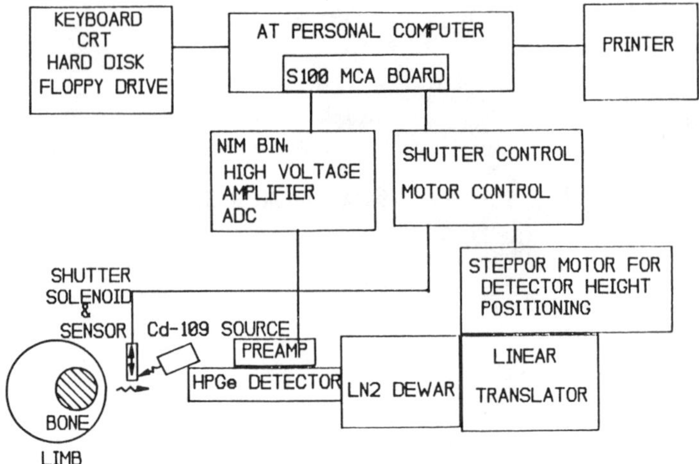

Fig. 1.   The Body Lead Analyzer is a complete self-contained system.

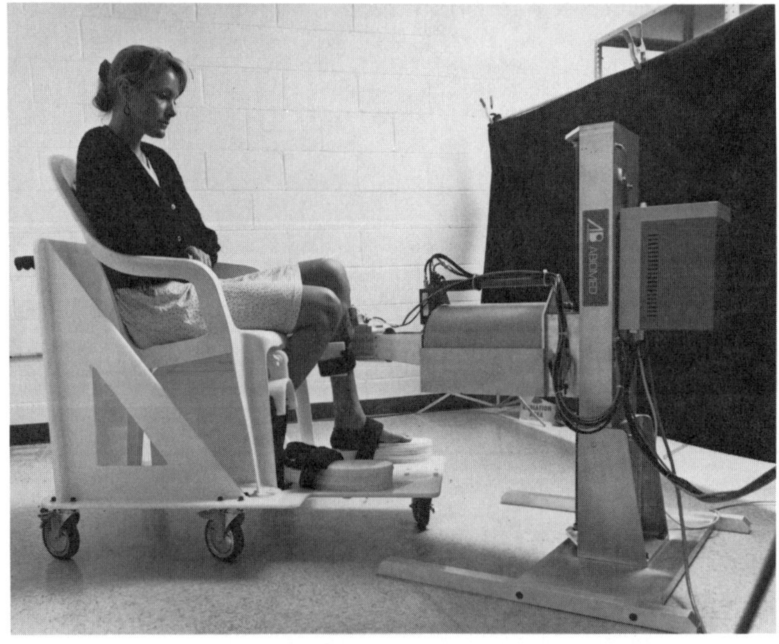

Fig. 2.   The patient sits in an adjustable chair that is lead free.

by 20 mm deep, high purity, intrinsic germanium (HPGe) detector. After amplification and shaping, the signal is coupled to a 450 MHz Wilkinson type ADC with integral pile-up rejection. The digitized signal is acquired by a Canberra model S100 multichannel analyzer located on a plug-in board in either an AT compatible or 386 computer.

System software operates in the Windows environment (Microsoft Corporation, Redmond, WA). Aside from Windows and the multichannel analyzer software, the XRF system uses the data base utility Btrieve (Novell, Inc., Provo, UT).

To take data on the patient, the user interacts with the computer via a mouse to select one of the following five lower limb sites: patella, proximal tibia metaphysis, mid-tibia diaphysis, medial maleolus, and calcaneus. The normal time for acquisition is 30 minutes. Analysis takes about 3 seconds and the result is reported on the screen under the site being measured and also stored in a patient data file. Results are reported as ppm lead/bone mineral with the estimated ppm uncertainty reported in parentheses.

METHODS

Systematic testing of the accuracy and precision of the ppm lead measurement was undertaken using lower limb phantoms built in-house. Previous researchers (Somervaille et al., 1986; Wedeen et al., 1987) have shown that K-XRF systems accurate for phantom models are also accurate for limbs. We, therefore, chose to methodically test our system with phantoms, for which we could create tibias with very low ppm lead and normal amounts of "bone mineral". Each phantom incorporated a simulated tibia molded from plaster-of-Paris with a central marrow cavity. Lead acetate was added to the plaster-of-Paris to obtain tibias with 0-1000 ppm lead/plaster-of-Paris dry weight. Systematic testing used only the 0-100 ppm phantoms. Water and polyethylene surrounded the tibias to simulate soft tissue. The overlying "soft tissue" was 4 mm thick in the front of the limb phantom at the site of the test measurements.

The accuracy of the system was measured by taking phantoms with 0, 4.2, 8.4, 12.6, 16.9, 25.3, 42.2, and 84.3 ppm lead and measuring each phantom multiple times. The multiplicity of measurements included moving the phantoms between measurements and also intentionally shifting the gain of the amplifier to the acceptable extremes of the calibration limits. In all, 53 measurements were taken in verifying the system accuracy. The analyzed peaks are the K-alpha$_1$ x-ray singlet, the K-beta$_1$, doublet, and the coherent scatter peak.

A variety of interferences must be dealt with in extracting the net peak areas to derive the ppm lead/plaster-of-Paris. Table I shows these interferences along with the quantities that scale them and their overall magnitudes. Beyond the interferences corrected for in previous work by other researchers, we also correct for: (1) the difference in attenuation between the 88 KeV coherent scattered gamma-ray and the 75 and 85 KeV lead x-rays; (2) the contribution to the coherent scatter peak made by the lead atoms themselves; (3) the apparent contribution to the coherent scatter peak which is actually the lead K-beta$_2$, fluorescence x-ray; and (4) the contribution to the coherent scatter peak made by the surrounding soft tissue. After applying the above corrections, we compared the theoretical peak area ratios to the observed ratios, expressing each in terms of ppm lead. The results are discussed in the next section.

The precision of the K-XRF measurement system was separated into two components, **systematic error** and **statistical error**, by the following

Table 1. Signal, background and interferences in extracting lead ppm

Signal	**Scales With**	
Pb fluorescence	Pb content	
Ca + P coherent scatter	Ca + P content	

Background	**Scales With**	
Compton scatter	Soft plus hard tissue mass	

Interference	**Scales With**	**Magnitude of Effect**
Soft tissue coherent scatter	Soft tissue mass	+5% on Ca + P coherence signal
Differential attenuation	Overlaying tissue	-3% on alpha fluorescence
Pb coherent scatter	Pb content	+.01%/ppm on Ca + P coherence signal
Pb beta-2,4 fluorescence	Pb content	+.04%/ppm on Ca + P coherence signal
Photo-electron bremsstrahlung	Ca content	-18 ppm on beta fluorescence typical

experiments. The 12.6 ppm phantom was measured repeatedly using data
collection times of 7.5, 15, 30, 60, 120, and 240 minutes and the standard
deviation (SD) of the measurements for each collection time separately
determined. The data of SD versus measurement time was then broken into a
systematic component and a statistical component by fitting the results to
a form:

$$SD=(sys + stat/min)^{1/2} \, .$$

Here, "sys" is the systematic error, "stat" is a statistical error
based on counts in the peaks, backgrounds, and an error constant and "min"
is the measurement time in minutes. Once this separation is done, one can
then predict the uncertainty of any one measurement by adding, in quadra-
ture, the predicted statistical error of that measurement to the measured
systematic error.

RESULTS AND DISCUSSION

Fig. 3 shows the results of the accuracy investigation. The best fit
linear regression line has a slope of 0.984 with a correlation coefficient
of 0.982. The ideal would be a slope of unity. Thus, we are accurate to
better than 2% over the range from 0-100 ppm lead. The zero intercept of
the regression line is -0.01 ppm lead, showing that the accuracy is maintained over the entire range and that there is no offset bias. The standard error of estimate is 3.75 ppm lead. This data was taken with 30 minute
measurements and a source strength of 125 mCi.

Fig. 4 presents the results of the precision experiments. Plotted in
the figure are: (1) the observed SD of 5 measurements at each of the measurement times (except for the 30 minute point which includes 21 measurements) with a 120 mCi source; (2) a prediction of that form for a 200 mCi
source strength (the strength of a newly replaced source); and (3) a prediction of that form for an 80 mCi source, which is the strength at which
we recommended source replacement. The measured systematic error is approximately 1 ppm lead.

From the same figure the standard error of estimate for a new source,
using a 30 minute collection time, is approximately 3.2 ppm lead for the
phantoms. It is known that in going from a plaster-of-Paris tibia to true
bone, the coherent scatter cross section per gram of material is greater
for bone mineral than for plaster-of-Paris. The net result of this is that
the SD error estimate for a single measurement increases on the order of

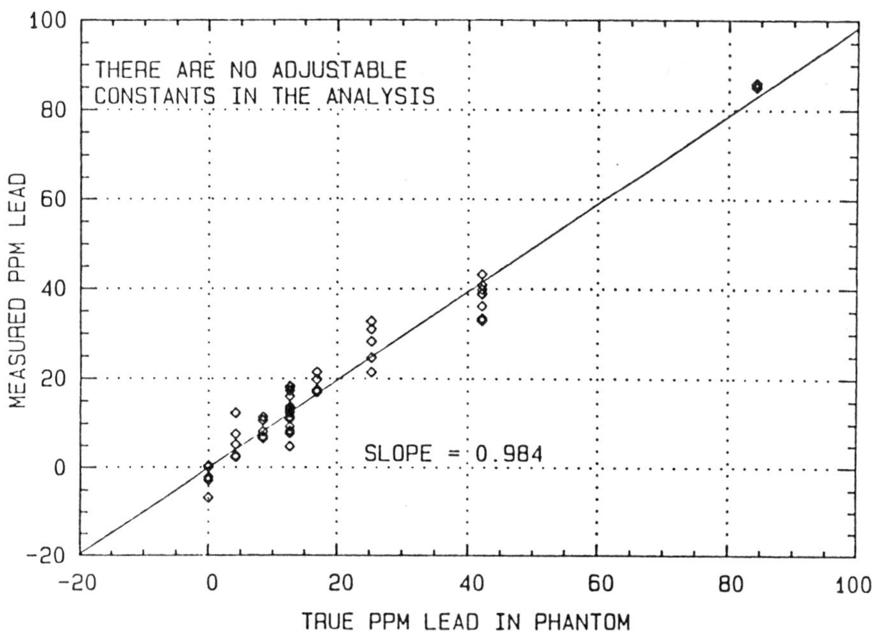

Fig. 3. Accuracy in the measurement of ppm is better than
2 percent.

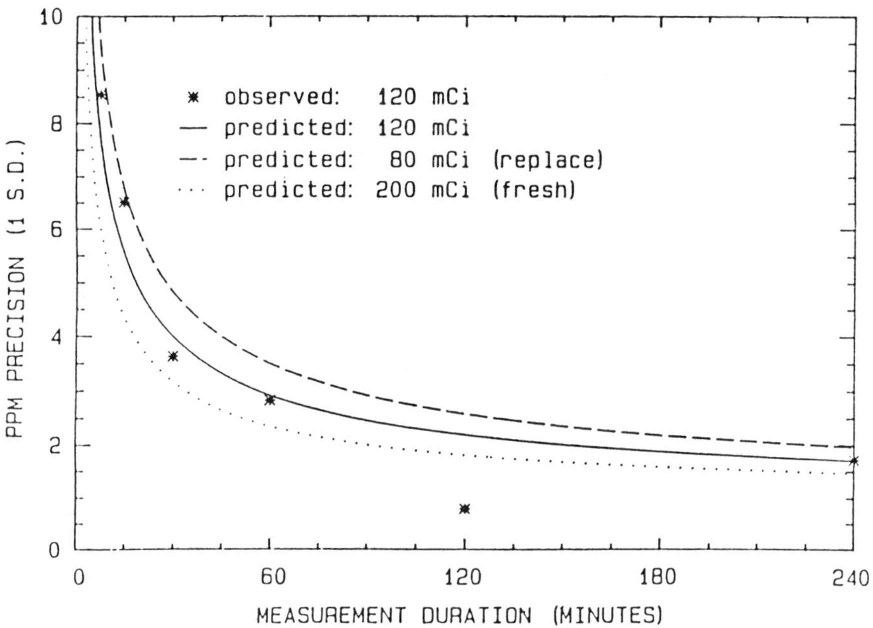

Fig. 4. The precision depends on source strength and the duration
of the measurement. With a new (200 mCi) source and a 30
minute collection time, the error estimate is 3.2 ppm.

25-50%. Thus, a nominal sensitivity number for the SD of a real tibia
measurement is 5 ppm.

Radiation exposure for the 30 minute measurement was taken on the
limb phantoms with thermoluminescent dosimeters placed at the "skin" surface, at the tibia surface, and in the marrow cavity. The on-axis skin
exposure is 2.80 mGy and the on-axis exposure in the marrow cavity is 0.88
mGy. Taking the integral of exposure per unit volume times the exposed
volume in the marrow cavity gives the energy absorbed by the red bone
marrow. Dividing this by the total body red marrow mass results in the
equivalent dose to the total red marrow organ. This value is 0.7 uSv. By
comparison, the typical skin exposure for a single dental bite-wing x-ray
is 4 mGy with a red bone marrow absorbed dose of 7-10 uSv (Sowby 1982).

The above described prototype clinical system is self calibrating,
analysis results are on-line and patient biographic data, raw spectrum data
and measurement results are stored automatically. This prototype is currently in clinical use at Brigham and Women's Hospital, Boston, MA.

ACKNOWLEDGEMENTS

Supported in part by NIEHS SBIR grant number 2R4403ES918.

REFERENCES

Ahlgren, L., Liden, K., Mattsson, S, Tejning, S., 1976, X-ray fluor.
    analysis of lead in the human skeleton in vivo, Scan J Work Envir, 2:
    82.
Ahlgren, L., Haeger-Aronsen, B., Mattsson, S. and Schutz, A. 1980, In vivo
    determination of lead in the skeleton after occupational exposure to
    lead, Brit J Ind Med, 37: 109.
Chettle, D.R., Scott, M.C. and Somervaille, L.J., Improvements in
    the precision of in vivo bone lead measurements, Physics in Medicine
    and Biology, In Press.
Jones, K.W. Schidlovsky, G., Williams, F.H., Wedeen, R.P. and Batuman,
    V., 1987, In vivo determination of tibial lead by K-X-Ray fluorescence
    with a Cd-109 source, In: "In vivo Body Composition Studies, Proceed-
    ings of an International Symposium" K.J. Ellils, S. Yasamura, W.D.
    Morgan, eds., Bocardo Press Limited, Oxford: 363.
Price, J., Baddeley, H., Kenardy, J.A., Thomas, B.J., Thomas, B.W., 1984,.
    In vitro x-ray fluorescence estimation of bone lead concentrations
    in Queensland adults, Br J Radiol, 57: 29.
Somervaille, L.J., Chettle, D.R. and Scott, M.C., 1985, In vivo measure
    ment of lead in bone using x-ray fluorescence, Phys Med Biol, 30: 929.
Somervaille, L.J., Chettle, D.R., Scott, M.C., Aufderheide, A.C., Wall
    gren, J.E., Wittmers, L.E. and Rapp, G.R., 1986, Comparison of two in
    vitro methods of bone lead analysis and the implications for in vivo
    measurements, Phys Med Biol, 31: 1267.
Somervaille, L.J., Chettle, D.R., Scott, M.C., Krishan, G., Browne, C.J.,
    Aufderheide, A.C. Wittmers, L.E., and Wallgren, J., 1987, XRF of lead
    in vivo: Simultaneous measurement of cortical and trabecular bone in
    a pilot study, In: "In vivo body composition studies", Proceedings of
    an International Symposium, K.J. Ellis, S. Yasumara and W.D. Morgan,
    eds., Bocardo Press Limited, Oxford: 325.
Sowby, F.D., ed., 1982, Protection of the Patient in Diagnostic Radiolo
    gy, In: "Annals of the ICRP", ICRP Pub. 34, Pergamon Press, N.Y.
Wedeen, R.P., Batuman, V., Quinless, F., Williams, F.H., Bogden,
    J., Schidlovsky, G. and Jones, K.J., 1987, Lead nephropathy: in vivo
    x-ray fluorescence for assessing body lead stores, In: "In vivo body
    composition studies", Proc. of an Inter. Sym., K.J. Ellis, S. Yasumu-
    ra and W.D. Morgan, eds., Bocardo Press Limited, Oxford: 357.

TRACE ELEMENTAL ANALYSIS IN BONE USING X-RAY MICROSCOPY[*]

R. S. Bockman, R. P. Warrell, Jr., and B. Levine
Department of Medicine, Cornell University Medical College
New York, New York 10021

J. G. Pounds, G. Schidlovsky, and K. W. Jones
Brookhaven National Laboratory
Upton, New York 11973

INTRODUCTION

The detection, quantification, and localization of trace elements
in biological tissues present major difficulties. Recent advances in x-
ray fluorescence now make it possible to quantify and localize naturally
occurring as well as therapeutic trace elements in a variety of tissues,
including bone. Studies of the metabolism and the pharmacology of these
high-Z trace elements are feasible. Using a synchrotron radiation
source, we used x-ray fluorescence to measure the naturally occurring
trace elements, iron, zinc, copper, and strontium, as well as gallium, a
new therapeutic agent that inhibits calcium loss from bone (Warrell et
al., 1984; Cournot-Witmer et al., 1987; Bockman et al., 1986; Warrell et
al., 1987; Repo et al., 1988). Our data show that gallium preferentially accumulates in the metabolically active regions of cortical and
trabecular bone, and leads to changes in iron and zinc levels within
bone.

METHODS

Six- to eight-week old female Sprague-Dawley rats were injected
intraperitoneally with 2.5 mg gallium nitrate/kg every other day for two
weeks. At the end of the injection schedule, the animals were killed,
the cleaned tibial bones were removed, frozen, and cut longitudinally
into 12-μm sections which were mounted on 7.3-μm polyamide film supported in cardboard slide mounts. X-ray fluorescence was conducted at
the Brookhaven National Laboratory National Synchrotron Light Source
(NSLS) (Jones et al., 1985; Hanson et al., 1987; Jones et al., 1988).
The synchrotron source provides an intense beam of collimated and

[*]This work was supported in part through PHS Grants CA38645,
CA42445, and CA29502 and NIH Biotechnology Research Resource Grant
P41RR01838. Development of the analytical instrumentation was supported
by the US Department of Energy, Office of Basic Energy Sciences,
Division of Chemical Sciences, Processes and Techniques Branch, Contract
DE-AC02-76CH00016.

Table 1. Percent Change in Trace Element Content in Bone Regions after
          Gallium Treatment

	Fe	Zn	Sr
Metaphysis	-47[1]	-27[2]	-10
Diaphysis	-74[2]	+32	-19

[1]Change compared to control bone significant, $p < 0.02$.
[2]Change compared to control bone significant, $p < 0.01$.

polarized x-rays. An energy-dispersive Si(Li) detector detects the x-ray emissions from the high-Z elements in the beam's path. To correct for variations in the sample, all data were normalized to calcium content, based on the gram weight ratio of the trace element to calcium.

RESULTS AND DISCUSSION

The concentration of calcium in the various bone regions was calculated using known bone mineral standards. More calcium (0.32 g/ml avg) was measured in the compact bone of the diaphyseal cortex than in the newly mineralized metaphyseal region (0.13 g/ml). This difference was highly significant, $p < 0.001$. The ratios of iron and zinc were significantly greater in the metaphyses compared to the diaphyses, $p < 0.001$. There was no difference in the ratios of strontium to calcium when the metaphyseal and diaphyseal bone regions were compared.

In vivo treatment with gallium nitrate resulted in detectable levels of gallium in the bone samples. Based on gallium-to-calcium ratios, there was three times the concentration of gallium in the metaphyseal bone compared to the diaphysis. A similar absolute difference was measured by atomic absorption in powdered bone samples from the metaphyses and diaphyses of similarly treated rats (Bockman et al., 1986; Repo et al., 1988).

The administration of gallium nitrate to the growing rats altered the concentration of the naturally occurring trace elements in iron and zinc in the various bone regions. Both iron and zinc concentration fell significantly in the metaphyseal region following such treatment. Only the iron concentration changed in the diaphyseal region, whereas no significant change in zinc was seen. No significant changes in strontium or copper were observed in either bone region following gallium treatment (Table 1). Each data point is the mean of 6-12 separate measurements from the bones of 1-2 rats. The variability was ±10%.

Scans were made across the diaphyseal cortex. No significant change in calcium concentration was observed as one scanned from the endosteal to periosteal surfaces of the diaphyseal cortex. This confirms the uniformity of mineral content in the compact bone. By contrast, there is a sharp drop in the ratio of gallium to calcium in the mid portion of this cortical region (Fig. 1). This finding suggests that gallium was preferentially accumulating in the periosteal and endosteal regions of the cortical bone.

SUMMARY

Following in vivo administration of gallium nitrate, the greatest concentrations of the therapeutic element gallium localized in the meta-

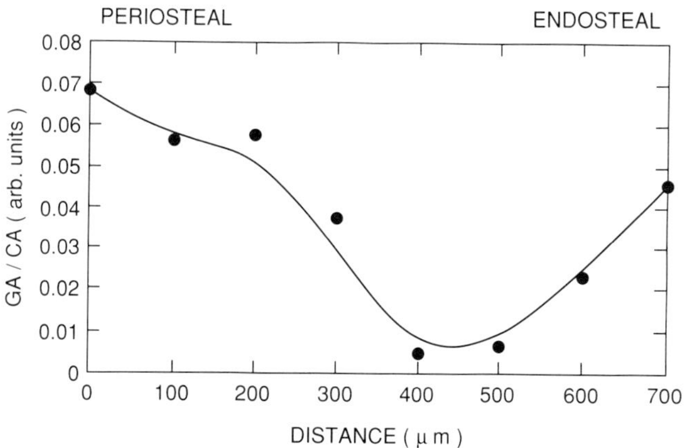

Fig. 1.   The ratio of elemental gallium to calcium is illustrated as a
function of anatomic localization in tibial bone of a rat.
XRM spectra were obtained as the beam scanned from the periosteal to the endosteal surface of the proximal portion of
the diaphyseal cortex of a rat treated in vivo with gallium
nitrate.

physis and at the endosteal and periosteal surfaces of the diaphysis.
These are the regions of greatest metabolic activity, where new bone
formation and remodeling are occurring. The lowest levels of gallium
were noted in the mid-cortical region of the diaphyseal shaft where bone
turnover is least. The accumulation of gallium in the metaphysis was
associated with a concomitant fall in iron and zinc. The gallium-
induced change in the metaphysis may reflect a subtle modulation of
metal dependent enzymes that are necessary for the active bone modeling
that occurs in this bone region. X-ray microscopy has provided the
first insights into the localization and possible mechanisms of action
of gallium in bone.

REFERENCES

Bockman, R. S., Boskey, A. L., Blumenthal, N. C., Alcock, N. W., and
      Warrell, R. P., Jr., 1986, Gallium increases bone calcium and
      crystallite perfection of hydroxyapatite, Calcif. Tis. Int.,
      39:376.
Cournot-Witmer, G., Plachot, J. J., Bourdeau, A., Lieberherr, M., and
      Balsan, S., 1987, Bone modeling in gallium-nitrate treated rats,
      Calcif. Tis. Int., 40:270.
Hanson, A. L., Jones, K. W., Gordon, B. M., Pounds, J. G., Kwiatek,
      W. M., Long, G. J., Rivers, M. L., and Sutton, S. R., 1987,
      Trace element measurements using white synchrotron radiation,
      Nucl. Instrum. and Meth., B24/25:400.
Jones, K. W. , Gordon, B. M., Hanson, A. L., Pounds, J. G., Rivers,
      M. L., and Schidlovsky, G., 1985, The NSLS photon microprobe--a
      resource for ultrasensitive trace element analysis on a
      microscopic level, EMSA Bulletin, 15(1):28.
Jones, K. W., Gordon, B. M., Hanson, A. L., Kwiatek, W. M., and Pounds,
      J. G., 1988, X-ray fluorescence with synchrotron radiation,
      Ultramicroscopy, 24:313.

Repo, M. A., Bockman, R. S., Betts, F., Boskey, A. L., Warrell, R. P.,
    Jr., 1988, Effect of gallium on bone mineral properties, <u>Calcif.
    Tis. Int.</u>, 43:300.
Warrell, R. P., Jr., Bockman, R. S., Coonley, C. J., Isaacs, M., and
    Staszewski, H. J., 1984, Gallium nitrate inhibits calcium
    resorption from bone and is effective treatment for cancer-
    related hypercalcemia, <u>J. Clin. Invest.</u>, 73:1487.
Warrell, R. P., Jr., Israel, R., Frisone, M., Snyder, T., Gaynor, J. J.,
    and Bockman, R. S., 1988, Gallium nitrate for acute treatment of
    cancer-related hypercalcemia: a randomized, double-blind
    comparison to calcitonin, <u>Ann. Intern. Med.</u>, 108:669.

FURTHER IMPROVEMENTS OF XRF ANALYSIS OF CADMIUM IN VIVO

U. Nilsson, L. Ahlgren, J-O. Christoffersson
and S. Mattsson

Department of Radiation Physics, Lund University
Malmö General Hospital, S-214 01 Malmö, Sweden

INTRODUCTION

Although clinically useful, current equipment and methods for X-ray
fluorescence (XRF) analysis in vivo should be improved in order to:
1) obtain better detectability of heavy elements, and thus extend the
measurements to new groups of persons and to new elements, 2) make the
equipment transportable and easier to handle in busy work-places or as
bedside instruments at hospitals when seriously ill patients are to be
examined. The stepwise introduction of new instruments and methods for in
vivo XRF analysis has resulted in considerable improvements of the
detectability for some elements but not for others.

The possibility of using XRF analysis for in vivo determination of
cadmium concentrations in the kidney cortex was shown by Ahlgren and
Mattsson (1981), who used 60 keV photons from a 11 GBq ^{241}Am source for
the excitation. However, the technique has a high detection limit, 20
μg/g for a kidney depth of 30 mm (distance skin to kidney surface) and a
measuring time of 30 minutes. An improved XRF technique was developed by
Christoffersson and Mattsson (1983), who used a partly plane polarised
X-ray beam for excitation and a triaxial geometry to reduce the
background. Together with the higher photon fluence rate available this
resulted in a lowered detection limit; approximately 9 μg/g for the same
conditions as mentioned above. By modifying the detector collimator so
that the detector field of view can cover the whole kidney, regardless of
the depth of the kidney, the detection limit has been further improved to
approximately 6 μg/g (Christoffersson et al., 1987). The aim of this work
was to investigate whether it is possible to decrease the detection limit
further by using a Si(Li) instead of a Ge detector for the registration
of the characteristic cadmium X-rays in the same way as was done by
Kaufman (1979) when analyzing iodine in vitro.

MATERIAL AND METHODS

The equipment and geometries used to produce the polarised photon
beam have been described elsewhere (Christoffersson and Mattsson, 1983).
To simulate an in vivo measurement, a kidney-simulating thin-walled
polymethyl methacrylate (PMMA) cylinder, containing a cadmium solution of
known concentration, was placed in a water-filled tank with PMMA walls.

A high-purity, planar Ge detector (FWHM=500 eV at 23.1 keV) and a
planar Si(Li) detector (FWHM=470 eV) were used. The detectors had

identical diameters, 16 mm, and thicknesses, 5 mm. Their positions, inside the protecting aluminium castings, were determined radiographically. By using a detector collimator of variable length (Christoffersson et al., 1987), an identical and optimum field of view of the detectors could be achieved. Both detectors were equipped with pulsed optical feedback pre-amplifiers and were connected to a 80 MHz Wilkinson ADC via the same linear amplifier (shaping time: 2 μs). The medial distance from the front of the polarised photon beam collimator to the centre of the kidney simulating cylinder was 70 mm, and the depth from the front of the detector collimator to the inside of the cylinder wall was 30 mm.

The variation of detectability with kidney depth was investigated using the described Si(Li) detector system. The kidney-simulating phantom was placed in the water tank at various depths, 30-70 mm.

From the pulse height distributions obtained, the number of counts in a region over the cadmium K_α peaks (23.1 keV) and in the corresponding

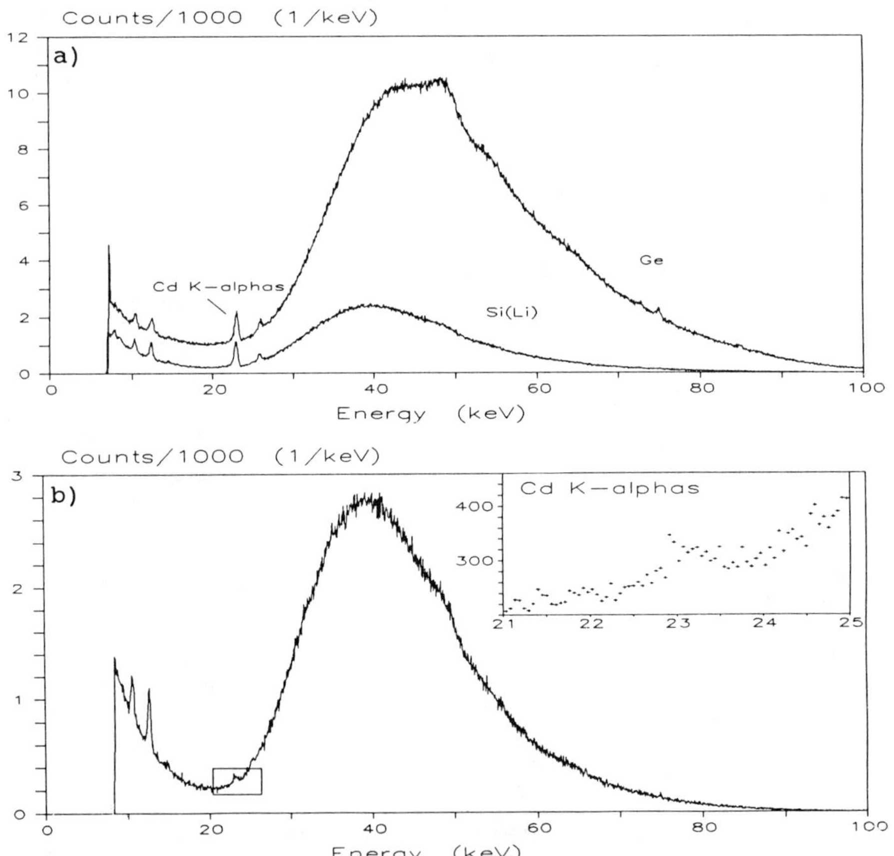

Fig. 1.   a) Pulse height distributions from the measurement on the cadmium-containing kidney phantom (200 μg/g).
b) Pulse height distribution from an <u>in vivo</u> measurement using the Si(Li) detector (measuring time 30 min). The cadmium concentration in the deeply lying kidney cortex (Table 1) was estimated to be (41 ± 11) μg/g.

Table 1. Measured Cadmium Concentrations in the Kidney Cortex of Two
          Non-occupationally Exposed Persons

Status	Age (y)	Kidney position Medial distance (mm)	Depth (mm)	Measured cadmium concentration in kidney cortex* (μg/g)
Former Smoker	43	68	34	7 ± 3
Smoker	63	82	46	41 ± 11

*The indicated uncertainty, 1 SD, is due to counting statistics and the
estimated uncertainty in determining the kidney position, ±3 mm (Ahlgren
and Mattsson, 1981).

background (regions to the left and right of the K_α peaks) were
determined, as was the number of counts in a region centred over the
distribution of incoherently scattered photons. Using data from these
distributions, the minimum detectable concentrations (MDC:s) of cadmium
corresponding to three standard deviations of the background have been
calculated for a measuring time of 30 minutes.

Two persons, who had not been occupationally exposed to cadmium, a
smoker and a former smoker, were measured for cadmium in the cortex of
the right kidney with the person sitting as described by Christoffersson
and Mattsson (1983). The measuring time was 30 minutes. The position of
the kidney was determined in the sitting position before the measurements
by an ultrasonic technique.

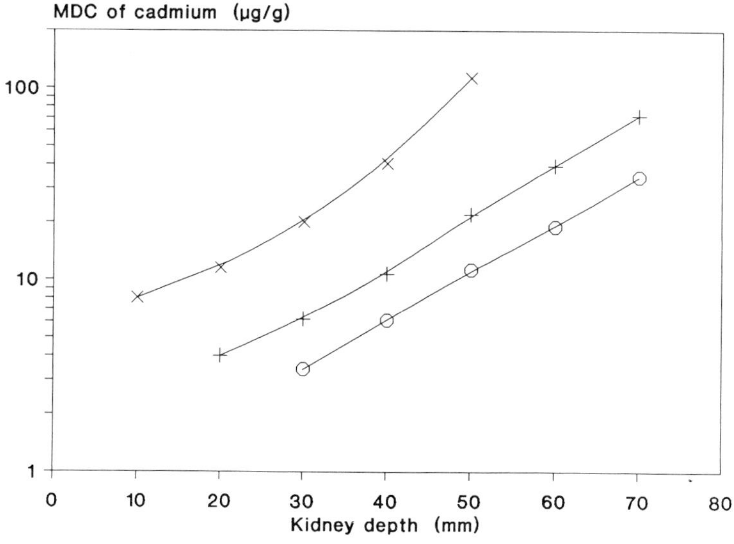

Fig. 2. The minimum detectable concentration (MDC) of cadmium in the
         kidney cortex for various kidney depths and a measuring time of
         30 minutes. The measurements were made using a cylindrical
         kidney phantom. The medial distance was 70 mm. x, [241]Am source
         (Ahlgren and Mattsson, 1981); +, polarised photons, Ge detector
         (Christoffersson et al., 1987); o, polarised photons, Si(Li)
         detector (this work).

299

RESULTS AND DISCUSSION

Pulse height distributions from the two detectors when measuring the same cadmium-filled phantom (200 μg/g), placed in a water tank, are given in Fig. 1a.

The evaluation of the pulse height distributions shows that an improvement in the MDC of cadmium by a factor of about 2 can be achieved for the investigated geometry if a Si(Li) detector is used instead of a Ge detector. The reason for this improvement can be explained by the difference in the variation of detection efficiency with photon energy for the two detector materials. Using the Si(Li) detector the count rate of incoherently scattered radiation (the most probable energy is between 40 and 50 keV, see Fig. 1a) is only 18 percent of the count rate using the Ge detector. According to manufacturer data, the full energy detection efficiency at a photon energy of 50 keV is approximately 25 percent for Si and 100 percent for Ge. Furthermore, the results show that with the use of the Si(Li) detector the net count rate in the cadmium K_α peaks is reduced to 79 percent and the background under these peaks to 21 percent. The full energy detection efficiency for the two detector materials is reported to be almost equal at 23.1 keV. The main reason for the decrease of the background under the characteristic cadmium K_α peaks, and consequently, an improvement in the MDC of cadmium, is the significant reduction of the detection efficiency for Si at a photon energy above approximately 30 keV. The experimental results indicate that the main source may be related to incomplete charge collection or other effects of the detector (Goulding, 1977). The contribution to this background from multiple scattered photons in the object seems to be of less significance (Kaufman, 1979).

The measured MDC:s of cadmium for various kidney depths are given in Fig. 2, which also shows the reported MDC values for the earlier XRF techniques. Using the Si(Li) detector, the MDC of cadmium can be lowered by approximately a factor of two for kidney depths between 30 and 70 mm.

The results of the _in vivo_ measurements are shown in Table 1 and Fig. 1b. The cadmium concentration in the kidney cortex was quantified from phantom measurements made with the same geometry.

CONCLUSION

Compared to occupationally exposed persons, the non-occupationally exposed show much lower levels of cadmium in the kidney cortex, approximately 10-20 μg/g (Elinder et al., 1976) in comparison to as much as 300 μg/g (Christoffersson et al., 1987).

As normal depths of the kidney are to be found in the range 40 to 50 mm (Christoffersson et al., 1987), the improved detection limit achieved using the Si(Li) detector implies that the majority of the public now can be screened for cadmium in the kidney cortex, regardless of the depth of the kidney, and the person's occupation and smoking habits.

REFERENCES

Ahlgren, L., and Mattsson, S., 1981, Cadmium in man measured _in vivo_ by X-ray fluorescence analysis, _Phys. Med. Biol._, 26:19.
Christoffersson, J-O., and Mattsson, S., 1983, Polarised X-rays in XRF analysis for improved _in vivo_ detectability of cadmium in man, _Phys. Med. Biol._, 28:1135.

Christoffersson, J-O., Welinder, H., Spang, G., Mattsson, S., and
    Skerfving, S., 1987, Cadmium concentration in the kidney cortex of
    occupationally exposed workers measured in vivo using X-ray
    fluorescence analysis, Environ. Res., 42:489.
Elinder, C.G., Kjellstrom, T., Friberg, L., Lind, B., and Linnman, L.,
    1976, Cadmium in kidney cortex, liver and pancreas from Swedish
    autopsies, Arch. Environ. Health, 31:292.
Goulding, F.S., 1977, Some aspects of detectors and electronics for X-ray
    fluorescence analysis, Nucl. Instr. Meth., 142:213.
Kaufman, L., 1979, Techniques for in vitro fluorescent excitation
    analysis of stable tracers, in: "Medical Applications of
    Fluorescent Excitation Analysis," eds., L. Kaufman, D.C. Price, CRC
    Press, Orlando.

RECENT DEVELOPMENTS IN THE PROMPT-GAMMA TOTAL BODY NITROGEN

MEASUREMENT FACILITY OF THE TORONTO GENERAL HOSPITAL

S.S. Krishnan, K.G. McNeill, J.E. Harrison

Toronto General Hospital, Departments of Physics and Medicine
University of Toronto, Toronto, Canada.

INTRODUCTION

The measurement of body nitrogen by neutron activation analysis
provides a measure of protein status, using the constant numerical
relationship of 6.25 between nitrogen and body protein. Measurements of
nitrogen balance have been used routinely for studying changes of total
body nitrogen (TBN). However, a method to measure total body protein
(TBP), itself, is required to assess protein status of an individual,
resulting from disease or from treatment. TBP has been clinically very
useful in several studies, e.g. obesity following calorie restriction
(Stallings et al., 1988), ascites following peritoneovenous shunting
(Blendis et al., 1986), renal failure following the start of treatment by
continuous ambulatory peritoneal dialysis (Schilling et al., 1985), cancer
(Shike et al., 1984) and malabsorption following total parenteral
nutrition (McNeill et al., 1982).

Biggin and his colleagues (Biggin et al., 1972) first used prompt
gamma analyses based on the thermal reaction, $^{14}N(n,y)^{15}N$, for in vivo
nitrogen measurement; a cyclotron was the source of neutrons. Later, other
neutron sources were used, for example ^{238}Pu-Be, ^{252}Cf, for such studies of
prompt gamma reaction (Mernagh, 1988; Vartsky et al., 1979; Beddoe et al.,
1984; Allen et al., 1987; Stroud et al., 1989), because such neutron
sources are cheaper and more reliable than the cyclotron.

METHODS

Principles of TBN Measurement

TBN is calculated using the following equations:

$$TBN = (N/H)_{pt} \times (H/N)_{st} \times (MN/MH)_{st} \times MH_{pt} \qquad \text{Eq. 1.}$$

where pt refers to the patient, st to a standard phantom, N and H are the
net nitrogen and the net hydrogen signals, respectively, and MN and MH are
the masses of nitrogen and hydrogen, respectively.

Hydrogen is used as an internal standard, taking the hydrogen
concentration as 10% of the total body weight reported for standard
reference man (International Commission on Radiological Protection); that
is, in the above equation, $MH_{pt} = 0.1 \times$ body weight.

    In our work, TBN is normalized to body size by the following
equation:

$$NI_{pt} = TBN_{pt}/TBN_n \qquad\qquad\qquad Eq.\ 2.$$

The nitrogen index, NI, is the ratio of the amount of nitrogen in a
patient to that of the mean value for healthy young adults of the same
size, based on height.

Irradiation Facility

    Fig. 1 shows our upgraded irradiation facility, the design of which
is essentially similar to our previous facility (Mernagh, 1977) except for
changes in outward appearance. However, we now use ^{252}Cf as neutron sources
(2 x 10 μg).

    The signals from each detector are routed through a preamplifier, an
amplifier, and a mixer-router to a multichannel analyzer (Nuclear Data
model ND66). The mixer-router routes the spectrum from each of the four
detectors into four separate 512-channel areas (of a total of 2048
channels). Thus, the signals from each detector are not affected by gain
drift from another detector. This arrangement avoids the need for having
complex and expensive electronic equipment for spectrum stabilization and
signal processing.

    In practice, the hydrogen signal is about 1000 times stronger than
the nitrogen signal. Therefore, the hydrogen signal is counted in 100's
and the nitrogen signal in 1000's, using a signal-gating arrangement. In
this arrangement, a single channel analyzer divides the total spectrum
from each detector into two areas, one roughly from 0 to about 5 MeV, the
other from about 5 to 10 MeV. During the first 100's, only the 0 to 5 MeV
spectrum, which includes the hydrogen signal, is counted. Following this,
the counting is done for 1000's during which the lower energy portion (0
to 5 MeV) is gated out and only the higher energy region, which includes
nitrogen but not hydrogen, is counted. This procedure improves the
signal-to-noise ratio for nitrogen from 5.0 to 5.9. Fig. 2 shows a

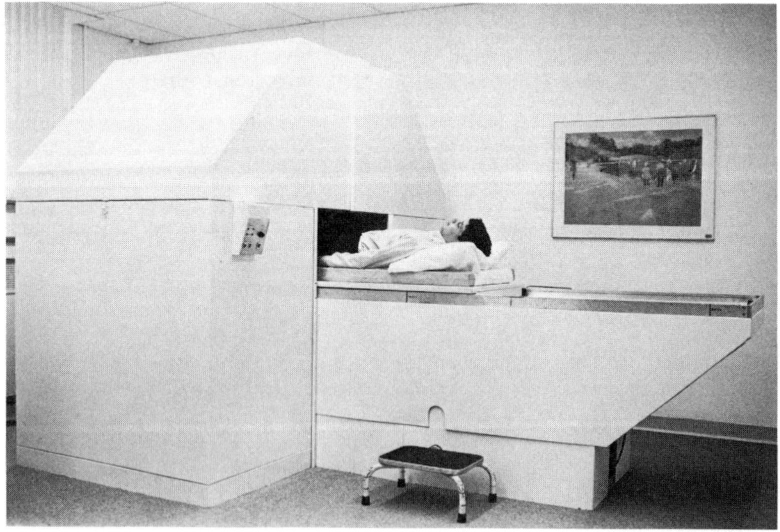

Fig. 1. The clinical facility for measurement of total body nitrogen.

typical TBN gamma ray spectrum of a patient using this counting
arrangement.

Analysis of the Gamma Spectrum

The nitrogen photopeak and first escape peak, i.e. the energy region
from 10 to 11 MeV, are integrated to provide the nitrogen signal count.
The channel numbers corresponding to this energy window is calculated by
linear regression, using the channel number of locations of the gamma
peaks at 2.2 MeV, 6.86 MeV, and 7.65 MeV from hydrogen, iron, and lead,
respectively. Using the hydrogen and nitrogen signals from the standard
urea phantom and the patient, together with the background and the body
weight of the patient, a computer calculates the percentage of body weight
which is nitrogen, the total body nitrogen in kg, and the nitrogen index
of the patient, using equations 1 and 2.

The data processing is done using an IBM-XT computer, attached to the
multichannel analyzer and custom-made computer software written in BASIC.

RESULTS

Factors Contributing to Background

Different neutron sources. Table 1 shows the data obtained from a
phantom containing 5.6 kg of nitrogen; ^{252}Cf neutron sources yield a lower
N background compared to ^{238}Pu-Be neutron sources of similar neutron yield.
This difference probably arises because the former source gives no gamma
photons, while the latter emits photons of average energy of about 4 MeV;
random summing of these photons with other photons may give more
background in the nitrogen region (McNeill et al., 1989).

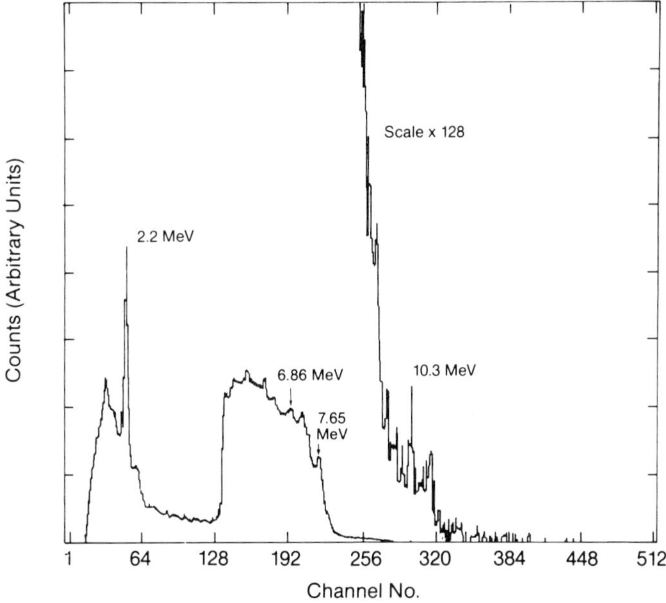

Fig. 2.   Typical gamma ray spectrum of a patient. The photopeaks of
hydrogen, iron, and lead are at 2.2, 6.87, and 7.65 MeV,
respectively. The first escape peak of nitrogen at 10.3 MeV
is identified.

Table 1. Nitrogen Signal-to-Background Ratio (S/B) with Different Neutron
         Sources

Phantom: 5.63 kg nitrogen as urea dissolved in 34.53 kg of water.
Counting time: 1000s

Source	Source Strength[*]	N signal (net)[*] (S)	Background[*] (B)	S/B
^{238}Pu-Be	2 x 10 Ci	9308	2780	3.45
^{238}Pu-Be	2 x 5 Ci	4592	1012	4.54
^{252}Cf	2 x 6 ug	5858	1044	5.61
^{252}Cf	1 x 6 ug[#]	3740	496	7.54

[*]sum of all four detectors.
[#]one source positioned below phantom without any source on top.

Additional considerations of the background signals in the nitrogen
and hydrogen regions. Various components within the area exposed to
neutrons may affect the nitrogen and hydrogen background counts. For
example, a mattress provided for the patient's comfort increased the
nitrogen and hydrogen background counts by about 10% (Table 2). A wooden
box that enclosed the plastic container (30 cm^3) of urea and water
standard increased the background for nitrogen by about 15% and for
hydrogen by about 5%. Heavy clothing on patients may also affect the
background counts.

The hydrogen and other non-nitrogen elements of the body might affect
the counts in the nitrogen region (McNeill et al., 1989). With our first
facility in which we used ^{238}Pu-Be sources, we found no difference in
nitrogen background counts, with or without a water phantom. The recent
change to ^{252}Cf sources in a similar collimated arrangement gave a
substantial reduction in the N background count (Table 1). Therefore, we
re-examined the effect of body water on nitrogen counts. A container with
35 kg of water caused a mean increase of 54 counts over the background
count of 909 counts from the empty container. (This 6% increase (Table 2)
is significant at p < 0.001). Sodium chloride added to the 35 kg of water
in a physiological concentration gave no further increase in counts in the
nitrogen region (Table 2). Therefore, we now use a water phantom for
measurement of the nitrogen background.

Reproducibility. The reproducibility of H/N ratio of our standard is
±2.6%. For a patient, the error of the hydrogen count is expected to be
similar to the standard hydrogen, ±2.33% (CV). However, the error in the
nitrogen counts will be expected to increase by about +3.0% (CV) (due to
lower counts in patients, compared to the standard phantom), and the
combined error of H/N for the patients will be about ±3.8% (CV). The
overall estimate of TBN (errors for H/N of patient and standard combined)
would be expected to give reproducibility of ±5% (CV) .

Other Potential Sources of Inaccuracy

The constant value of 10% of body weight for hydrogen concentration
in the body is reasonable for most normal subjects, but is not accurate
for the very obese or in subjects with extreme water retention (or
oedema). Even in these very abnormal situations, it is estimated that the
error would be less than 5% (McNeill et al, 1979). The validity of the N/H
measurement also depends on the assumption that both elements are
similarly distributed throughout the body. This assumption has been found
to be correct in normal humans and in healthy pigs by several
investigators (Vartsky, 1976; McNeill, 1979; Waana, 1986; Thompson,

Table 2. Nitrogen Background from Different Sources.

	No.	H mean count (SD)	N mean count (SD)
Air background			
1. no mattress	6	69301 (333)	909 (21)
2. mattress	8	77555 (540)	978 (16)[*]
Water background			
3. water	4	234344 (583)	963 (7.5)[*]
4. water in wooden box	4	246495 (636)	1109 (14)[**]
5. NaCI, water in wooden box	4	250531 (551)	1109 (13)
Mattress (2-1)		8254	69
Wooden box (4-3)		12151	146
Water (3-1)			54
NaCl in water (5-4)			0

[*]significant difference from 1 (p<0.01)
[**]significant difference from 3 (p<0.01)

private communication, 1989) but, again, may be subject to error in the very obese people (Stroud et al, 1989) or subjects with localized water retention, e.g. ascites.

CONCLUSION

An improved facility for prompt gamma nitrogen measurement has been built at the Toronto General Hospital. This facility has improved the precision of measurements compared to our previous facility, and the new one is comfortable for patients. Two sources, each of 10 μg ^{252}Cf, are used. The overall precision of total body nitrogen measurement of a patient is about ±5% (CV), and the dose received per measurement is about 0.2mSv (Effective Dose Equivalent; QF=10) for a 20-minute irradiation.

REFERENCES

Allen, B.J., Blagojevic, N., McGregor, B.J., Parsons, D.E., Gaskin, K., Soutter, V., Waters, D., Allman, M., Stewart, P., and Tiller, D., 1987, In vivo determination of protein in malnourished patients, In: "In Vivo Body Composition Studies", K.J. Ellis, S. Yasumura, and W.D. Morgan, eds., IPSM, London.

Beddoe, A.H., Zuidmeer, H., and Hill, G.I., 1984, Prompt gamma in vivo neutron activation facility for measurement of total body nitrogen in the critically ill, Phys. Med. Biol., 29:371.

Biggin, H.C., Chen, N.S., Ettinger, K.V., Fremlin, J.H., Morgan, W.D., Nowotony, R., and Chamberlain, M.J., 1972, Determination of nitrogen in living patients, Nature (New Biology), 236:187.

Blendis, L.M., Harrison, J.L., Russell, D.M., Miller, C., Taylor, B.R., Greig, P.D., and Langer, B., 1986, The effects of peritoneovenous shunting on body composition, Gastroenterology, 90:127.

International Commission on Radiological Protection, 1975, Report of the Task Group on Reference Man. Pergamon Press, Oxford.

McNeill, K.G., Borovnicar, D.J., Krishnan, S.S., Wang, H., Waana, C., and Harrison, J.E., 1989, Investigation of factors which lead to the background in the measurement of nitrogen by IVNAA, Phys. Med. Biol., 34:53.

McNeill, K.G., Harrison, J.E., Mernagh, J.R., Stewart, S., and
    Jeejeebhoy,K.N., 1982, Changes in body protein, body potassium and
    lean body mass during total parenteral nutrition, _Parent. Ent. Nutr._,
    2:106.
McNeill, K.G., Mernagh, J.R., Jeejeebhoy, K.N., Wolmen, S.L., and
    Harrison, J.E., 1979, In vivo measurement of body protein based on
    the determination of nitrogen by prompt gamma analysis. _Am. J. Clin._
    Nutr., 32:1955.
Mernagh, J.R., 1977, In vivo neutron activation analysis and capture gamma
    ray analysis in man, Ph.D. thesis, University of Toronto, Toronto,
    Ontario, Canada.
Schilling, H., Wu, G., Pettit, J., Harrison, J.E., McNeill, K.G.,
    Shepherd, F.A., Feld, R., Evans, W.K., and Jeejeebhoy, K.N., 1984,
    Changes in body composition in patients with small-cell lung cancer -
    the effect of total parenteral nutrition as an adjunct to
    chemotherapy, _Ann. Intern. Med._, 101:303.
Shike, M., Russell, D., Detsky, A.S., Harrison, J.E., McNeill, K.G.,
    Shepherd, F.A., Feld, R., Evans, W.K., and Jeejeebhoy, K.N., 1984,
    Changes in body composition in patients with small-cell lung cancer -
    the effect of total parenteral nutrition as an adjunct to
    chemotherapy. _Ann. Intern. Med._, 101:303.
Stallings, V.A., Archibald, E.H., Pencharz, P.B., Harrison, J.E., and
    Bell, L.E., 1988, One year follow-up of weight, total body potassium
    and total body nitrogen in obese adolescents treated with the protein
    sparing modified fast, _Am. J. Clin. Nutr._, 48:91.
Stroud, D.B., Borovnicar, D.J., Lambert, J.R., McNeill, K.G., Marks, S.J.,
    Rassool, R.P., Rayner, H.C., Strauss, B.J.G., Tai, E.H., Thompson,
    M.N., Wahlqvist, M.L., Watson, B.A., and Wright, C.M., Clinical
    studies of total body nitrogen in an Australian hospital. This
    volume.
Vartsky, D.V., 1976, Ph.D. thesis, University of Birmingham, Birmingham,
    U.K.
Waana, C., 1986, In vivo measurement of small quantities of nitrogen by
    neutron activation analysis, M.Sc. thesis, University of Toronto,
    Toronto, Ontario, Canada.

PERFORMANCE OF THE DELAYED- AND PROMPT-GAMMA NEUTRON

ACTIVATION SYSTEMS AT BROOKHAVEN NATIONAL LABORATORY

F. A. Dilmanian,[1] D. A. Weber,[1] S. Yasumura,[1] Y. Kamen,[2]
L. Lidofsky,[2] S. B. Heymsfield,[3] R. N. Pierson, Jr.,[3]
J. Wang[3], J. J. Kehayias,[1*] and K. J. Ellis[1+]

[1]Medical Department, Brookhaven National Laboratory,
Upton, N.Y.; [2]Department of Applied Physics and Nuclear
Engineering, Columbia University, New York, N.Y.;
[3]St. Luke's/Roosevelt Institute of Health Sciences,
College of Physicians and Surgeons, Columbia University,
New York, N.Y.

INTRODUCTION

Brookhaven National Laboratory (BNL) is one of the major facilities
pioneering the development of in vivo neutron activation (IVNA) techniques
for body composition studies. The IVNA facility at BNL includes a delayed-
and prompt-gamma neutron activation system (DGNA and PGNA), as well as an
inelastic neutron scattering facility (INS). The BNL DGNA system was first
fully established in the 1960's by Cohn et al. (1969). It is composed of a
total-body neutron activation facility (TBNAF) and a whole body counter
(WBC), and is used to measure total body sodium, phosphorus, chlorine, and
calcium. Body potassium is measured by counting endogenous ^{40}K with the
whole body counter. The PGNA system to measure total body nitrogen (TBN)
was developed by Vartsky et al. (1979), and the INS system to measure total
body carbon (TBC) was instituted by Kehaiyas et al. (1987). The DGNA and
PGNA facilities have been upgraded and modified since they were first
built.

THE DGNA SYSTEM

The WBC was upgraded in 1987. The detector configuration was modified
by replacing round detectors with 32 rectangular NaI(Tl) detectors, each
with a volume of 10.2 cm x 10.2 cm x 45.7 cm; 16 detectors are positioned
above and 16 detectors below the subject (Fig. 1). The outcome of these
changes was a four-fold increase in the sensitivity of the WBC for the
detection of 3.1 MeV gamma rays from ^{49}Ca. This increase corresponds to

---

*Present address: USDA Human Nutrition Research Center, Tufts
University, Boston, MA.

+Present address: Children's Nutrition Research Center at Baylor
College of Medicine, Houston, TX.

Advances in In Vivo Body Composition Studies
Edited by S. Yasumura *et al.*, Plenum Press, New York, 1990

the capability of the system to measure total body ^{49}Ca with a precision of
±2.5% (coefficient of variation [CV]) at a level of activity of 37 Bq (one
nCi) for a data acquisition time of 15 minutes. The absolute photopeak
sensitivity of the WBC for the 3.1 MeV gamma rays from ^{49}Ca is about 5%
(Fig. 2). An important characteristic of any WBC system is its background
radiation. The well shielded room in which the BNL WBC is positioned
reduces the background radiation underneath the ^{40}K peak (E_γ = 1461 keV) by
a factor of seven. As a result, the background, on average, is less than
50% of the total counts in human ^{40}K studies, before the background
spectrum is subtracted from the patient spectrum.

The TBNAF part of the DGNA system consists of 14 ^{238}PuBe sources of 1.55
TBq (42 Ci) each, positioned symmetrically above and below the patient. To
slow down the fast neutrons emitted from the source, a moderator made of
2.0-cm thick polyethylene plates covers the patient's body. The patient is
activated for five minutes, and counted in the WBC for 15 minutes,
starting three minutes after activation. The activation dose at the skin
is less than 3 mSv using an RBE = 10 for neutrons (Cohn and Palmer, 1974).

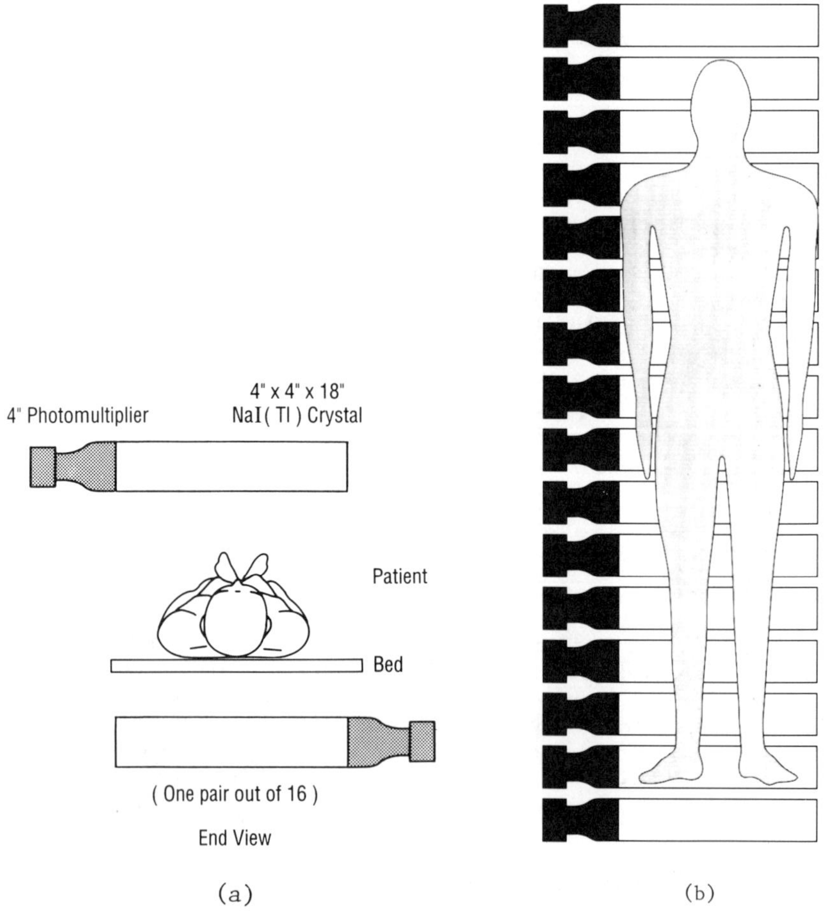

Fig. 1. Axial view of the 32-Detector WBC (a); and its top view (b),
with a subject of 175 cm height included as reference. Only
the lower bank of the detectors is shown in the top view.

An important characteristic of any WBC is the dependence of its response on the patient positioning. Fig. 3 shows the response dependence measured with a ^{22}Na point source in the longitudinal and the lateral directions, respectively. According to these results, the WBC sensitivity for 1.75 MeV gamma rays emitted from a point source varies by 0.3% for a 5 cm longitudinal displacement in the center of the WBC. The variation in sensitivity for lateral displacement is 0.4% for a 2-cm displacement.

THE PGNA SYSTEM

The PGNA system includes 2.85 TBq (77 Ci) of ^{238}PuBe as a neutron source, and two 15.2 cm x 15.2 cm NaI(Tl) detectors. The ^{238}PuBe source, moderated with 5 cm of heavy water (D_2O), is located below a motor-driven platform for the patient. The detectors are positioned above the patient, out of the direct path of the neutron beam. Prompt-gamma measurements are made at five 20-cm increments along the trunk of the body with the subject in a supine position. The process is repeated with the patient in the prone position.

TBN is determined by measuring the prompt-gamma yield from ^{14}N neutron capture (E_γ = 10.83 MeV), relative to that of ^{1}H neutron capture (E_γ = 2.23 MeV). The technique is based on calibrating the count ratio N/H with an anthropomorphic phantom containing known amounts of hydrogen and nitrogen, and determination of the subject's total body hydrogen (TBH) independently. The method to measure TBN is described in detail by Vartsky et al. (1979; 1984). TBH is obtained from the formula (Vartsky, 1979):

$$TBH = 0.11\ TBW + 0.07\ TBPr + 0.12\ TBF$$

where TBW is total body water measured by tritiated water dilution, TBPr is the total body protein, obtained from TBN through TBPr = 6.25 TBN in a recursive manner, and TBF is total body fat obtained from the formula:

$$TBF = body\ weight - (TBW + TBPr + BMA)$$

where BMA is bone mineral ash estimated from TBK.

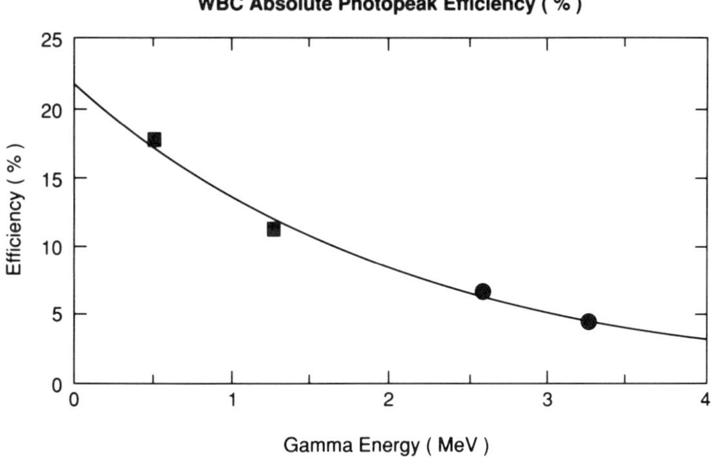

WBC Absolute Photopeak Efficiency (%)

Fig. 2. Absolute efficiency of the WBC as a function of gamma energy, measured with ^{22}Na (■) and ^{56}Co (●) sources.

The skin dose to the patient in a PGNA study is less than 0.26 mSv
(Vartsky et al., 1979). The principal application of the system has been
in determination of the total body protein as derived from the TBN
measurement (Vartsky et al., 1984).

PERFORMANCE OF THE DGNA FACILITY

Recent Improvements in the System

Several changes were made in the software and the quality control
procedures of the DGNA system in the last two years to improve its
performance. These improvements include: a) introduction of an energy-
dependent gamma-ray attenuation correction algorithm, b) development of an
improved spectral analysis algorithm to correct for spectral shifts, and c)
improved quality-control routines to monitor changes in the activation and
detection systems. Extensive phantoms studies using Alderson and other
anthropomorphic phantoms show that the reproducibility of the upgraded
system was significantly improved. Table 1 compares the reproducibility of
the system before and after the changes in software and quality control
were made.

PERFORMANCE OF THE PGNA SYSTEM

Recent Improvements in the System

Major changes were made in the hardware and the software of the PGNA
system during the past two years to improve the precision and accuracy of
the TBN measurements. These improvements include: a) the introduction of a
full-charge pulse integration amplifier and a pile-up rejection circuit,
b) the addition of faster ADCs to provide shorter dead time and finer pulse
digitization, c) improved shielding of the detectors for better protection
against slow and fast neutrons, and d) the modification of data analysis
program to include an improved background subtraction algorithm and an
improved internal energy calibration. As a result, the system's

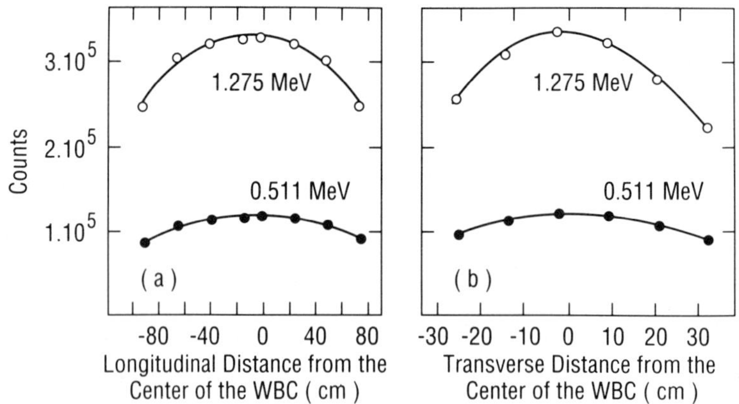

Fig. 3. The WBC response invariance test for source positioning.
Counts from the two gamma lines of ^{22}Na were measured along
the WBC longitudinal and transverse axes.

Table 1. Reproducibility of the DGNA System

Element	Reproducibility with phantoms (Coefficient of variation, %)	
	Before changes	After changes
Sodium	±2.5	±2.0
Phosphorus	±4.0	±2.6
Chlorine	±2.5	±2.5
Potassium	±3.3	±1.5
Calcium	±1.0	±0.8

The system reproducibility for TBK measurement in humans is currently ±1.8%.

reproducibility (S.D./mean) for nitrogen measurements in an Alderson phantom (5 measurements each day, for 5 successive days) is ±2.7%. The previous value, obtained for patient studies, was ±3.5% (Ellis et al., 1982).

Sensitivity of the System's Response to the Amount of D_2O

A well-known problem in designing a PGNA system is to establish the optimal pre-moderation of neutrons between the source and the patient. Too little pre-moderation will allow the fast neutrons to pass through the near side of the patient's body with little interaction with the hydrogen and the nitrogen nuclei in their path, while too much pre-moderation will limit the depth of penetration of the neutrons to the narrow front section of the patient's body. We tested the effect of changing the D_2O thickness of the pre-moderator using a ground-beef phantom.

The ground-beef phantom was prepared from 63 kg of frozen ground beef, distributed into seven hermetically sealed plastic bags of about 35 cm x 35 cm x 10 cm in dimension, with about 9 kg of beef in each bag. The beef in different bags had differing amounts of fat, i.e. bag number 1 had only lean meat, bag number 7 contained only fat, and the fat content of the bags 2,3,4,5, and 6 were 16%, 28%, 40%, 52%, and 67%, respectively.

The measurements were carried out in sets of two, one in which the three bags with the leanest tissue were placed closer to the pre-moderator, with the three bags with the most fat on top; in a second experiment, the position of the lean and fat bags were reversed. For each set, the measurements were repeated for 0-,1-,3-, and 5-cm high D_2O in the pre-moderator. The following results were obtained:

1. For all values of the pre-moderator thickness, TBN values were the highest when the lean bags were placed close to the pre-moderator.

2. The ratio of the TBN measured with lean bags positioned below to the TBN measured with these bags positioned above, increased with the increasing thickness of the pre-moderator.

3. The TBN results averaged for the two measurements with the lean
bags above and below show a decreasing trend with the increasing
thickness of the pre-moderator.

It should be noted that although TBN is normalized to H, the measured
TBN varies with variation in distribution of thermal neutron flux in a non-
homogeneous N and H mixture. The reason is that in such a mixture the
measured N/H count ratio is dominated by the N/H concentration at the site
of the peak of the thermal neutron flux, which differs from the mean N/H
concentration ratio in the whole subject.

On the basis of these effects, the neutron beam appears to be most
effective in the lower layer of the target, with an insufficient number of
neutrons reaching the upper bag, especially for larger thicknesses of D_2O
pre-moderator. Thus, the 5-cm height of D_2O used currently in the
container of the pre-moderator is optimized for small body thicknesses.

<u>A Correction for Body Thickness</u>

Dependence of the response of the PGNA system on patient's body size is
a combination of several effects, including changes in the thermal neutron
distribution in the body, changes in the detector background due to
neutrons and gamma-rays from the source, changes in the solid angle of
prompt gamma rays to the detector, and changes in the attenuation of the
gamma rays in their way out of the patient's body. However, the first
factor is probably the dominant one, overwhelming the others. Since the
correlation between the values for TBN and TBK in healthy people is close
to 0.95 (Ellis et al., 1982), and since the TBK measurement is less
sensitive to the body size than the TBN measurement, comparison between the
results of TBN and our TBK in the same patients can be used to evaluate the
adequacy of the correction for body thickness to the TBN results.

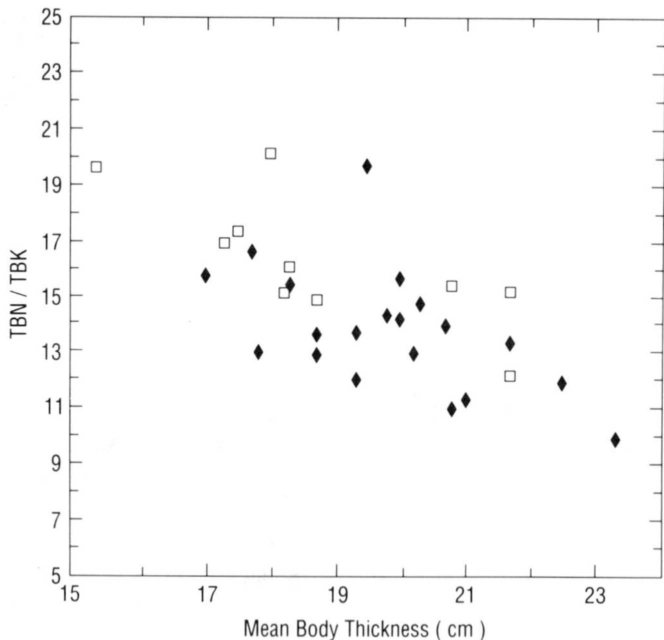

Fig. 4. TBN/TBK results as a function of the mean body thickness of the
upper body for 30 healthy subjects. ♦ : male volunteers,
□ : female volunteers.

TBN and TBK were measured in 30 healthy volunteers (20 male and 10 females), all of whom weighed in excess of 50 kg of weight. Figure 4 shows the results of TBN/TBK plotted versus the mean body thickness, which was the average of three body thickness measurements of the shoulder, trunk, and abdomen.

The decrease in the TBN results as a function of the mean body thickness suggests that the existing correction in the TBN software for body thickness is not adequate for the wide range of body sizes included in the measured subjects. Since the effect that requires correction is mostly the change in the distribution of thermal neutrons in the body, use of Monte Carlo simulations to predict that distribution for a given body shape can provide the correction. Another solution, that can be used in conjunction to simulations, is to use anthropomorphic phantoms of different sizes and shapes to calibrate the system for a variety of body configurations.

ACKNOWLEDGEMENTS

The authors wish to thank A.F. LoMonte and M. Schmaeler for their assistance with the measurements, H.D. Lee for help with the computer systems, J.F. Gatz for assistance with the hardware, and A.D. Woodhead for valuable comments. A summer student, A. Daw, contributed to the software development; his efforts are appreciated. We appreciate the help of A.L. Ruggiero in preparing this manuscript. This research was supported in part by the U.S. Department of Energy under Contract DE-AC02-76CH00016. Accordingly, the U.S. Government retains a nonexclusive, royalty-free license to publish or reproduce the published form of this contribution, or allow others to do so, for U.S. Government purposes.

REFERENCES

Cohn, S. H., Dombrowski, C. S., Pate, H. R., and Robertson, J. S., 1969, A whole-body counter with an invariant response to radionuclide distribution and body size, _Phys. Med. Biol._ 14:645.

Cohn, S. H., and Palmer, H. E., 1974, Recent advances in whole body counting: a review," _J. Nucl. Med. Biol._ 1:155.

Ellis, K. J., Yasumura, S., Vartsky, D., Vaswani, A. N., and Cohn, S. H., 1982, Total body nitrogen in health and disease: effects of age, weight, height, and sex, _J. Lab. Clin. Med._ 99:917.

Kehayias, J. J., Ellis, K. J., Cohn, S. H., and Weinlein, J. H., 1987, Use of a high repetition rate generator for in vivo body composition measurements via neutron inelastic scattering, _Nucl. Instr. and Meth._ B24/25, 1006.

Vartsky, D., Ellis, K. J., and Cohn, S. H., 1979, In vivo measurement of body nitrogen by analysis of prompt gammas from neutron capture, _J. Nucl. Med._ 20:1158.

Vartsky, D., Ellis, K. J., Vaswani, A. N., Yasumura, S., and Cohn, S. H., 1984, An improved calibration for the _in vivo_ determination of body nitrogen, hydrogen and fat, _Phys. Med. Biol._ 29:209.

HIGH PRECISION IN-VIVO NEUTRON ACTIVATION ANALYSIS: A NEW ERA

FOR COMPARTMENTAL ANALYSIS IN BODY COMPOSITION

Richard N. Pierson, Jr.,[1] Jack Wang,[1] Stephen B. Heymsfield,[1]
F. Avraham Dilmanian,[2] and David A. Weber[2]

[1]St. Luke's-Roosevelt Hospital Center, Columbia University
[2]and Brookhaven National Laboratory, Upton, New York

INTRODUCTION

Since the high-water mark days of the 1960s when the combination of
newly available isotope dilution spaces and the mathematics of
compartmental analysis seemed on the verge of revolutionizing physiologic
research (Bassingthwaighte, 1970), there has been a general decline in
the applications of body composition methods to clinical medicine. In
retrospect, concept outstripped competence, as the spaces that could be
measured accurately, such as blood volume, turned out to be easier to
cure (by transfusion) than to measure, and the mathematical expression of
complex solutions for ferrokinetics and iodine kinetics through multiple
compartments could rarely be turned to clinical benefit. The reasons
were several, but important among them, a "respectable" standard error of
measurement of $\pm 5\%$ was combined with a clinical uncertainty, at least as
large, as to how "normal" could be defined. The terrain has now changed,
in part because some new body compartments can be measured, in part
because we have developed age-, sex-, and race- specific definitions of
"normal", but, in largest measure, because new precisions of measurement
have been achieved, especially in the techniques of in-vivo neutron
activation analysis. We shall consider some specific examples of their
application. The benefits we envisage from high-precision in-vivo
neutron activation (IVNA) derive largely from understanding the
interdependence of the body compartments, and, therefore, from the
development of a series of models which interrelate the compartments.

The Body Compartments: Structure and Function

Our basic model for the body (Fig. 1) defines four functional
compartments: the body cell mass (containing the intracellular fluid),
the extracellular fluid (ECW), the inert skeletal structures, and the
adipose tissue mass (ATM), which contains neutral lipid, or fat, and its
cellular host the adipocyte (Pierson and Wang, 1977). The adipose tissue
mass is an organ, and is a preferable compartmental entity to its
subcompartment, fat, for the purposes of our model (Wang and Pierson,
1976). These four compartments are both convenient and useful; convenient
because they can be measured, and useful because each is related to one
of four functions: metabolic activity, transport, structure, and energy
storage.

The model invites a second level of description: definition of the
interdependencies which link the compartments by the ratios of their
masses. Our model builds from the body cell mass, the only living
portion of the body where metabolism is carried out in cells which
oxidize substrates to carry out specialized functions.

The body cell mass (Moore et al, 1963) requires, and in a sense
controls, the volume of the extracellular water mass, the ECW, which is
tightly related in composition and volume to the former by homeostatic
mechanisms. The ECW is a bathing solution which buffers the internal
milieu from the outer environment, and transports the substrates of
metabolism by convection and diffusion from the outer boundaries of the
organism to its specialized cells. Under normal conditions the mass of
the ECW is remarkably well correlated with the mass of the body cells;
however, in almost all illnesses the ECW increases in volume to
compensate for a variety of incipient failures in one or another organ
system (Pierson and Wang, 1987).

The skeletal and support structures are metabolically inert and
tough, providing a framework and envelope for more fragile organs and
systems. The skeleton, the densest structure in the body, weighs a fixed
7% of body weight, (when faced with earth's gravity); this bone mass
melts down rapidly in the weightless environment of space, a change which
has nothing to do with nutrition, and everything to do with gravity and

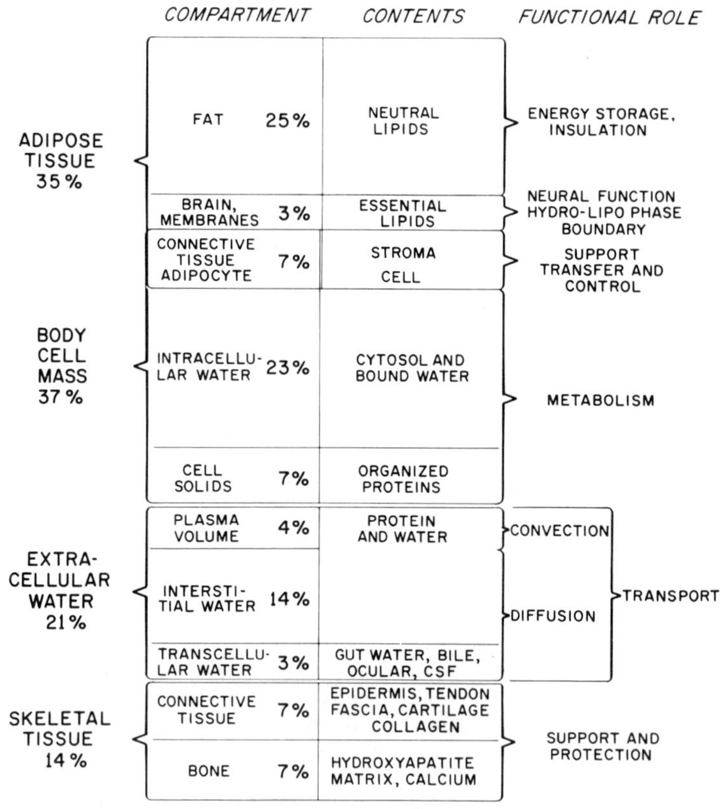

Fig. 1. A four-compartment functional model for body composition.
Reprinted with permission from: Pierson, R.N. Jr., Wang,
J., 1977 Body Composition. Spencer, R., Ed., Nuclear
Medicine Handbook. CRC Press, Cleveland, 161-189.
Copyright CRC Press, Inc., Boca Raton, FL.

Table 1. Compartments and Methods

Measured Component	Compartment	Method	Precision
Carbon	Fat (77%) Protein (52%) CHO (42%)	IVNA Inelastic Scattering	± 3.0 %
Nitrogen	Protein	IVNA Prompt Gamma	± 2.7 %
Sodium, total	Extracellular Cation; Bone (10%)	IVNA Delayed Gamma	± 2.5 %
Sodium, exchangeable	ECW	Indicator dilution	± 5.0 %
Phosphorus	Muscle, Bone	IVNA Delayed Gamma	± 3.0 %
Chloride	ECW	IVNA	± 2.5 %
Potassium	Body Cell Mass	Whole Body Count	± 1.5 %
Calcium	Bone	IVNA Delayed Gamma	± 0.8 %
HTO (or D_2O) Space	Total Body Water	Indicator dilution	± 1.5 %
Radiosulfate Space	ECW	Indicator dilution	± 4.0 %

weight bearing (Carter, 1982).) For every gram of bone, there is
approximately one additional gram of collagen, skin, cartilage, and other
interstitial tissues which connect and attach the other organs.

<u>Elements, Compartments, and the Precisions of the New Measurements</u>

Table 1 lists the measurements of body composition required for our
model. These precisions (repeatability on replicate tests) are not, at
present, accuracies (correct, by comparison with a "gold" standard"), a
more difficult target which will engage a major part of our attention
during the next decade. The data listed here are fundamental to our
argument, if we are to extend the application of these measurements to
clinical issues more effectively than at present. These numbers for
precision represent the initial conditions, which, in turn, will set the
boundaries to what we may accomplish. As these precision levels are
further improved, the boundaries will move back, and the value to our
projects and patients will be improved.

When only a single measurement is available, there can only be a
two-compartment model, the second quantity being obtained by subtraction
from body weight. When more compartments are measured, the compartments
may be discrete, nested, or overlapping, and families of models (Fig. 2)
will be invoked (Moore et al, 1963; Burkinshaw et al, 1978; Cohn et al,
1984). When two or more methods measure a single compartment and differ,

both benefits and dilemmas arise; which method should one believe, and how should one interpret the results? How accurate must the measurement be? For any particular measurement, the choice of method is determined by the goal of the study. Since ultimate "gold standard" calibrations (as in cadaver analysis) are rarely possible, the accuracy of a body composition measurement can be questionable. Comparison measurements have been useful, most notably in challenging several traditional assumptions, and in extending body composition methods to studies of illness.

<u>New Data Challenge the Traditional Assumptions of Constancy</u>

The extension of studies to older, fatter, leaner, and ill subjects disclosed the inadequacy of traditional assumptions which related a component by a constant to the lean body mass (Forbes et al, 1961). Body

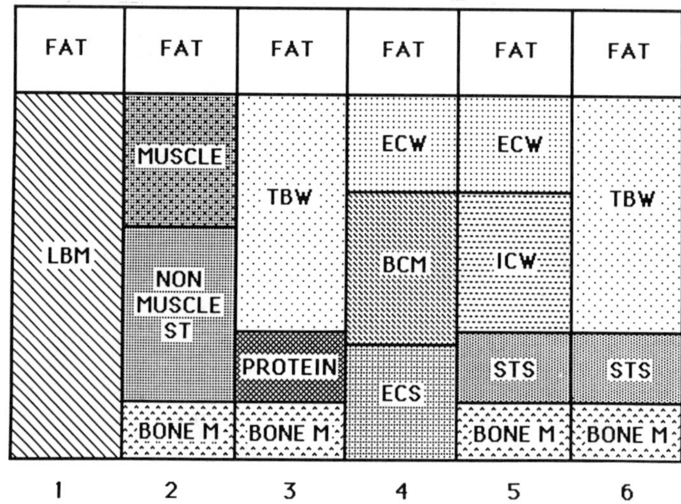

Fig. 2. Traditional two compartment, five compartment, and four compartment models, each based on sets of measurable parameters. STS is soft tissue solids, BCM is body cell mass, and M is mineral.

water (Pace and Rathbun, 1945), body density (Behnke, 1942), and body potassium (Forbes, 1961) each have been used to define the lean body, for "normal", "average", and "young" persons; indeed, the hydration, density, and K contents of the lean body are reasonably constant in young, healthy people. However, these assumptions of constancy fall apart with increasing fatness, increasing age, and with most systemic illnesses, all of which show, in various degrees, increased hydration, and lower density and K concentrations than in the lean body; depending on the degree of change this can result in errors of from 5 to over 25 percent in estimates of body fat (Pierson 1976, 1982, 1987; Kotler 1985). This has

Table 2. Correlation Between Methods for Body Fat Measurement n=13

	TBC	TBN	DPA	UWW	TOBEC	TBK	TBW	DUR-NIN	STEIN-KAMP	BIA
Fat Mean %	31.3	28.7	27.3	27.4	26.5	28.7	26.2	25.2	32.9	26.8
SD	7.0	9.5	8.0	8.6	5.9	6.8	7.3	8.4	7.0	7.1
TBC										
TBN	0.75									
DPA	0.83	0.95								
UWW	0.74	0.88	0.90							
TOBEC	0.70	0.92	0.92	0.75						
TBK	0.60	0.87	0.87	0.85	0.74					
TBW	0.83	0.84	0.82	0.78	0.74	0.68				
DURNIN	0.88	0.73	0.75	0.65	0.65	0.46	0.78			
STEINKAMP	0.83	0.90	0.95	0.89	0.80	0.84	0.84	0.79		
BIA	0.59	0.61	0.78	0.73	0.66	0.64	0.57	0.48	0.79	

been well demonstrated by IVNA and dual photon absorptiometry (DPA)
(Mazess 1984). The changing hydration, density, and average
intracellular potassium can be estimated precisely in subjects of varying
fatness, and of all ages and illnesses, with these techniques.

Newly Precise Data Invite New Models: in Search of Accuracy

As more elements and, therefore, more compartments became
measurable, conceptually totally orthogonal analytic techniques have
appeared. ("Orthogonal" modeling refers to the collection of
measurements from a variety of "angles", each technique depending on
different physical measurements, based on different assumptions.) One
such set of techniques is inherently anatomic; CT, NMR, and dual photon
absorptiometry. These methods collect data in two-dimensional slices of
anatomy, and, by reconstruction, develop three dimensions of information,
with the capacity to define the various components in each serial
section. The systematic summing of the regional signals for fat, bone,
muscle, or other signal-producing components can be applied for an organ,
or for the entire body, a technique elegantly demonstrated in the CT
reconstructions by Alpsten (1989).

Let us consider the measurement of fat and its slightly larger
organ, the adipose tissue mass. Many methods have been proposed for
measuring fat and the ATM. Table 2 shows the average of body fats in 13
normal subjects measured by ten different techniques. Remarkable
differences in the results depend on the method; fat as percent of weight
varied from 25 to 31. Correlations between methods provide another
comparison of the techniques; thus DPA measurements best predict the
element-based analyses of fat from C, N, K, and TBW (Cohn et al, 1984),
and also the many-skinfold regimen of Steinkamp et al (1965), but are
less predictive of the Durnin skinfold (Durnin and Womersley, 1974) and
BIA techniques (Segal et al, 1985). However only 13 subjects, aged from
24 to 94, of both sexes, were studied. A much larger number of subjects
will be required to provide a definitive comparison of these methods.
Choosing an optimal method for measuring fat depends on the purpose of
the study; the choices vary widely in complexity and expense.

By cross correlation of measurements of different elements, the
compartments of fat, body cell mass, or skeletal mass, for example, may

be more reliably defined. The extra effort of making multiple
measurements for a single parameter is justified, particularly when an
accurate estimation of a difficult-to-assess compartment is required.
The orthogonal strategy may provide the dual benefits of giving a result
with fewer assumptions and add insight into the nature of the mass which
is being studied. Although each measurement technique has different
precisions, assumptions, and sources of error, the multiple method
approach offers a rationalization of the sources of error by exposing
faulty assumptions. Thus, the orthogonal use of several measurements put
less reliance on constancy assumptions, and hence are more likely to be
accurate.

Accuracy and precision are different goals. While much of our
current thrust is to seek accuracy, some clinical circumstances require
serial single dimension measurements of a compartment to measure changes
as a result of an intervention. Measurements must be precise, not
necessarily accurate, to measure change, particularly over a short time.

The preliminary results of some current work will illustrate some
dilemmas in multiple-element studies. We are systematically evaluating
measurements of body composition in white, black, and Asian adults in
cohorts of 24 men and women at approximately each decade of age from 20
to 100. Preliminary results provided insight into the relationships
between compartments, and also drew our attention to calibration issues.

The relationship of exchangeable to total sodium has potential
significance in cardiology in terms of hypertension's possible
relationship to sodium metabolism. A subset of this issue relates to the
size and availability of the bone-bound sodium compartment. Figure 3
shows the relationship of total sodium measured by IVNA to exchangeable

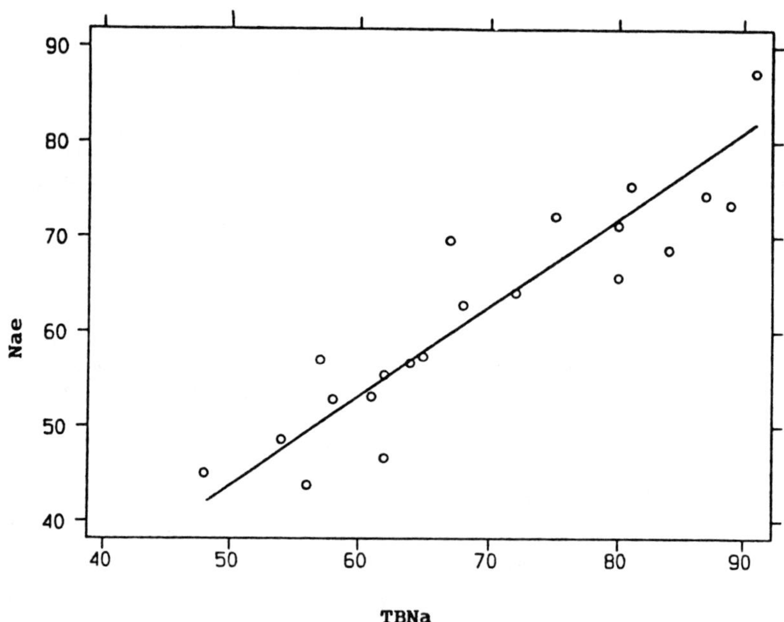

Nae = 0.86TBNa + 2.6, r^2 = 0.8,   SEE = 4.9
Intercept is not different from zero

Nae/TBNa = 0.895 $\pm$ 0.069, ranged 0.75 to 1.04

Fig. 3. Na_e (exchangeable sodium) correlates with TBNa (total body
sodium) in 21 normal subjects.

Table 3. Comparison Between TBCa by DPA and TBCa by IVNA (n=39)

	Obs. #	Mean (g)	SD	Min.	Max.
TBCa by DPA	40	961	256	465	1643
TBCa by IVNA	40	800	181	454	1184
Fat % DPA	40	25	9	7	47
Age	40	55	21	23	94

sodium measured by ^{24}Na. The scatter in this data is assumed to be
primarily due to low precision in the measurement of Na_e. The difference
between the Na_T and Na_e is non-ionized sodium bound in bone, previously
estimated at 15 to 25% of the total, based on direct chemical
measurements of residual sodium in bones. The measured ratio of bound to
total sodium was 0.14 ± 0.047 in these 21 subjects. Exchangeable sodium
is higher in the obese, as the high-sodium ECW is expanded in adipose
tissue. Hypertension is associated with obesity, with a global
coefficient value of 0.3.

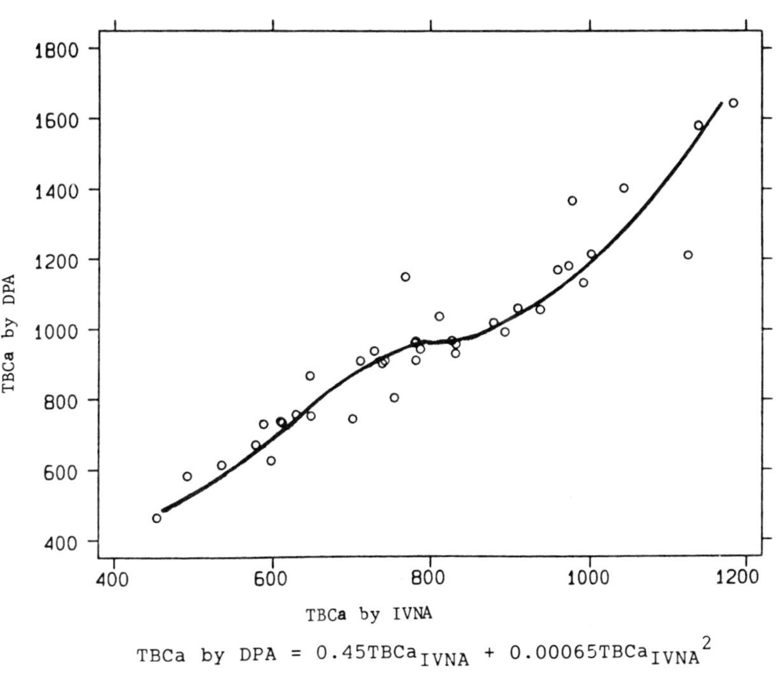

$$\text{TBCa by DPA} = 0.45\text{TBCa}_{IVNA} + 0.00065\text{TBCa}_{IVNA}^2$$
$$+ 51.1\text{Fat\%} + 1.6\text{Fat\%}^2 + 0.02\text{Fat\%}^3 - 1.14\text{Age} - 321$$
$$r^2 = 0.97, \quad SEE = 47$$

Fig. 4. Partial regression of TBCa by IVNA vs TB bone mineral by DPA.

Body Calcium by IVNA or DPA: Cognitive Dissonance?

Table 3 shows measurements of body calcium by IVNA and by dual photon absorptiometry in the first forty subjects in our study. These estimates of calcium are more than 15% apart, a finding which raises new questions. The correlation coefficient is 0.97 after multiple regression adjustment for fat and age, as shown in Figures 4. Therefore, the differences between the results by these two orthogonal techniques depend on anthropometric variables which affect, in different ways, the attenuation of incident neutrons, the exiting photons, or both, in the IVNA procedure. The solution to these differences lies in calibrations which account for all of the anthropological determinants. We plan to use phantoms which model the distributions of bone and soft tissue. In addition, we will apply mathematical simulations through the Monte Carlo method (Kalos and Whitlock, 1986), which may allow interpolations to be made between the calibration points established for the physical models. Some aspects of radiation dosimetry and radiation therapy planning will be directly affected. The refinement of all techniques in body composition research which use neutron activation analysis require development of a rational and explicit physical basis. The tools for that development include comparisons between orthogonal techniques.

Other techniques include assays based on indicator dilution theory, conservation of mass, and compartmental analysis. These techniques add the ordinates of physiology to the static measurements of mass: turn-over, intermediary metabolism, and the mechanisms of change in body compartments. Useful tracers exist for almost all biological materials, usually in non- radioactive as well as radioactive forms.

SUMMARY

We have focused on two topics in body composition research. The first is basic: we must quantitatively define the interrelationships of the body compartments to one another in health before studying these relationships in disease. The precisions now available can enable us to get to do this work, over the full range of age, race, and occupational diversity. The second direction is more modest and more technical, but nonetheless important. It is the calibration and validation of surrogate measurement methods suitable for field research conditions, doctors' offices, intensive care units, etcetera. Our methods have advanced in fundamental ways. We have an obligation to see them applied effectively.

ACKNOWLEDGEMENTS

We appreciate the statistical analysis of John C. Thornton and the technical assistance of Yakov Kamen. The study was supported in large part by NIH NIDDK-37352, and in part by the U.S. Department of Energy under Contract DE-AC02-76CH00016.

REFERENCES

Alpsten, from Toronto conference; quantitative CT paper

Bassingthwaighte, J. B., 1970, Blood flow and diffusion in mammalian organs, _Science_, 167:1347-1353

Bassingthwaighte, J. B., 1988, Physiological heterogeneity: Fractals link determinism and randomness in structures and functions, _Newsletter of the International Physiological Society_ 3:5-9

Behnke, A. R., Feen, B. G., and Welham, W. L., 1942, Specific gravity of healthy men, J. Am. Med. Assoc., 118:495-498.

Burkinshaw, L., Hill, G. L., and Morgan, W. D., 1978, Assessment of the distribution of protein in the human body in in vivo neutron activation analysis, in: "Nuclear Activation Techniques in the Life Sciences, 1978," (Proc. Symp. Vienna, 1978), 787-798.

Carter, D.R., 1982, The relationship between in vivo strains and cortical bone remodeling, Crit. Rev. Biomed. Eng., 8:1-28.

Cohn, S.H., Vaswani, A.N., Yasumura, S., Yuen, K., and Ellis, K. J. 1984, Improved models for determination of body fat by in-vivo neutron activation. Am. J. Clin Nutr. 40:255-259.

Durnin, J. V. G. A., and Womersley, J., 1974, Body fat assessed from total body density and its estimation from skinfold thickness: measurements on 481 men and women aged from 16 to 72 years, Br. J. Nutr., 32:77-97.

Forbes, G. B., Gallup, J., and Hursch, J.B., 1961, Estimation of total body fat from ^{40}K content, Science, 133:101-102 Kotler, D. P., Wang, J, and Pierson, R. N. Jr., 1985, Studies of body composition in patients with the aquired immuno- deficiency syndrome, Am. J. Clin. Nutr., 42:1255-1265.

Kalos, M. H., and Whitlock, P. A., 1986, "Monte Carlo Methods", vol. 1, J. Wiley & Sons, New York.

Mazess, R.B., Peppler, W.W., and Gibbons, H., 1984, Total body composition by dual-photon (^{153}Gd) absorptiometry. Am. Jour. Clin Nutrition, 40:834-839.

Moore, F.D., Olesen, K. H., McMurrey, J. D., Parker, H. V., et al, 1963, The body cell mass and its suporting environment: Body composition in health and disease. W. B. Saunders Company, Philadelphia-London.

Pace, N., and Rathbun, E. N., 1945, Studies on body composition, body water and chemically combined nitrogen content in relation to fat content, J. Biol. Chem., 158:685-691.

Pierson, R. N. Jr., Wang, J., Frank, W, et al. 1976, Alcohol affects intracellular potassium, sodium, and water distribution in rats and man, Currents in Alcoholism, 1:161-178.

Pierson, R.N. Jr., and Wang, J., 1977, Body Composition. Spencer, R. Ed., Nuclear Medicine Handbook. CRC Press, 161-189.Pierson, R.N. Jr., Wang, J., Thornton, J.C., and Van Itallie, T.B., 1982, Body potassium by 4-Pi counting: an anthropometric correction. Am. Jour. Physiol., 246:F234-239.

Pierson, R. N. Jr., Wang, J., 1987, The quality of the lean body mass: Implications for clinical medicine, in: "In Vivo Body Composition Studies," eds. K. J. Ellis, S. Yasumura, and W. D. Morgan, The Institute of Physical Sciences in Medicine, London. 123-130.

Segal, K. R., Gutin, B., Preston, E., Wang, J., and Van Itallie, T. B., 1985, Estimation of human body composition by electrical impedance methods, a comparative study, J. Appl. Physiol., 58(5):1565-1571.

Steinkamp, R. C., Cohen, N. L., Siri, W. Z., Sargent, W., and Walsh, H.E., 1965, Measures of body fat and related factors in normals - II, J. Chron. Dis., 18:1292-1307.

Wang, J., and Pierson, R.N. Jr., 1976, Disparate hydration states of adipose and lean tissue require a new model for body water distribution in man. J. Nutr. 106:1687- 1693

DUAL PHOTON ABSORPTIOMETRY: VALIDATION OF

MINERAL AND FAT MEASUREMENTS

Steven B. Heymsfield,[1] Jack Wang,[1] Mary Aulet,[1]
Joseph Kehayias,[2] Steven Lichtman,[1] Yakov Kamen,[3]
F. Avraham Dilmanian,[3] Robert Lindsay,[4] Richard N. Pierson, Jr.[1]

[1]Department of Medicine, Luke's-Roosevelt and [4]Helen Hayes
Hospitals, Columbia University College of Physicians and
Surgeons, New York; [3]Brookhaven National Laboratory, Upton,
New York; and [2]USDA Human Nutrition Research Center on
Aging, Boston, Massachusetts

INTRODUCTION

Photons passing through human tissue undergo attenuation in
relation to the specific chemical substances with which they interact.
Soft tissues, consisting largely of water and organic compounds, reduce
photon flux to a lesser extent than bone mineral, which contains the
intermediate-Z element calcium. By selecting two appropriate photon
energies and recording their attenuation, the investigator can solve
simultaneous equations that subdivide body mass into two components: soft
tissue and bone mineral ash (Heymsfield, et al., 1989b). Systems are now
available that use either a 153-Gadolinium (^{153}Gd) source that generates
photons with energies at 100 KeV and 44 KeV or filtered X-rays with
energies at about 70 KeV and 40 KeV. Instruments that rely on these
sources are referred to as dual photon absorptiometers (DPA) and dual
energy X-ray absorptiometers (DEXA), respectively.

Dual photon systems are also capable of further resolving soft
tissue into fat and fat-free components (Heymsfield et al., 1989b). This
is accomplished by use of the R_{ST}, a ratio of attenuations through soft
tissue (ST) pixels at the two energy levels (Mazess et al., 1989). The
R_{ST} is linearly and inversely correlated with the proportion of soft
tissue fat. The patient's percentage fat can be calculated either by
internal system calibrations or by using phantoms of known fat content.
The latter approach involves serial scans of beef-lard mixtures of known
composition, and then using this information to derive the subject's fat
mass (Heymsfield et al, 1989b; Wang et al., in press).

Fig. 1 shows the dual photon model of body composition. Body weight
is divided into three components, total body bone ash (TBBA), fat-free
soft tissue (FFST), and fat. Total fat-free body mass (FFM) is the sum of
TBBA and FFST.

The aim of this paper is to describe the joint efforts by our group
to validate the estimates of body composition derived by dual photon

Advances in In Vivo Body Composition Studies
Edited by S. Yasumura *et al.*, Plenum Press, New York, 1990

systems. Our initial studies largely involved DPA (Lunar DP4, Madison, Wisconsin), although our discussion will also include the more recently developed DEXA.

MINERAL MEASUREMENTS

     The skeleton consists of mineral, protein, water, and small amounts of other compounds (Woodard, 1964). Dual photon systems estimate bone mineral ash, the skeletal fraction remaining after prolonged heating at high temperatures (Heymsfield et al., 1989b). The calcium content of bone ash in humans is reasonably constant at 34%-38% of dry mineral weight. The primary focus in our laboratory was to validate TBBA derived by DPA against total body calcium (TBCa), which is found almost entirely within bone mineral. An accurate and precise method of measuring TBCa, delayed gamma neutron activation analysis, is available at one of our Centers (Brookhaven National Laboratory, Upton, NY). This delayed gamma system is described by Cohn et al. (1980).

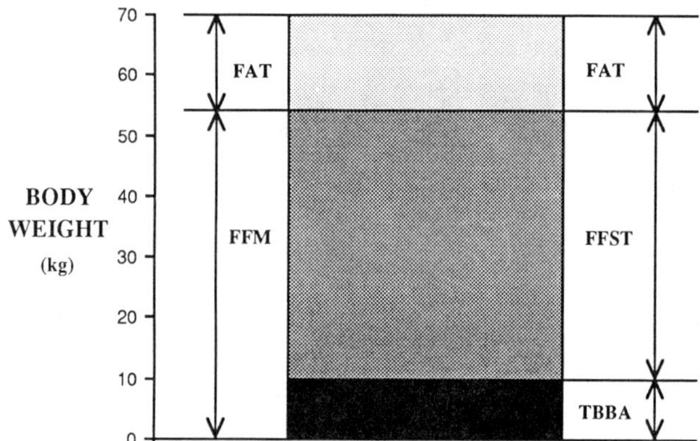

Fig. 1.   Three-compartment DPA model of body composition. F = fat, FFST = fat free soft tissue, and TBBA = total body bone ash. FFM = fat free body mass (FFST + TBBA).

     Fig. 2 presents our study of TBBA versus TBCa in 31 healthy non-obese adults between the ages of 24 and 93 years (X+SD=58+20 years)(Heymsfield et al., 1989b; Heymsfield et al., in press). The correlation between bone ash and whole body Ca was highly significant (TBBA=3.08 x TBCa - 432, SEE = 238 g, r = 0.94, p<0.001). The ratio of TBCa to TBBA was 0.380 for men, 0.377 for women, and 0.379 for pooled subjects. These results, which confirm earlier initial findings by Mazess et al (1981), support the validity of bone mineral estimates provided by DPA.

     An additional point worthy of mention is the relation between bone ash and bone mineral. Mineral, when heated to over 500°C and ashed, loses crystalline water and other volatile components that reduce its mass by

4-5%. We therefore use the limited data available to convert bone ash to osseous mineral (M_o) as (Brozek et al., 1963):

$$M_o = TBBA \times 1.0436.$$

There are two reproducibility issues of concern in relation to bone ash measurements. First, each instrument should be highly reproducible from day to day so that small changes in bone mineral over time are readily detected. The second concern relates to the variety of instruments available that differ in technological features and manufacturer. Optimally, each instrument should provide equivalent estimates of bone ash in the same subject. We performed two pilot studies aimed at examining each of these issues.

In the first study four subjects underwent between four and six DPA scans daily, and the TBBA results were used to calculate a coefficient of variation (CV) (Table 1). The range in CV's was between 0.65% and 1.69%, with an average of 1.34% for the four subjects. We are presently extending this study to DEXA, which reportedly has a CV one-half that of DPA. For comparison, the CV for TBCa estimated by neutron activation, derived on anthropomorphic phantoms, is 0.8%.

The second study involved a cross-comparison between two DPA instruments followed by a comparison of DPA to DEXA (Table 2). Eight subjects underwent DPA scan on our system (#1, Lunar DP4) followed by a scan on a second DPA system (#2, Norland 2600, Helen Hayes Hospital, Havestraw, NY). The mean bone mineral results differed by 2.6% and the regression line of DPA-1 versus DPA-2 was almost identical to the line of identity (Fig. 3). The next phase involved eleven subjects who were first studied using DPA-1 and then underwent study using DEXA (DPX, Lunar). Again the difference in TBBA detected by the two systems was small (X=2.7%), with a near-equivalent regression line and line of identity (Fig. 4).

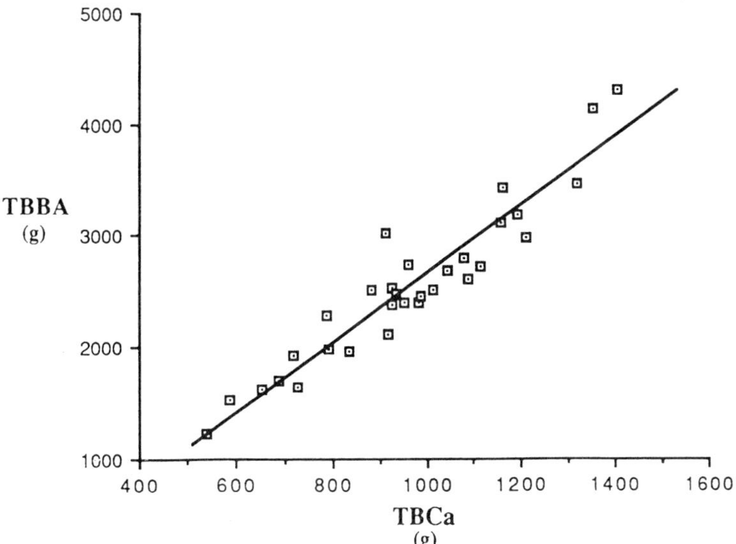

Fig. 2.   TBBA on the ordinate versus total body calcium (TBCa) on the abscissa (TBBA = 3.08 TBCa - 432, SEE = 238 g, r = 0.94, p<0.001) (Heymsfield et al., in press).

Table 1. Coefficient of Variation (%) in Subjects Undergoing
          4-6 Daily Studies

	Subjects					
	1	2	3	4	5	$\overline{X}$
DPA						
TBBA	-	1.60	1.69	1.43	0.65	1.34
% Fat	-	2.65	3.73	3.27	4.41	3.52(3.22)+
UWW *						
% Fat	-	5.79	4.18	7.23	-	5.73

* UWW = underwater weighing.
+ Means for 3 subjects with % fat measured using DPA and UWW.

Hence, these initial studies suggest that bone mineral estimates by
DPA and DEXA in normal weight, healthy subjects are precise, reproducible
between instruments, and compare favorably to the more complex, costly,
and less accessible neutron activation method as a means of evaluating
bone mineral mass in vivo.

These encouraging results represent only an introductory evaluation
of bone mineral estimates by dual photon studies. Many technical and
biological concerns prevail, including the effects of body thickness,
bone marrow fat, and ectopic calcifications on bone mineral results.

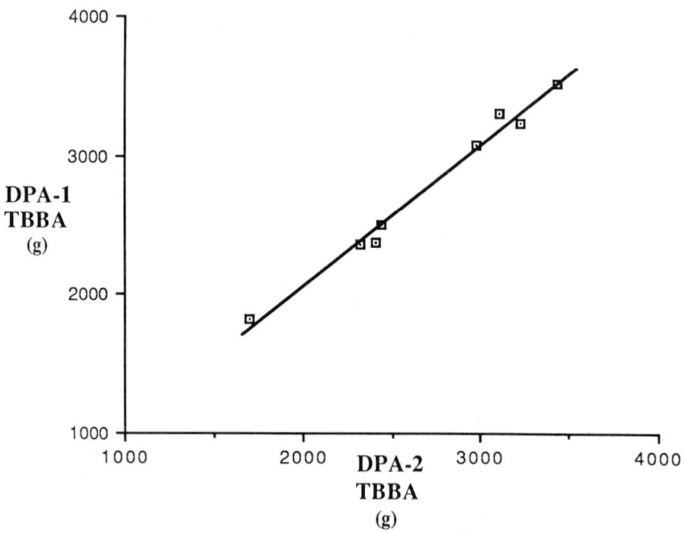

Fig. 3. TBBA measured by DPA-1 (Lunar DP4) on the ordinate versus TBBA
         derived by DPA-2 (Norland 2600) on the abscissa (DPA-1 = 1.02
         DPA-2 + 18.5, r=0.99, p<0.001).

Table 2. Comparison of Bone Ash (TBBA) Between Dual Photon Systems

	DPA-1	DPA-2	% Δ
TBBA (n=8)			
X̄	2771	2700	2.6%
SD	555	541	

	DPA-1	DPX	% Δ
TBBA (n=11)			
X̄	3385	3292	2.7%
SD	648	607	

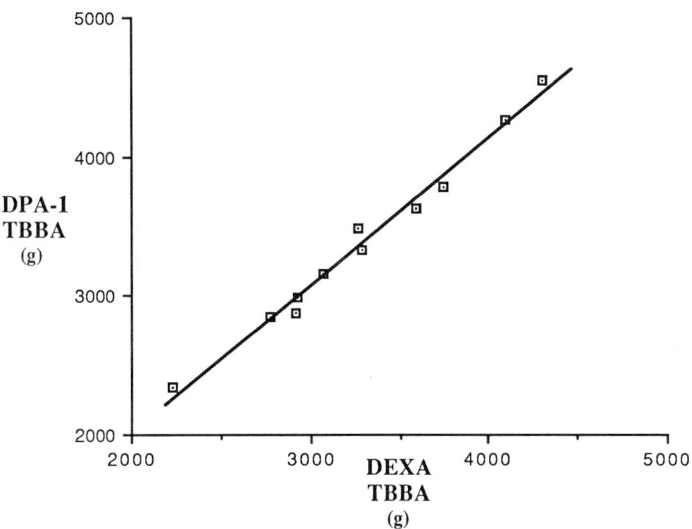

Fig. 4. TBBA measured by DPA-1 (Lunar DP4) on the ordinate versus TBBA
        derived by DEXA (Lunar DPX) on the abscissa (DPA-1= 1.08 DPX +
        55.2, r=0.99, p<0.001).

Table 3. Results of Phase I Studies (X±SD)

	Males (n=18)	Females (n=13)	Total (n=31)	r(p)**	SEE(kg)
Age	55.2 ±18.8	62.5 ±22.5	58.3 ±20.4	-	-
BMI	23.9 ±2.2	22.8 ±3.1	23.4 ±2.6	-	-
% fat *					
DPA	20.8 ±5.1	32.3 ±7.3	25.6 ±8.3	-	-
TBW	18.9 ±7.3	28.6 ±7.9	23.0 ±8.9	0.91 (0.001)	3.6
UWW	21.6 ±5.8	31.0 ±8.9	25.6 ±8.5	0.88 (0.001)	4.0
Model A	24.1 ±7.3	33.0 ±9.3	27.9 ±9.2	0.90 (0.001)	3.8
Model B	22.6 ±6.2	32.0 ±8.3	26.5 ±8.5	0.93 (0.001)	3.2

* % fat by DPA, 3H2O dilution (TBW), underwater weighing (UWW),
Model A = combined TBW, TBN, TBCa, TBK, TBNa, and TBCl, and
Model B = combined TBW, UWW, and DPA TBBA.
** DPA % fat versus corresponding % fat.

BODY FAT

The studies of DPA-estimated fat at our Center involved two phases
that differed in methodology and subject number. The first phase involved
an extensive series of body composition measurements on the healthy
non-obese subjects mentioned above (Heymsfield et al., 1989b; Heymsfield
et al., in press). In this analysis we compared fat derived by DPA to fat
estimated from the following: total body water (TBW), hydrodensitometry
(Db), and combined methodologies [Model A: TBW, TBCa, total body nitrogen
(TBN), TBK, TBNa, and TBCl; and Model B: TBW, TBBA, Db] (Heymsfield et
al., 1989a; Heymsfield et al., in press). In the combined methods TBW was
established by tritiated water dilution; TBCa, TBNa, and TBCl by delayed
gamma neutron activation (Cohn et al., 1980); TBN by prompt gamma neutron
activation (Vartsky et al., 1979); and TBBA by DPA (Heymsfield et al.,
1989b).

Table 3 presents the results of the phase I studies. Fat estimated
by DPA was on average similar to, and highly correlated with percentage
fat derived by conventional methodologies (all p<0.001). Fig. 5 presents
the results for DPA percentage fat versus percentage fat estimated by
Model A. This multicompartment neutron activation model is unique in that
measured components (TBW, TBCa, TBK, TBNa, TBCl, and TBN) are completely
independent from DPA and that assumptions involved are unrelated to age,
gender, and ethnicity.

In phase II we evaluated 317 healthy, non-obese Caucasians between

Table 4. Results of Phase II Studies (X±SD)

	Males (n=111)	Females (n=206)	Total (n=317)	r(p)**	SEE(kg)
Age	53.7 ±17.1	52.7 ±18.5	53.1 ±18.0	-	-
BMI	24.8 ±2.7	22.7 ±2.8	23.5 ±3.0	-	-
% fat *					
DPA	20.9 ±5.9	30.9 ±8.0	27.4 ±8.9	-	-
TBW	21.1 ±8.4	29.4 ±7.8	26.5 ±9.0	0.79 (0.001)	5.4
UWW	22.7 ±6.1	29.1 ±7.8	26.9 ±9.0	0.81 (0.001)	3.2
Model B	22.0 ±6.7	29.6 ±7.3	27.0 ±8.0	0.88 (0.001)	1.4

 * % fat by DPA, 3H2O dilution (TBW), underwater weighing (UWW), and
    Model B = combined TBW, UWW, and DPA TBBA.
 ** DPA % fat versus corresponding % fat.

the ages of 20 and 93 years. This large subject pool not only included
more individuals than in phase I, but the age distribution was younger
and more evenly balanced between males and females (Table 4). The
protocol in this study was similar to phase I described above, with the

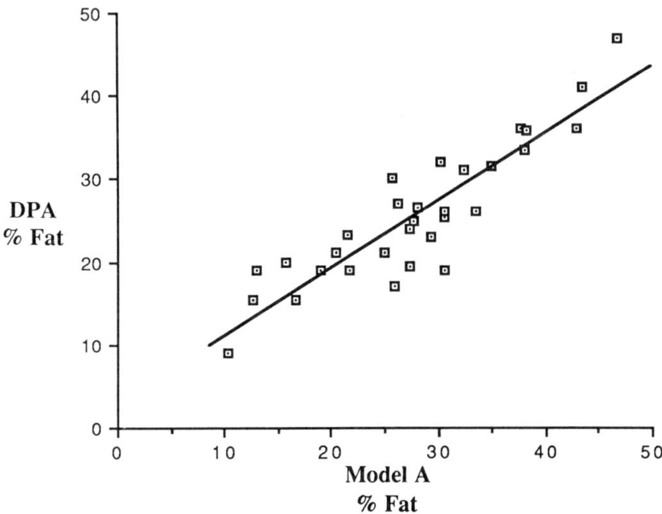

Fig. 5. Percentage fat estimated by DPA on the ordinate versus
         percentage fat derived by multicompartment neutron activation
         Model A on the abscissa [DPA Fat (%) = 0.81,
         Model A Fat + 3.1, SEE = 3.8%, r = 0.90, p<0.001].

exception that neutron activation was not included. As in the earlier phase I, results showed good agreement between DPA percentage fat and fat derived by conventional methodologies. Fig. 6 presents the correlation between DPA % fat and % fat calculated from our multicompartment model B, which is based on TBW, body density, and TBBA ($r=0.88$, $p<0.001$). This model has theoretical advantages over the other two-compartment models presented in Table 4, which fail to adjust fat estimates based on individual differences in hydration and bone mineral.

These studies indicate that DPA is capable of partitioning soft tissues into fat and fat-free components in a manner similar to that provided by conventional methodology.

As with bone mineral measurements, there exists a concern with fat estimated by DPA in the same individual over time and between different absorptiometers. Repeated scans over time of our meat phantoms shows good reproducibility. For example, phantoms #3 and #4 on daily DPA scans have a soft tissue attenuation ratio (R_{ST}) CV of 0.35% and 0.30%, respectively. These CVs translate to ~ 2% on the fat scale for R_{ST}, which ranges between about 1.2 and 1.4 for 100% and 0% fat. When these two meat phantoms were repeatedly studied on the newer DEXA (DPX), their R_{ST} CVs decreased by about 50% to 0.16% and 0.17%, respectively.

Repeated between-day studies for % fat are also available for DPA on the previously mentioned four subjects (Table 1). A parallel comparison of the CV for % fat was made using underwater weighing in three of these individuals. The mean CVs for % fat by DPA and underwater weighing were 3.22% and 5.73%, respectively for the three subjects undergoing both studies. The mean % fat in this group was 28.2% by DPA, so the CV represents about 1% total body fat. We anticipate improved precision with the newly introduced DEXA.

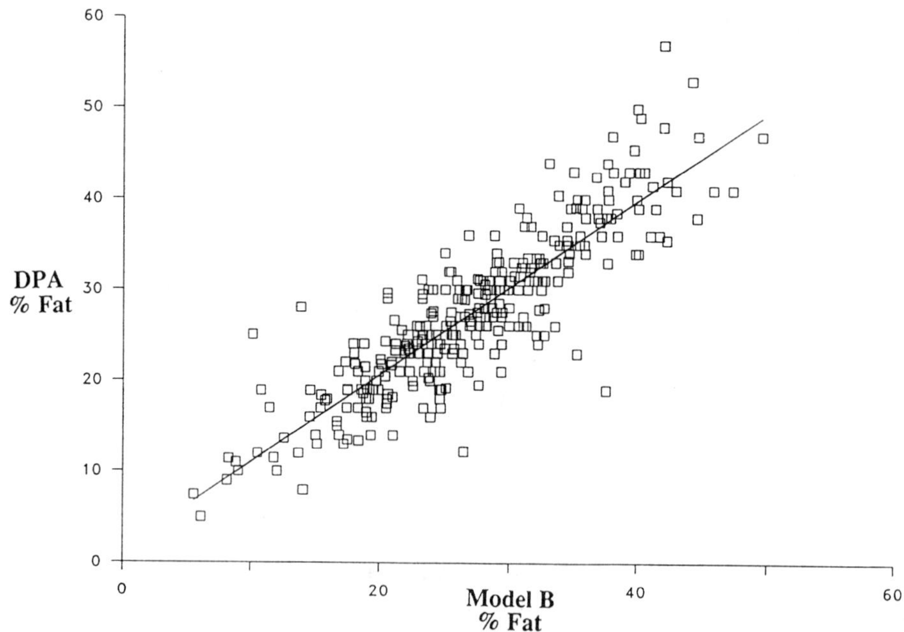

Fig. 6. Percentage fat estimated by DPA on the ordinate versus percentage fat derived by multicompartment Model B on the abscissa [DPA Fat (%) = 0.91 model B Fat + 1.5, SEE = 3.2%, r=0.93, p<0.001].

The question of between-instrument comparability is an important one that as yet does not have a definitive answer. Our initial studies can be briefly summarized, and they indicate that discrepancies exist between fat estimated by meat phantoms and by internal instrument calibrations, between different instruments, and between DPA and DEXA. These observed differences are worthy of further analysis, and extensive comparative studies are now underway at our Center.

As with bone mineral estimates, future studies need to examine the effects of body size and other conditions, such as state of hydration, on percent fat estimates by DPA.

APPENDICULAR SKELETAL MUSCLE

The fat-free, non-osseous portion of the extremities is almost entirely skeletal muscle, with skin and bone marrow comprising less than 5 percent of soft lean appendicular mass (Heymsfield et al., in press). As DPA software can isolate the extremities, this affords the opportunity to quantify a large portion (70-75%) of skeletal muscle mass in vivo. Our approach was to first develop a simple model of extremity composition. Limb bone ash was converted to skeletal weight by assuming ash comprises 55% of defatted fresh bone mass (Heymsfield et al., in press). Extremity fat was estimated by the beef phantom calibration procedure described earlier (Heymsfield et al., 1989b). Appendicular skeletal muscle was then calculated as total limb mass minus fat and bone mass.

The validity of the DPA method for appendicular skeletal muscle mass was evaluated on subjects of the phase I and three additional healthy, lean individuals (n=34) by comparing DPA results to TBK, TBN, and muscle calculated from anthropometric estimates and combined methodologies (Heymsfield et al., in press). The calculated muscle estimates involve use of two models of skeletal muscle mass proposed by Burkinshaw (1985; Burkinshaw et al., 1987). One model requires measurement of TBK and TBN, and the second model is based on TBK and fat-free body mass. Our approach to the second model was to calculate fat-free body mass as body weight minus DPA fat.

Total (upper and lower) DPA appendicular skeletal muscle mass was highly correlated with TBK (r=0.94, p<0.001; Fig. 7) and summed upper plus lower anthropometric limb muscle plus bone cross sectional areas (r=0.92, p<0.001). Significant, but weaker correlations were observed between DPA limb muscle and TBN (r=0.78, p<0.001), and between DPA limb muscle and estimates of whole body skeletal muscle mass based on Burkinshaw's two models (both r=0.82, p<0.001). These initial results are encouraging and suggest that DPA may also be useful as a means of quantifying a large proportion of skeletal muscle mass in vivo.

APPLICATIONS AND CONCLUSION

The availability of total body bone mineral measurements outside the few neutron activation centers that can quantify TBCa is an important advance in the field of body composition. The traditional two-compartment model of fat and fat-free body mass can be expanded to further subdivide lean tissues into FFST and TBBA. Moreover, by combining DPA with other methods, such as ^3H_2O dilution, the investigator could potentially evaluate four (H_2O, mineral, protein, fat) or more body compartments. The door is thus open to answering countless interesting and important questions in human biology previously limited by our inability to quantify skeletal mineral mass.

Similarly, the ability to quantify appendicular skeletal muscle mass, given the limitations of other available muscle-measuring techniques, is a novel and important feature of dual photon systems. The combined capacity to estimate skeletal muscle and bone mineral simultaneously in the appendages further expands the range of questions answerable by these methodologies.

Finally, although fat can be measured using other techniques, dual photon systems offer a new approach independent from classical assumptions or models related to body composition.

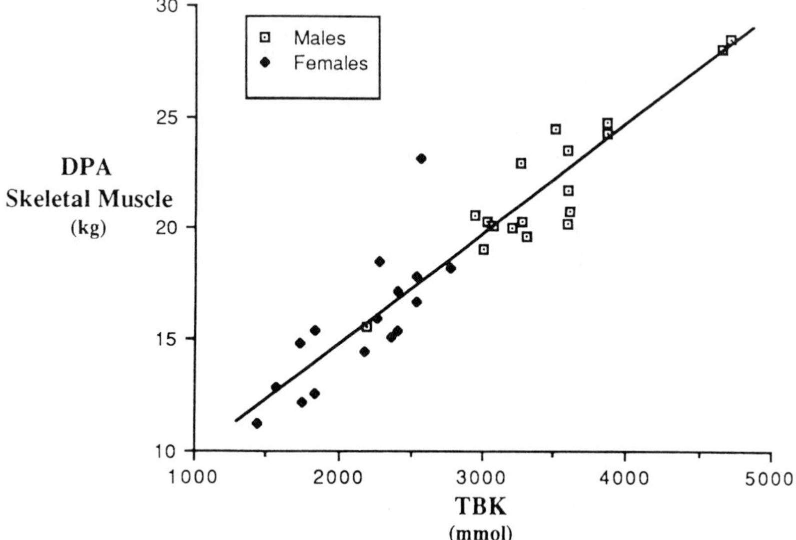

Fig. 7. DPA total appendicular skeletal muscle mass on the ordinate versus TBK on the abscissa (DPA muscle (g) = 0.0005 x TBK + 5.0, SEE = 1.6 kg, r = 0.94, p<0.001) (Heymsfield et al., in press).

ACKNOWLEDGEMENTS

Supported in part by NIH grants AR 39191, DK 41000 and DK 37352, and U.S. DOE contract DE-A CO2-76CH00016.

REFERENCES

Brozek, J., Grande, F., Anderson, T., and Keys, A., 1963, Densitometric analysis of body composition: revision of some assumptions, Ann. N.Y. Acad. Sci., 110:113.
Burkinshaw, L., 1985, Measurement of human body composition in vivo, in: "Progress in Medical Radiation Physics," Vol. 2, C. G. Orton, ed., Plenum Publishing Corporation, New York.
Burkinshaw, L., 1987, Models of the distribution of protein in the human body, in: "In Vivo Body Composition Studies," K. J. Ellis, S. Yasumura, and W. D. Morgan, eds., Institute of Physical Sciences in Medicine, London.
Cohn, S.H., Vartsky, D., Yasumura, S., Sawitsky, A., Zanzi, I., Vaswani, A., and Ellis, K.J., 1980, Compartmental body composition based on total body nitrogen, potassium, and calcium, Am. J. Physiol. 239:E524.

Heymsfield, S.B., Lichtman, S., Baumgartner, R.N., Wang, J., Kamen, Y., Aliprantis, A., and Pierson, R.N., Human body composition: comparison of two improved four-compartment models that differ in expense, technical complexity, and radiation exposure, Am J Clin Nutr, in press.

Heymsfield, S.B., Smith, R., Aulet, M., Bensen, B., Lichtman, S., Wang, J., Pierson R.N., Appendicular skeletal muscle mass: measurement by dual photon absorptiometry, Am J Clin Nutr, in press.

Heymsfield, S.B., Wang, J., Kehayias, J., Heshka, S., Lichtman, S., and Pierson, R.N., 1989a, Chemical determination of human body density in vivo: relevance to hydrodensitometry, Amer J Clin Nutr, 50:1282.

Heymsfield, S.B., Wang, J., Kehayias, J.J., and Pierson, R.N., 1989b, Dual photon absorptiometry: comparison of bone mineral and soft tissue mass measurements in vivo with established methods, Am. J. Clin. Nutr., 49:1283.

Mazess, R.B, Peppler, W.W, and Chestnut, C.H, 1981, Total body bone mineral and lean body mass by dual photon absorptiometry. II. Comparison with total body calcium by neutron activation analysis, Calcif.Tiss. Int., 33:361.

Mazess, R.N., Peppler, W.W., and Gibbons M., 1984, Total body composition by dual-photon (153Gd) absorptiometry, Am. J. Clin. Nutr., 40:834.

Vartsky, D., Ellis, K.J., and Cohn, S.H., 1979, In vivo quantification of body nitrogen by neutron capture prompt gamma-ray analysis, J Nucl Med 20:1158.

Wang, J., Heymsfield, S.B., Aulet, M., Thornton, J.C., and Pierson, R.N., 1989, Body fat from body density: underwater weighing vs dual photon absorptiometry, Am. J. Physiol., in press.

Woodard, H.Q., 1964, The composition of human corticol bone, Clin. Orthop., 37:187.

MEASUREMENT OF BODY FAT BY NEUTRON INELASTIC SCATTERING:

COMMENTS ON INSTALLATION, OPERATION AND ERROR ANALYSIS

Joseph J. Kehayias, Steven B. Heymsfield[*], F. Avraham
Dilmanian[+], Jack Wang[*], David M. Gunther, and Richard
N. Pierson[*] Jr.

USDA Human Nutrition Research Center on Aging at
Tufts University, Boston, MA 02111
[*]St. Luke's-Roosevelt Hospital, Columbia University
College of Physicians and Surgeons, New York, NY
10025
[+]Medical Department, Brookhaven National Laboratory
Upton, NY 11973

INTRODUCTION

The existing techniques for the _in vivo_ measurement of
body fat can be classified into three categories:
    Subtraction of lean body mass from total weight. Body
lean mass is estimated by measurements of total body potassium
(TBK), total body water (TBW), or total body nitrogen (TBN).
Under the assumption that lean tissue has a constant content
of potassium or water, TBK or TBW measurements can produce an
estimate of body lean mass. The constant content assumption
does not always hold; therefore, more detailed techniques have
been developed for evaluating the individual components of
lean tissue. These techniques include the measurement of
total body nitrogen (TBN) to evaluate body protein (Biggin et
al., 1972 and Vartsky et al., 1979), and total body calcium
(TBCa) for bone ash (Chamberlain et al., 1968, Nelp et al.,
1970 and Cohn et al., 1974). After the lean body mass is
determined, fat mass is derived by subtraction of this value
from total body mass. As most of the body is composed of lean
tissue, the fat evaluation by subtraction suffers from an
error propagation disadvantage which is more pronounced for
lean individuals. This sensitivity to measurement errors is
demonstrated in Fig. 1 where the bracket shows the total error
in the evaluation of lean body mass. This total error
includes both systematic and random measurement errors, as
well as deviations from the assumptions upon which the
specific model relies.
    Measurement of a physical property of the body. Fat
content is related directly or indirectly to some physical
properties of the body such as body density, impedance,
photon attenuation, and electrical conductivity. By measuring
these physical parameters, body composition results can be

derived. The corresponding techniques are underwater
weighing, bioelectrical impedance, dual photon absorptiometry
(DPA), and total body electrical conductivity (TOBEC). We
include all skinfold and other anthropometric techniques in
this category. Some of these techniques, such as underwater
weighing (Heymsfield et al., 1989) are exceptionally sensitive
to measurement errors while others depend upon body
composition assumptions which may significantly vary with age
and health condition of the subject.

Neutron inelastic scattering. This relatively new
technique (Kyere et al., 1982; Kehayias et al., 1987) is used
to measure _in vivo_ body carbon. Body fat and protein are the
main contributors to total body carbon (TBC); bone ash and
carbohydrates contribute less than 3%. The contribution of
protein is typically 15-30%. Therefore, a measurement of
total body carbon can be associated directly with body fat if
a reasonable adjustment is made for the contribution of carbon
due to body protein. TBC is measured by neutron inelastic
scattering, while protein is evaluated directly by measuring
TBN, or indirectly by measuring TBK. The accuracy of the
protein evaluation is not critical, since only a small
fraction of its measurement error propagates to the estimate
of fat.

THE TECHNIQUE

Total body carbon is measured _in vivo_ through the
reaction $^{12}C(n,n'\gamma)^{12}C$ by detecting the 4.44 MeV gamma rays
from the inelastic scattering of fast neutrons. The neutrons

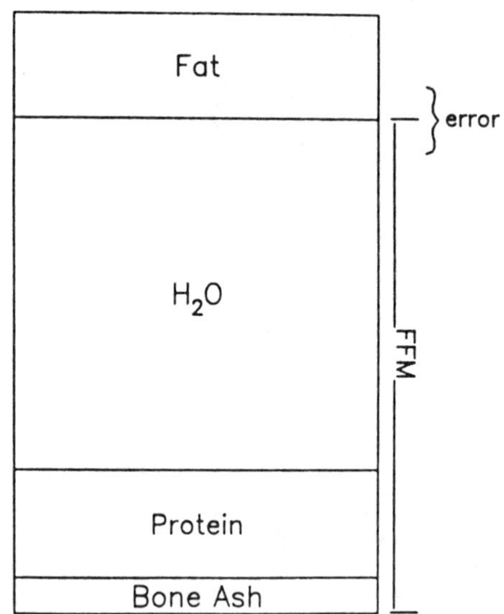

Fig. 1.   Schematic representation of a four-component model
          for body composition. The bracket indicates the
          total error in determining the fat free mass
          (FFM). When fat is estimated by subtraction from
          body weight the same error-bar represents a large
          relative error for fat.

are generated by a miniature D-T accelerator neutron source
(Shope et al., 1981) as a product of the deuterium-tritium
fusion reaction. The patient is scanned over the neutron
source at constant speed for 20 min. The total radiation
exposure to the skin is less than 0.16 mSv, which allows
multiple studies of the same volunteer.

Two carbon scanners exist based on a small D-T neutron
source for _in vivo_ TBC measurements. The prototype facility
was built at Brookhaven National Laboratory, New York. A
second scanner is near completion at the USDA Human Nutrition
Research Center on Aging at Tufts University, Boston. This
scanner will use two simultaneously operated D-T neutron
generators to deliver a more uniform neutron beam to the
patient. In both the Brookhaven and the Tufts version, the
neutron source is positioned under a scanning bed, shielded
with steel. Two NaI(Tl) detectors are positioned on each side
of the patient to detect gamma rays. The detectors are
shielded from the neutron cloud with borax ($Na_2B_4O_7 \cdot 10H_2O$) and
lead. The details of the miniature accelerator design and the
fast data acquisition electronics have been described
elsewhere (Kehayias et al., 1987). Table 1 shows the
instrumental characteristics and estimated cost for the
facility.

Analysis of Spectral Data

Fig. 2 shows an energy spectrum of gamma rays from a
normal volunteer. There are two factors which complicate the
spectral analysis process:

1. The detectors do not maintain their energy calibration

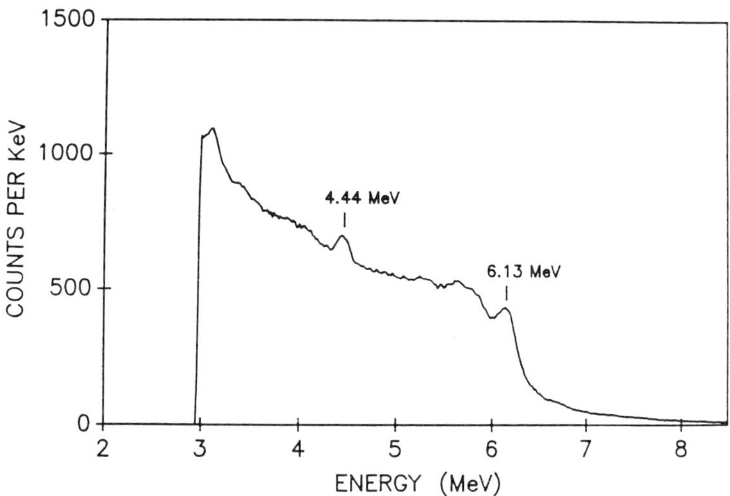

Fig. 2.   Energy spectrum of gamma rays from a normal
          volunteer.

Table 1. Instrumental characteristics of the neutron
generator scanning facility

Neutron Source	sealed D-T accelerator
Neutron output	approximately 10^8 n/s
Operation	pulsed source at 10 KHz
Duty cycle	10 %
Neutron burst duration	10 μs
Detectors	two NaI(Tl) (15.2x15.2cm)
Shielding (patient)	steel, cadmium and lead
Shielding (detectors)	borax, lead
Scanning time	20 min
Skin dose to patient	< 0.16 mSv

Replacement costs (thousands of 1989 US dollars)

sealed D-T tube	8
electronics (accelerator)	35
electronics (data acquisition)	22
scanning bed	9
detectors	11
shielding	6
computer	5

as both the neutron irradiation level and their counting rate
change substantially when the neutron generator is turned on
and off. To minimize this effect, energy calibration is
performed before each measurement.

2. An accurate evaluation of the background under the
carbon peak is essential for the accuracy of the measurement.
The escape peaks on the left of the main carbon peak interfere
with a simple "left background" measurement.

With these problems in mind, a simple library-fitting
technique was used for the analysis. All energy spectra were
first compressed to 512 channels, so as to have the same
energy calibration. The compression algorithm determines the
centroids of the 4.44 MeV carbon peak and the 6.13 MeV oxygen
peak for each spectrum, and adjusts the horizontal gain and
offset so that all spectra result with identical energy
calibration. This function corrects for small drifts in
detector gain. Then, two spectra, one from a phantom rich in
carbon and one from a water phantom, are used as a base for
fitting the patient's spectral data. We found that this
technique is less sensitive to variations in the performance
of the detectors and the neutron output of the D-T generator.

Table 2. Characteristics of studied volunteers

		N	Age (yrs)	Weight (Kg)	BMI (Kg/m^2)
Group I	men	6	40-94	58.2-73.0	21.5-25.5
	women	8	24-85	52.6-71.2	21.0-26.1
Group II	men	3	62-77	55.8-91.2	22.2-28.0
	women	4	26-60	49.4-68.5	19.8-22.8

The amount of body carbon for each individual is derived by comparing the total number of detected carbon gamma rays to a series of measurements with anthropomorphic phantoms. Fat is then derived by using the equation:

$$FAT\ (Kg) = [TBC - (3.75 \times TBN + 0.05 \times TBCa)]/0.77$$

where TBC, TBN and TBCa are the measured amounts of body carbon, nitrogen and calcium in kilograms. The terms in parenthesis represent corrections due to the carbon contributions of protein and glycogen, both estimated from TBN, and the carbon contribution due to bone ash. It is assumed that fat contains 77% carbon by weight (Elia and Livesey, 1988).

RESULTS

Our experience with the neutron inelastic scattering technique was gained from two experimental protocols, one with 14 and the other with 7 healthy, non-obese volunteers. The volunteers underwent several other measurements of body composition so that cross validation could be made. We have completed the experiments with the first group of volunteers (Kehayias et al., unpublished results). Table 2 shows the principal characteristics of the subjects which participated in the two studies. The main difference between the two groups is in the way the carbon and nitrogen spectral data were collected and analyzed. The carbon measurements of group I were performed with high flux (5×10^7 n/s) and a constant output of fast neutrons from the neutron generator. This mode of operation offers a reproducibility of 3%. Group II measurements were performed with a neutron output approximately three times lower, as tritium was depleted from the aging neutron generator target. The data in this group carry a higher reproducibility error (6%) because of lower counting statistics and instability in the neutron flux. On the other hand, the Brookhaven nitrogen facility was improved for the group II measurements, providing a more accurate estimate of lean body mass. The improvement in the nitrogen measurement has only a small effect on the total error of fat as calculated from TBC, because protein serves as a relatively small correction to total body carbon. TBN measurements

offer, however, a significant improvement when fat is
calculated by subtraction from body weight. Fig. 3 shows the
data on fat for group II calculated from total body water and
nitrogen by using the model proposed by Cohn et al. (1984)
plotted against fat derived by carbon. The equation used is:

FAT(Kg) = (body weight - TBW - body protein - bone ash)

where body protein = TBN x 6.25 and bone ash = TBCa/0.34.

Fig. 4 demonstrates how two other techniques, used to
predict fat, correlate with the carbon fat measurements.
Considering the TBC technique as the standard, we notice that
the TBW method seems to work better for men, while the dual
photon absorptiometry (Heymsfield et al., 1989) predicts fat
better for women, with the exception of the very lean.

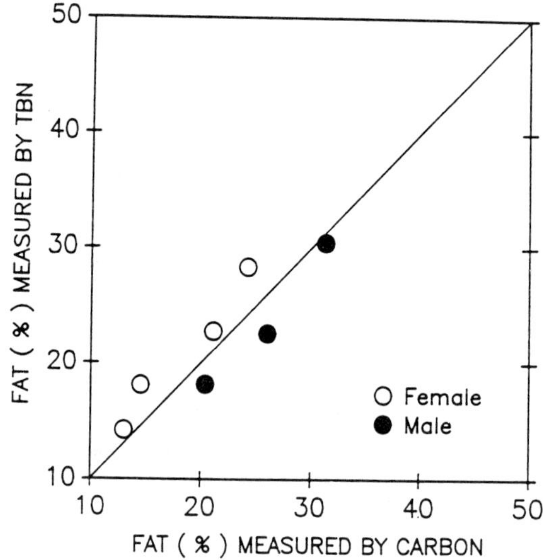

Fig. 3.   Fat content for group II subjects calculated from
           TBN, TBW and TBCa using the Cohn model (see text)
           plotted against fat measured by neutron inelastic
           scattering.

Based on the operation of the facility with the two
groups of volunteers we concluded:

1. Constant neutron output should be maintained
throughout the life of the neutron generator, which is
estimated to 200-300 hours of operation. This aim is achieved
by adjusting the accelerating voltage to compensate for the
depletion of target material.

2. Accurate data analysis is best achieved with energy calibrated spectra fitted to a basis of standards.

3. The accurate estimation of protein is not critical for the TBC technique, as only a third of its error propagates to the final estimate of fat. On the contrary, if fat is derived by subtraction from body weight the accurate determination of lean body mass becomes very critical.

All indirect methods for determining body fat lack independent validation. Each of these methods is applicable

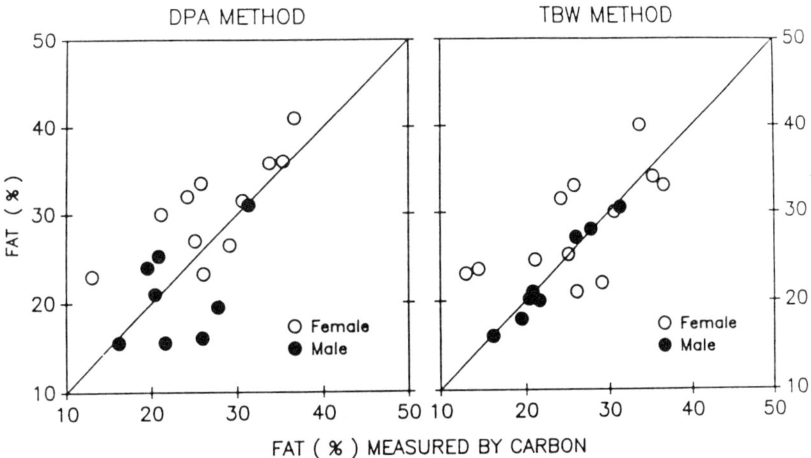

Fig. 4.　Fat content for both groups of volunteers measured by DPA (left) and TBW (right) plotted against fat content derived from TBC.

only to particular classes of subjects mainly because of limitations in the validity of the assumptions upon which they rely. The TBC technique, being model-independent and less sensitive to errors in measurement, qualifies for studies of subjects of any age and condition of health. The TBC technique will also serve as the measure for validating and determining the limitations of less direct techniques.

ACKNOWLEDGEMENTS

This research was supported by US Department of Energy grant No DE-AC02-76CH000016, NIH grants DK37352 and DK26687 and US Department of Agriculture grant No 53-3K06-5-10.

REFERENCES

Biggin, H. C., Chen, N. S., Ettinger, K. V., Fremlin, J. H., Morgan, W. D., Nowotny, R. and Chamberlain, M. J., 1972, Determination of nitrogen in living subjects. Nature 236:187.

Chamberlain, M. J., Fremlin, J. M., Peters, D. K., Philp., H., 1968, Total body calcium by whole body neutron activation analysis: new technique for the study of bone disease, Brit. Med. Jour. 2:581.

Cohn, S. H., Ellis, K. J. and Wallach, S., 1974, In vivo neutron activation analysis: clinical potential in body composition studies, Amer. J. Med. 57:683.

Cohn, S. H., Vaswani, A. N., Yasumura, S., Yuen, K. and Ellis, K. J., 1984, Improved models for determination of body fat by in vivo neutron activation, Am. J. Clin. Nutr. 40:255.

Elia, M. and Livesey, G., 1988, Theory and validity of indirect calorimetry during net lipid synthesis, Am. J. Clin. Nutr., 47:91.

Heymsfield, S. B., Wang, J., Heshka, S., Kehayias, J. J. and Pierson, R. N., 1989, Dual-photon absorptiometry: comparison of bone mineral and soft tissue mass measurements in vivo with established methods, Am. J. Clin. Nutr. 49:1283.

Heymsfield, S. B., Wang, J., Kehayias, J. J., Heshka, S., Lichtman, S. and Pierson, R. N., 1989, Chemical determination of human density in vivo: relevance to hydrodensitometry, Am. J. Clin. Nutr. 50:1282.

Kehayias, J. J., Ellis, K. J., Cohn, S. H. and Weinlein, J. H., 1987, Use of a high repetition rate neutron generator for in vivo body composition measurements via neutron inelastic scattering, Nucl. Instr. and Meth. in Phys. Res. B24/25:1006.

Kehayias, J. J., Heymsfield, S. B., LoMonte, A. F., Wang, J. and Pierson, R. N., In vivo determination of body fat by neutron inelastic scattering, Unpublished results.

Kyere, K., Oldroyd, B., Oxby, C. B., Burkinshaw, L., Ellis, R. E. and Hill, G. L., 1982, The feasibility of measuring total body carbon by counting neutron inelastic scatter gamma rays, Phys. Med. Biol. 27:805.

Nelp, W. B., Palmer, H. E., Murano, R., Pailthorpe, K., Hinn, G. M., Rich, C., Williams, J. L., Rudd, T. G., Denney, J. D., 1970, Measurement of total body calcium (bone mass) in-vivo using total body neutron activation analysis, Jour. Lab. Clin. Med. 76:151.

Shope, L. A., Berg, R. S., O'Neal, M. L. and Barnaby, B. E., 1981, Operation and life of the Zetatron: a small neutron generator for borehole logging, IEEE Trans. Nucl. Sci. NS-28:1696.

Vartsky, D., Ellis, K. J. and Cohn, S. H., 1979, In vivo measurements of body nitrogen by analysis of prompt gammas from neutron capture, J. Nucl. Med. 20:1158.

PROMPT GAMMA MEASUREMENTS OF NITROGEN AND

CHLORINE IN NORMAL VOLUNTEERS

S J S Ryde, W D Morgan, D W Thomas[*], J L Birks,
C J Evans[*], P A Ali, and H Jenkins

Swansea In Vivo Analysis Research Group
Department of Medical Physics and Clinical
Engineering, Singleton Hospital, Swansea, SA2 8QA,
[*]Department of Physics, University College of
Swansea, Swansea, SA2 8PP, Wales, UK.

INTRODUCTION

The Swansea _in vivo_ neutron activation analysis (IVNAA)
instrument, which is based on a 4 GBq Cf-252 neutron source,
has been developed primarily for the prompt-gamma measurement
of total body (TB) Ca, N, Cl, H and C, and partial body Cd,
using semiconductor (HPGe) and scintillation (NaI) gamma ray
detectors (Ryde et al., 1987).

All of the total body elements listed can be measured
simultaneously using high resolution HpGe detectors, but when
TBCa is not of prime interest, the combination of H,N,Cl and,
to a lesser extent C, can be satisfactorily analysed using
NaI(Tl) detectors and at a much smaller dose to the subject.

This paper describes a study of normal volunteers, (i) to
validate clinically the method for TBN (and hence protein)
measurement, and (ii) to evaluate the calibration of the
instrument for TBCl by comparison with stable bromide dilution
measurements of extracellular fluid volume (ECFV).

STUDY POPULATION

Twenty-four normal, healthy volunteers with stable body
weight during the 3 months prior to the study were recruited
(Table 1). All subjects provided written informed consent in
accordance with the study protocol approved by West Glamorgan
District Ethical Committee.

METHODS

The features and calibration of the instrument associated
with the measurement of TBN and Cl have been described elsewhere (Ryde et al., 1989). The subjects were scanned prone and

Table 1. Subjects In The Study Population. Values Are
         Mean ± SD (Range)

	n	Age (y)	Height (m)	Weight (kg)
Male	12	40.1 +/-8.5 (24.7-55.7)	1.80 +/-0.06 (1.69-1.89)	75.34 +/-11.6 (59.3-104.2)
Female	12	42.7 +/-12.5 (23.4-61.4)	1.57 +/-0.08 (1.42-1.69)	56.68 +/-5.4 (44.5-65.2)

supine from the shoulder to knee above the neutron source and
beneath two 152 x 152 mm NaI(Tl) detectors. The absorbed
radiation dose equivalent to the skin was 0.4 mSv (QF=10).
The net counts due to gamma rays in the H (2.08 to 2.39 MeV),
Cl (5.46 to 5.82 and 5.99 to 6.21 MeV), and N (9.65 to 11.14
MeV) regions of interest were determined. Body habitus correction for N and Cl counts was achieved by using the hydrogen
internal standardisation technique (Vartsky et al., 1979), in
which body H was independently assessed by measurement of total
body water (TBW) (see following section). The precision
(coefficient of variation) of the technique determined from
repeated measurements of an anthropomorphic phantom is 1.6% and
5.1% for the N/H and Cl/H ratios, respectively (Ryde et al.,
1989).

TBN was calculated using the equations of Vartsky et al
(1984) and total body protein (TBProt) obtained by multiplying
TBN by 6.25. The fat-free mass (FFM) was assumed to consist of
protein, water, glycogen and minerals, where the mass of glycogen was given by M_G = 0.0091 x TBW/0.73 (Beddoe et al., 1984)
and the mineral compartment was derived from predicted TBCa
(Ellis, 1989). Total body fat (TBFat) was then calculated as
the difference between body weight and the FFM.

To determine TBCl it was necessary to correct the gamma
ray counts in the Cl region of interest for interferences from
N and O (Ryde et al., 1989); the former via measured TBN and
comparison with a standard phantom, and the latter by assuming
the following stochiometric relationship, TBO = 0.889(TBW) +
0.11(TBF) + 1.44(TBN) + 0.4(TBCa/0.208), and subsequent comparison with a standard phantom.

The ECFV was derived from the TBCl measurements by the
following equation (Yasumura et al., 1983)

$$ECFV[l] = 0.9 \ (TBCl[mol]/plasma \ Cl[mol/l]) \qquad (1)$$

Total Body Water (TBW)

Subjects were fasted overnight but allowed a normal fluid
intake until the start of the test. Urine and blood samples
were taken immediately before an oral dose of 3.7 MBq tritiated
water (THO). Further blood and urine samples were collected at
4 and 5 hours, during which time no food or drink was taken.
Plasma and urine samples were assayed by liquid scintillation
counting using a spiked sample for quench correction. An

Table 2.   Comparison of Measured TBN in Healthy Swansea Volun-
teers to Predicted Values from IVNAA Centres; and to
Predicted Values of TBN from Measured TBK in the Same
Subjects Using the Swansea Whole-body Counter

Centre	Measured TBN/ Predicted TBN	SEM	Linear Regression Analysis	
			correlation coefficient, r	%CV of Swansea data about regression
Brookhaven (Ellis et al., 1982)	1.015	+/-0.019	0.940	9.1%
Leeds (Burkinshaw et al., 1981)	0.996	+/-0.023	0.944	11.6%
Brookhaven - from TBK (Ellis et al., 1982)	0.967	+/-0.016	0.960	7.4%

aliquot of each sample was evaporated to dryness to correct for
contained solids. This method gives excellent agreement for ^{3}H
dilution volumes derived from either urine or plasma samples,
and the mean value was taken for TBW calculation. TBW was
obtained by muliplying the derived tritium dilution space by
0.97 to correct for non-aqueous exchange of hydrogen.

Extracellular Fluid Volume (ECFV)

Concurrently with the administration of THO each subject
received a maximum oral dose of KBr corresponding to 0.047g
Br^{-}/kg body weight. Aliquots (2.5ml) of plasma and urine were
analysed for Br content by X-ray fluorescence (XRF). The XRF
instrument is based on a 5.2 GBq Cd-109 source and achieves a
detection limit (3 SD's of the background) of 2 µg/ml for a
2000s counting time. ECFV was calculated as the bromide
dilution space (at 5 hours) multiplied by an overall correction
factor of 0.9 (Forbes, 1987).

If it is assumed that bromide and chloride equilibrate in
the same compartment, then TBCl can be calculated as the product of the Br space and the plasma Cl^{-} concentration.

RESULTS AND DISCUSSION

Nitrogen

The measured TBN for each subject was compared with the
predicted value based on age, height, weight, and sex (Table 2)
(Ellis et al., 1982; Burkinshaw et al., 1981). A prediction of
TBN based upon measured TBK using the Swansea whole-body counter was also calculated (Table 2). The Swansea data show

Table 3.   Comparison of TBCl Measured in Swansea by IVNAA and
            Derived From Br Dilution Space (Figures in Brackets)
            to Predicted Values From Other Centres

Centre	Measured TBCl/ Predicted TBCl	SEM	Linear Regression Analysis	
			Correlation Coefficient, r	%CV of Swansea data about regression
Brookhaven (Ellis, 1989)	0.988 (1.016)	+/-0.034 (+/-0.018)	0.823 (0.937)	16.8% (8.6%)
Leeds (Siwek et al., 1987)	1.022 (1.054)	+/-0.036 (+/-0.025)	0.816 (0.870)	17.2% (11.4%)

encouraging agreement with the predicted values. The ratio of
mean TBN/body weight was 2.59% +/-0.09 (SEM) and 2.17% +/-0.07
(SEM) for males and females, respectively, and probably reflects the differences in fat between the sexes. If fat is
excluded, the ratio of mean TBN/FFM (g/kg) was 32.7 +/-0.4
(SEM) and 33.2 +/-0.4 (SEM) for males and females, respectively.

Both measures of TBN (i.e. as a percentage of body weight
and FFM) are comparable to values for normal subjects reported
elsewhere (Beddoe and Hill, 1985), thereby providing confirmation that the TBN measurement system is valid.

Furthermore, the mean values of the hydration coefficient
of the FFM for males and females of 0.727 +/-0.002(SEM) and
0.719 +/-0.002(SEM) respectively are in excellent agreement
with the currently accepted values for normal human subjects.

<u>Chlorine</u>

The IVNAA measured TBCl for each subject was compared with
a predicted value based on sex, weight, and height (Ellis,
1989) and with a value based on sex, age, and height (Siwek et
al., 1987) (Table 3). The table also shows (in brackets) the
ratio of TBCl values derived from Br dilution space, TBCl(Br),
to the respective predicted values. The correlation coefficient between measured and predicted TBCl is higher for TBCl(Br)
than for TBCl(IVNAA), which probably reflects the imprecision
of individual IVNAA measurements. Although the means of the
measured-to-predicted values suggest that there may be a small
difference between the two methods for TBCl measurement, the
difference is not significant (p>0.05).

Linear regression analysis was used to examine the relationship between TBCl(IVNAA) and TBCl(Br). The analysis yields
a correlation coefficient of r=0.881 and a regression line
slope of 0.99 +/- 0.12 (SE). The coefficient of variation of
the data about the regression line is 12.8%. The TBCl(IVNAA)
correlates better with TBCl(Br) than with either of the predictor equations (Table 3), but this may reflect the normal
biological variations inherent in these equations (for example,

the standard error estimate of the Brookhaven and Leeds equations is approximately 7 and 13g, respectively).

Linear regression was also used to examine the measurement of ECFV by two techniques, viz ECFV(IVNAA) derived from TBCl(IVNAA) and equation 1, and ECFV(Br) derived from the Br dilution space. The correlation coefficient (r=0.881) and slope of the regression line (0.98 +/- 0.12 (SE)) are almost identical to the result given for TBCl(IVNAA) upon TBCl(Br) in the preceeding discussion. This is presumably because the error associated with the measurement of plasma Cl is small. The mean ratio of ECFV(IVNAA) to ECFV(Br) for the subject group is 0.970 +/- 0.026 (SE).

In conclusion, the results from TBCl measurements using IVNAA compared with predicted values and those derived from Br dilution space are very encouraging. Although the imprecision of the TBCl(IVNAA) measurement is a limitation on an individual basis, it is possible to overcome this by increasing the scanning time and/or detection efficiency. Finally it is noted that the measurement of TBCl by stable KBr dilution is a valuable addition to the IVNAA method.

OVERALL CONCLUSION

The combination of prompt gamma IVNAA measurements of N and Cl with total body water estimation, provides unique information on the composition of the fat-free mass and on the partition between intra- and extracellular water. By incorporating predicted values of glycogen and minerals into the model for FFM, a value for TBFat is obtained (by difference from body weight) that is independent of assumptions of tissue density and/or tissue hydration.

ACKNOWLEDGEMENTS

We are grateful to the Medical Research Council for supporting the construction of the instrument, and to colleagues R. Wyatt, M.G. Hughes, I. Hainsworth and R.R. Ghose for their support and assistance. We also thank the Directors of SIVARG, A. Sivyer, J. Dutton, and B.N.C. Littlepage for initiating and overseeing the work. A Wellcome Trust Travel Grant was awarded for one of us (SJSR) to attend this Conference. DWT is in receipt of an SERC Research Studentship.

REFERENCES

Beddoe, A.H., and Hill, G.L., 1985, Clinical measurement of body composition using in-vivo neutron activation analysis, J. Parenteral and Enteral Nutr., 9 : 504.
Beddoe, A.H., Streat, S.J., and Hill, G.L., 1984, Evaluation of an in-vivo prompt gamma neutron activation facility for body composition studies in critically ill intensive care patients : results on 41 normals, Metabolism, 33 : 270.
Burkinshaw, L., Morgan, D.B., Silverton, N.P., and Thomas, R.D., 1981, Total body nitrogen and its relation to body potassium and fat free mass in healthy subjects, Clin. Sci., 61 : 457.
Ellis, K.J., 1989, Reference Man : variations due to body size, age, sex and race, Biol. Trace Element Res., in press.

Ellis, K.J., Yasumura, S., Vartsky, D., Vaswani, A.N. and Cohn, S.H., 1982, Total body nitrogen in health and disease : effects of age, weight, height, and sex, J. Lab. Clin. Med., 99 : 917.

Forbes, G.B., 1987, "Human Body Composition", Springer-Verlag, New York.

Ryde, S.J.S., Morgan, W.D., Sivyer, A., Evans, C.J., and Dutton, J., 1987, A clinical instrument for multi-element in-vivo analysis by prompt, delayed and cyclic neutron activation using Cf-252, Phys. Med. Biol., 32 : 1257.

Ryde, S.J.S., Morgan, W.D., Evans, C.J., Sivyer, A., and Dutton, J., 1989, Calibration and evaluation of a Cf-252 based neutron activation instrument for the determination of nitrogen in-vivo, Phys. Med. Biol., 34 : 1429.

Siwek, R.A., Wales, J.K., Swaminathan, R., Burkinshaw, L., and Oxby, C.B., 1987, Body composition of fasting obese patients measured by in-vivo neutron activation analysis and isotopic dilution, Clin. Phys. Physiol. Meas., 8 : 271.

Vartsky, D., Prestwich, W.V., Thomas, B.J., Dabek, J.T., Chettle, D.R., Fremlin, J.H., and Stammers, K., 1979, The use of body hydrogen as an internal standard in the measurement of nitrogen in-vivo by prompt neutron capture gamma-ray analysis, J. Radioanal. Chem., 48 : 243.

Vartsky, D., Ellis, K.J., Vaswani, A.N., Yasumura, S., and Cohn, S.H., 1984, An improved calibration for the in-vivo determination of body nitrogen, hydrogen and fat, Phys. Med. Biol., 29 : 209.

Yasumura, S., Cohn, S.H., and Ellis, K.J., 1983, Measurement of extracellular space by total body neutron activation, Am. J. Physiol., 244 : R36.

DETERMINATION OF TOTAL BODY CALCIUM BY PROMPT GAMMA NEUTRON

ACTIVATION ANALYSIS: ABSOLUTE <u>IN VIVO</u> MEASUREMENTS

SJS Ryde, WD Morgan, JE Compston+, AJ Williams*, CJ
Evans**, A Sivyer and J Dutton**

Swansea In Vivo Analysis Research Group
Department of Medical Physics and Clinical
Engineering, Singleton Hospital, Swansea SA2 8QA.
+Department of Pathology, University of Wales
College of Medicine, Heath Park, Cardiff CF4 4XN.
*Department of Renal Medicine, Morriston Hospital,
Swansea SA6 6NL. **Department of Physics, Univer-
sity College of Swansea, Swansea SA2 8PP, Wales, UK.

INTRODUCTION

A method for the measurement of total body calcium (TBCa)
by prompt gamma neutron activation analysis has previously been
described (Ryde et al., 1987). This involves irradiation with
Cf-252 neutrons and on-line detection of 6.420 MeV gamma-rays
resulting from neutron capture in Ca-40. It therefore differs
significantly from all other measurements of TBCa, which hither-
to have employed the delayed activation method.

The ultimate aim of such measurements is, of course, to
make absolute measurements of total body calcium. While the
calibration method necessary for such measurements was being
developed, however, it was considered worthwhile to initiate
clinical studies by undertaking serial measurements. Provided
the body habitus does not change significantly with time, such
serial measurements of 'calcium counts' in the same individual
can provide useful clinical information.

The purpose of this paper is to describe recent progress
in achieving an absolute calibration of the system and to
report the results obtained so far for various clinical groups
involved in the serial studies.

THE INSTRUMENT

Ryde et al. (1987) gives a detailed description of the
neutron activation instrument. The instrument contains a
nominally 4 GBq (200 µg) Cf-252 neutron source and is capable
of performing both total and partial body multi-element analysis. Total body measurements of Ca, N, Cl, C and H are obtained simultaneously by scanning the subject between the collimated

Table 1. Known and Measured Parameters of the Anthropomorphic
          Phantoms.

Phantom	Weight/height[2] (kg/m[2])	Known quantities		Measured quantities	
		TBCa (g)	TBCl (g)	TBCa (g)	Discrepancy[a] (%)
1	18.4	581 +/-6	47 +/-1.6	566 +/-53[b]	-2.6
2	20.2	967 +/-10	113 +/-2.9	1034 +/-67	+6.9
3	30.9	1213 +/-13	105 +/-4.8	1197 +/-89	-1.3
4	24.3	1010	97	973 +/-47	-3.7

[a]Defined as TBCa (measured-known)/TBCa (known) x 100%
[b]Errors are based on counting statistics and the known errors
 in the TBCl content

neutron source and two 20% relative efficiency high resolution
HpGe detectors. The scanning geometry therefore offers the
possibility of obtaining head-to-toe profiles of element distribution in addition to a total body content. The *in vitro*
measurement precision (CV) for TBCa is 2.6% for a dose equivalent to the skin of 6.4 mSv (QF = 10) (Ryde et al., 1989a).

SEQUENTIAL CLINICAL MEASUREMENTS

     To date over 100 subjects have been scanned from shoulder-
to-knee in the prone and supine positions for a total scanning
time of less than one hour per subject. In practice, the dose
equivalent is less than 5 mSv to the skin surface, but the mid-
body dose equivalent is estimated to be less than 2.5 mSv.

     Several clinical studies are in progress to determine
changes in TBCa with time. These include patients with Crohn's
disease, women receiving oestrogen implant therapy, patients on
renal dialysis and others with various metabolic bone disorders
including osteomalacia. In most cases, TBCa is only a part of
a wider measurement protocol, which includes quantitative compu-
terised tomography (QCT) of the lumbar spine, single photon
absorptiometry (SPA) of the radius and quantitative histomorphometry of an iliac crest bone biopsy. The longitudinal data
are not yet complete although preliminary results from those
subjects with Crohn's disease are presented elsewhere (Compston
et al., 1989). The following sections, therefore, refer only to
single measurements of bone status without any reference to
changes therein.

ABSOLUTE MEASUREMENTS

     The method that has been adopted for absolute measurement
relies on the use of total body chlorine (TBCl) as an 'internal
standard' for calcium (Ryde et al., 1989b) and using the ratio
of Ca/Cl counts to correct for body habitus. The technique has
been tested by measuring 4 anthropomorphic phantoms (Table 1),
(Bewley, 1988; Oxby and Brooks, 1979) in which the contents of

Ca and Cl were known. The mean discrepancy (from the modulus
of the individual values) is 3.6% and comparable with measurements by the delayed activation technique of three of these
phantoms at other centres (Bewley, 1988).

The absolute calibration technique has been used to assess
TBCa in the Crohns, oestrogen implant, and renal dialysis patient groups. For each patient a predictor equation based on
sex, height and weight was used to determine TBCl (Ellis, 1989).
TBCa was then calculated using the measured Ca/Cl counts ratio.
The values so obtained were compared with those expected on the
basis of sex, height, weight and age (Figure 1) (Ellis, 1989).

Patients were admitted to the implant group on the basis
of menopausal symptoms and previous hysterectomy. The TBCa
measurements were undertaken before therapy commenced. The
Crohns patient group have a number of risk factors, including
corticosteroid therapy, malabsorption, malnutrition and amenorrhoea. Other investigations (e.g. QCT, SPA) in this group have
yielded results which have, in general, been below the values
observed in age and sex matched normal subjects (Compston et
al, 1987).

It is not unreasonable, therefore, that the TBCa of both
groups might be below that expected in a normal population as
indicated in Figure 1. However, some caution is necessary, as
the use of predictor equations for TBCa and TBCl may only be
strictly applicable to the North American population from which
they were derived. Caution is also necessary in interpreting
the results from the renal dialysis group in whom several
different types of metabolic bone disease may be present.

Nevertheless, initial results are encouraging and work is
in progress to measure the TBCl directly using a stable bromide
dilution technique (Ryde et al., 1989c) and to consider further
the effects of body habitus upon the Ca/Cl ratio.

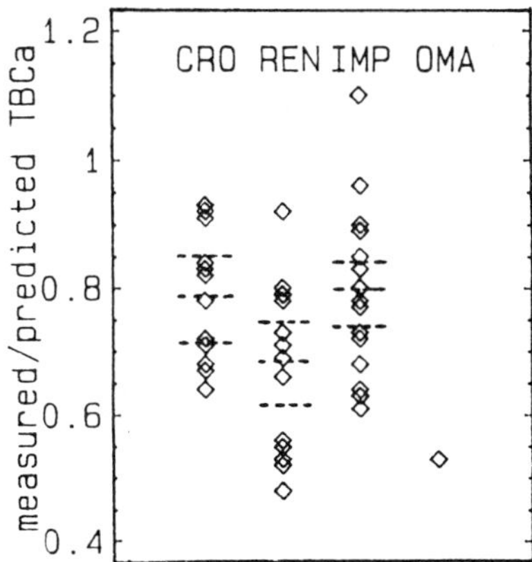

Fig. 1. The comparison of the measured-to-
         predicted TBCa in 3 patient groups
         and in one osteomalacia subject.
         The dashed lines indicate mean +/-95%
         confidence interval. (CRO = Crohns,
         REN = renal, IMP = implant,
         OMA = osteomalacia).

SUMMARY

Sequential TBCa measurements using the prompt gamma ray technique are firmly established and a number of patient groups are under investigation. A technique has been proposed for the absolute measurement of TBCa based upon the use of TBCl as an internal standard. Further work is needed to establish the validity of the method, but the initial results are encouraging.

ACKNOWLEDGEMENTS

We are grateful to the Medical Research Council for supporting the construction and calibration of the instrument. The support of the National Kidney Research Fund, Organon, and the contribution of colleagues within the Department of Medical Physics and Clinical Engineering is also gratefully acknowledged. A Wellcome Trust Travel Grant was awarded for one of us (SJSR) to attend this Conference. DK Bewley and CB Oxby are thanked for the loan of their anthropomorphic phantoms.

REFERENCES

Bewley, D.K., 1988, Anthropomorphic models for checking the calibration of whole-body counters and activation analysis systems, _Phys. Med. Biol._, 33 : 805.

Compston, J.E., Evans, W.D., Crawley, E.O., Judd, D., Motley, R., Evans, C., and Rhodes, J., 1987, Osteoporosis in patients with inflammatory bowel disease, _in_: "In-Vivo Body Composition Studies", K.J. Ellis, S. Yasumura, and W.D. Morgan, eds., Institute of Physical Sciences in Medicine, London.

Compston, J.E., Ryde, S.J.S., Motley, R.J., Crawley, E.O., Evans, W.D., and Morgan, W.D., 1989, Longitudinal study of total body calcium measurements in patients with inflammatory bowel disease: correlations with quantitative CT and single photon absorptiometry, _these proceedings_.

Ellis, K.J., 1989, Reference Man: variations due to body size, age, sex and race, _Biol. Trace Element Res._, in press.

Oxby, C.B., and Brooks, K., 1979, A versatile variable anthropomorphic phantom, _Phys. Med. Biol._, 24 : 440.

Ryde, S.J.S., Morgan, W.D., Sivyer, A., Evans, C.J., and Dutton, J., 1987, A clinical instrument for multi-element _in vivo_ analysis by prompt, delayed and cyclic neutron activation using Cf-252, _Phys. Med. Biol._, 32 : 1257.

Ryde, S.J.S., Morgan, W.D., Cobbold, S., Sivyer, A., Dutton, J., Evans, C.J., Compston, J.E., and Motley, R., 1989a, Measurement of total body calcium by prompt gamma neutron activation analysis: instrument performance and preliminary clinical experience, _in_: "Osteoporosis and Bone Mineral Measurement", E.F.J. Ring, W.D. Evans, and A.S. Dixon, eds., Institute of Physical Sciences in Medicine, York.

Ryde, S.J.S., Morgan, W.D., Compston, J.E., Evans, C.J., Sivyer, A., and Dutton, J., 1989b, Absolute measurements of total-body calcium by prompt gamma neutron activation analysis using a Cf-252 source, _Biol. Trace Element Res._, in press.

Ryde, S.J.S., Morgan, W.D., Thomas, D.W., Birks, J.L., Evans, C.J., Ali, P.A., and Jenkins, H., 1989c, Prompt gamma measurements of nitrogen and chlorine in normal volunteers, _these proceedings_.

DISTRIBUTION OF BODY WATER IN RATS

S. Yasumura,[1,3] D. Glaros,[4] J. Kalef-Ezra,[4]
J. Xatzikonstantinou,[4] A. F. LoMonte,[3]
J. K. Yeh,[2] and R. I. Moore[3]

[1]SUNY-Health Sciences Center, Brooklyn, NY 11203
[2]Winthrop University Hospital, Mineola, NY 11501
[3]Brookhaven National Laboratory, Upton, NY 11973 and
[4]University of Ioannina, Ioannina, Greece

INTRODUCTION

Textbook values for the size of the extracellular water compartment
(ECW) expressed as a fraction of total body water (TBW): ECW/TBW, range
from .27 to .45 (Berne and Levy, 1988). These discrepancies are disturbing since the size of the ECW and intracellular (ICW) is important to the
decision-making process in the management of patients with altered fluid
balance, and also important in considering the pharmacokinetics of the
distribution of drugs in the body.

The differences in the reported values for ECW/TBW are due to differences in the choice of compounds (tracers) used in the dilution
technique to determine ECW. Some tracers underestimate the ECW because
they fail to penetrate some spaces such as dense connective tissue,
whereas other tracers overestimate the ECW because they enter some cells.
Since there are no substances that distribute uniformly and exclusively
within cells, the intracellular water (ICW) is derived from the difference between the TBW and ECW, and is, therefore, subject to the same
uncertainties.

In 1983, we measured the ECW in normal human subjects by neutron
activation analysis (NAA) based on total body chloride (TBCl) (Yasumura
et al., 1983). The calculated ICW/TBW by this technique averaged .54 in
males and .50 in females. We made a similar study in rats where the
measurements could be verified by use of phantom standards equal in size
and weight to the rats, and by additionally comparing the NAA results
with chemical analysis of the carcass.

We used rats of two different age groups, immature and mature
animals, and they were subjected to extremes of physical activity. Our
primary objective was to compare the ECW/TBW values with human data, and
to verify the accuracy of the <u>in vivo</u> measurements for TBCl, TBW, and
total body potassium (TBK) by direct analysis of the carcass.

METHODS

Female Sprague-Dawley rats (Charles River Labs) were fed a standard
pellet chow diet (Rodent Lab Chow 5001, Ralston Purina). In the first
study, the "young" rats were five weeks old at the start of the experiment; in the second study, one-year old rats were used. The animals were
divided into three groups: exercised; immobilized by bilateral sciatic
denervation; and controls. Rats in the exercised group were trained to
run on a flat-bed treadmill. The running speed was gradually increased
over one week to 25 m/min for 60 min/day, five days per week, for 10
weeks.

Total body potassium (TBK) was measured by ^{40}K-counting using NaI(Tl)
detectors (Harshaw/Filtrol), each with dimensions of 4 x 4 x 16 inches,
that were located above, below, and one on each side of the rat. The
accuracy and precision of a one-hour count of the ^{40}K peak was 3% (coefficient of variation [CV]) for rats weighing 300 g (Yasumura et al.,
1987). Total body chloride (TBCl) was measured by neutron activation
analysis using Californium-252 (Glaros et al., this volume). Total body
water (TBW) was measured using tritiated water (HTO) (New England
Nuclear). From the measured values, derived values were calculated for
the percentage of extracellular water (ECW), intracellular water (ICW),
and the concentration of intracellular potassium (ICK) as described by
Yasumura et al. (1983):

ECW = 0.9 x TBCl (mEq)/plasma Cl (Meq/l)

ICW = TBW - ECW

ICK = 0.97 x TBK/ICW

Eight rats from the first study ("young" rats) were individually
homogenized using a Brinkman polytron unit. Samples, each weighing
100-200 mg, were used for analysis of water content by desiccation, and
potassium and chloride content by ion selective electrodes. Each determination was repeated on a minimum of three aliquots of each homogenized
carcass.

RESULTS

The average intracellular water fraction of the total body water was
between 54 and 60 percent for all rats, except the older rats that were
immobilized. The ICW/TBW ratio of this group was slightly, but significantly lower than that of control rats. Intracellular potassium was
similar in all groups of rats with an average range of 145 to 165 mEq/l
(Table 1).

For the eight "young" rats homogenized for _in vitro_ analysis, the
average TBCl was 2% lower, and the TBK 5% lower than values obtained by
NAA and whole body counting; however, the differences were not statistically significant. Also, water values obtained by the desiccation method
were slightly (2%), but not statistically, lower than the results derived
from the HTO dilution.

DISCUSSION

The ICW/TBW ratios in rats are similar to those found in human
subjects (Yasumura et al., 1983). In young rats, neither exercise nor

Table 1. Effect of 10 Weeks of Exercise (Exer.) or Immobilization
(Immob.) on the Distribution of Body Water Expressed as
Intracellular Water Percent of the Total Body Water (% ICW/TBW),
and Intracellular Body Potassium Concentration (ICK in mEq/l),
in Young and Year-Old Rats

| | YOUNG RATS | | | YEAR-OLD RATS | | |
	Contr.	Exer.	Immob.	Contr.	Exer.	Immob.
% ICW/TBW	60±2	59±2	59±2	58±2	54±3	51±1*
ICK mEq/l	145±7	150±3	145±8	149±3	165±5	149±12

Values represent the mean of 6 to 7 rats per group ± standard error (SE).
*Significantly lower than controls, p<.05.

immobilization alters this ratio. In older animals, however, immobilization is associated with a decrease in the ICW/TBW ratio. This finding
is similar to that observed in women over 60, where a general decline in
lean body mass (LBM) and an increase in body fat is consistent with a
relative reduction in the ICW volume. This decrease in the ICW/TBW ratio
is explained, in part, by the fact that adipose tissue water is about 14%
by weight and of that, 11/14 is extracellular (Wang and Pierson, 1976).
Thus, as the adipose tissue mass increases the ECW fraction also
increases.

The intracellular potassium concentration (ICK), as derived by 0.97
TBK/ICW, falls in the range of 145-165 mEq/l for all rats and is, again,
similar to the ICK values noted in normal human subjects. Also, the ICK
values derived in these studies agree with the value of 150 mEq/l generally quoted in the literature (Moore et al., 1963). Muscle tissue is
known to have an ICK greater than 160 mEq/l (Valentin et al., 1973), and
since muscle is the dominant potassium-containing tissue in the body, the
ICK of other tissues would have to be exceedingly low to reduce the total
body average ICK to a level much below 150 mEq/l. On this basis, it
seems unlikely that the ICW in normal humans or rats is any larger than
that reported here. However, this argument depends on the accuracy of
the measured values.

The animal experiments presented here are important because the TBW,
TBK, and ECW can be measured by more than one technique. The TBW and
TBK, measured by isotope dilution and whole body counting, were only
slightly higher than values determined by dehydration and chemical analysis of the rat carcass; and the ICW/TBW ratios were nearly identical.
These results provide evidence that the _in vivo_ techniques used in both
the rat and human studies are accurate.

The _in vivo_ methods used in these studies have major advantages over
carcass analysis. Since _in vivo_ methods can be repeated over time,
longitudinal studies can be conducted in the same animal. Neutron
activation and whole body counting are less labor-intense and less

time-consuming than carcass analysis. The precision and accuracy of the
in vivo techniques are at least as good as, and generally better than
that obtained from chemical analyses. The chemical method, in particular, is prone to errors in extrapolation due to non-uniformity of tissue
sampling, errors that become greatly magnified when total body values are
calculated.

The relative ICW/TBW reported in the present experiments is similar
to that established by Edelman and Liebman in their classic review
(1959). We conclude that the ICW/TBW ratio in both rats and humans is
between .50 and .60; the lower values are observed in those individuals
and rats in which the ratio of adipose/lean mass is increased.

ACKNOWLEDGEMENTS

This research was supported in part by the U.S. Department of Energy
under Contract DE-AC02-76CH00016. Accordingly, the U.S. Government
retains a nonexclusive, royalty-free license to publish or reproduce the
published form of this contribution, or allow others to do so, for U.S.
Government purposes.

The research described in this report involved animals maintained in
animal care facilities fully accredited by the American Association for
Accreditation of Laboratory Animal Care.

REFERENCES

Edelman, I. S., and Leibman, J., 1959, Anatomy of body water and
    electrolytes, _Am. J. Med._, 27:256.
Moore, F. D., Olesen, K. H., McMurray, J. D., Parker, H. V., Ball, M. R.,
    and Boyden, C. M., 1963, "The Body Cell Mass and its Supporting
    Environment," Saunders, Philadelphia, PA.
Berne, R.M., and Levy, M., eds., 1987, "Physiology", Mosby, St. Louis.
Valentin, N., and Olesen, K. H., 1973, Measurement of muscle tissue water
    and electrolytes, _Scan. Clin. Lab. Invest._, 32:155.
Wang, J., and Pierson, R. N., Jr., 1976, Disparate hydration and adipose
    and lean tissue require a new model for body water distribution in
    man, _J. Nutr._, 106:1687.
Yasumura, S., Kiebzak, G. M., Ellis, K. J., LoMonte, A. F., Zhang, R.,
    Yuen, K., and Cohn, S. H., 1987, Body composition studies in aging
    rats, _in_: "In Vivo Body Composition Studies," K. J. Ellis, S.
    Yasumura, W. D. Morgan, Eds., Institute of Physical Sciences in
    Medicine, York, England.
Yasumura, S., Cohn, S. H., and Ellis, K. J., 1983, Measurement of
    extracellular space by total body neutron activation, _Am. J.
    Physiol._, 244:R36.

Rn-222 AND Rn-222 PROGENY IN THE HUMAN BODY:

EFFECT ON IN VIVO K-40 MEASUREMENTS

Glenn Lykken[1,2], William Ong[2], Dale Patrick[3],
and Lucian Wielopolski[4]
[1]USDA, ARS, Human Nutrition Research Center, [2]UND Dept
Physics, [3]UND Safety Office, Grand Forks, ND and [4]Dept
Radiation Oncology, SUNY, Stony Brook, NY

INTRODUCTION

In vivo measurements of total body potassium (TBK) have been shown to
be affected by ^{222}Rn and its decay products, mainly Pb-214 and Bi-214
(Arvidsson and Hagar, 1976, Lykken et al., 1983). Corrections for
contributions to K-40 window counts using a matrix inversion method
resulted in measured TBK differences of a few percent in the presence of
radon progeny (Rn-prog) when repeated measurements were performed on
subjects (Lykken et al., 1987). This method had no provision for
estimation of accuracy, which was reduced with photopeak overlap and/or
large relative gamma ray activities. Library least-squares analysis
(LLSA), as developed for multidetector whole body counting systems by
Wielopolski and Cohn (1984), uses data from each channel and provides an
estimate of accuracy through the reduced chi square (red. chi-sq.)
measure of fit.

METHODS

A group of measurements were obtained to examine the accuracy of TBK
measurement by LLSA in the presence of Rn-prog. TBK in a free-living
male subject who had a radon concentration of 16 kBq/m^3 in his bedroom
was measured approximately one hour after he left his residence. One
week later, another TBK measurement was performed approximately nine
hours after he had left his residence.

The TBK in a second free-living male subject was measured
approximately 15 minutes after one hour exposures to radon-laden
atmospheres containing from 14 to 25 kBq/m^3 Rn-222. These atmospheres
were initially unfiltered and later electronically filtered. Baseline
data were obtained prior to each exposure. A comparative series of
exposures and whole body counts were obtained after an electrostatic air
filter and an ion generator had been in operation in the test chamber for
6 days. The subject wore a mask designed to filter dust, fumes, and mist
in these exposures. The Rn-222 concentration was 14 kBq/m^3 with
approximately 0.07 WL.

A shower and hair shampoo preceded all individual data collection.

RESULTS AND DISCUSSION

The subject with high radon concentrations in his home was analyzed for TBK and Rn-prog using LLSA. In the first measurement, respective TBK and Rn-prog were calculated to be 169 g and 4.8 kBq; in the second measurement, 172 g and 0.9 kBq (Fig. 1). TBK in the second male subject was documented to be 155 ± 6 g (mean ± SD) in 11 measurements with a Rn-prog range of 0.15 to 5.2 kBq and a reduced chi-square value from 1.08 to 3.0.

Linearity is assumed with the LLSA method. However, this rarely is the case. Inability to distinguish between two nearly coincident, monoenergetic gamma events, and consequent pulse pile-up, occurs even with very low activity sources. This results in a distorted spectrum that can not be precisely fitted to a library standard (LS) unless the LS also includes this effect. More accurate TBK values would be obtained if the LS were generated under a similar source activity and a "best fit" from a family of library standards was chosen.

Accuracy of TBK measurement was reduced when Rn-prog was greater than 2.6 kBq; presumably due to pulse pile-up. However, the LLSA TBK accuracy was greater than that obtained through a window method with excessive photopeak overlap (Respective TBK mean ± standard deviation: 170 ± 2 g; 178 ± 13 g, see Fig. 1, region centered at 1.46 MeV).

Radon and Rn progeny were stored in the body and measurable activity was found in a free-living subject even after considerable time had elapsed (see Fig. 1).

These studies demonstrate that LLSA as applied to whole body counting can be used to accurately measure TBK in the presence of interfering radioisotopes, provided appropriate library standards are chosen.

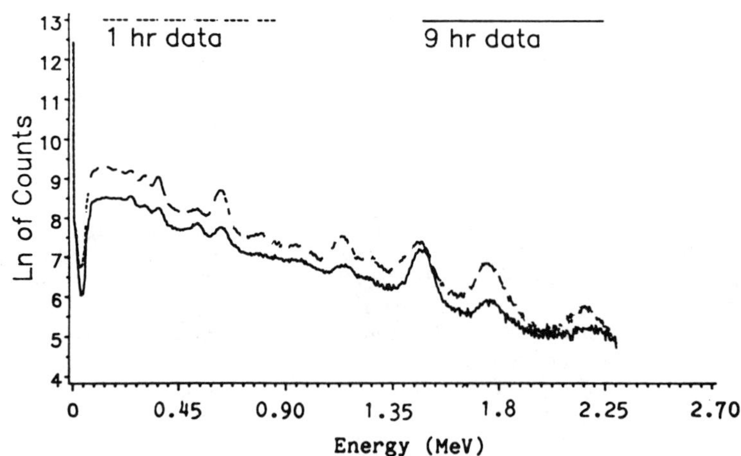

Fig. 1. Whole body counter spectra from a free-living male subject. One hr data: TBK = 169 g, Rn prog. = 4.8 kBq, red. chi-sq. = 3.8; 9 hr data: TBK = 172 g, Rn prog. = 0.9 kBq, red. chi=sq = 2.6. See text for details.

REFERENCES

Arvidsson, B., and Hager, A., 1976, Whole body counting of infants: A
    longitudinal study of total body potassium, in: "The Adipose Tissue
    in Obese and Nonobese Children, A Clinical and Metabolic Study," A.
    Hager, ed., Linköping University Medical Dissertation, Linköping,
    Sweden.
Lykken, G.I., Lukaski, H.C., Bolonchuk, W.W., and Sandstead, H.H., 1983,
    Potential errors in body composition as estimated by whole body
    scintillation counting, J. Lab. Clin. Med., 101:651.
Lykken, G.I., Speaker, K.K., and MacKichan, A.K., 1987, Estimation of
    total body potassium in the presence of interfering radio-isotopes,
    in: "In Vivo Body Composition Studies," K.J. Ellis, S. Yasumura,
    and W.D. Morgan, eds., The Institute of Physical Sciences in
    Medicine, London.
Wielopolski, L., and Cohn, S.H., 1984, Application of the library least-
    squares analysis to whole-body counter spectra derived from an array
    of detectors, Med. Phys., 11:528.

APPLICATIONS OF BIOELECTRICAL IMPEDANCE ANALYSIS: A CRITICAL REVIEW

Henry C. Lukaski

USDA, ARS
Grand Forks Human Nutrition Research Center
Grand Forks, N.D. 58202, U.S.A.

INTRODUCTION

Bioelectrical impedance analysis (BIA) is a relatively new method for the assessment of body composition. Because this approach is safe, noninvasive, rapid, portable, inexpensive, and easy to use, it may be amenable for laboratory, clinical, and field assessment of human body composition. The purpose of this presentation is to summarize the current uses of this method and to evaluate its validity.

The BIA method is based upon the conduction of an applied electrical current in the body. Introduction of a constant, low level, alternating current (AC) results in a frequency-dependent opposition to the flow of current in the organism. The fat-free tissues, which contain the preponderance of water and electrolytes in the body, are highly conductive and represent a low impedance pathway. In contrast, bone and triglyceride or fat are nonconductive and have a very high impedance to current flow. In general, at low frequencies, the current passes through the extracellular fluids, while at higher frequencies it penetrates all water compartments (Nyboer, 1972; Pethig, 1987).

The hypothesis that BIA can be related to conductive tissue volume is derived from the physical principle that the impedance of an isotropic conductor is related to its length, configuration, cross-sectional area, and the signal frequency used. Using a constant signal frequency and assuming a relatively constant conductor configuration, the bioelectrical impedance to the flow of alternating current can be related to the volume of the conductor (Hoffer et al., 1969): $Z = \rho L^2/V_2$ where Z is impedance in ohms, ρ is volume resistivity in ohm·cm, L is conductor length in cm, and V is volume in liters. Although there may be difficulties in applying this general relationship to a system with complex geometry and bioelectrical characteristics as the human body, this paradigm has been used to generate mathematical models relating impedance measurements to body composition variables.

Bioelectrical impedance variables have been measured using either two or four surface electrodes. Applications of the bipolar and tetrapolar approaches are summarized below.

<u>Bipolar Technique</u>

   <u>Total body water and extracellular fluid volume</u>. Thomasset (1963,
1965, 1962) used two stainless steel needle electrodes placed subdermally
on the dorsal surfaces of one hand and the contralateral foot.
Transverse, whole body impedance was determined in response to a 10 μA AC
administered at two frequencies (1 kHz and 100 kHz). Whole body
impedance at 1 kHz was correlated (n = 65: r = 0.71, p <0.001) with
height^2/radiobromide space, and impedance at 100 kHz was related (n = 44;
r = 0.93; p <0.0001) to height^2/tritium dilution space). Jenin et al.
(1975) used these data and developed a relationship to predict the
relative distribution of water (e.g., total body water/extracellular
fluid volume or TBW/ECF) based upon the ratio of high to low impedance
values. They observed that this ratio was greater than 1.3 in healthy
volunteers, but it declined to much less than 1.3 in pathological cases.

   This bipolar approach of Thomasset and his colleagues has some
practical limitations. Use of needle electrodes can cause both volunteer
discomfort and induce electrochemical reactions that alter the measured
impedance values.

<u>Tetrapolar Technique</u>

   The disadvantages of the bipolar approach hindered its acceptance and
use outside of Thomasset's laboratory and stimulated interest in the
development of a four-electrode technique. The major advantage of the
tetrapolar method was to minimize contact impedance or skin-electrode
interactions.

   <u>Total body water</u>. Hoffer et al. (1969) used four surface electrodes,
two placed on the dorsal surface of the right hand and two placed on the
flexor surface of the left foot, and administered 100 μA AC, at 100 kHz
in 20 healthy volunteers and 34 patients in whom TBW was estimated from
tritium dilution space. In both groups of subjects, height^2/impedance
(Ht^2/Z) was the best predictor (r = 0.92 and 0.93; p <0.001) of TBW.
These preliminary findings indicated the potential of the four-electrode
technique to assess TBW.

   Other investigators used tetrapolar impedance to develop models for
estimating TBW. However, the current source was 800 μA at 50 kHz, and
the impedance plethysmograph partitioned Z into its components of
resistance (R) and reactance (Xc), which represent the sum of in-phase
and out-of-phase electrical vectors, respectively. Using this method,
Lukaski et al. (1985) demonstrated significant inverse relationships
between TBW and Z, R, and Xc (r = -0.86, -0.86, and -0.55, respectively)
in 37 men. Because R and Z are equally well-correlated with TBW (r =
0.99) and Xc is not as strongly related to Z (r = 0.70), these
investigators decided to use R as the criterion impedance variable for
prediction of TBW and fat free mass (FFM). Thus, the highest correlation
coefficient (r = 0.98, p <0.0001) was found between Ht^2/R, where R is the
lowest R measured across the ipsilateral and contralateral electrode
arrangements and TBW. Other studies by Segal et al. (1985) and Van Loan
and Mayclin (1987) confirmed the prominent role of Ht^2/R as a predictor
of TBW (r = 0.912 and 0.933, respectively).

   While these reports indicate the potential of BIA to predict TBW,
they do not provide an estimate of validity or accuracy of this method.
To address these issues, cross-validation studies using two independent
samples are required. Kushner and Schoeller (1986) tested the validity
of an impedance model derived in a sample of 40 healthy men and women,
then cross-validated it in a group of 18 obese patients. Their model

successfully predicted TBW (r = 0.95, p <0.01) with absolute errors
(e.g., differences between observed and predicted values) of only 0.6-
1.0 L, and relative errors (absolute error x 100%/mean value) of 2-2.3%.

Lukaksi and Bolonchuk (1988) conducted a double cross-validation of
the impedance method to predict TBW in a sample of 110 healthy, lean and
obese men and women that were randomly divided into two groups. Models
for prediction of TBW were developed for each group and were used to
predict TBW in the other group. It was determined that the relationships
between observed and predicted values were similar to the line of
identity, thus establishing validity of the impedance method in normal
people. Also, the errors of prediction, as indicated by the standard
error of the estimate (SEE) of the relationship between observed and
predicted values, were 1.48-1.58 L. Thus, these observations indicate
the validity of BIA to predict TBW among individuals without electrolyte
disturbances.

Extracellular fluid space. The tetrapolar approach permits
determination of R and Xc, which may be an index of extracellular fluid
space (ECF). In two clinical studies, patients undergoing either renal
dialysis (Nyboer and Sedensky, 1974) or furosemide diuresis (Subramanyan
et al., 1980) had impedance variables measured before and after therapy.
After fluid loss, Xc showed the largest relative change (50% increase),
whereas R and Z exhibited the smallest relative change (10-15% increase).
Because both of the interventions stimulated fluid loss, and this loss
was presumably due to a change in the extracellular fluid space, it was
hypothesized that Xc may predict ECF.

In an abstract, Segal et al. (1987) reported that Xc was a good
predictor (r = 0.721, p <0.0001) of extracellular water (ECW) determined
by radiosulfate space in a sample of 62 men and women. The best
predictor of ECW was Ht^2/Xc (r = 0.793), while Ht^2/R was the best
predictor (r = 0.875) of TBW. To test the hypothesis that Xc is an
independent predictor of TBW, the influence of TBW on predictors of ECF
was removed statistically. It was determined that Xc (r = -0.595) was a
stronger (p <0.05) predictor of ECW than R (r = -0.327).

These preliminary findings were confirmed and extended in a recent
double cross-validation study (Lukaski and Bolonchuk, 1988). The
variables Ht^2/R and Ht^2/Xc were the best predictors of TBW (determined as
deuterium dilution space) and ECF (estimated as corrected bromide space),
respectively. Because ECF was correlated (r = 0.90; p <0.0001) with TBW,
the effect of TBW was removed from ECF and candidate predictor variables
of ECF using partial correlation analysis. The results indicate that R
and Ht^2/R (r = -0.12 and 0.14) were not as important (p <0.005)
predictors of ECF as were Xc and Ht^2/Xc (r = -0.45 and 0.50). Also, the
cross-validation trial showed that observed and predicted values were
significantly correlated (r = 0.91-0.94; p <0.0001) and were distributed
on lines not different than the line of identity. These results,
therefore, support the hypothesis that Xc is an independent predictor of
ECF and validate the BIA method to assess fluid compartments in humans.

The physiological significance of the impedance variables is not well
understood. Using the tetrapolar approach, Nyboer (1970) clearly
demonstrated that Z and R were dependent upon both electrolyte
concentration and fluid volume. Recently, Meguid et al. (1988) used a
single-cell model and showed linear relationships between R and saline
volume, and among R osmolarity and sodium concentration. These findings
are consistent with the reported observations that R is highly correlated
with in vivo determinations of TBW in lean and obese humans.

Our understanding of the biological meaning of Xc is less complete. In theory, Xc is a measure of the capacitance of the cell membrane in response to the applied electrical current (Nyboer, 1970). In the functioning cell, this capacitance is probably dependent upon the ionic distribution across the membrane. Thus, the preliminary finding that Xc is independently related to the radiosulfate and corrected bromide spaces suggests that Xc is an index of fluid and/or electrolyte partitioning (e.g., the extracellular to intracellular ratio) in the body. This attractive hypothesis remains to be tested in clinical populations.

<u>Fat free mass</u>. Investigators recognized that water and electrolytes, the major component of FFM, are also the primary factors responsible for conduction of an electrical signal in the body, and attempted to develop mathematical relationships between FFM and impedance variables.

Nyboer et al. (1983) conducted a pilot study and related impedance measures to body composition estimates in a sample of college-aged men and women. Using densitometric data, but lacking accurate residual lung volume data, these investigators developed preliminary statistical relationships between Ht^2/R and body composition variables. Lukaski et al. (1985) used the same tetrapolar method and reported that R was a good predictor of densitometrically-determined FFM and total body potassium or TBK (r = -0.86 and -0.79), but the best predictor was Ht^2/R (r = 0.98 and 0.96) in a homogenous sample of men. Segal et al. (1985), Van Loan and Mayclin (1987) and Chumlea et al. (1988) also concluded that Ht^2/R was a strong predictor of FFM in adults.

There have been a few attempts to cross-validate the BIA method in adults. Lukaski et al. (1986) studied the validity of this method in 114 men and women aged 18-50 yr. It was determined that regression equations for prediction of FFM developed separately for men and women were valid in the other group and that errors of prediction (e.g., SEE) were low, 2.20-2.30 kg. To broaden the application of this finding, another cross-validation study was conducted in a heterogenous sample of 312 adults aged 18-73 yr (Lukaski and Bolonchuk, 1987). Stepwise multiple regression analysis identified the best model for estimating FFM. The validity of this model was tested and found to give predictions of FFM not different from the densitometric reference method. Also, the SEE of the regression between observed and predicted values was 2.29 kg. When percent body fat (%BF) was calculated from body mass and FFM predicted using the derived impedance model, it was highly correlated (r = 0.96; p <0.0001) with and similar to reference values. It was also shown that a standard skinfold prediction (Durnin and Womersley, 1974) had a less reliable (r = 0.91; p <0.05) prediction of %BF relative to densitometry than did the impedance approach.

It has been reported that %BF may influence the validity of the BIA method to assess FFM. Segal et al. (1985) reported a significant relationship (r = -0.796) between residual FFM scores, calculated as the difference between observed and predicted values calculated using the BIA manufacturer's equation, and %BF. In contrast, only a weak correlation (r = -0.32) was found between residual FFM scores and %BF in a cross-validation study using an impedance prediction formula derived and validated in one laboratory (Lukaski, in press).

Recently, the tetrapolar BIA method has been used to derive models for estimating FFM of children. Cordain et al. (1988) studied 16 girls and 14 boys aged 9-14 yr and reported significant correlations between Ht^2/R and TBK (r = 0.92) and densitometrically-determined FFM (r = 0.83). Houtkooper et al. (1989) developed an impedance model for predicting FFM

determined from body mass and %BF calculated using body density and TBW
in a sample of 94 boys and girls. An equation containing Ht^2/R and body
mass had a coefficient of determination of 0.93 and an SEE of 2.0 kg.
Stepwise multiple regression identified the best equation (R^2 = 0.94, SEE
= 1.9 kg) as containing Ht^2/R, chest circumference, hip skeletal width,
and Xc. Cross-validation using adult-based prediction equations
indicated concordance between observed and predicted FFM values using an
equation developed by Lukaski et al. (1986).

   Weight loss. Some recent studies addressed the use of BIA to
estimate body composition changes during human weight reduction. Gray
(1988) measured bioelectrical impedance variables daily in a group of six
obese women over a two-week fast. Body water was determined at the start
and end of the fast. Average weight loss was 10 kg. A good correlation
was noted between the changes Ht^2/R and TBW (r = 0.94; p <0.02). A
significant relationship was also found between the changes in measured
TBW and changes in Ht^2/R (r = 0.94, SEE = 0.7L). Thus, it appears that
BIA is capable of estimating acute TBW changes after a short-term fast.

   Deurenberg et al. (1989a) studied composition changes after
consumption of a very low calorie diet for two days. They reported
similar values of change in FFM using densitometry and BIA (1.2 and 0.7
kg) calculated using a prediction equation by Lukaski et al. (1986). In
another study, Deurenberg et al. (1989b) reported changes in the body
composition determined using BIA and densitometry in a group of 13 women
participating in an 8-week weight reduction. Mean weight loss was 10 kg.
Relative to densitometry, FFM change was underpredicted (p <0.02) by BIA
(2.3 vs 0.6 kg). It cannot be determined whether the method is reliable
to assess changes in FFM because TBW was not measured in conjunction with
densitometry after weight reduction.

EVALUATION

   Descriptions of applications of the BIA method indicate a variety of
possible uses. However, an understanding of the limitations and
procedures necessary for proper use of this method is needed.

Reliability of Measurement

   The reproducibility of the impedance measurement determines its value
in biological applications. An initial step in assessing instrument
reliability is determination of technical accuracy. Van Loan and Mayclin
(1987) reported an error of 0.5-1% using this method to determine the R
of calibrated resistors ranging from 100-680 ohm measured twice daily
over 11 days.

   Studies of biological variability of human volunteers measured
repeatedly under similar conditions indicate small intrasubject
deviation. Lukaski et al. (1985) found a 2% (range: 0.9-3.4%)
coefficient of variation in R values measured in 14 men during five days.
Similar values were reported by other investigators (Segal et al., 1985;
Van Loan and Mayclin, 1987).

Physical Factors

   If instrument reliability is high, then factors affecting the
biological stability of BIA variables require identification. Some
factors affecting the validity of impedance predictions of body
composition have been identified (Lukaski, 1987). Relative to
measurements made under controlled conditions, dehydration, incorrect
electrode placement and cool skin increase R and Xc values and decrease

estimates of FFM. Conversely, preceding exercise, use of a conductive
surface for BIA measurements and warm skin can decrease impedance
variables and spuriously increase predictions of FFM. Thus, avoidance of
alcohol and exercise for the preceding 24 hours seems prudent for valid
BIA measurements.

Errors in Reference Composition Model

One criticism of the BIA method is that it may overestimate %BF in
lean individuals and underestimate %BF in obese volunteers (Segal et al.,
1985; Hodgdon and Fitzgerald, 1987; Segal et al., 1988). We tested this
statement (unpublished data) in an independent sample of 120 men and
women aged 18-60 yr using our previously validated equation for
estimation of FFM (Lukaski and Bolonchuk, 1987). Predicted FFM was
similar to observed values. The slope (1.006) and intercept (-0.219) of
the relationship between observed and predicted values were not different
than those of the line of identity. The observed and predicted values
were significantly correlated (r = 0.99) with a SEE of 0.97 kg. The
correlation between residual FFM scores and %BF was low (r = 0.293). To
assess whether errors in estimating FFM were dependent upon %BF, the %BF
data were partitioned into quartiles and FFM values were predicted using
our equation (Lukaski and Bolonchuk, 1987). Observed and predicted FFM
values in each quartile were related using regression analysis and the
coefficients are presented in Table 1. The only significant difference
between observed and predicted FFM was found in the fourth quartile of
%BF. With increasing %BF, the slopes of the lines relating observed and
predicted FFM began to deviate from unity. Interestingly, divergence
from unity was coincident with an increase in the fraction of women in
each ascending quartile of fatness. Perhaps reduced bone mineral density
of the women, an observation more commonly noted in women than men, is
contributing to the error between observed and predicted values. Thus,
the two compartment model may be inadequate in establishing the validity
of BIA method in heterogenous samples of volunteers.

Table 1. Errors in Prediction of Fat Free Mass
Relative To Body Fatness

	% Body Fat Quartiles			
	1	2	3	4
n	31	27	31	31
Fat Free Mass Residuals, kg	0.34	0.28	-0.38	-0.43[a]
Slope	0.99	0.98	1.03[b]	1.09[b]
Intercept	0.76	1.04	-1.03	0.17
F/M[c]	4/27	12/15	22/9	25/6

[a]p <0.05;   [b]Significantly different than 1, p <0.05

[c]F/M = ratio of female to male volunteers

<u>Whole Body Versus Segmental Impedance Measurements</u>

Recently, some investigators (Smith, 1987; Patterson, 1989)
questioned the appropriateness of measuring impedance variables using the
wrist and ankle because of possible error due to the changes in cross-
sectional area between the measurement sites. These investigators
proposed that body segments be measured and mathematical models developed
for prediction of whole body compositional variables. Whether this
approach can provide improved prediction accuracy remains to be
determined.

Other support for the use of regional or segmental impedance
measurements for assessment of human body composition also is available.
Chumlea et al. (1988) estimated specific resistivity of body segments of
123 children and adults. They found that the specific resistivity of the
trunk, including the thorax and the abdomen, was significantly greater
among these individuals with increased %BF, total body fat and
subcutaneous adipose tissue estimated from skinfold thickness. In
addition, Baumgartner et al. (1988) reported that the trunk phase angle,
defined as the arc tan (Xc/R), was an independent predictor of %BF when
age, mean skinfold thickness and body mass index were controlled
statistically. These findings suggest that regional body bioelectrical
impedance measurements may be useful in assessing regional body
composition.

CRITIQUE

Current applications of the tetrapolar BIA method have focused solely
on healthy individuals for development and validation of prediction
models for assessment of human body composition. It is reasonable,
therefore, to question the relative importance of impedance variables
relative to simple anthropometric measurements, such as body weight and
standing height, to predict body compositional components. These data
are summarized in Table 2. As expected, body weight and height are
significantly correlated with each body composition variable.
Bioelectrical R, however, is a significantly better predictor of TBW,
total body potassium and FFM than are height and weight. The best
predictor of these body composition variables is Ht^2/R. These findings
indicate the robust statistical capacity of R to predict indices of FFM
in healthy people.

Table 2. Relationships Between Body Composition Estimates and Body and
Bioelectrical Impedance Measurements in 250 Healthy Adults.

	Total Body Water	Total Body Potassium	Fat Free Mass	Fat Mass
Height	0.73[b]	0.74[b]	0.75[b]	0.42[a]
Weight	0.77[b]	0.78[b]	0.78[b]	0.52[a]
Resistance	-0.90[c]	-0.85[c]	-0.88[c]	0.40[a]
Reactance	-0.50[a]	-0.49[a]	-0.54[a]	0.32[a]
Height²/Resistance	0.98[d]	0.96[d]	0.99[d]	0.38[a]

[a,b,c,d]Values in a column with different superscripts are different.

[a]p <0.05; [b]p <0.01; [c]p <0.001; [d]p <0.0001.

A critical and unresolved question is the validity of the tetrapolar
BIA method to estimate body composition among patients with altered fluid
and electrolyte status. Unfortunately, data are not available on this
important issue. Among such individuals, the TBW/FFM and the
intracellular to extracellular fluid volume may be altered. It is
reasonable to speculate that BIA may be useful in estimating TBW;
however, the extrapolation to FFM, assuming a relatively constant
TBW/FFM, probably will yield spurious results (Cohn, 1985). One
preliminary report (Katch et al., 1986) indicated significant differences
between %BF values estimated by densitometry and BIA in 24 cardiac and
pulmonary patients. Whether this discrepancy can be attributed to
inadequate reference methods, prediction model or a single frequency
impedance method is not known.

Resolution of this question may require the use of advanced
technology in both reference body composition assessment and BIA
measurement. Use of multicompartmental, compositional models that
account for altered fluid distribution are needed to avoid reliance upon
the assumption of a relatively constant TBW/FFM. Also, development and
availability of multifrequency BIA plethysmographs may facilitate the
ability to discriminate fluid and water distribution and, hence, improve
estimation of body cell mass and FFM. Until these technical advances are
used, it is not possible to evaluate the tetrapolar BIA method to assess
body composition of individuals with altered physiological status.

SUMMARY

Tetrapolar BIA offers great potential for noninvasive assessment of
human body composition because it is safe, portable and easy to use.
Additional research is needed in clinical studies among patients in whom
body composition and fluid status are impaired. However, appropriate
reference methods are required to establish the validity of the BIA
method. Critical evaluation of the regional measurement of bioelectrical
impedance variables for the assessment of human body composition is
needed.

REFERENCES

Baumgartner, R.N., Chumlea, W.C., and Roche, A.F., 1988, Bioelectric
    impedance phase angle and body composition, Am. J. Clin. Nutr.,
    48:16.
Chumlea, W.C., Baumgartner, R.N., and Roche, A.F., 1988, Specific
    resistivity used to estimate fat free mass from segmental measures of
    bioelectrical impedance, Am. J. Clin. Nutr., 48:7.
Cohn, S.H., 1985, How valid are bioelectrical impedance measurements in
    body composition studies? Am. J. Clin. Nutr., 42:889.
Cordain, L., Whicker, R.E., and Johnson, J.E., 1988, Body composition
    determination in children using bioelectrical impedance, Growth
    Devlop. Aging, 52:37.
Deurenberg, P., Weststrate, J.A., and Van Der Kooy, K., 1989a, Body
    composition changes assessed by bioelectrical impedance measurements,
    Am. J. Clin. Nutr., 49:401.
Deurenberg, P., Weststrate, J.A., and Hautvast, J.G.A.J., 1989b, Changes
    in fat free mass during weight loss measured by bioelectrical
    impedance and by densitometry, Am J. Clin. Nutr., 49:33.
Durnin, J.V.G.A., and Womersley, J., 1974, Body fat assessed from total
    body density and its estimation from skinfold thicknesses:
    measurements on 481 men and women aged 16-72 years, Brit. J. Nutr.,
    32:77.
Gray, D.S., 1988, Changes in bioelectrical impedance during fasting, Am.
    J. Clin. Nutr., 48:1184.

Hodgdon, J.A., and Fitzgerald, P.I., 1987, Validity of impedance
    predictions at various levels of fatness, <u>Human Biol.</u>, 59:281.
Hoffer, E.C., Meadow, C.K., and Simpson, D.C., 1969, Correlation of whole
    body impedance with total body water, <u>J. Appl. Physiol.</u>, 27:531.
Houtkooper, L.B., Lohman, T.G., Going, S.B., and Hall, M.C., 1989,
    Validity of bioelectrical impedance for body composition assessment
    in children, <u>J. Appl. Physiol.</u>, 66:814.
Jenin, P., Lenoir, J., Roullet, C., Thomasset, A.L., and Ducrot, H.,
    1975, Determination of body fluid compartments by electrical
    impedance measurements, <u>Aviat. Space Environ. Med.</u>, 46:152.
Katch, F.I., Solomon, R.T., Shayevitz, M., and Shayevitz, B., 1986,
    Validity of bioelectrical impedance to estimate body composition in
    cardiac and pulmonary patients, <u>Am. J. Clin. Nutr.</u>, 43:972.
Kushner, R.F., and Schoeller, D.A., 1986, Estimation of total body water
    by bioelectrical impedance analysis, <u>Am. J. Clin. Nutr.</u>, 44:417.
Lukaski, H.C., 1987, Bioelectrical impedance analysis, <u>in</u>: "Nutrition
    '87," O.A. Levander, ed., American Institute of Nutrition, Bethesda,
    MD.
Lukaski, H.C., 1989, Use of bioelectrical impedance analysis to assess
    human body composition: a review, <u>in</u>: "Nutritional Status
    Assessment of the Individual," G.L. Livingston, ed., Food and
    Nutrition Press, Inc., Trumbull, CT.
Lukaski, H.C., and Bolonchuk, W.W., 1987, Theory and validation of the
    tetrapolar bioelectrical impedance method to assess human body
    composition, <u>in</u>: "In Vivo Body Composition Studies," K.J. Ellis, S.
    Yasumura, W.D. Morgan, eds., Institute of Physical Sciences in
    Medicine, London.
Lukaski, H.C., and Bolonchuk, W.W., 1988, Estimation of body fluid
    volumes using tetrapolar bioelectrical impedance measurements, <u>Aviat.
    Space Environ. Med.</u>, 59:1163.
Lukaski, H.C., Bolonchuk, W.W., Hall, C.B., and Siders, W.A., 1986,
    Validation of tetrapolar bioelectrical impedance method to assess
    human body composition, <u>J. Appl. Physiol.</u>, 60:1327.
Lukaski, H.C., Johnson, P.E., Bolonchuk, W.W., and Lykken, G.L., 1985,
    Assessment of fat free mass using bioelectrical impedance
    measurements of the human body, <u>Am. J. Clin. Nutr.</u>, 41:810.
Meguid, M.M., Campos, A.C.L., Lukaski, H.C., and Kiell, C., 1988, A new
    single-cell in vitro model to determine volume and sodium
    concentration changes by bioelectrical impedance analysis.
    <u>Nutrition</u>, 4:363.
Nyboer, J., 1970, "Impedance Plethysmography," 2nd Ed., C.C. Thomas
    Publ., Springfield, IL.
Nyboer, J., 1972, Workable volume and flow concepts of biosegments by
    electrical impedance plethysmography, <u>T.-I.-T. J. Life Sci.</u>, 2:1.
Nyboer, J., Liedtke, R.J., Reid, K.A., and Gessert, W.A., 1983,
    Nontraumatic electrical detection of total body water and density in
    man, <u>Med. Jadertina</u>, 15 (Suppl):381.
Nyboer, J., and Sedensky, J.A., 1974, Bioelectrical impedance during
    renal dialysis, <u>Proc. Dialysis Transplant. Forum</u>, 4:214.
Patterson, R., 1989, Body fluid determinations using multiple impedance
    measurements, <u>IEEE Engin. Med. Biol.</u>, 8:16.
Pethig, R., 1987, Dielectric properties of body tissues, <u>Clin. Physiol.
    Meas.</u>, 8 (Suppl A):5.
Segal, K.R., Gutin, B., Presta, E., Wang, J., and Van Itallie, T.B.,
    1985, Estimation of human body composition by electrical impedance
    methods: a comparative study, <u>J. Appl. Physiol.</u>, 58:1565.
Segal, K.R., Kral, J.G., Wang, J., Pierson, R.N., and Van Itallie, T.B.,
    1987, Estimation of body water distribution by bioelectrical
    impedance, <u>Fed. Proc.</u>, 46:1334 (abst).

Segal, K.R., Van Loan, M., Fitzgerald, P.I., Hodgdon, J.A., and
    Van Itallie, T.B., 1988, Lean body mass estimation by bioelectrical
    impedance analysis: A four-site cross-validation study, <u>Am. J. Clin.
    Nutr.</u>, 47:7.
Smith, D.N., 1987, Body composition by tetrapolar impedance measurements:
    correlation or con?, <u>Seventh ICEBI Conference</u>, Austria.
Subramanyan, R., Manchanda, S.C., Nyboer, J., and Bhatia, M.L., 1980,
    Total body water in congestive heart failure: a pre- and post-
    treatment study, <u>J. Assoc. Phys. Ind.</u>, 28:257.
Thomasset, A., 1962, Bioelectrical properties of tissue impedance
    measurements, <u>Lyon Med.</u>, 207:107.
Thomasset, A., 1963, Bioelectrical properties of tissues, <u>Lyon Med.</u>,
    209:1325.
Thomasset, A., 1965, Electrochemical volume of measurement of
    extracellular liquids: biophysical significance of impedance at one
    kilocycle, <u>Lyon Med.</u>, 214:131.
Van Loan, M., and Mayclin, P., 1987, Bioelectrical impedance analysis: is
    it a reliable estimator of lean body mass and total body water?,
    <u>Human Biol.</u>, 59:299.

THE USE OF SEGMENTAL BIOELECTRIC IMPEDANCE IN ESTIMATING BODY COMPOSITION

Wm. Cameron Chumlea, Richard N. Baumgartner, and
Carol O. Mitchell

The Division of Human Biology, Department of Pediatrics, Wright
State University School of Medicine, Dayton, OH, 45435 and The
Department of Home Economics, Memphis State University,
Memphis, TN, 38152.

INTRODUCTION

The use of measures of bioelectric impedance to estimate body composition is based upon the greater conductivity of electricity by fat-free mass compared to that of adipose tissue or bone. Bioelectric impedance is typically measured in the whole body between the wrist and ankle on one or both sides of the body and used in combination with stature (S^2/R) to predict body composition (Lukaski et al., 1985, 1986; Segal et al., 1985; Kushner and Schoeller, 1986; Guo et al., 1987; Jackson et al., 1988). However, these whole body techniques are prone to sources of error because differences in lengths and cross-sectional areas of the body alter the uniformity of the current field which affects whole body impedance values (Smith, 1987; Baker, 1989). Segmental bioelectric impedance measures may provide body composition information obscured in the whole body approach (Chumlea et al., 1987; Smith, 1987; Patterson et al., 1988; Patterson, 1989). Chumlea et al. (1987, 1988) showed that estimates of specific resistivities for the arm, leg and trunk can be used to calculate fat-free mass directly. Also, Baumgartner et al. (1988) demonstrated that the phase angle for the trunk was correlated significantly with percent body fat for the whole body. Patterson and coworkers (1988, 1989) report that combined segmental impedance measurements produce more reliable determinations of total body water and its changes than the whole body impedance method.

An index of the conductive volume of a body segment can be determined from its length (L), its bioelectric resistance (R) and its specific resistivity (ρ) according to the formula $V = \rho L^2/R$. In normal individuals, this conductive volume should approximate the segment volume of fat-free mass and is smaller than the amount of the total segment volume which includes adipose tissue. As a result, it is theoretically possible to derive an index of the amount of the total subcutaneous adipose tissue in the body by subtracting the conductive volume of each body segment (arms, legs and trunk) from the corresponding total volume. Simplistically, the inner volume of fat-free mass is subtracted from the total volume of each corresponding body segment (excluding the head) leaving the outer volume of subcutaneous adipose tissue. These summed remainders produce an index of the amount of total subcutaneous adipose tissue for the body. However, the extracellular fluid content of subcutaneous adipose tissue varies among individuals and increases with the amount of adipose tissue (Pierson et al., 1982). This increased hydration could cause overestimates of fat-free mass from bioelectric impedance at 50 kHz and, subsequently, underestimate subcutaneous adipose tissue in individuals with large amounts of adiposity.

Settle et al. (1980) and, more recently, Smith (1987) and Patterson (1989) have reported that the whole body resistance is determined primarily by the resistances of the arm and/or the leg in normal individuals. Therefore, it should be possible to predict whole body composition accurately from measurements of the length and resistance of the arm or leg only, assuming that fat/lean ratios are similar throughout the body. Compared to whole body impedance techniques, this approach would be particularly useful in the elderly, or in other chair- or bed-fast individuals for whom an accurate measurement of stature is difficult to obtain (Chumlea et al., 1985).

The purpose of the present study was to investigate the use of segmental measures of bioelectric impedance to estimate body composition and to compare predictions of body composition based on the lengths and resistances of the arm and leg with those obtained using the whole body method.

SAMPLE AND METHODS

The study sample consisted of 52 white men, ages 18 to 63 years and 74 white women, ages 18 to 58 years. These normal individuals are participants in the Fels Longitudinal Study (Roche et al., 1982) and were not selected by any criteria relating to their body composition. The sample included a wide range of levels of adiposity (3 to 53% fat).

The anthropometry was collected using standardized procedures closely similar to corresponding techniques in the Anthropometric Standardization Reference Manual (Lohman et al., 1988). These measurements included weight, stature, sitting height, acromial height, arm length, and circumferences of the abdomen, thigh, calf, and upper arm. All measurements were made by two observers, and the means of paired measurements were used in the analyses. The circumferences were used to calculate the cross-sectional areas of the abdomen, calf, and upper arm. Leg length was derived as the difference between stature and sitting height, and trunk length was calculated as acromial height minus leg length.

Measures of the resistance of the arm, leg, trunk, and whole body were taken on the right side of the body using a BIA 101 impedance analyzer (RJL Systems, Detroit, MI). These techniques have been described in detail previously (Chumlea et al., 1988). In brief, the participant lay supine on an examining table with the arms and legs relaxed but not touching the body. Source and receiving electrodes were placed at the conventional locations on the anterior surface of the foot and ankle and on the posterior surface of the hand and wrist for whole body measurements. For measurements of the leg, one pair of electrodes was placed at the conventional locations on the foot and ankle and the other pair was placed on the anterior midline of the proximal thigh, with the receiving electrode in the same plane as the gluteal crease and the source electrode 5.0 cm proximal to the receiving electrode. Trunk impedance was measured by placing one pair of electrodes on the proximal thigh at the same locations used for measurement of the leg, but with the source electrode 5.0 cm distal to the receiving electrode. The other pair of electrodes was placed with the receiving electrode over the sternal notch and the source electrode on the anterior midline of the neck 5.0 cm cranial to the receiving electrode. The impedance of the arm was measured by attaching one pair of electrodes at the conventional locations on the hand and wrist and the other pair on the anterior surface of the shoulder, with the measuring electrode over the mid-point of a line between the acromial process and the axilla and the source electrode 5.0 cm medial to this mid-point. Trunk impedance measurements were taken at midrespiration.

Fat-free mass (FFM) was estimated from body density measured by underwater weighing. Underwater weights were measured to the nearest 2.0 gm using a steel chair suspended into a tank of water at 34°C from 4 load cells connected to a digital processor. Each participant was weighed underwater ten consecutive times, and the mean of the last three weights was used to compute body density. Residual volume was measured on land to the nearest 0.01 L by a nitrogen-washout method. The percentage of body weight as fat (%BF) was computed from body density using Siri's equation (1961). Fat-free mass was calculated as the product of body weight and (1-%BF/100).

RESULTS

The men tended to have larger and heavier bodies but less adipose tissue than the women who had larger resistances for all measurements than the men (Table 1).

Table 1. Physical Characteristics of the Study Population

	Men		Women	
	Mean	SD	Mean	SD
Age	33.9	12.6	35.6	11.2
Weight (kg)	79.8	15.6	63.0	13.3
Stature (cm)	179.6	8.8	165.0	5.9
Body Density (gm/cm^3)	1.053	0.019	1.031	0.019
Leg Length (cm)	85.80	5.23	76.85	3.75
Trunk Length (cm)	62.38	3.22	58.32	3.15
Arm Length (cm)	70.10	3.79	62.92	2.51
Arm Circ. (cm)	31.3	3.8	28.7	4.0
Abdominal Circ. (cm)	90.9	12.8	82.6	12.4
Thigh Circ. (cm)	51.4	5.3	49.5	5.9
Calf Circ. (cm)	37.3	3.3	35.6	3.4
Triceps Skf. (mm)	11.6	7.4	18.5	7.1
Subscapular Skf. (mm)	15.4	7.8	16.1	8.4
Lateral Calf Skf. (mm)	7.6	2.8	12.0	3.7
Resistances (Ω)				
Whole Body	476	65	601	66
Leg	251	29	299	34
Trunk	83	17	93	27
Arm	224	26	303	38

In estimating an index of the amount of subcutaneous adipose tissue for the participants, the total segment volume of the arm, leg and trunk were calculated using the formula for a cylinder $V_i = (C_i^2/4\pi) \times L_i$, where C_i and L_i were the corresponding segment circumferences and lengths respectively. The conductive volumes of the arm, leg, and trunk were calculated using the formula, $\rho_i L_i^2/R_i$, where L_i, and R_i, were the corresponding lengths and the resistances and ρ_i was the muscle-specific resistivity for each segment.

Theoretically, the muscle-specific resistivity for a segment, or for the body as a whole, should be independent of the level of adipose tissue. Its coefficient of variation among individuals should not be substantially greater than that for fat-free mass. Practically, it is very difficult to calculate a muscle specific resistivity from simple anthropometric measurements and a segment resistance (Rush et al., 1963). In the present study, arm muscle area (AMA) was calculated (Heymsfield et al., 1982) and used with the resistance of the arm to calculate the specific resistivity of the arm by the formula $\rho_a = R_a(\text{AMA})/L_a$. The minimum value (175 Ω) from the total sample distribution of the muscle-specific resistivities for the arm was used to calculate the conductive volume of the arm for each individual. For each individual, the muscle specific resistivity for the leg and trunk was adjusted for the effects of adipose tissue on bioelectric impedance measurements based upon the assumption that differences in the segment volume-resistivities would be associated with differences in the resistances and the proportions of the segments. Thus, for each participant, the adjusted muscle-specific resistivity for the leg or trunk was calculated using the following formula: $\rho_i = \rho_a R_i (L_a A_i) / R_a L_i A_a)$. In this formula, R_i, L_i, and A_i, were the respective resistances, lengths, and cross-sectional areas of the leg and trunk. Sex-specific, sample means of the adjusted muscle-specific resistivities for the leg and trunk were used to compute the conductive volumes of these segments for each individual.

The conductive volumes were subtracted from the corresponding total segment volumes for each individual. The right-side volumes were assumed equal to those of the left for the arms and legs, based upon reports of the absences of significant lateral differences (Lohman et al., 1988). The volumes of subcutaneous adipose tissue from the body segments (2 arms, 2 legs, and trunk) were summed and multiplied by the reported density of fat (0.9 gm/ml/1000) to compute an index of the amount of the total subcutaneous adipose tissue (SATI) for each individual in kilograms. These data are presented in Table 2.

In the men and women, the means for the subcutaneous adipose tissue index (SATI) were slightly larger than those for total body fat (TBF) computed from hydrodensitometry, but these differences were statistically significant ($p<0.05$). Sex-specific regressions of SATI on %BF from hydrodensitometry were computed to determine if the association might be non-linear due to increased extracellular hydration at higher percentages of body fatness. In the women, a second degree polynomial best described the data ($R^2 = 0.70$; SEE = 6.4 kg). In the men, a third degree polynomial best described a more complex relationship than in the women ($R^2 = 0.74$; SEE = 6.4 kg). Despite these differences, correlation coefficients between total body fat and anthropometric variables were similar to corresponding coefficients between SATI and the same variables in the men and the women (Fig. 1).

Theoretically, the arm, leg, and trunk should act as a series of resistors, and their separate resistances should be additive (Patterson et al., 1988). In each sex in the present study, however, the sum of the arm, leg and trunk resistances was about 16% greater than the resistance for the whole body. Nearly all of the resistance of the whole

Table 2. Means and Standard Deviations for Body Composition Variables

	Men		Women	
	Mean	SD	Mean	SD
*%BF (%)	23.2	9.1	29.6	8.4
*FFM (kg)	61.8	6.9	42.0	4.7
*TBF (kg)	19.1	8.2	18.4	7.8
**SATI (kg)	23.4	12.0	20.7	11.7

* from hydrodensitometry, ** from impedance

body (95%) in each sex was determined by the resistances of the arm and the leg. Since so much of the total resistance was accounted for by these body segments, their impedance and lengths could be used to estimate fat-free mass for the whole body.

Regressions of fat-free mass on S^2/R_w for the whole body and on the conductive volume (L_i^2/R_i^2) for the arm and leg are presented in Table 3. In each sex, the regression of fat-free mass on S^2/R_w had the highest R^2 and lowest standard error of estimation (SEE). However, the regression of fat-free mass on the conductive volume

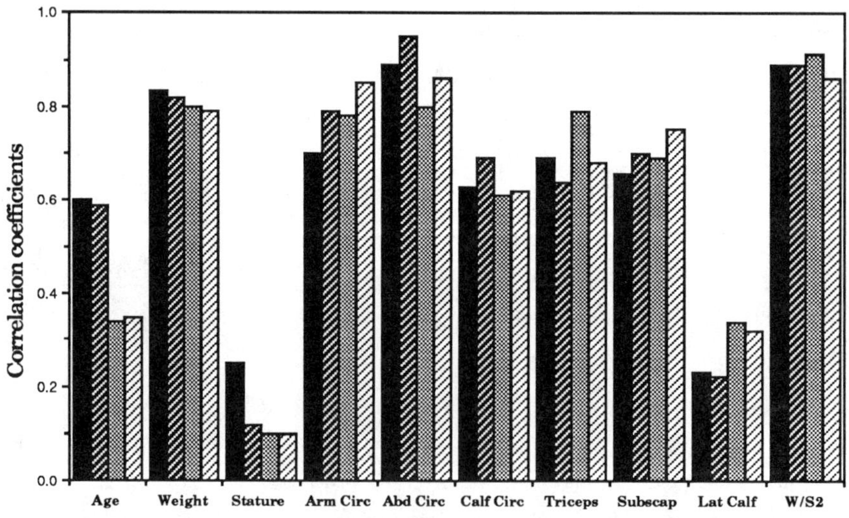

Fig. 1. Correlations with anthropometric variables. ■ TBF-men, ▨ SATI-men, ▧ TBF-women, and ▨ SATI-women.

Table 3. Regressions of Fat-Free Mass on Bioelectric Impedance Variables

	Intercept	slope (95%CI)	R^2	SEE (kg)
S^2/R_w (cm^2/Ω)				
Men	16.35	0.68 (0.55-0.81)	0.65	4.10
Women	8.60	0.74 (0.64-0.85)	0.73	2.80
Combined Sexes	4.40	0.84 (0.80-0.89)	0.90	3.65
L_a^2/R_a (cm^2/Ω)				
Men	22.77	1.80 (1.41-2.20)	0.58	4.48
Women	18.04	1.87 (1.41-2.32)	0.49	3.90
Combined Sexes	14.63	2.15 (2.01-2.30)	0.86	4.25
L_l^2/R_l (cm^2/Ω)				
Men	34.21	0.96 (0.66-1.26)	0.40	5.36
Women	20.88	1.09 (0.74-1.44)	0.36	4.37

of the arm, produced R^2's that were almost as high, and SEE values that were nearly as low as for the regression of S^2/R_w. In these normal individuals, the conductive volume of the leg was almost as good as the arm for predicting fat-free mass. The loss of accuracy in the prediction of fat-free mass using the conductive volume of the arm rather than stature and whole body resistance was only about 0.6 kg. The R^2's for the prediction fat-free mass from S^2/R_w and L_a^2/R_a were 0.90 and 0.86, and the SEE's were 3.65 and 4.25 kg, respectively.

The utility of bioelectric impedance of the arm for estimating fat-free mass for the whole body was supported in an independent sample of 33 healthy, elderly women, 72 to 92 years of age. These participants were voluntary residents in a retirement community and were determined to be free of renal disease, heart failure, severe hypertension, and diabetes. Only those participants within 20% of their ideal body weight were included. The bioelectric impedance and the length of the right arm were measured in each woman. Total body water was determined by measures of deuterium isotope dilution in urine analyzed using infrared absorption. Fat-free mass was computed as 73% of the volume of total body water.

In this sample of elderly women, the conductive volume of the arm (L_a^2/R_a) had as high a correlation with total body water (r = 0.74) as did S^2/R_w (r= 0.75). The conductive volume of the arm was also significantly correlated with fat-free mass estimated from anthropometry (r = 0.45) using the Durnin and Womersley equation (1974). In this small sample of healthy, elderly women, the applicability of using the conductive volume of the arm to estimate body composition would appear to be as good as that for S^2/R_w.

DISCUSSION

The calculation of SATI from segmental bioelectric impedance provides an indicator of the amount of subcutaneous adipose tissue in the body that is separate from other conventional methods. The mean amounts of SATI in the men and the women were statistically larger than corresponding amounts of total body fat from hydrodensitometry, but the SATI does appear to be a valid index of the amount of total subcutaneous adipose tissue. The correlations between SATI and anthropometric measurements are almost the same as corresponding correlations with TBF. Also, at zero values of SATI in regressions on %BF from hydrodensitometry, there is still some small percentage of body fat remaining, which should be expected. The calculation of SATI from bioelectric impedance is probably affected by increased levels of extracellular hydration in adipose tissue. This is also suggested in the polynomial regressions of SATI on %BF from hydrodensitometry. With increasing percentages of body fatness, the increase in SATI was less than at lower percentages. The ability of bioelectric impedance to detect differences between intra- and extracellular water would be greatly improved by the availability of multiple frequency bioelectric impedance analyzers.

The results of this study confirm reports that the resistance of the whole body in normal individuals is primarily determined by the resistances of skeletal muscle in the arm and the leg. Indices of the conductive volume of the arm can be used to estimate body composition in place of S^2/R_w with only a marginal loss in accuracy with the assumption that the composition is representative of the rest of the body. It is likely that indices of the conductive volumes of shorter segments of the limbs, such as the forearm or the lower leg can be used to estimate fat-free mass with similar accuracy.

No effort was made in the present analyses to improve predictions by including age or additional bioelectric or anthropometric variables. The addition of these variables might have reduced the errors of prediction, but it also would have made subsequent comparisons among models more complex. The attitude was also taken that the addition of variables in multiple regression equations increases the sample-specificity of the predictions.

The finding that the whole body resistance is less than the sum of the resistances for the body segments by an amount approximately equal to the resistance of the trunk suggests that the trunk contributes little or nothing to whole body resistance when measured using the conventional method with electrodes on the hand and the foot. This fact has been reported by several other investigators who are critical of the utility of the whole body impedance method in predicting body composition (Settle et al., 1980; Smith, 1987; Patterson et al., 1988, 1989). This marginal contribution of the trunk to whole body resistance may be attributable to the effects of anisotropy, or the dependence of resistance on the direction of flow of the current in the conductor (Rush et al., 1963).

The present study demonstrates that an index of the amount of subcutaneous adipose tissue can be calculated from the bioelectric resistances and lengths of the arm, leg and trunk. Also, fat-free mass can be predicted from the length and bioelectric resistance of the arm. The segmental bioelectric impedance approach can be adapted for estimating the body composition of subjects who have limited mobility and for whom accurate measurements of stature cannot be obtained. The indices and equations in the present study, however, should be used with caution until they have been cross-validated.

Supported by Grant HD-12252 from the National Institutes of Health, Bethesda, MD, and Ross Laboratories, Columbus, OH.

REFERENCES

Baker, L. E., 1989, Principles of the impedance technique, I. E. E. E. Eng. Med. Biol. 8:11.

Baumgartner, R. N., Chumlea, W. C., and Roche, A. F., 1988, Bioelectric impedance phase angle and body composition, Am. J. Clin. Nutr. 48:16.

Chumlea, W. C., Roche, A. F., and Steinbaugh, M. L., 1985, Estimating stature from knee height for persons 60 to 90 years of age, J. Am. Geriat. Soc. 33:116.

Chumlea, W. C., Baumgartner, R. N. and Roche, A. F., 1987, Segmental bioelectric impedance measures of body composition, in: "In Vivo Body Composition Studies," K. J. Ellis, S. Yasumura, and W. D. Morgan eds., The Institute of Physical Sciences in Medicine, London.

Chumlea, W. C., Baumgartner, R. N., and Roche, A. F., 1988, The use of specific resistivity to estimate fat-free mass from segmental body measures of bioelectric impedance, Am. J. Clin. Nutr. 48:7.

Durnin, J. V. G. A., and Womersley, J., 1974, Body fat assessed from total body density and its estimation from skinfold thickness: measurements on 481 men and women aged 16 to 72 years, Br. J. Nutr. 32: 77.

Guo, S., Roche, A. F., Chumlea, W. C., Miles, D. S., and Pohlman, R. L., 1987, Body composition predictions from bioelectric impedance, Hum. Biol. 59:221.

Heymsfield, S. B., McManess, C., Smith, J., Stevens, V., and Nixon, D. W., 1982, Anthropometric measurements of muscle mass: revised equations for calculating bone-free arm muscle area, Am. J. Clin. Nutr. 36:680.

Jackson, A. S., Pollock, M. S., Graves, J. E., and Mahar, M. T., 1988, Reliability and validity of bioelectrical impedance in determining body composition, J. Appl. Physiol. 64:529.

Khaled, M. A., McCutcheon, M. J., Reddy, S., Pearman, P. L., Hunter, G. R., and Weinsier, R. L., 1988, Electrical impedance in assessing human body composition: the BIA method, Am. J. Clin. Nutr. 47:789.

Kushner, R. F., and Schoeller, D. A., 1986, Estimation of total body water by bioelectrical impedance analysis, Am. J. Clin. Nutr. 44:417.

Lohman, T. G., Roche, A. F., and Martorell, R., 1988, "Anthropometric Standardization Reference Manual," Human Kinetics Books, Champaign, IL.

Lukaski, H. C., Johnson, P. E., Bolonchuk, W. W., and Lykken, G. I., 1985, Assessment of fat-free mass using bioelectrical impedance measurements of the human body, Am. J. Clin. Nutr. 41:810.

Lukaski, H. C., Bolonchuk, W. W., Hall, C. B., and Siders, W. A., 1986, Validation of tetrapolar bioelectrical impedance method to assess human body composition, J. Appl. Physiol. 60:1327.

Patterson, R., Ranganathan, C., Engel, R., and Berkseth, R., 1988, Measurement of body fluid volume change using multisite impedance measurements, Med. Biol. Engineer. Comput. 26:33.

Patterson R., 1989, Body fluid determinations using multiple impedance measurements, I. E. E. E. Eng. Med. Biol. 8:16.

Pierson, R. N., Wang, J., Colt, E. W., and Neumann, P., 1982, Body composition measurements in normal man: the potassium, sodium, sulfate and tritium spaces in 58 adults, J. Chronic Dis. 35:419.

Roche, A. F., Siervogel, R. M., Chumlea, W. C., Reed, R. B., Eichorn, D., McCammom, R. M., 1982, Serial changes in subcutaneous fat thicknesses of children and adults, Mono. Paediat. 17:1.

Rush, S., Abildskov, J. A., and McFee, R., 1963, Resistivity of body tissues at low frequencies, Circul. Res. 12:40.

Segal, K. R., Gutin, B., Presta, E., Wang, J., and Van Itallie, T. B., 1985, Estimation of human body composition by electrical impedance methods: a comparative study, <u>J. Appl. Physiol.</u> 58:1565.

Settle, R. G., Foster, K. R., Epstein, B. R., and Mullen, J. L., 1980, Nutritional assessment: whole body impedance and body fluid compartments, <u>Nutr. Cancer</u> 2:72.

Siri, W. E., 1961, Body composition from fluid spaces and density, analysis of methods. in: "Techniques for Measuring Body Composition." J. Brozek, and A. Henschel, eds., Nat. Acad. Sci., Nat. Res. Council, Washington, DC.

Smith, D. N., 1987, Body Composition by tetrapolar impedance measurements - correlation or con? <u>Proc. VIIth Int. Conf. Electrical Bio-Impedance</u>. pp 1-15, Portschach, Austria.

TRIM : AN ELECTROMAGNETIC BODY COMPOSITION ANALYSER

P.J. Chadwick and N.H. Saunders

Swansea In Vivo Analysis Research Group (SIVARG)
Department of Physics
University College of Swansea
Swansea SA2 8PP, U.K.

INTRODUCTION

An extensive range of techniques is now available to assess the body composition of humans. At one extreme, sophisticated and expensive imaging procedures like Computed Tomography and Magnetic Resonance scanning can in principle enable the relative proportions of different kinds of tissue in an individual to be estimated, while at the other, well-established anthropometric measurements can simply and inexpensively yield useful if less precise data based on the average parameters of a population or group.

Within the last decade a number of electrical and electromagnetic techniques have emerged which provide information about the body in terms of a broad two-compartment model comprising fat (non-conducting) tissue and fat-free (conducting) tissue. Perhaps the simplest, in principle, is the measurement of bioelectrical impedance via electrodes positioned, for example, on the wrist and ankle in a standard four probe configuration. The current is typically less than 1 mA at a frequency of 50 kHz (Lukaski et al., 1985). An electrical imaging procedure also capable in principle of yielding body composition data is applied potential tomography or electrical impedance tomography (Brown et al., 1987 and 1988). In this case, alternating current at up to 100 kHz is passed through pairs of electrodes selected successively from an array attached to the body surface. Potential differences between the other electrodes in the array can then be used to reconstruct a resistivity image, albeit one of limited resolution at the moment.

In this paper we wish to describe an electromagnetic body composition analyser which, from substantial work in the United States (see, for example, Van Loan and Mayclin, 1987), holds considerable promise as a rapid, safe, reliable, non-invasive and inexpensive predictor of lean body mass. It takes advantage of the fact that the configuration of eddy currents induced in the body when subjected to an external electromagnetic field will reflect the distribution of conducting and non-conducting tissue. The electromagnetic field is generated by a large coil into which the subject is inserted in either a whole body or a scanning mode. Electrical parameters of the coil which respond to the eddy currents, and which can be monitored, include impedance, as in the American TOBEC programme, and resonance frequency as used in the present work using the Tissue Resonance Impedance Monitor (TRIM).

Table 1. Correlation Between the TRIM Variable and Lean Body Mass (LBM)
          From Other Body Composition Data, (Study 1)

	Variable	Skinfold	^{40}K	Anthropometry	Impedance
Combined n=45		0.92	0.84	0.96	0.87
Men n=22	TRIM	0.87	0.74	0.92	0.78
Women n=23		0.83	0.75	0.96	0.69

TISSUE RESONANCE IMPEDANCE MONITOR

The electrical properties of a solenoidal coil carrying a high frequency
current and containing material of arbitrary geometry and non-uniform electrical conductivity, dielectric coefficient and magnetic susceptibility, are
difficult to calculate in general. The problem is simplified if the inserted
material is modelled by a uniform cylinder coaxial with the coil, but even
in this case it is important to consider the wavelength of the current in
relation to the dimensions and other parameters of the coil. The ordinary
solenoid approximation corresponds to a wavelength much greater than the
length of wire in the coil and solutions for the changes in its inductance
and resistance are given, for example, by Scott (1930). These solutions
indicate that for typical tissue conductivities the best discrimination
between tissue types will occur at frequencies of 1 - 10 MHz. At these
frequencies, however, the ordinary solenoid approximation may not be valid.
Thus, the coil designed for the present body composition work has a cylinder
diameter of approximately 1 m and length 1.8 m and is helically wound in
such a way that, at the frequency used (about 2 MHz), the current wavelength in the wire is comparable with the total wire length while being
greater than the length of a single turn. The electric and magnetic field
configurations in such a structure are quite complex but can be approximately
calculated using the sheath helix waveguide model (Pierce, 1950; Sensiper,
1955; Chute and Vermeulen, 1981). The three components of the $\underline{B}$ and $\underline{E}$
fields in the natural resonant modes of the coil have been evaluated.
Experiments have been carried out to measure some of these field components
and reasonable agreement has been found with the predicted values. The
change in resonance frequency on totally inserting cylindrical phantoms of
various radii filled with saline solution of a range of concentrations has
also been calculated and compared with experimental values. The coil
impedance has been similarly considered and reasonable agreement with
experiment found, bearing in mind the simplified nature of the sheath helix
model. In the system constructed, the second natural resonance at about
2 MHz was found to be particularly sharp and was generally used in the body
composition measurements.

The procedure followed to monitor lean body mass was to measure the
change in this resonance frequency on inserting the subject in a supine
position wholly inside the coil. Generally this was repeated five times
and the average taken. Several TRIM variables could then be evaluated,
involving different combinations of the resonance frequency change and
anthropometric data such as the height, (see below). As already noted,
the geometrical and electrical complexity of the human body precludes any
of these TRIM variables being analytically and exactly linked with say lean
body mass. As in the existing TOBEC programme however, some link is to be
anticipated on physical grounds, and this expectation can be tested by
examining the correlation between a TRIM variable and other currently
accepted measures of lean body mass. This is the programme followed in the
body composition measurements outlined below.

Table 2. Correlation Between the TRIM Variable and LBM From Other Body
         Composition Data, (Study 2: 12 men, 12 women)

Variable	Skinfold	^{40}K	Anthropometry	Impedance	TBW (^{3}H$_2$0)	TBNitrogen (NAA)
TRIM	0.95	0.82	0.99	0.89	0.92	0.86

SURVEY OF RESULTS

The TRIM analyser has been included in two independent studies of normal
healthy volunteers within the context of the body composition research programme of the Swansea In Vivo Analysis Research Group (SIVARG). The first
involved 45 subjects (22 male, 23 female) ranging in age from 17 to 71 and
in body weight from 44 to 104 kg. The other analytical techniques employed
were skinfold thickness, anthropometric measurements, whole body potassium 40,
and whole body bioelectrical impedance using a commercial body composition
analyser marketed by Holtain Ltd. The second study is currently in progress
and so far has covered 24 volunteers (12 men, 12 women, age range 23-61,
weight range 44-104 kg). A more extensive range of techniques is being used
with total body water determined by isotope dilution (^{3}H$_2$0), extra-cellular
water using KBr, and total body nitrogen and chlorine by IVNAA being included
alongside the techniques of the first study. In both cases, appropriate
protocols were drawn up and approved, and the informed consent of the
volunteers obtained.

Tables 1 and 2 summarise some of the results of the two studies. Correlation coefficients are given between a helical waveguide or TRIM variable
and lean body mass estimated from the other techniques. In Table 1, separate
correlation coefficients are shown for the men and women as well as for the
combined groups. Several different definitions of the TRIM variable were
examined, the simplest being T, the percentage change in resonance frequency.
Also considered were T^2, $\sqrt{T}$ and $\sqrt{hT}$, where h is the body height. The latter
variable typically changed by about 35% over the members of the volunteer
groups, the range in $\sqrt{T}$ alone being about 22%. The TRIM variable used in
the Tables is $\sqrt{hT}$ which often yielded the highest correlation coefficients.
Further analysis is in progress and will be reported elsewhere, but already
the results support the promise established by the TOBEC programme, at least
for normal healthy volunteers. Validation of the technique with other groups
is planned.

ACKNOWLEDGEMENTS

The work reported was carried out with the financial support of the
Welsh Scheme for the Development of Health and Social Research. Valuable
assistance with the electrical measurements was given by S. Al-Zeibak.
Other data, some reported separately at this Symposium, were supplied by
the following SIVARG colleagues to whom grateful acknowledgement is made:
P.A. Ali, J.L. Birks, C.J. Evans, H. Jenkins, W.D. Morgan and S.J.S. Ryde.

REFERENCES

Brown, B., Barber, D., and Tarassenko, L., eds, 1987, Special issue
    on electrical impedance tomography - applied potential tomography,
    Clin. Phys. Physiol. Meas., 8 (Supp. A).
Brown, B., Barber, D., and Jossinet, J., eds, 1988, Special issue
    on electrical impedance tomography - applied potential tomography,
    Clin. Phys. Physiol. Meas., 9 (Supp. A).

Chute, F.S., and Vermeulen, F.E., 1981, A visual demonstration of
     the electrical field of a coil carrying a time-varying current,
     I.E.E.E. Trans. Educ., E-24(4) : 278.
Lukaski, H.C., Johnson, P.E., Bolonchuk, W.W., and Lykken, G.I., 1985,
     Assessment of fat-free mass using bioelectrical impedance measure-
     ments of the human body, Am. J. Clin. Nutr., 41: 810.
Pierce, J.R., 1950, "Traveling-Wave Tubes", D. Van Nostrand Co., Inc.,
     New York.
Scott, K.L., 1930, Variation of the inductance of coils due to the
     magnetic shielding effect of eddy currents in the cores, Proc.
     I.R.E., 18(10) : 1750.
Sensiper, S., 1955, Electromagnetic wave propagation on helical
     structures. (A review and survey of recent progress), Proc.
     I.R.E., 43 : 149.
Van Loan, M., and Mayclin, P., 1987, A new TOBEC instrument and
     procedure for the assessment of body composition : use of
     Fourier coefficients to predict lean body mass and total body
     water, Am. J. Clin. Nutr., 45 : 131.

THE ASSESSMENT OF THE BODY COMPOSITION IN THE ELDERLY BY

DENSITOMETRY, ANTHROPOMETRY AND BIOELECTRICAL IMPEDANCE

Paul Deurenberg , Karin van der Kooy, and
Joseph G.A.J. Hautvast

Department of Human Nutrition, Agricultural
University, Bomenweg 2   6703 HD Wageningen,
The Netherlands

INTRODUCTION

Body composition was assessed in a group of 165 men and women, aged
37-83 years, by densitometry, anthropometry, and bioelectrical
impedance.

The percentage of body fat (BF) and the fat free mass (FFM) were
calculated from body density, using Siri's equation (BF_S%, FFM_S) in
which the density of the FFM is assumed to be 1.100 kg/l, and using an
adapted equation (BF_A%, FFM_A). In this equation, the density of the FFM
was recalculated from the chemical composition of the FFM. Based on data
on bone mineral loss (Mazess, 1982) and loss of active cell mass
(Forbes, 1976) with advancing age, the calculated density of the FFM was
found to be lower at older ages in women but not in men (Deurenberg et
al., 1989a). A larger relative amount of water in the FFM in the obese
state also results in a decreasing density of the FFM with increasing
amounts of body fat (Deurenberg et al., 1989b).

RESULTS

Applying both corrections of Siri's formula to the data of this
population, the adapted formulas resulted in a lower body fat percentage
compared to Siri's formula (Table 1). When predicting FFM or BF% from
body impedance or from simple anthropometric variables (body mass index
BMI), the explained variance was mostly higher and the predicting error
was always lower when impedance or anthropometric characteristics were
related to the body fat or fat-free mass obtained by the adapted
formula:

$$FFM_S= 0.382*H^2/R - 0.135*Age + 0.216*BW + 4.5*Sex +$$
$$0.107*H - 3.95 \ (r^2 = 0.92, \ SEE = 2.84 \ kg, \ CV = 5.9 \ \%)$$

$$FFM_A= 0.340*H^2/R + 0.288*BW - 0.116*Age + 3.5*Sex +$$
$$0.097*H - 3.79 \ \ (r^2 = 0.94, \ SEE = 2.54 \ kg, \ CV = 5.1 \ \%)$$

$$BF_S\%_{MALES} = 1.078*BMI + 0.150*Age - 6.84 \ (r^2 = 0.34,$$
$$SEE = 4.46\%)$$

Table 1. Age and Physical Characteristics of the Subjects (Mean ± SD)

	Male (85)			Female (80)		
Age (years)	57.8	±	12.6	57.4	±	11.8
Weight (kg)	78.6	±	9.7	65.6	±	9.1
Height (cm)	177.6	±	7.7	164.6	±	6.3
BMI (kg/m^2)	24.9	±	2.3	24.2	±	3.2
BF$_S$ (%)	28.7	±	5.4	38.7	±	7.8
FFM$_S$ (kg)	55.9	±	7.4	39.8	±	4.8
BF$_A$ (%)	27.3	±	4.9	35.6	±	6.9
FFM$_A$ (kg)	57.0	±	7.4	41.9	±	4.8
Impedance (Ω)	449	±	36	553	±	43

$$BF_A\%_{MALES} = 0.970*BMI + 0.136*Age - 4.67 \ (r^2 = 0.34, \ SEE = 4.01\%)$$

$$BF_S\%_{FEMALES} = 1.146*BMI + 0.333*Age - 8.14 \ (r^2 = 0.72, \ SEE = 4.16\%)$$

$$BF_A\%_{FEMALES} = 1.028*BMI + 0.283*Age - 5.56 \ (r^2 = 0.72, \ SEE = 3.74\%)$$

[H = body height; R = body impedance; BW = body weight; BMI = body mass index (kg/m^2); Age (years); Sex (male=1, female=0)]

When applying the fat-specific prediction formulas of Segal et al. (1988) for FFM from body impedance, or the age-specific prediction formulas of Womersley & Durnin (1977) for BF% from BMI to this population, BF% was only slightly better predicted from body impedance.

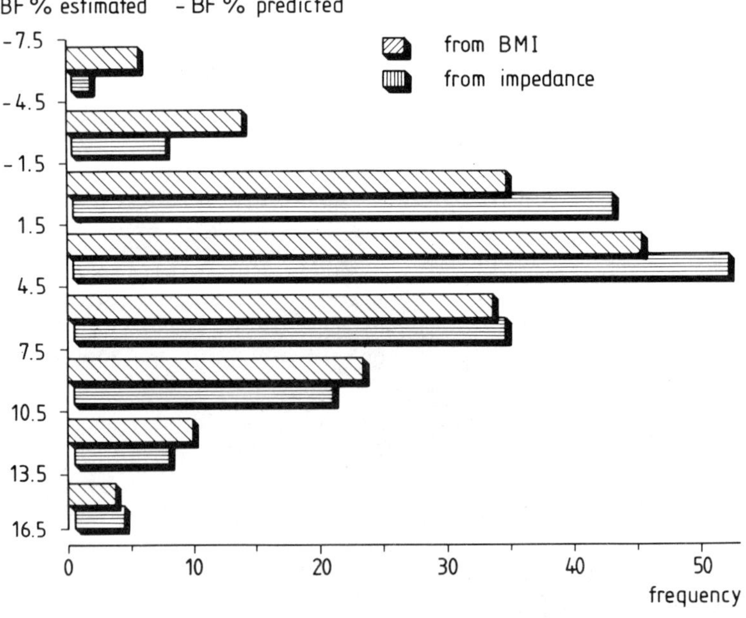

Fig. 1.   Histogram of differences between densitometric BF% and predicted BF% by body mass index or body impedance.

However, both the impedance method and the anthropometric method resulted in large estimation errors in individuals (Fig. 1). In 4.2%
(impedance), respectively 4.7% (BMI) of the subjects, the predicted BF%
differed more than 10% from the value determined by densitometry.

CONCLUSION

The estimation of the body composition in elderly subjects by
bioelectrical impedance may be an adequate method to predict body
composition at a population level only as far as age-specific formulas
are used. However, at a population level the method is hardly better
compared to simple anthropometric methods. In 5 percent of the subjects
the difference between predicted value and observed value was
very large. Therefore, it can be questioned whether the bioelectrical
impedance method is suitable to assess the body composition at an
individual level, at least in subjects from middle age onwards.

REFERENCES

Deurenberg, P., Weststrate, J.A., and van der Kooy, K., 1989a, Is an
    adaptation of Siri's formula for the calculation of body fat
    percentage from body density necessary?, <u>Eur. J. Clin. Nutr.</u>,
    43:559.
Deurenberg, P., Leenen, R., van der Kooy, K., and Hautvast,
    J.G.A.J., 1989b, In obese subjects the body fat percentage
    calculated with Siri's formula is an overestimation, <u>Eur. J. Clin.
    Nutr.</u>, 43:569.
Forbes, G.B., 1976, The adult decline in lean body mass, <u>Hum. Biol.</u>,
    48:161.
Mazess, R.B., 1982, On aging bone loss, <u>Clin. Orthop. Res.</u>, 165:239.
Segal, K.R., van Loan, M., Fitzgerald, P.I., Hodgdon, J.A., and van
    Itallie, T.B., 1988, Lean body mass estimation by bioelectrical
    analysis: a four site cross-validation study, <u>Am. J. Clin. Nutr.</u>,
    47:7.
Womersley, J., and Durnin, J.V.G.A., 1977, A comparison of the skinfold
    method with extent of 'overweight' and various weight-height
    relationships in the assessment of obesity, <u>Brit. J. Nutr.</u>, 38:271.

A METHOD FOR *IN VIVO* DETERMINATION OF CARBON AND OXYGEN USING PROMPT
GAMMA RADIATIONS INDUCED BY 14.7-MeV NEUTRONS

C. L. Hollas, L. E. Ussery, K. B. Butterfield,
and R. E. Morgado

Advanced Nuclear Technology Group
Los Alamos National Laboratory
Los Alamos, NM 87545 USA

INTRODUCTION

Quantification of body fat is important in studying obesity and
other diseases involving nutritional assessment.[1] The *in vivo*
determinations of total-body carbon and oxygen can be used to obtain
clinical information about a person's fat content. Biggin and Morgan[2]
showed that the ratio of total-body oxygen to total-body carbon is a
sensitive indicator of the percentage of fat tissue in the human body.
Kyere et al.,[3] measured total-body carbon by counting the number of
4.438-MeV gamma rays emitted from the ^{12}C nuclei after excitation by
14-MeV neutrons. At Brookhaven National Laboratory, Kehayias et al.,[4]
developed a similar technique for measuring total-body carbon with a beam
of 14-MeV neutrons from a pulsed neutron generator. We present here a
variation on these two methods that permits the simultaneous
determination of both carbon and oxygen, with the potential for a
significantly reduced dose of neutron radiation.

EXPERIMENTAL APPARATUS AND TECHNIQUE

In the experimental apparatus (see Fig. 1), a small accelerator
provides a beam of ~150-keV deuterons that strikes a tritiated titanium
target, producing 14.7-MeV neutrons and 3.7-MeV alpha particles in
correlated pairs. An alpha detector within the vacuum envelope of the
accelerator detects alpha particles from the reaction and allows us to
define a beam of neutrons that leave the vacuum chamber through the wall
opposite the alpha detector. These neutrons travel with a velocity of
5.2 cm/ns. The detection of an alpha particle also provides a start
signal to a time-to-amplitude converter (TAC). The neutron associated
with the detected alpha particle can interact with a nucleus within the
object under investigation and produce prompt gamma radiation. A gamma
radiation detector then measures the energy of the radiation, and also
provides a stop signal to the TAC. The time difference between the
arrival of the fast timing signals from the alpha detector and gamma
radiation detector determines the point of interaction within the object.
In a typical measurement, the time interval in which events are accepted
is less than 10 ns, which corresponds to a distance of less than 52 cm.
This narrow time-window in which good events are recorded allows us to
reduce greatly the unwanted background of gamma radiation in the measured
spectra. For example, the measured background is at least a thousand
times lower than that achieved with the pulsed beam method[4] (which
records events within a 12-μs time window). The energy spectrum of gamma
radiation that is obtained is characteristic of the elemental composition
of the object.

THE ALPHA PARTICLE DETECTOR

The alpha particle detector is a Bicron BC-400 fast plastic
scintillator, 50 μm thick and 1.9 cm in diameter. The scintillator is
located 10 cm from the intersection point of the deuteron beam with the
tritium target, at an angle of 135 deg with respect to the deuteron
beam's direction, and is viewed by an EMI 9813B photomultiplier. The
detector has excellent time-resolution characteristics (<700 ps) at high
data rates >1 MHz.

THE GAMMA RADIATION DETECTOR

The gamma radiation detector is a NaI(Tl) cube, 10.2 cm on each
side, located ~80 cm from the tritium target, and shielded from direct
neutrons by a tungsten- and borated-polyethylene-shadow shield. In
addition, a 10.2-cm-thick lead cave shields the detector on four sides
from background gamma radiation. This detector has moderate energy
resolution (7% at 660 keV) and good timing characteristics (~2 ns).

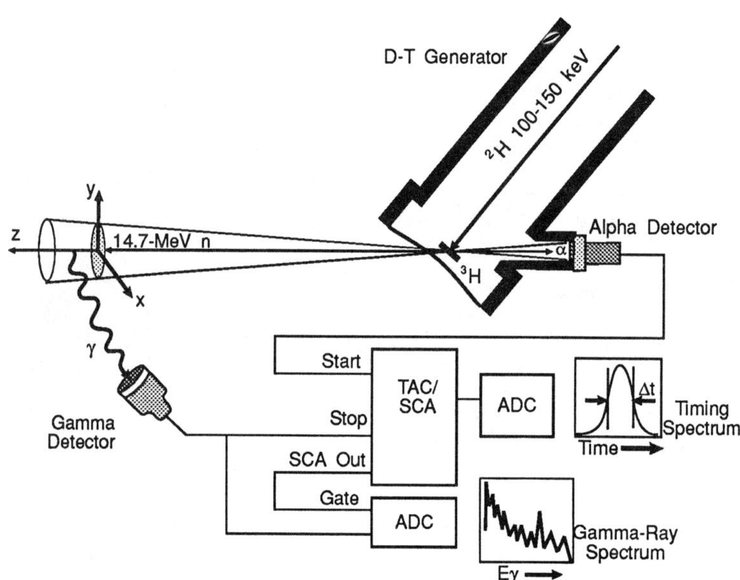

Fig. 1. Schematic representation of the technique.

THE ELECTRONICS OF THE DATA ACQUISITION SYSTEM

Figure 2 shows a schematic of the electronics of the data
acquisition system. The anode signals from the photomultiplier tubes for
both the alpha and gamma detectors are discriminated, using constant
fraction discriminators. Resulting fast NIM logic pulses then are
presented to a fast coincidence module. A delay is added to the
coincidence input signal from the alpha detector to ensure that the input
signal arrives after the gamma detector signal. Events are in
coincidence if they arrive within a 40-ns time interval. The coincidence
output triggers the start of the TAC and minimizes the number of unwanted
events that initiate a time measurement. A second branch of the logic
signal from the gamma detector is delayed and then provides the stop
signal for the TAC. The amplitude signal from the TAC is then presented
to a CAMAC analog-to-digital converter (ADC).

A linear-gate module passes gamma-ray energy signals for those gamma rays that meet coincidence requirements. This linear analog signal is then presented to a second CAMAC ADC module. The arrival of the signal from the TAC at the CAMAC ADC initiates a data transfer of the two values of the CAMAC ADCs into the memory of a VMEbus computer system. The values in the ADCs are used as address pointers to a location in a two-dimensional array. The value of the array element at this lcoation is then incremented by one. After many events and data transfers, a two-dimensional data array of energy vs time is generated, which allows projections of either time or energy spectra to be easily produced. A peak in the time spectrum is always visible, corresponding to gamma rays being emitted by the object within the neutron beam at ~1 m from the tritium target. Gamma-ray energy spectra are selected from only those events associated with this time peak.

GAMMA-RAY SPECTRA

Figure 3 shows the gamma-ray spectrum that was recorded when carbon in the form of a graphite sheet was placed in the neutron beam 1 m from the tritium target. At this distance, the diameter of the neutron beam (defined by the alpha detector) is ~19 cm. Carbon is identified by a gamma radiation of 4.438 MeV from the first excited state of ^{12}C. The gamma-ray spectra recorded for oxygen, in the form of water, is shown in Fig 4. Oxygen is identified by gamma radiation primarily at 6.129 MeV from the second excited state of ^{16}O.

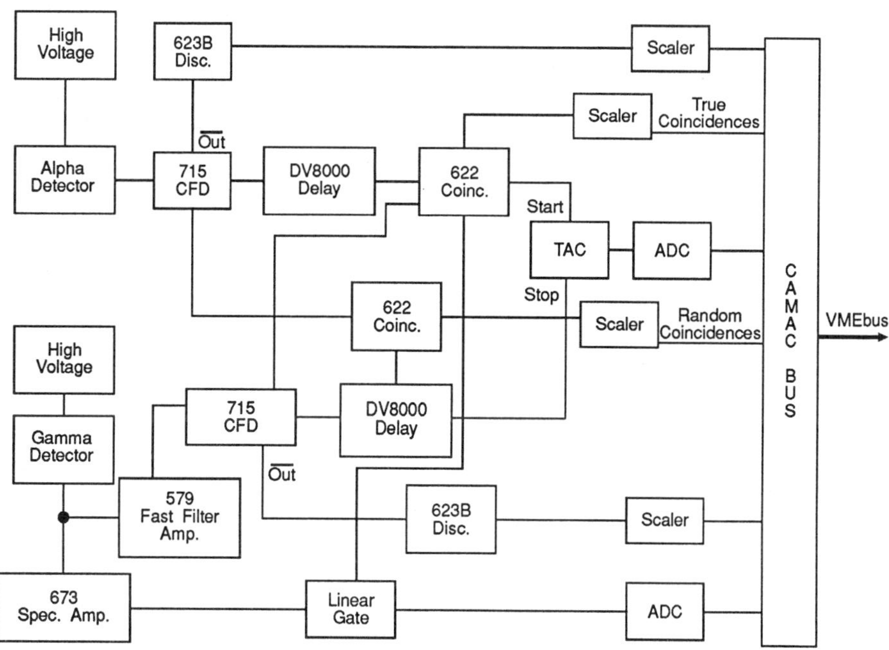

KEY: 623B - LeCroy leading edge discriminator
      622 - LeCroy 2-fold logic unit
      222 - LeCroy dual gate-and-delay generator
      DV8000 - Ortec octal delay box
      715 - Phillips constant fraction discriminator
      673 - Ortec spectroscopy amplifier
      579 - Ortec fast filter amplifier

Fig. 2. Block diagram of the data acquisition electronics.

THE PHANTOM TARGET

A small phantom target (total mass: ~8 kg) representative of the carbon, oxygen, nitrogen, and hydrogen composition of Reference Man[5] was assembled from ethylene glycol, urea, and water. The composition is given in Table 1. This phantom sample then was placed at 1 m from the tritium target and irradiated for 20 min to provide a total dose of 4.2×10^5 neutrons/cm^2. Assuming a quality factor of 10, this exposure corresponds to a dose of ~50 mrem, or 0.5 milliseiverts. Figure 5 shows

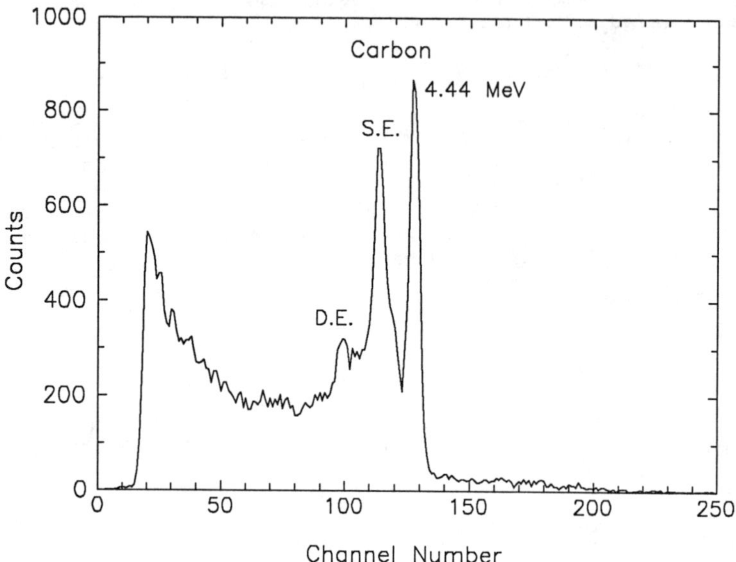

Fig. 3. Gamma-ray energy spectrum for carbon.

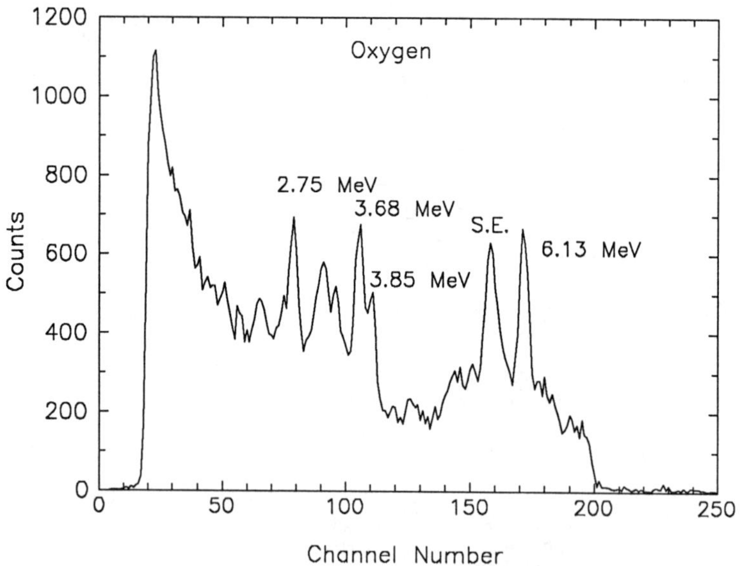

Fig. 4. Gamma-ray energy spectrum for oxygen.

Table 1. Composition of Phantom Target

Material	Quantity	Element	Percent by Weight
Ethylene Glycol	4 l	O	64.5
Urea	425 g	C	22.9
Water	3 l	N	2.5
		H	10.1

the gamma-ray spectrum obtained. The statistical precision for the number of gamma rays from carbon and oxygen is ± 3.0 % and $\pm 2.5\%$, respectively.

CONCLUSION

We can make some conclusions and extrapolations from these preliminary measurements. The carbon and oxygen content of a 70-kg subject could be determined to a statistical precision of better than $\pm 3\%$ with a total neutron radiation dose of ~5 mrem by using a scanning method of measurement. In this method, the subject is shielded from the neutrons not contained within the defined cone. The total radiation dose of 50 mrem used in the study above is distributed over ten 7-kg sections, each section receiving approximately a 5-mrem dose. With the addition of multiple gamma-ray detectors, reductions in the statistical error, the measuring time, and the total radiation dose are readily achievable. Thus, with this method, the total radiation dose would be low enough that longitudinal studies on individuals would be possible during the course of treatment and recovery.

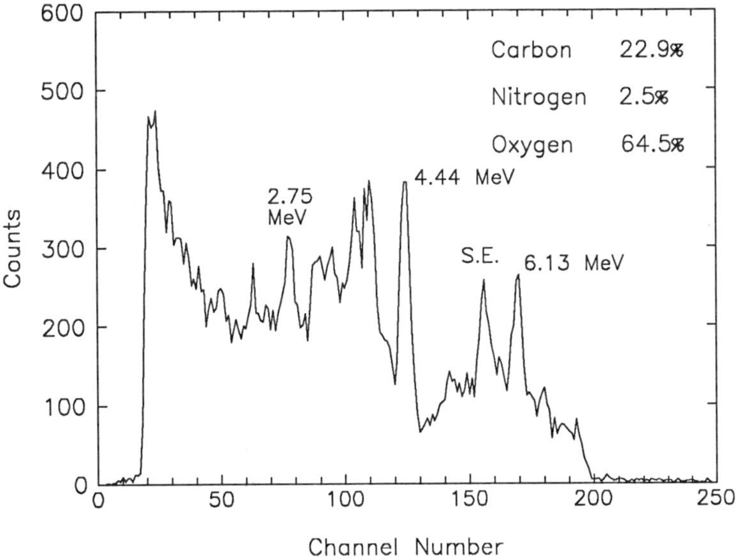

Fig. 5. Gamma-ray spectrum from phantom target representing Reference Man.

REFERENCES

1. S. H. Cohn, New concepts of body composition, in: "*In Vivo* Body
    Composition Studies," K. J. Ellis, S. Yasamura, and W. D. Morgan,
    eds., The Institute of Physical Sciences in Medicine, London,
    pp. 1-14 (1987).

2. H. C. Biggin and W. D. Morgan, The measurement of tissue composition
    by neutron activation analysis, <u>International Journal of Nuclear
    Medicine and Biology</u> 4:133 (1977).

3. K. Kyere, K. B. Oldroyd, C. B. Oxby, L. Burkinshaw, R. E. Ellis, and
    C. L. Hill, G. L., The feasibility of measuring total body carbon by
    counting neutron inelastic scatter gamma rays, <u>Phys. Med. Biol.</u>,
    27:805 (1982).

4. J. J. Kehayias, K. J. Ellis, S. H. Cohn, S. Yasamura, and J. H.
    Weinlein, Use of a pulsed neutron generator for *in vivo* measurement
    of body carbon, in: "*In Vivo* Body Composition Studies," K. J. Ellis,
    S. Yasamura, and W. D. Morgan, eds., The Institute of Physical
    Sciences in Medicine, London, pp. 427-435 (1987).

5. International Commission on Radiological Protection, Report on the
    Task Group of Reference Man, Publication 23, Oxford, Pergamon (1975).

DEVELOPMENT OF A TOTAL BODY CHLORINE ANALYSER USING

A BISMUTH GERMANATE DETECTOR SYSTEM AND A ^{252}Cf NEUTRON SOURCE

N. Blagojevic, B. J. Allen, and A. Rose

Australian Nuclear Science and Technology Organisation
Private Mail Bag 1
Menai NSW 2234 Australia

INTRODUCTION

The initial method for _in vivo_ determination of total body chlorine (TBCl) was total body neutron activation followed by whole body counting of residual products. The technique relied on the exposure of the subject to a sufficient fluence of neutrons to activate body sodium and chlorine. Subsequently, ^{38}Cl and ^{24}Na were quantified and correlated to the extracellular water (ECW) previously determined by tracer dilution analysis (Yasumura et al., 1983). The major drawbacks of this technique are the following:

a)   the radiation exposure during the activation process is approximately 280 mrem (2.8 mSv);
b)   the method is time consuming as it consists of two separate operations, namely activation and a counting step;
c)   the unit cost is high because it requires two separate facilities and a number of high flux neutron sources; and
d)   the patient must be quickly moved from one facility into another to be counted as ^{38}Cl, the chlorine activation product, which has a half-life of only 37.3 minutes.

A better alternative involves the use of a prompt _in vivo_ neutron activation method. An attempt was made to calculate TBCl from the NaI spectra obtained using this type of system; however, errors involved were large (Beddoe et al., 1987). The major difficulty was the high background produced by prompt neutron capture reactions within the sodium iodide (NaI) detectors used in the facility. Although the detectors are shielded, a small proportion of fast neutrons penetrate and are captured by iodine within the crystal structure of the detector. When the cascade gamma rays are all absorbed in the NaI crystal, a full energy peak of 6.8 MeV is observed, together with a spectral line shape which masks the observation of the most intense chlorine gamma rays at 6.6, 6.1 and 5.7 MeV. Despite the background problems the prompt activation system has several advantages:

a)   the radiation dose is very low, 20 to 30 mrem depending on the patient's size (Allen et al., 1985);
b)   the scan is completed within 15 minutes (Allen et al., 1987); and
c)   the results are obtained with a single facility.

To circumvent the problem of high background in the chlorine region, a bismuth germanate (BGO) detector was added to the existing Total Body Nitrogen (TBN) unit, specifically to detect the chlorine gamma rays. The BGO scintillator has the advantage of high counting efficiency, and more importantly, it is relatively insensitive to the neutron background. The low thermal capture cross-section of Bi (36 mb) and oxygen (<0.2 mb), coupled to that of Ge (2450 mb) gives a thermal neutron capture probability per gramme mole which is 13.4% of that calculated for NaI. We then reasoned that this detector might be more suitable for TBCl determinations (Fig. 1). The disadvantage of BGO is the reduced gamma ray resolution, typically 15% at ^{137}Cs (661 kev). However, because of the complexity of the chlorine

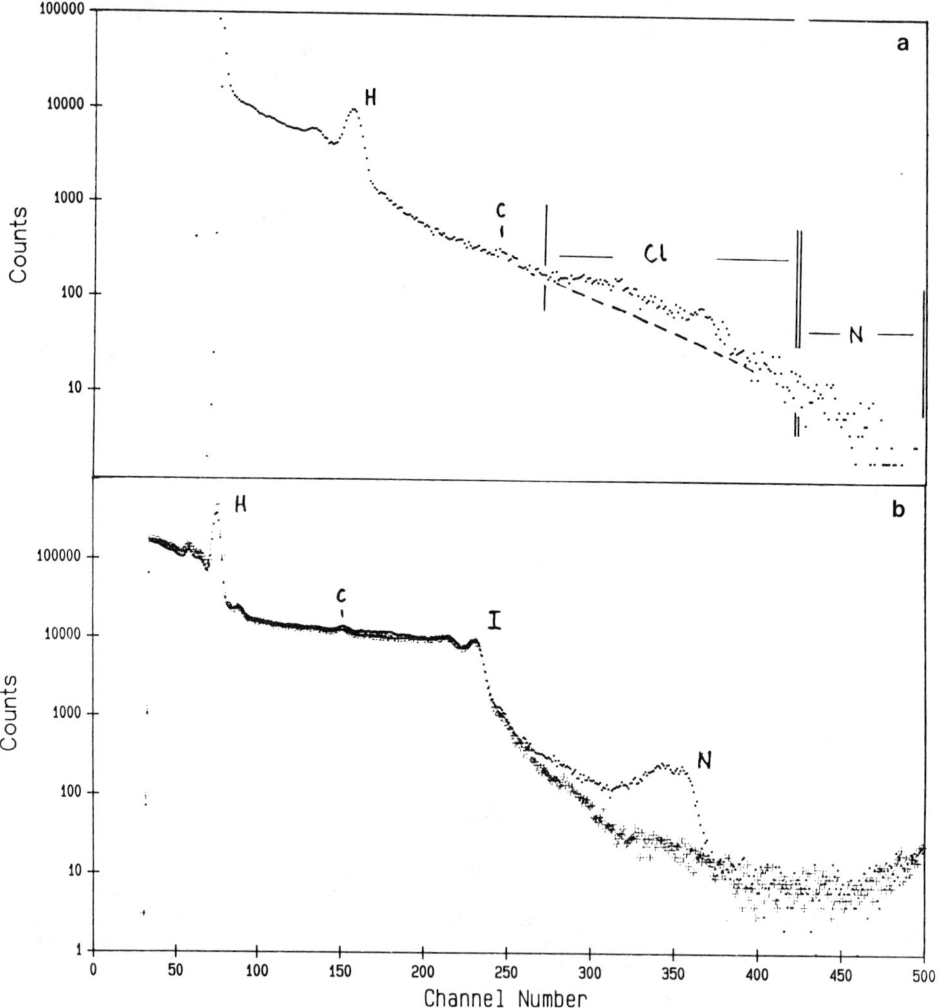

Fig. 1.   a) Typical BGO subject spectrum. No significant peaks due to scintillator-neutron interaction are produced in or above the chlorine region of interest (5 to 9 MeV). b) Typical phantom solution NaI spectrum with a 6.8 MeV iodine capture peak.

spectrum in the region of the most intense gamma rays (Bird et al., 1973), mainly 5.7, 6.1, 6.6, 7.4 and 7.9 MeV, and the small size of the available detector, we integrated the whole region of interest (ROI) between 5 and 9 MeV. Similarly, we integrated the nitrogen ROI between 9 and 11 MeV. The objective of this study is to show that the background in these ROI's is independent of body habitus, such that a linear system is achieved.

The TBCl may be converted into extracellular water volume (ECW) by the TBCl (g)/[Cl-plasma](g/L) ratio since the Cl$^-$ ions are mainly confined to the extracellular space. The chlorine plasma concentration can be obtained from a single blood sample. The only assumption is that the plasma chloride is representative of the chloride concentration in ECW. By including the total body water (TBW) measurement with the ECW analysis, the intracellular water (ICW) can be calculated by subtracting ECW from the TBW. The principle aim of this study, therefore, is to investigate the possibility of using the BGO spectrometer to determine TBCl in human subjects and to do so simultaneously with the TBN analysis.

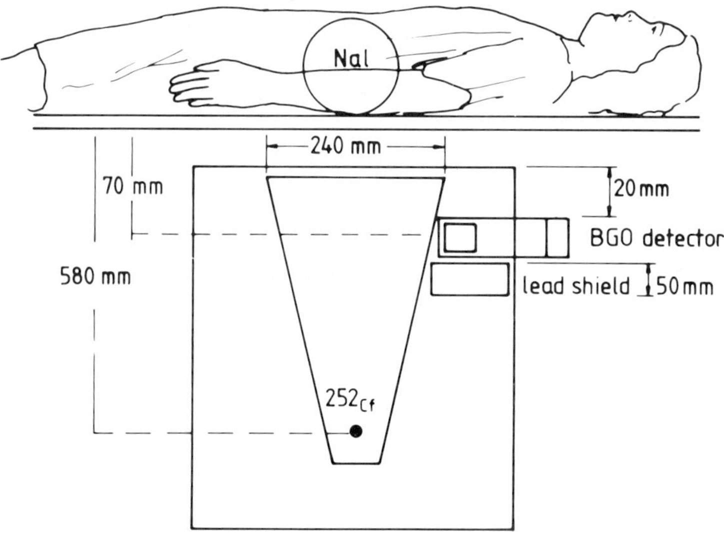

Fig. 2.   Position of the BGO detector relative to the subject and existing NaI detectors.

METHOD

One 50 mm diameter by 50 mm cylindrical BGO detector was incorporated into the ANSTO TBN facility. The detector was placed approximately 100 mm below the scanning bed, centrally along the scanning axis of the subject; the detector was imbedded in borated paraffin and shielded from the neutron source gamma radiation by 50 mm of lead (Fig. 2).

The detector was connected to a Canberra, Series 40 multichannel analyser through a preamplifier and a Canberra 2020 shaping amplifier. The spectral data were collected using an PC/AT compatible computer.

A number of experiments were performed to determine the following data:

a)   the system calibration for the Cl net yields and the Cl/H ratio;
b)   the independence of Cl/H ratio on body habitus; and
c)   the possible spectrum interference from neutron inelastic scattering
     in oxygen.

The system was calibrated using a series of sodium chloride solutions with varying quantities of nitrogen as urea. The nitrogen solution concentrations were multiples of the tissue equivalent of nitrogen (2.6 % wt/v). Our purpose was to measure possible interference from high-energy (10.8 Mev) nitrogen prompt neutron capture gamma-peaks. Solutions were placed in identical five litre containers and positioned in a constant geometry in the neutron beam. The container size represented the smallest subject (small child) geometry the TBN unit was likely to measure. The spectrum was divided into two major regions of interest (ROI), representing the chlorine region (5 to 9 MeV), and nitrogen region (9 to 11 MeV). Another ROI was the well defined peak of the hydrogen (2.2 MeV) gamma ray.

Oxygen Inelastic Scattering Experiment

Oxygen inelastic scattering gamma rays are possibly the only interference expected in the chlorine analysis window (Ryde et al., 1986). However, the use of ^{252}Cf neutron source should minimise this reaction because the average neutron energy of 2.2 MeV is well below the 6.1 MeV threshold. The oxygen effect was investigated by measuring the gamma spectrum produced with varying volumes of water in a fixed geometry. The data also serve as a correction for the background due to the subject's thickness.

Correction for Subject's Thickness

The effect of the subject's thickness on the Cl net yield was measured using the tissue equivalent of chlorine and nitrogen in varying volumes in

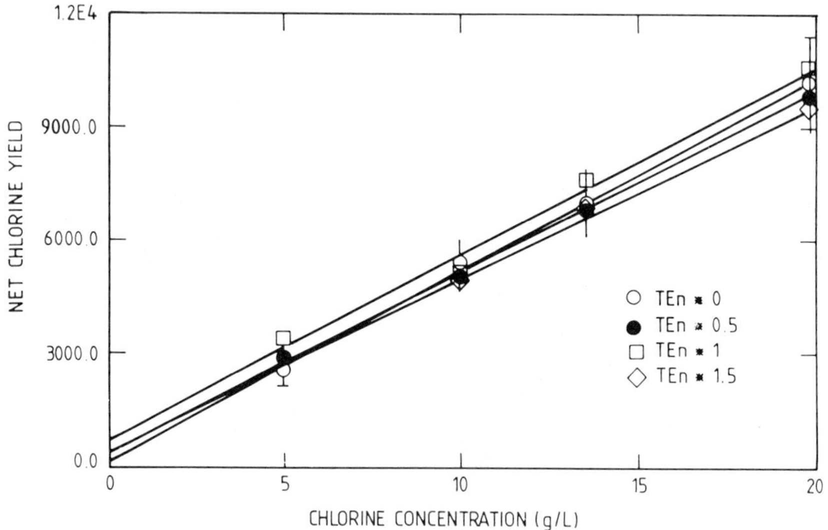

Fig. 3.   Chlorine net yield plotted against chlorine concentration (g/L) for the nitrogen tissue equivalent (TEN) solutions (range 0.5 to 1.5 times the TEN). Standard error of estimate is 2.7 % for the tissue equivalent solution.

fixed lateral geometry. The solution was placed in a 30 by 30 cm perspex
box and measured after each addition of 2 litres of solution. Chlorine and
hydrogen ratios were calculated from net chlorine yields and hydrogen areas
and plotted as a function of the subject's thickness in cm. Two factors are
significant here; the different attenuation coefficients for the ~6 MeV Cl
and 2.2 MeV hydrogen gamma rays, and the thermal neutron fluence. Both
these factors vary with thickness of the phantom, the latter reducing the
effect of the former.

RESULTS

Calibration

The system calibration for Cl net yield and Cl/H ratio as a function
of chlorine concentration exhibited a linear relationship, with the
correlation coefficients of $r = 0.998$ and $r = 0.999$, respectively (Figs 3
and 4). The standard error of estimate for each line (2.7 % and 1.2 %
respectively) showed that the Cl/H ratio is the better parameter to use
in the TBCl analysis. The relationship is independent of nitrogen
concentration and, therefore, we conclude that the contribution of the
nitrogen high-energy spectrum is not significant. Slight variations in the
slope and intercepts in the yield calibration are within the error of
analysis.

Oxygen Interference

One of primary requirements in any analytical procedure is to
establish constant, or at least, predictable and reproducible background
data in order to calculate the contribution to the sample measurement.
Although the elements in the BGO detector are of a low neutron capture
cross section, there is a possibility that inelastic neutron scattering in
borated paraffin neutron shields, subject tissue, and the detector itself
might produce a sufficiently large background to interfere with the

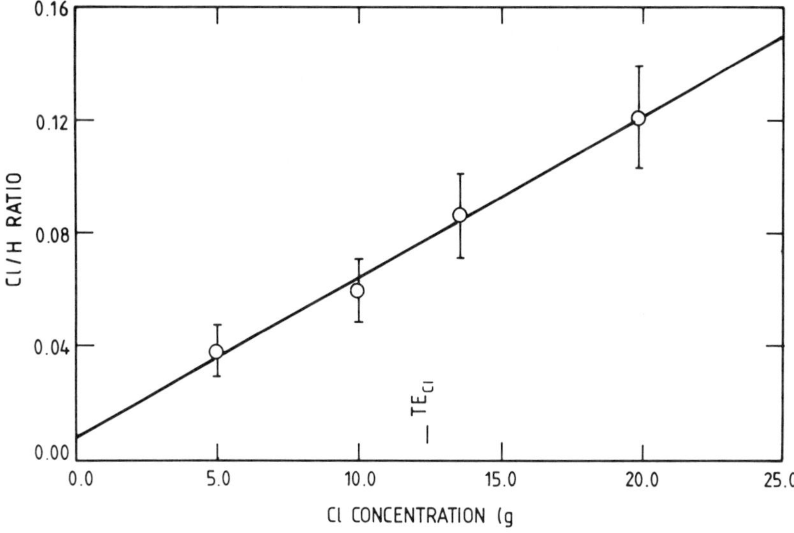

Fig. 4.   Chlorine to hydrogen habitus independent ratio vs chlorine
concentration. Standard error of estimate = 1.2 %; r = 0.999 for
the tissue equivalent solution.

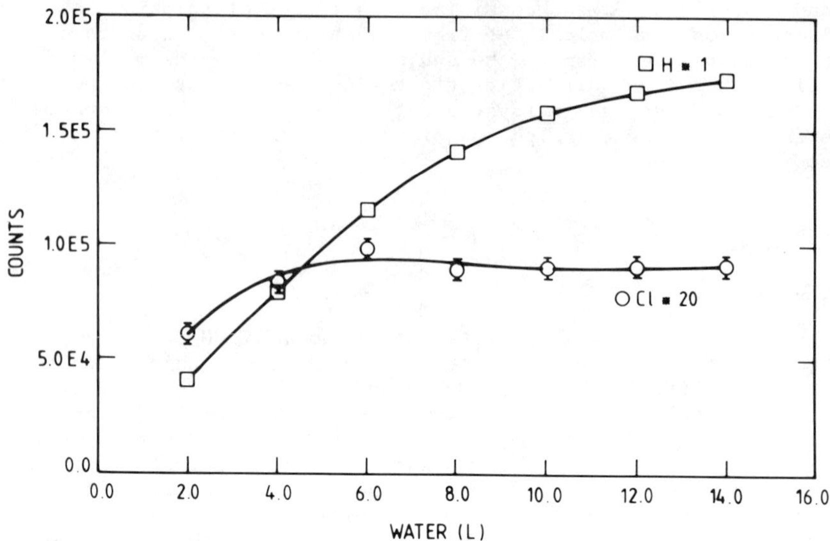

Fig. 5.   Cl background as a function of phantom thickness. The background is essentially constant for subject thickness above 5 cm, thus we conclude that the oxygen neutron scattering reaction does not make a significant contribution to the chlorine region of interest. The hydrogen peak reflects the build up in yield for large volumes of water.

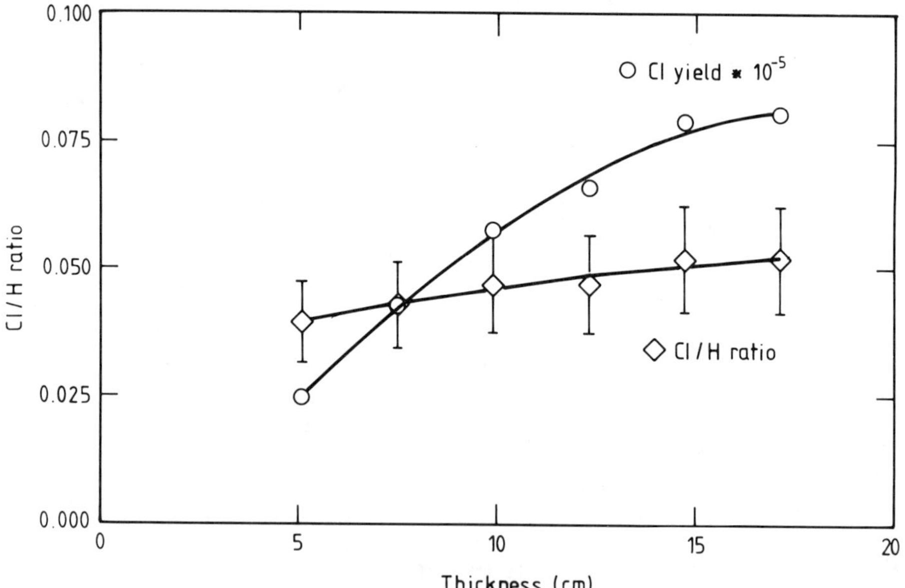

Fig. 6.   Chlorine to hydrogen ratio correction as a function of phantom thickness is less then 10 % for thickness above 10 cm.

analysis. Fig. 5 shows that the water background in the chlorine ROI as a
function of sample volume is constant within error above 4-5 l or 5 cm
thickness and appears to be independent of the quantity of oxygen present
in the neutron beam. The background component of oxygen-neutron inelastic
scattering is thus not of significance in the TBCl analysis for subject
thickness above 5 cm.

<u>Subject Thickness</u>

Subject geometry, such as thickness or the anterior-posterior diameter
(APD), is an important parameter in the TBCl analysis. Due to a variety of
sizes and shapes of subjects, certain corrections must be made for each
geometry. However, irrespective of the geometry variations, the primary
requirement for the TBCl analysis is that the Cl/H ratio should show a slow
variation with thickness. Our experience with children (mean APD = 15.4
cm, SD = 1.2, min = 12.0, n = 49 for age <= 10 years) is that this ratio
should be constant for the APD values of 10 cm and above. Fig. 6 shows that
the Cl/H ratio change for the APD range of normal children is less then 10
% and is well within the error of measurement with the small BGO detector.
This small effect provides an empirical correction for patient data.

CONCLUSION

A BGO detector added to the conventional TBN unit can extend the
useful analytical range of the system. To determine ECW, bromide or another
type of chloride analogue is normally required. This requirement leads to
protracted analytical procedures to determine the tracer concentration in
serum samples and then to calculate the chloride space from the dilution
factor. The proposed system directly measures TBCl with high accuracy,
without making assumptions regarding metabolism and distribution of
tracers. Yasumura et al. (1983) have shown a high degree of correlation
between TBCl determined by ECW and tracer derived values. Our technique
quantifies TBCl at the same time as the subject is scanned for total body
protein. The technique relies on the neutron insensitivity of BGO
detectors to minimise the background in the chlorine region of interest,
thus gaining maximum sensitivity.

The number of body composition parameters which can be determined in a
conventional TBN system are limited to TBN, TBH, and perhaps TBC depending
on the type of neutron source(s) used. Our system extends this by adding
TBCl to the list. Current prices and the relatively low resolution of BGO
detectors limit their use to an add-on or a secondary system. However, in
the future it may be economical to construct the entire system using the
BGO scintillator, and eliminate the NaI detectors.

REFERENCES

Allen, B. J., Baily, G.M., and McGregor, B.J., 1985, Dose equivalent
    distribution in the AAEC total body nitrogen facility. Proc. Fourth
    Australian Conference on Nuclear Techniques of Analysis, Sydney,
    AINSE.
Allen, B. J., Blagojevic, N., McGregor, Parsons, D. E., Gaskin, K.,
    Soutter, V., Waters, D., Allman, M., Stewart, P., and Tiller, D.,
    1987, In-vivo determination of protein in malnourished patients, <u>in</u>:
    "In vivo Body Composition Studies", K.J. Ellis, S. Yasumura, and
    W.D. Morgan, eds., Institute of Physical Sciences in Medicine,
    England.
Beddoe, A. H., Streat, S. J., and Hill, G. L., 1987, Measurement of total
    body chlorine by prompt gamma in vivo neutron activation analysis,
    <u>Phys Med Biol</u>, 32:191.

Bird, J. R., Allen, B. J., Bergqvist, I., and Biggerstaff, J. A., 1973,
    Compilation of KeV-neutron-capture gamma-ray spectra, <u>Nucl Data Tab</u>,
    11:433.
Ryde, S. S., Morgan, W. D., Evans, C. J., Dutton, J., Sivyer, A., McNeil,
    E., and Sandhu, S., 1987, A new multi-element system using Cf-252
    calibration and performance, <u>in</u>: "In vivo Body Composition Studies",
    K.J. Ellis, S. Yasumura, and W.D. Morgan, eds., Institute of Physical
    Sciences in Medicine, England.
Yasumura, S., Cohn, S. H., and Ellis, K. J., 1983, Measurement of
    extracellular space by total body neutron activation, <u>Am J Phys</u>,
    13:R36.

THE IN VIVO PRECISION OF BROADBAND ULTRASONIC ATTENUATION

W.D. Evans, E.A. Jones, and G.M. Owen

Department of Medical Physics and Bioengineering
University Hospital of Wales
Cardiff CF4 4XW, Wales.

INTRODUCTION

Broadband ultrasonic attenuation (BUA) is a relatively new technique for
the measurement of bone mineral content (BMC) in the os calcis (Langton et
al., 1984). This is a site of mainly trabecular bone and it has been
suggested that BUA values reflect bone structure as well as mineral density
(Palmer and Langton, 1987). Recent clinical studies (Baran et al., 1988)
indicated the high diagnostic sensitivity of the technique in both spine and
hip fracture patients. It has the additional advantages of being cheaper
than other methods of BMC measurement and not using ionising radiation.

However, the precision (reproducibility) of BUA is relatively poor
compared with photon absorptiometry; Poll et al. (1986) reported a
coefficient of variation (COV) of 3.9% for repeated measurements on normal
volunteers over a period of several weeks, whereas Evans et al. (1988) found
a value of 4.6% under similar circumstances. In this study, the possible
sources of imprecision were investigated in a systematic manner.

SUBJECTS AND METHODS

Investigations were performed on a group of 10 normal volunteers (4
females) with a mean age of 35 years (range 24-68 years).

The experiments were performed using a Walker Sonix UBA1001 bone mineral
analyser. In this apparatus, the heel is positioned centrally in a water
tank between two ultrasound transducers, one of which acts as a transmitter
and the other as a receiver. Measurement of attenuation is repeated
automatically until three consecutive consistent readings are achieved; the
BUA is taken as the mean of these three readings. Before immersion, the
dimensions of the foot are measured with anthropometric calipers so that the
appropriate number of plastic spacers (thickness 1.5 mm) may be placed below
and behind the foot to ensure that the axis of the transducers passes through
the centre of the os calcis.

Precision was estimated by measuring each member of the group of 10
volunteers twice within a period of two weeks. In addition, one male
volunteer (age 38 years) was measured 10 times over an 8-hour period
(short-term) and a 6-week period (long-term). In each case, both feet were

measured and the mean BUA calculated. The investigations were performed
using ordinary water drawn from the hospital supply, and repeated using
hospital water from which air had been removed by prior boiling in the
laboratory. For the one male volunteer, the COV was calculated by expressing
the standard deviation of 10 measurements as a percentage of the mean BUA
value. For the group studies, on the other hand, the average COV was
estimated from the paired measurements in each volunteer as described by
Hosie and Smith (1986).

In the male volunteer, the following effects were examined using boiled
water: the effect of immersion time by repeated measurements over a 45-minute
period; the effect of temperature variation by adding cold water, keeping the
water level constant and correcting for immersion time; the effect of water
level by removing water, keeping the temperature constant and correcting for
immersion time; the effect of lateral position of the foot by moving it from
one side of the tank to the other and measuring the distance from one
transducer by a pulse-echo technique (repeated 4 times), and finally the
effect of moving the foot by systematically changing the number of plastic
inserts below and behind the foot.

RESULTS

With ordinary water, a COV of 4.6% was obtained for the 10 volunteers;
for the one male volunteer the short-term precision was 1.2% and the
long-term value 3.2%. The use of boiled water did not have a consistent
effect on precision but reduced measurement time - the median number of
readings required to achieve consistency was reduced from 4.4 to 3.7.

For the right foot, BUA increased linearly at a rate of 0.03%/min. with
immersion time, but decreased at a rate of 0.04%/min. for the left foot.
However, this effect makes a small contribution to precision with a maximum
variation of only 0.2% for 5 minutes of immersion. BUA remained fairly
constant as the water temperature increased from about 29 to 35 C but
decreased above this value. It is estimated that this may account for a
variation of about 2% if the temperature is maintained in the range 30 to 35
C. On the other hand, changing the water level by several centimetres did
not significantly affect BUA values: in fact, no changes were observed until
the transducers were exposed. As the foot was moved towards the side of the
tank, BUA was observed to decrease, a movement of 1 cm contributing a maximum
variation of about 2%. A relatively large effect was noted by changing the
number of plastic inserts; a movement of 1.5 mm in both directions may cause
a variation of up to 11%, although values were constant to within about 5%
for a small area in the centre of the os calcis.

DISCUSSION AND CONCLUSIONS

In a group of normal volunteers, the precision of BUA is about 4% but it
may be better in some individuals. These results are similar to those which
have been previously reported. Although the use of boiled water does not
improve precision, it is recommended since measurement times are reduced.
The effects of immersion time, water temperature, water level and lateral
position are relatively small and may be minimised by careful positioning of
the foot and control of the measurement conditions. However, it is important
to use the same number of plastic inserts for repeated measurements on a
given individual (provided that there are no significant changes in foot size
and shape). In addition, efforts should be made to prevent forward movement
of the foot in the tank. Other factors which may affect precision (such as
orientation of the foot) remain to be investigated.

ACKNOWLEDGEMENT

The Walker Sonix UBA1001 bone mineral analyser was purchased through the generosity of the Women's Royal Voluntary Service.

REFERENCES

Baran, D.T., Kelly, A.M., Karellas, A., Gionet, M., Price, M., Leahey, D., Stauterman, S., McSherry, B., and Roche, J., 1988, Ultrasound attenuation of the os calcis in women with osteoporosis and hip fractures, Calcif. Tissue Int., 43:138.
Evans, W.D., Crawley, E.O., Compston, J.E., Evans, C., and Owen, G.M., 1988, Broadband ultrasonic attenuation and bone mineral density, Clin. Phys. Physiol. Meas., 9:163.
Hosie, C.J., and Smith, D.A.S., 1986, Precision of measurement of bone density with a special purpose computed tomography scanner, Br. J. Radiol., 59:345.
Langton, C.M., Palmer, S.B. and Porter, R.W., 1984, The measurement of ultrasonic attenuation in cancellous bone, Eng. Med., 13:89.
Palmer, S.B., and Langton, C.M., eds., 1987, "Ultrasonic Studies of Bone", IOP Publishing, Bristol.
Poll, V., Cooper, C. and Cawley, M.I.D., 1986, BUA in the os calcis and SPA in the distal forearm: a comparative study, Clin. Phys. Physiol. Meas., 7:375.

FEASIBILITY STUDIES IN THE IN VIVO MEASUREMENT OF IRON IN

SYNOVIAL MEMBRANE

F.E. McNeill[1], D.M. Franklin[1], D.R. Chettle[1],
R.E. Ellis[2], S.P. Pittard[2], M.C. Scott[1] and
W. Vennart[2]

[1]Medical Physics Group, School of Physics and
Space Research, University of Birmingam, P.O.
Box 363, Birmingham B15 2TT, England

[2]Department of Physics, University of Exeter,
Stocker Road, Exeter EX4 4QL, England

INTRODUCTION

Rheumathoid arthritis is a widespread and currently incurable
disease. Its most obvious symptom is inflammation of the synovial joints;
it also results in a reduced life expectancy. Certain principles of
structure are common to all synovial joints (Fig. 1). They are formed by
two bones in apposition, the ends being capped with a smooth surface of
hyaline cartilage. The bones are held together by a sleeve, formed from
an outer fibrous membrane and an inner synovial membrane; the space
created between the bones is filled with synovial fluid, which acts as a
lubricant. It is known that subjects with rheumatoid arthritis have
increased levels of iron in the synovial membrane (Muirden, 1966) and mean
levels rise from 15 g/g to 350 g/g dry tissue weight (Senator and
Muirden, 1968). In addition the mass of the synovium increases
dramatically, the membrane thickening from a few cells up to 1-10 mm.

Some of the iron found in the membrane is free, but most of it is
contained in the iron storage molecule ferritin (Muirden, 1966). Ferritin
consists of an almost spherical protein shell, called apoferritin, of
approximate molecular weight 460000, surrounding a ferrihydrite core. The
molecule can store up to 4500 iron atoms, but normally contains about
2000. It has been suggested that these increased levels are not just an
index of the disease, but could be in part causative (Blake and Bacon,
1981). Free iron and fully loaded ferritin can catalyse oxidative free
radical reactions, which degrade the lubricating properties of synovial
fluid, and ultimately erode the cartilage and bone (Blake et al, 1981).
Measurement of iron levels can be made on synovial biopsies, but the use
of this invasive technique makes it difficult to assess the effect of iron
chelation therapy or anti-inflammatory drugs on the course of the disease.

Investigations of possible in vivo techniques for measuring iron in
synovial membranes have been made. Brief studies have been undertaken
using magnetic susceptometry (SQUID), x-ray fluorescence, nuclear
resonance fluorescence, and neutron inelastic scattering. The two which

offered the greatest potential, nuclear magnetic resonance and neutron activation, are reported here.

METHODS

Nuclear Magnetic Resonance - Relaxation Times

Measurements were made of both spin-lattice (T_1) and spin-spin (T_2) gross proton relaxation times in dilute protein solutions. When a system of protons is immersed in a static magnetic field, B_0, there is a net magnetisation, M_0, in the direction of the applied field (the z-axis). The application of a 90° pulse of r.f. radiation at the resonance frequency tips the magnetisation into the x-y plane. The return of M_z and M_{xy} to their equilibrium values are characterized by the time constants T_1 and T_2, respectively. In aqueous solutions the dominant form of relaxation is generally due to the interactions between proton spin (dipole-dipole). However, paramagnetic molecules greatly accelerate relaxation because their magnetic moments and associated fields are much larger. In this context it should be noted that free iron is paramagnetic, and that Mossbauer studies have shown that the mineral core of ferritin is also paramagnetic.

In this study three types of ferritin were used. Two were commercially available horse spleen ferritin from Sigma and Boehringer, each thought to have approximately 2000 iron atoms per protein molecule. The third was Boehringer apoferritin reloaded to have 700 iron atoms per protein molecule. It had been estimated that a rheumatoid synovium would contain about 0.15 g/l ferritin, so solutions of each of the three types of ferritin were made in a range of protein concentrations from 0.03 g/l to 1.0 g/l. A reference sample of the solvent used, HEPES pH 7 buffer, was also included. The solutions were all made under a nitrogen atmosphere to prevent oxidation of the iron.

The measurements were performed on the 20MHz NMR machine at the University of Exeter. The samples consisted of 0.5 ml of each solution sealed in glass test tubes. They were measured at body temperature, 37° +/- 0.5°C, relaxation times being highly dependant on temperature. T_1 was measured by the pulse sequence known as inversion recovery, whilst T_2 was measured using the Carr-Purcell sequence.

Neutron Activation

Samples were irradiated on the University of Birmingham Dynamitron with 14 MeV neutrons and iron detected by the $^{56}Fe(n,p)^{56}Mn$ reaction, which has a 113 mb cross-section at this energy. ^{56}Mn has a half-life of 2.58 hours and emits 847 keV γ-rays in 99% of decays. Low levels of iron were expected, requiring a high gamma detection efficiency. Samples were therefore counted at the low background Whole Body Monitor at the Queen

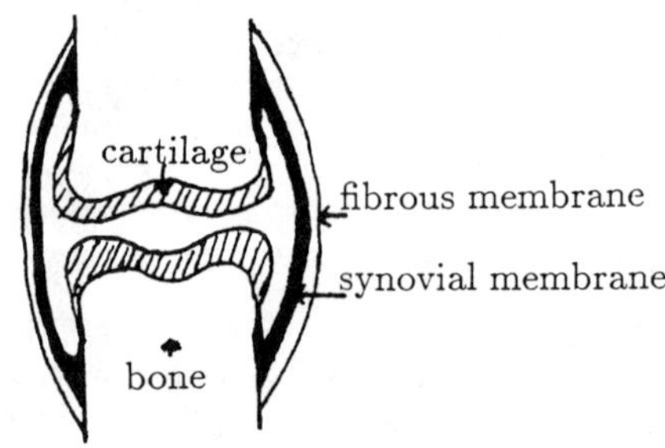

Fig. 1 Features common to synovial joints.

Elizabeth Hospital, the eight 5" diameter x 4" depth NaI (Tl) detectors
being re-arranged in two rings of four around the samples to maximise the
counting efficiency.

Phantoms were constructed to simulate the size, mass and elemental
composition of a human knee, with additional levels of iron. The samples
were made in two sets. One set had a mock 'bone' inserted; the other did
not. The 'bone' was calcium orthophosphate, the iron content of which was
unknown and was therefore determined by comparing the two sets of
phantoms. Iron was added to the phantoms in the form of FeCl, which was
acidified with HCl to prevent precipitation of the iron from the solution.
NaCl was also added to the phantoms in quantities expected from the ICRP
Reference Man levels in the appropriate tissues. Since calcium, potassium
and phosphorous were present in the bone, elements which occur in tissue
and which could produce γ-ray's contributing to the background under the
iron peak were accounted for.

The γ-ray spectra for each sample from each of the NaI crystals were
analysed by a standard fitting procedure. The resulting iron peak areas
were corrected for variations in dose and timing.

The major limiting factor on an in vivo system is the dose which can
be administered. It is assumed that there is no radiosensitive tissue in
an adult joint and that the mass of a knee joint is 1.5 kg and that of a
subject is 70 kg (ICRP, 1974). To keep the irradiation levels to within
WHO category II, the maximum permissible local dose is thus 0.2 Sv, whole
body dose equivalent 5 mSv.

In addition to the two phantoms described above, two knee joints were
obtained at necropsy. Both were removed because of circulatory failure;
neither had been diagnosed as rheumatoid arthritic, but one had had severe
joint dysfunction for many years. The phantoms and knees were measured at
a high dose, 1.4 Sv, to assess the iron contents of the knees and the
'bone' components of the phantoms. Measurements were also made at 0.2 Sv
to determine the practical system sensitivity.

RESULTS AND DISCUSSION

<u>Nuclear Magnetic Resonance</u>

The variations of T_1 and T_2 with protein concentration are shown in

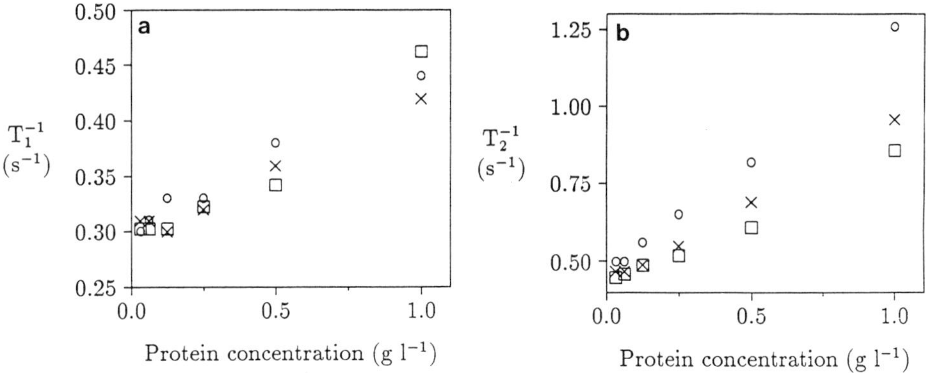

Fig. 2   Inverse relationships of a) spin-lattice relaxation time (T_1) and
(b) spin-spin relaxation time (T_2) with protein concentration   o -
Boehringer; x - Sigma; □ - reloaded ferritin.

Figs. 2a,b for the three different types of ferritin sample. Fig. 2a
shows an inverse relationship that is common to all three ferritins,
implying that the spin-lattice relaxation depends predominantly on the
protein concentration. In contrast, the inverse relationship between
spin-spin relaxation time and protein concentration is distinct for each
of the ferritin samples. The reloaded ferritin produces the lowest slope
and the Sigma ferritin is thought to be intermediate because of the
effects of possible contamination. The Boehringer material is guaranteed
cadmium free, but traces of this metal in Sigma ferritin are known to
inhibit iron uptake. The gradient of $1/T_2$ versus concentration therefore
appears to depend on the level of iron loading.

The results above are not these of a paramagnetic material, because
they are not of sufficient magnitude. At present time they can offer no
suitable explanation for observed relationship.

Neutron Activation

The results from the high dose measurements showed that the γ-ray
spectra from the 'with bone' phantom spectra and the knees were very
similar, although the knees did contain slightly higher amounts of Cl, Ca
and K (Fig. 3). Calibration lines for a single detector are shown in Fig.
4. The iron content results are listed in Table 1. The level of iron in
knee(2) is not unexpected as it compares with a normal level of 80-100 mg
in normal muscle and bone, but the absence of iron from knee (1) is rather
surprising. However, this subject (1) had suffered the joint dysfunction,
so there had almost certainly been bone erosure and muscle wastage.

The calibration lines were used to determine a Minimum Detection
Limit (MDL). This was defined as being twice the error of the peak given
a concentration of zero. The Minimum Detectable Concentration (MCD) was
calculated as being the MDL divided by the calibration line. The MDC was
calculated for each detector and these were combined to give the MDC of
the system as a whole.

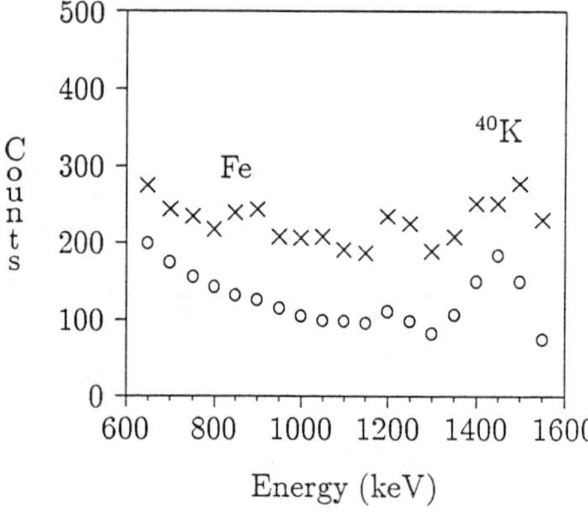

Fig 3.   Single detector γ-ray
spectra from phantom with
'bone' (given +ve offset
of 100 counts per channel)
and necropsy knee.

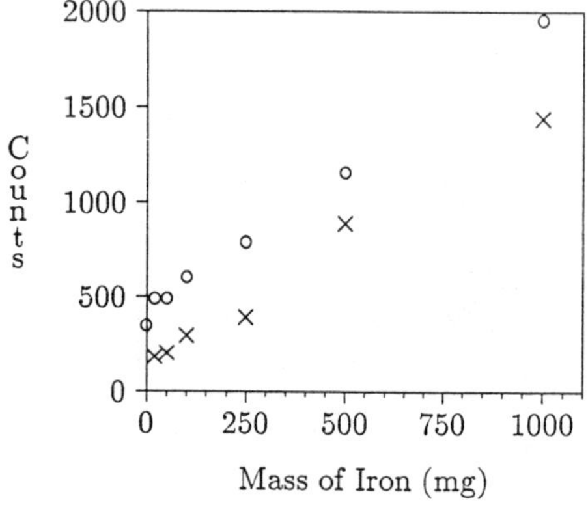

Fig. 4   Single detector calibration
lines from phantoms without
(x) and with (o) mock bone,
dose 1.4 Sv.

Table 1. Iron Content of Samples

Sample	Mass of iron (mg)	Error (mg)
mock bone	126.8	9.5
knee (1)	1.6	7.1
knee (2)	86.8	7.6

Based on the phantom measurements, the MDC for a dose of 1.4 Sv was
12.9 mg. When errors from actual leg measurements were used, it was
calculated to be 14.7 mg (leg 1) or 15.3 mg (leg 2).

Similar calibration lines and knee results were obtained for the
lower dose (0.2Sv) irradiations. The MDC's were determined as above. For
phantoms containing mock bone the result was 59.5 mg; for leg (1) 58.0 mg,
and for leg (2) 49.5 mg.

CONCLUSION

The nuclear magnetic resonance measurements show that iron in the
form of ferritin in solution does affect the relaxation times. However,
more measurements would be needed to quantify the results. Whether this
technique could ultimately lead to in vivo measurements is questionable,
but it could provide valuable research information on the composition of
synovial fluid in varying disease states.

The work on neutron activation has resulted in a dramatic improvement
in detection limit compared to previous systems designed to measure liver
iron (Fletcher, 1988). The gain of a factor of twenty is partly
attributable to a higher permissible local dose, but principally arises
from the use of a monoenergetic fast neutron beam in a low scatter
environment. The latter factor also greatly reduces any interference from
manganese in tissue. However, it is still uncertain whether this
measurement will be clinically useful in the study of rheumatoid disease
in joint. This is because although the total iron present in a knee joint
is likely to be measurable, its distribution amongst synovium, muscle and
bone cannot be distinguished.

The development work on neutron activation has, at the very least,
pointed the way to new, more sensitive measurements of iron stores.

ACKNOWLEDGEMENTS

This study has been possible because of grants from the Medical
Research Council (who supported F.E. McN.) and The Oliver Bird Trust (who
supported D.M.F.); their grants are acknowledged with gratitude. The
in-vitro ferritin studies were performed in collaboration with A. Treffrey
and P. Harrison, of the University of Sheffield. The post-mortem knee
studies were performed in collaboration with M. Simms, of Selly Oak
Hospital, Birmingham. This collaboration is also gratefully acknowledged.
Discussion and collaboration with D.R. Blake of the London Hospital and
his colleagues have been vital to the development of this project, and
their support and enthusiasm has been much appreciated.

REFERENCES

Blake, D.R. and Bacon, P.A., 1981, Short reports: Synovial fluid ferritin
in rheumatoid arthritis: an index or cause of inflammation? Brit.
Med. J., 282:189.
Blake, D.R., Hall, N.D., Bacon, P.A., Dieppe, P.A., Halliwell, B., and
Gutteridge, J.M.C., 1981, The importance of iron in rheumatoid
disease, Lancet, (ii) 1142.
Fletcher, J.G., 1988, Ph.D. thesis, Department of Physics, University of
Birmingham.
Muirden, K.D., 1966, Ferritin in synovial cells in patients with
rheumatoid arthritis, Ann. Rheum. Dis., 27:38.
Report on the Task group on reference man, 1974, ICRP Publication 23.
Pergamon Press N.Y.
Senator, G.B., and Muirden, K.D., 1968, Concentration of iron in synovial
membrane, synovial fluid, and serum in rheumatoid arthritis and
other joint diseases, Ann. Rheum. Dis., 27:49.

IN VIVO NMR SPECTROSCOPY

Peter S. Allen

Department of Applied Sciences in Medicine

University of Alberta, Edmonton, Alberta   T6G 2G3

INTRODUCTION

In vivo nuclear magnetic resonance (NMR) spectroscopy has an assured
role to play in medical research because it enables one to interrogate
non-invasively a range of biochemical processes, both in the resting state
and then serially when they are changing in response to some induced
stress. The stress may be physical or chemical, and the careful observation by NMR of the response which stress evokes, coupled with the recovery
from it, provides a most fruitful source of information on the biochemical
mechanisms involved. In diagnostic medicine, on the other hand, two roles
can be envisaged for the routine use of NMR spectroscopy. The first is as
a monitor of some particular marker, the status of which indicates the
presence or absence of some particular pathology. For example, the substantially depressed levels of N-acetylaspartate in several types of brain
tumours in humans are in marked contrast to its level in normal brain
tissue (Luyton et al., 1988; Bruhn et al., 1988). When such markers can be
found, and much clinical research effort is currently being devoted to
their discovery, non-invasive diagnosis by means of NMR spectroscopy
becomes much more straightforward. The second role is as an adjunct to
therapy, initially evaluating the biochemical status of the disease to
enable the appropriate therapy to be administered, and subsequently
monitoring the biochemical changes that reflect the efficacy or otherwise
of that therapy, prior to changes in the clinical symptoms. Cancer
treatment by means of chemotherapy seems to offer a promising opportunity
for this latter role.

In order to carry out biochemical analysis by means of NMR spectroscopy, one requires a nuclear probe with a magnetic moment. A number of
the stable nuclear species having intrinsic magnetic moments do occur
naturally in the body. Moreover, several of these species are present in
sufficient concentrations to make their application to in vivo NMR spectroscopy a viable proposition. These include the nuclei of the hydrogen
atom, ^1H; the carbon atom, ^{13}C; the sodium atom, ^{23}Na; and the phosphorus
atom, ^{31}P. Because of the different strengths of their magnetic moments
(bigger is better) and because their concentrations in tissue are also
different (again, bigger is better), the use of some nuclei as biochemical
probes is much more demanding experimentally than that of others. For
example, when the most concentrated biologically occurring molecular
species, namely water, is observed by means of the largest biologically
occurring magnetic moment, that of the proton ^1H, it gives rise to the most
intense biological NMR signal. This signal can then be exploited to

produce magnetic resonance images that provide diagnostic quality structural information with an in-plane spatial resolution of $\leq$1mm, or in other words provide an excellent NMR signal-to-noise ratio from a tissue volume on the order of a microlitre. When one's goal is to obtain biochemical information rather than the structural information provided by water proton images, the concentrations of the metabolite species that one wishes to observe are at least a factor of 10^4 less than that of water. To compensate for this loss in concentration, one must therefore increase one's signal acquisition volume to several millilitres, or greater, in order to derive signal from a similar number of nuclei. The volumes (several millilitres) quoted here correspond to NMR experiments carried out in a magnetic field strength of 1.5T. Higher magnetic field strengths give rise to a greater sensitivity, and hence, smaller acquisition volumes.

As the technique of in vivo NMR spectroscopy has evolved, the exploitation of the different observable nuclei seems to have developed in such a way that specific roles are assigned to each nuclear species. For example, because of the crucial function that the phosphate group fulfills in energy metabolism, spectroscopy with ^{31}P has evolved as the principal tool to use when monitoring the energy status. Similarly, ^{23}Na is used to monitor the functioning of various membrane pumps that control sodium concentration. Hydrogen and carbon are ubiquitous in organic molecules, and, in principle, their nuclei offer a much greater potential for biochemical problem solving than either ^{31}P or ^{23}Na. However, each of them has its own peculiar difficulties, to be outlined below, which prevent it from becoming the diagnostic panacea one might hope for.

In the following sections, I shall first explore the constraints which distinguish in vivo NMR spectroscopy from conventional high resolution NMR spectroscopy and make it, as a result, a more limited analytical tool. Secondly, I shall describe some of the successful spectroscopic applications of the various nuclei.

SOME CONSTRAINTS OF IN VIVO NMR SPECTROSCOPIC ANALYSIS

The principal objective of in vivo NMR spectroscopy is to carry out the chemical analysis of the biochemicals present in a preselected region of tissue. This is similar to the role that high resolution NMR plays in analytical chemistry. However, when applied in vivo, NMR suffers from several constraints that limit its power, relative to that enjoyed when it is practiced in vitro. For example, conventional in vitro high resolution NMR can take advantage of the high sensitivities and spectral resolutions obtained by working at magnetic field strengths up to 14.1T. Because of the wide magnet bore diameter required for in vivo work, field strengths are much lower, typically 1.5T for clinical NMR units and up to 4.7T for animal NMR units. Moreover, signal averaging to enhance the effective sensitivity is much more limited in vivo, where patients can only tolerate about one hour of data acquisition compared with many hours or even days that are acceptable for inanimate solutions. Correspondingly, concentrations must be that much greater for in vivo observation. An acceptable rule of thumb is that concentrations must be on the order of several millimolar for in vivo observation.

When applied in vivo, it is necessary to incorporate into the NMR excitation pulse sequence some means of defining the boundaries of the tissue region from which the spectrum will be acquired (several millilitres in volume or greater) and also of excluding signal from outside these boundaries. No such encumbrance is encountered with in vitro chemical analysis, where the measurements are averages over the whole sample. The "localization" of the spectral acquisition region requires that some parameter involved in the NMR experiment be made spatially dependent. This is done by incorporating well-controlled spatial variations in either or both of the static magnetic field, B_o, or the radio frequency magnetic field, B_1. These controlled spatial variations are called gradients. The use of either B_0 or B_1 gradients has its corresponding advantages and disadvantages. For example, B_0 gradients enable well-controlled shapes such as cuboids to be defined with ease and lead to the application of single voxel (Ordidge et al., 1986) or multiple (Meyer et al., 1988) voxel signal acquisition routines. However, when used in conjunction with selective excitation (worst for single voxel acquisition), B_0 gradients give rise to registration errors in the excited cube location of different spectral lines. These registration errors can be thought of as fuzzy boundaries which, at 1.5T and employing a gradient strength of $5mTm^{-1}$ ($0.5Gcm^{-1}$), are typically 0.75cm thick for the ^{31}P spectrum. This boundary thickness is to be compared with typical cube sides of 3cm used in human brain spectroscopy acquired with large circumscribing radio frequency (RF) coils. The use of B_1 gradients, on the other hand, does not lead to such fuzzy boundaries. However, B_1 gradients are normally applied with RF surface coils and are only appropriate at limited depths. When employed as the sole means of localization (Bendall and Gordon, 1983), B_1 gradient methods can also lead to inappropriate shapes of the region of acquisition, for example, curved regions that resemble the shape of a concavo-convex lens. The use of B_1 gradients has, nevertheless, been used with success in the case of an exteriorized kidney (Stewart et al., 1988), where it enabled ^{31}P spectral bands, approximately 2.5mm wide, to be stepped through the kidney and reveal the variations in the concentrations of certain phosphorus metabolites; and also in the case of rat skeletal muscle (Challis et al., 1988), where it enabled changes in the heterogeneity of the muscle to be observed from contiguous 4mm thick slices.

Although the "localization" technique referred to above needs to be applied for any nucleus observed in vivo, "spectral editing" is a further constraint that is important mainly for protons and ^{13}C. In principle, the proton spectrum should reflect the presence of each of the numerous proton-bearing metabolites that occur in excess of millimolar concentration. In practice, it is not always easy to distinguish these metabolite signals because of the occurrence of very large signals from the water protons and from the lipid protons that result from the high concentration of both water and fat in the body. Although in vitro NMR often encounters solvent suppression problems, it is rarely faced with the accommodation of two inordinately large unwanted signals. Techniques do exist for suppressing the water signal and for extracting metabolite signals hidden under the lipid band, and this is called "spectral editing." However, in editing the proton spectrum one usually has to tailor one's editing strategies to suit the particular metabolite of interest, and this precludes the simultaneous observation of all proton-bearing metabolites. Currently, much emphasis is being placed on the development of editing techniques that will enable lactate levels to be observed (Sotak et al., 1988). The spectral editing encountered with the observation of ^{13}C stems not from difficulties caused by water or fat, but from the very low natural abundance of the ^{13}C isotope (1%) that leads to exceedingly weak signals. As a result, editing techniques that transfer information from the ^{13}C spin system to the more easily observed proton spin system (Rothman et al., 1985), or alternatively remove the splittings of ^{13}C spectral lines caused by neighboring protons, are often employed to enhance the observability of the carbon nucleus.

SOME APPLICATIONS OF IN VIVO NMR SPECTROSCOPY

<u>^{31}P Spectroscopy</u>

^{31}P spectroscopy has been used extensively to evaluate the status of the energy metabolism in tissue. Most applications have been in skeletal muscle (see Fig. 1 and below for a more detailed discussion), where it is relatively straightforward to stress the system and observe the subsequent metabolic response, and where the problems of spatial localization are

reduced. Nonetheless, the more challenging experimental problems of determining both the spatial and temporal variations in energy metabolism within
the muscle of the beating heart are being successfully pursued by
Robitaille et al. (1989), where heart rate and metabolism can be perturbed
by drugs.

Another area of high activity is the application of ^{31}P spectroscopy to
cancer. By applying it to cancer, one might hope that, in principle, the
^{31}P spectrum from a tumour would reflect its degree of hypoxia and so be a
predictor of the efficacy of radiotherapy, while at the same time providing
a measure of its intracellular pH that would be important in assessing the
likely efficacy of certain chemotherapeutic agents such as BCNU. The
fulfillment of this hope is being hindered by the variable heterogeneity of
tumours. The individual necrotic, hypoxic, and well-perfused regions of
tumour tissue each give rise to markedly different ^{31}P spectra, and, in
general, the global average spectrum from a large piece of tumour is very
difficult to disentangle.

Returning to skeletal muscle where the potential of in vivo ^{31}P
spectroscopy has been exploited most, one could catalogue several areas in
which this technique has provided information not readily available by
other means. For example, with regard to our basic understanding of
metabolite reactions, it was demonstrated quite early (Brown et al., 1980;
Rees et al., 1989) that the concentration of free ADP that participates in
the creatine kinase reaction is very much less than the total ADP that is
measured by chemical analysis of muscle extracts. With regard to the
diagnosis of muscle disease, deficiencies in enzymes such as phosphofructokinase (Argov et al., 1987) and myophosphorylase (Ross et al., 1981) have
been diagnosed by ^{31}P spectroscopy because it could detect, respectively,
the blockage of glycolysis and abnormal accumulation of phosphorylated
sugars; and a mild ischemic-exercise-induced alkalosis (instead of
acidosis) coupled with fairly normal changes in metabolite concentrations.
With regard to nutritional therapy, ^{31}P spectroscopy of the working muscle
has been shown to be sensitive to nutritional status (Lunt et al., 1986;
Phang et al., 1987; Pichard et al., 1988).

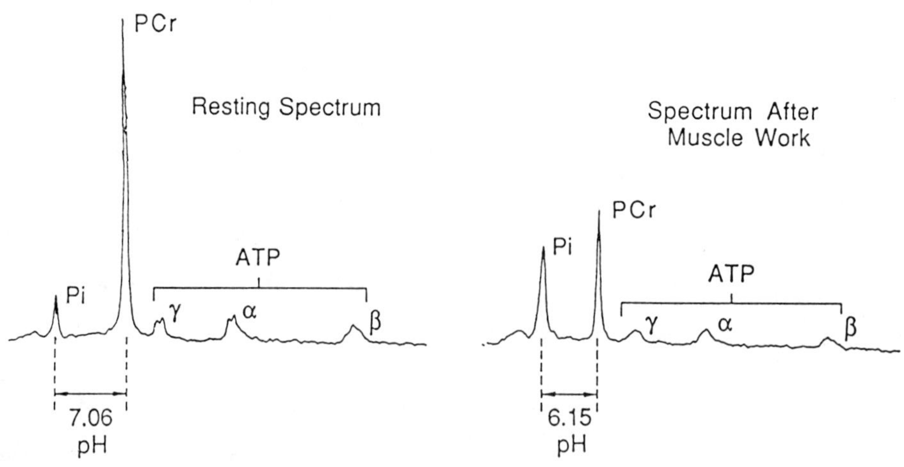

Fig. 1. ^{31}P Spectra of Skeletal Muscle.

[31]P spectroscopy of exercising skeletal muscle has also been used to explore the body's adaptive response to hypoxia (Matheson et al., in press). Our understanding of the body's natural response to certain forms of stress can often be enhanced by studying sub-populations that have adapted to this stress so that treatment can be optimized when these same forms of stress are encountered in disease. For example, chronic hypoxia is an accompanying feature of several disease conditions and, to help understand the body's preferred adaptive response to hypoxia, a group of 6 high altitude-adapted Andean Indians were evaluated during a joint project in several laboratories at the University of British Columbia and the University of Alberta. One component of the evaluation in Alberta was to measure the response of their muscle metabolism to exercise by means of in vivo ^{31}P spectroscopy. These measurements were carried out under two different levels of oxygen supply, both prior to any acclimatization to sea level and following an acclimatization period of 6 weeks. The two levels of oxygen supply corresponded to sea-level air, on the one hand, and air found at their native altitude (14,000 ft.), on the other.

Immediately after arrival from Peru (within 72 hours of leaving 14,000 ft.), ^{31}P spectroscopy showed distinct differences in the fatigue values of pH and Pi/PCr between Andean Indians and lowlanders. The lowlanders became significantly more acidoxic for a similar change in Pi/PCr at fatigue. After six weeks at sea level the Peruvians had acclimatized somewhat to sea level conditions, but were still significantly less acidoxic at fatigue than lowlanders. Additionally, the high altitude-adapted natives did not show a significant difference in metabolism between the conditions of sea level normoxia or hypoxia equivalent to 14,000 ft.

It was clear from these NMR experiments (and several additional experiments carried out during the overall project) that the Andean natives exhibit the lactate paradox (West, 1986). In lowlanders, this character- istic is an acclimatization. It requires about 10-14 days at altitude to be expressed and it is lost along a similar time course on descent. Andean natives, in contrast, were found to express the lactate paradox even 6 weeks after descent. It thus seems to be a fixed metabolic feature which one might term the "perpetual lactate paradox." Since it is an expression of oxidative metabolism which seems fixed and does not acclimate, we also tentatively assume that it is a developmental or genetic adaptation. While the adaptive significance of the lactate paradox at high altitude is well recognized (excessive lactate production and accumulation under these conditions are counter-productive), its mechanism has been the subject of some controversy for over 50 years. West (1986) postulated that high altitude-adapted individuals should display a more labile intracellular muscle pH. He supposed that an inordinately large drop in muscle pH for a given level of work would serve to inhibit glycolysis and so account for the lactate paradox. Our in vivo spectroscopy measurements conflict with that supposition. The skeletal muscle in Quechua Indians displays more stable intracellular pH than in lowlanders, and this difference is retained during work performed while breathing hypoxic air. Switching from normoxia to breathing hypoxic air did not significantly alter the ^{31}P-NMR spectrum of high energy phosphate metabolites in working skeletal muscles of Quechua Indians, implying that acute adjustments were able to fully compensate for the reduced oxygen availability in the inspired air.

Instead of differences in muscle pH homeostasis, it appears that differences in energy coupling stabilization lie at the heart of the lactate paradox in high altitude-adapted muscle. At this time we infer that the tighter energy coupling observed in muscles of Andean natives is a constrained or fixed characteristic. This component in the final link in the path of oxygen to tissues is thus tentatively assumed to be a develop- mental or genetic adaptation.

Proton Spectroscopy

Because of its high sensitivity, proton spectroscopy probably offers the greatest promise for evaluating different forms of metabolism with NMR, notwithstanding the substantial spectral editing that has to be undertaken to observe millimolar concentration metabolites. A discussion of the analysis of metabolism by in vivo proton spectroscopy can be found in the review by Avison et al. (1986). In terms of body composition, the most obvious contribution that proton NMR can make is to provide a measure of the proportions of fat and water in a predefined volume. The methyl and methylene proton resonances from its fattyacyl chains give rise to a broad signal band (approximately 0.5 to 2 ppm) separated sufficiently from the wings of the intense water peak (at 4.8 ppm) that both signals can be well resolved. The spatial resolution of such measurements is much better (cube of side of few mm) than for the observation of millimolar metabolites because of the greater concentrations of water and fat.

^{13}C Spectroscopy

Because of the low natural abundance of the ^{13}C isotope, the ^{13}C NMR signals are very weak without isotropic enrichment. This tends to limit the usefulness of in vivo natural abundance ^{13}C spectroscopy at 1.5T to an exploration of the fat composition of human tissue, although with animals more stringent experimental conditions are possible that allow the observation of metabolites, as for example the observation of glucose and glycogen metabolism in rat liver at 8.5T (Reo et al., 1984). A positive aspect of the low signal strength is that the resonances from the much lower concentration metabolites are mainly below the observational threshold at 1.5T and do not complicate the observed spectrum. Despite the disadvantage of a low signal intensity, ^{13}C spectroscopy has some advantages over proton spectroscopy for investigating fat composition. The first is the much larger chemical shift dispersion, which spreads the aliphatic fatty acid resonances in the carbon spectrum over a range between 15 and 170 ppm, whereas in the proton spectrum they may fall in a range between 0.5 and 7.5 ppm. The second is the absence of a dominating water peak in the carbon spectrum. Therefore, it is easy with ^{13}C (Fig. 2) to distinguish

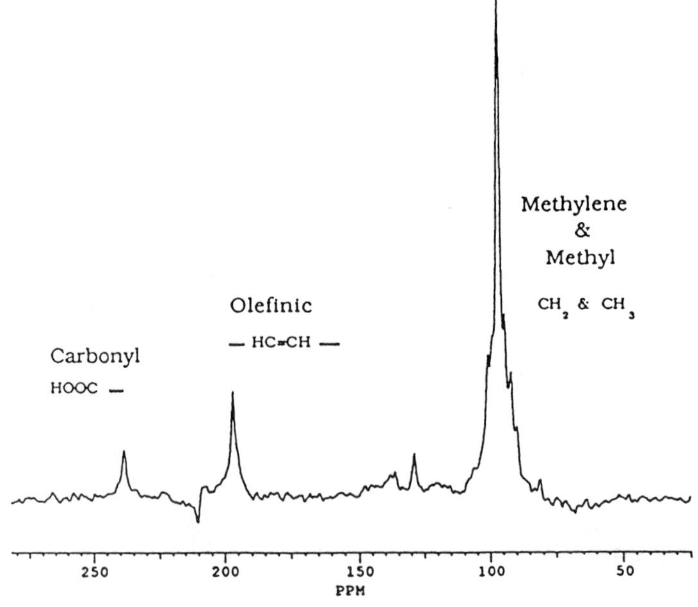

Fig. 2. ^{13}C Spectrum of Rat Muscle.

the carbonyl carbon, which reflects the total fat concentration, from the olefinic carbon signal that reflects the unsaturated fats. Rough comparisons of total saturation to mono-unsaturation may therefore be possible (Dimand et al., 1986).

Nevertheless, as with all NMR of biological systems, the spectral resolution in vivo is inferior to that observable from excised tissue or tissue extracts in vitro (Sillerud et al., 1986), and the ability to discriminate between precursor and product is difficult to make in vivo, even with the help of isotope enrichment of ^{13}C or deuteron labelling of the precursors. An additional and striking example of the resolution and sensitivity differences between in vivo and in vitro experiments can be found in the efforts of Sillerud and co-workers (Sillerud et al., 1988) to use ^{13}C spectroscopy to discriminate between normal and neoplastic tissue.

REFERENCES

Argov, Z., Bank, W. J., Maris, J. , Leigh, J. S. Jr., and Chance, B., 1987, Muscle energy metabolism in human phosphofructokinase deficiency as recorded by ^{31}P nuclear magnetic resonance spectroscopy, <u>Ann. Neurol.</u>, 22:46.
Avison, M. J., Hetherington, H. P., and Shulman, R. G., 1986, Applications of NMR to studies of tissue metabolism, <u>Ann. Rev. Biophys. Biophys. Chem.</u>, 15:377.
Bendall, M. R., and Gordon, R.E., 1983, Depth and refocusing pulses designed for multipulse NMR with surface coils, <u>J. Magn. Reson.</u>, 53:365.
Brown, T. R., Chance, E. M., Dawson, M. J., Gadian, D. G., Radda, G. K., and Wilkie, D. R., 1980, The activity of creatine kinase in frog skeletal muscle studied by saturation transfer nuclear magnetic resonance, <u>J. Physiol</u>. (London), 305:84P.
Bruhn, H., Frahm, J., Gyngell, M. L., Merboldt, K. D., Hänicke, W., and Sauter, R., 1988, Localized proton spectroscopy of tumors in vivo: patients with primary and secondary cerebral tumors, Proc. VII Ann. Meeting of Soc. Mag. Res. Med., San Francisco.
Challiss, R. A. J., Blackledge, M. J., and Radda, G. K., 1988, Spatial heterogeneity of metabolism in skeletal muscle in vivo studied by ^{31}P NMR spectroscopy, <u>Am. J. Physiol.</u>, 254:C417.
Dimand, R. J., Moonen, C. W., Bradbury, E. M., and Cox, K. L., 1986, Non-invasive determination of linoleic acid content of adipose tissue by carbon-13 topographic magnetic resonance, <u>Clin. Res.</u>, 35:228A.
Lunt, J. A., Allen, P. S., Brauer, M., Swinamer, D. , Treiber, E. O., Belcastro, A., Eccles, R., and King, E. G., 1986, An evaluation of the effect of fasting on the exercise-induced changes in pH and P_i/PCr from skeletal muscle, <u>Magn. Reson. Med.</u>, 3:946.
Luyten, P. R., den Hollander, J. A., Segebarth, C., and Baleriaux, D., 1988, Localized ^{1}H-NMR spectroscopy and spectroscopic imaging of human brain tumors in situ, Proc. VII Ann. Meeting of Soc. Mag. Res. Med., San Francisco.
Matheson, G. O., Hochachka, P. W., Hanstock, C. C., Ellinger, D., Gheorghiu, D., Man, S. F. P., McKenzie, D., Parkhouse, W., Stanley, C.A., and Allen, P. S., Intracellular pH and phosphogen metabolism during exercise to fatigue in altitude-adapted natives during deacclimation, <u>J. Respir. Physiol</u>, in press.
Meyer, R. A., Fisher, M. J., Nelson, S. J., and Brown, T. R., 1988, Evaluation of Manual Methods for Integration of in vivo phosphorus NMR spectra, <u>NMR in Biomedicine</u>, 1:131.
Ordidge, R. J., Connelly, A., and Lohman, J. A. B., 1986, Image-selected in vivo spectroscopy (ISIS). A new technique for spatially selective NMR spectroscopy, <u>J. Magn. Reson.</u>, 66:283.

Phang, P. T., Swinamer, D. L., Eccles, R., Lunt, J. A., Allen, P. S., Jeejeebhoy, K. N., Belcastro, A., and King, E. G., 1987, Muscle function testing and ^{31}P-NMR spectroscopy in fasted healthy volunteers, J. Crit. Care, 2:156.

Pichard, C., Vaughan, C., Smik, R., Armstrong, R. L., and Jeejeebhoy, K. N., 1988, Effect of dietary manipulations (fasting, hypocaloric feeding and subsequent refeeding) on rat muscle energetics as assessed by nuclear magnetic resonance spectroscopy, J. Clin. Invest., 82:895.

Reo, N. V., Siegfried, B. A., and Ackerman, J. J. H., 1984, Direct observation of glycogenesis and glucagon stimulated glycogenolysis in the rat liver in vivo by high field carbon-13 surface coil NMR, J. Biol. Chem., 259:13664.

Rees, D., Smith, M. B., Harley, J., and Radda, G. K., 1989, In vivo functioning of creatine phosphokinase in human forearm muscle, studied by ^{31}P NMR saturation transfer, Magn. Reson. Med. 9:39.

Robitaille, P. M., Merkle, H., Sublett, E., Hendrich, K., Lew, B., Path, G., From, A. H. L., Bache, R. J., Garwood, M., and Ugurbil, K., 1989, Spectroscopic imaging and spatial localization using adiabatic pulses and applications to detect transmural metabolite distribution in the canine heart, Magn. Reson. Med., 10:14.

Ross, B. D., Radda, G. K., Gadian, D. G., Rocker, G., Esiri, M., and Falconer-Smith, J., 1981, Examination of a case of suspected McArdle's syndrome by ^{31}P nuclear magnetic resonance, N. Engl. J. Med., 304:1338.

Rothman, D. L., Behar, K. L., Hetherington, H. P., den Hollander, J. A., Bendall, M. R., Petroff, O. A. C., and Shulman, R. G., 1985, ^{1}H-observe/^{13}C-decouple spectroscopic measurements of lactate and glutamate in the rat brain in vivo. Proc. Natl. Acad. Sci. USA, 82:1633.

Sillerud, L. O., Han, C. H., Bitensky, H. W., and Francendese, A. A., 1986, Metabolism and structure of triacylglycerols in rat epididymal fat pad adipocytes determined by ^{13}C nuclear magnetic resonance, J. Biol. Chem., 261:4380.

Sillerud, L. O., Halliday, K. R., Griffey, R. H., Fenoglio-Preiser, C., and Sheppard, S., 1988, In vivo ^{13}C NMR spectroscopy of the human prostate, Magn. Reson. Med., 8:224.

Sotak, C. H., Freeman, D. M., and Hurd, R. E., 1988, The unequivocal determination of in vivo lactic acid using two-dimensional double-quantum coherence-transfer spectroscopy, J. Magn. Reson., 78:355.

Stewart, C. A., Hanstock, C. C., and Allen, P. S., 1988, The application of depth-pulse localized ^{31}P NMR spectroscopy to monitor tumor metabolism and response to chemotherapy in the rat kidney, Magn. Reson. Med., 7:100.

West, J., 1986, Lactate during exercise at external altitude, Fed. Proc., 45:2953.

BODY COMPOSITION BY DUAL-PHOTON ABSORPTIOMETRY AND DUAL-ENERGY
X-RAY ABSORPTIOMETRY

R.B. Mazess, H.S. Barden, and J.A. Hanson

Department of Medical Physics, University of
Wisconsin, Madison, and Lunar Radiation Corporation
Madison, Wisconsin 53713, U.S.A.

INTRODUCTION

Dual-energy projection methods have been used over the past
decade for non-invasive measurement of bone. Dual-photon
absorptiometry (DPA) using a 153-Gd radionuclide source is used
clinically in about 1000 centers to measure regional bone
mineral density (BMD), particularly of the spine and proximal
femur (Barden and Mazess, 1989). The same method has been
applied to measurement of the total body bone mineral content
(BMC) and BMD (Peppler and Mazess, 1981; Mazess et al., 1984b;
Nuti et al., 1987; Gotfredsen et al., 1984a, 1984b, 1986).
Total body scans using DPA require about 60-80 minutes; the
radiation dose is low (1 mrem). Total body BMC correlates
highly with actual skeletal mass and with total body calcium by
neutron activation analysis _in vivo_ (Mazess et al., 1981;
Heymsfield et al., 1989) because calcium is a constant fraction
(about 38%) of the mineral component, or calcium hydroxyapatite.
The typical precision error (1 SD) for total body BMD is under
1% (Nijs et al., 1988; Mazess et al., 1988; Gallagher et al.,
1987). In the past year, new instrumentation was developed in
which the two monoenergetic peaks formerly supplied by the
radionuclide source are provided by an x-ray generator and K-
edge filter (Mazess et al., 1989; Sartoris and Resnick, 1989).
This new methodology is called dual-energy x-ray absorptiometry
(DEXA). DEXA provides total body scans in a fraction of the
time required for DPA (10 vs 70 minutes), yet the precision,
spatial resolution, and radiation dose (<0.02 mrem) are better.

METHODS

Both DPA and DEXA can be used for body composition
analysis. Body composition results are a by-product of total
body scans for bone mineral; the relative lean/fat composition
is given in areas where there is no bone present (Peppler and
Mazess, 1981). A quantity called the R-value is calculated for
each pixel where bone is absent. The R-value (ratio of low to
high-energy attenuation in soft-tissue) ranges from 1.30 to 1.50
for DPA and 1.20 to 1.40 with DEXA based on measurements in fat
and water. In a typical projection of the total body, about

half of the 12,000 pixels contain bone and soft-tissue, and half contain soft-tissue alone (Fig. 1).

Dual-energy absorptiometry has been called the "gold-standard" of body composition because it allows measurement that is largely independent of compartmental assumptions, with a precision error that is very small. The accuracy of DPA composition analysis has been compared to hydrostatic density and other methods (Mazess et al., 1984a; Heymsfield et al., 1989). While the accuracy of total body soft-tissue composition from DPA is quite good (r>0.95), the precision error has been relatively high; the SD for repeat measurements of percent body fat is about ± 2% to 3%. DEXA provides comparable composition analysis with 1% precision.

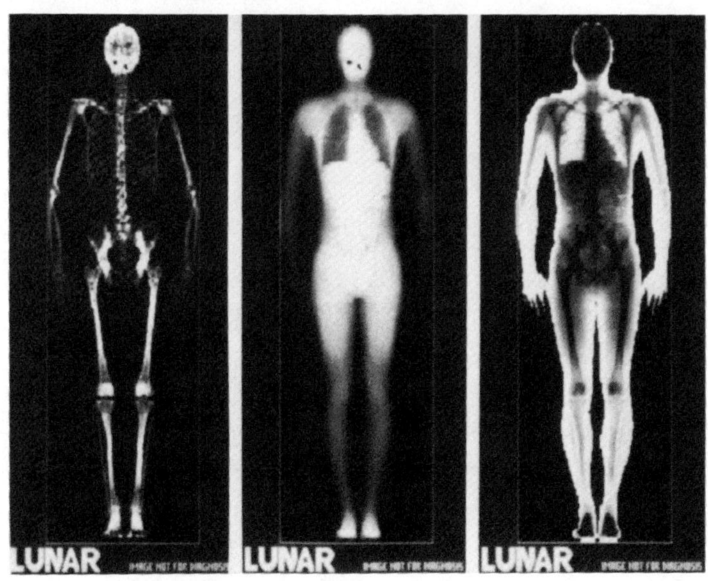

Fig. 1. The images from the screen of the DPX scanner show the skeleton, soft-tissue, and both components together in a normal individual.

We have used both DPA and DEXA for measurement of both the skeletal phase and soft-tissue composition. We also compared the DEXA results to those obtained with DPA. The DPA measurements using 153-Gd were made with the DP4 scanner, and DEXA measurements were obtained with the DPX scanner (LUNAR Radiation Corp., Madison, Wisconsin). The DP4 energies are 44 and 100 keV from 153-Gd, while the DPX energies are 40 and 70 keV (effective) using an 80 kVp x-ray beam filtered with K-edge material (350 mg/cm^3 of cerium). The change of R-value for percentage alcohol is shown in Fig. 2 (Note that the calibration line for alcohol and plastics differs from that of the soft-tissue). The relative invariance of R-value versus effective thickness for water and lard is shown in Fig. 3; the small change with thickness can be corrected in software. We observed stability of R-value results on a phantom made of polyoxymethylene (Delrin) over one year (Fig. 4); the precision error of .002 is equivalent to an uncertainty of 1% in %FAT.

428

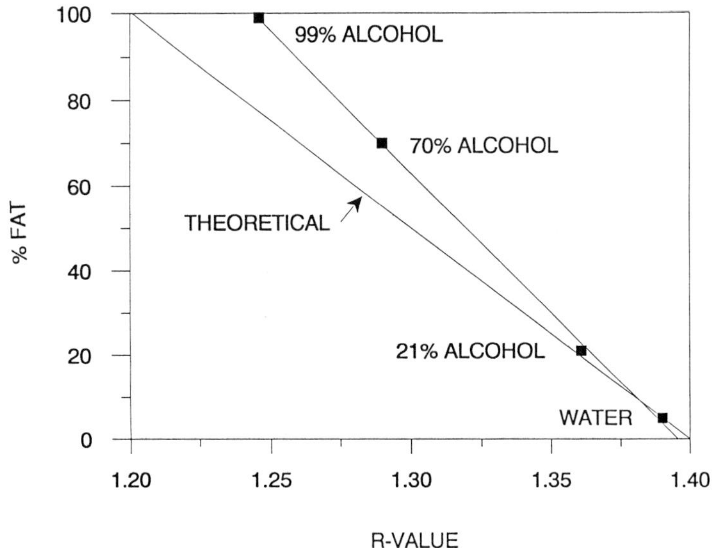

Fig. 2.   The R-value using DEXA is directly proportional to the %FAT. The regression line for alcohol concentration by weight deviates from the theoretical line for soft-tissue.

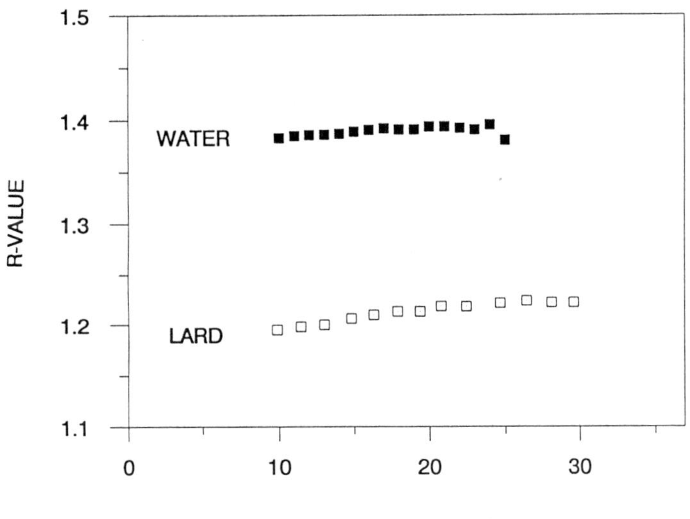

Fig. 3.   The R-value using DEXA is plotted versus effective thickness (water-equivalent) rather than thickness; fat (lard) is less dense and attenuates less than water. There is a small increase of R-value for a large change in thickness.

CORRELATION OF DPA AND DEXA

Measurements were made on 15 subjects ranging in composition from 5% to 40% body fat with DPA and DEXA. The correlation between the DPA and DEXA results for R-value was high, both for total body and regional composition (r=0.94, SEE=.003). The precision of DEXA measurements was evaluated by doing repeat scans on 12 subjects. The reproducibility of DEXA was .002 for total body, and .003 for regional R-values, respectively, which corresponds to an error in percent fat of about 1%. We confirmed that the soft-tissue mass plus the bone mass determined with DEXA corresponds to body weight (r>0.99). There was no systematic offset when actual weights were used, but at higher values the reported weights slightly underestimated DEXA weights. The DEXA method permits a more precise measurement of body composition than DPA, with much faster determination (10 or 20 minutes).

MEASUREMENTS IN ANOREXIA

Dual-photon absorptiometry (DPA) was used to measure the bone and soft-tissue in 10 anorexic women ranging in age from 18 to 27 years. The average weight of the women was 44 kg and the fat content of soft-tissue was about 3.3 kg or about 7.5% fat. The anorexic women averaged 15 to 17 kg below young controls (20 to 40 years) in weight. Of this difference, about 12 to 14 kg was fat, 2 kg was lean tissue, and 0.6 kg was bone mineral.

The total body bone mineral amounted to about 1900 g or approximately about 4.9% $\pm$ 0.5% of lean tissue mass. This value is well under the typical value of 5.9% found in young normal subjects. Spinal BMD averaged 25% below the value in young normal women, or about equivalent to that of 75-year old women. Femoral BMD was reduced only 13% on the average.

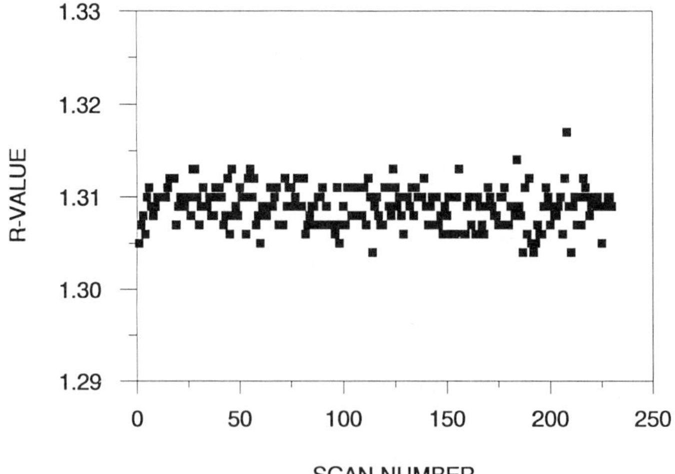

Fig. 4. The precision of R-value using DEXA determinations (n=230) over a 1-year period on a phantom of polyoxymethylene.

The anorexic patients had marked osteopenia both in absolute
terms and relative to lean tissue. Most densitometric techniques assume that the skeleton is a constant fraction of lean
tissue. This assumption causes an overestimation of the percent
body fat in anorexic patients and, in other women with
osteopenia, by about 5%-10% (Baker et al., 1961). For women
with an average fat content of 25%-30% this is a relatively
large error, and it is even higher in anorexics or thin elderly
subjects.

CONCLUSION

Dual-energy projection methods allow a new approach to
quantitation, not only of total body, but of regional composition. Bone, fat, and lean tissue can be directly measured.
This approach overcomes many of the assumptions made using other
composition approaches. In contrast to inductance, impedance
and dilution methods, DEXA and DPA are not dependant upon
electrolyte concentrations or compartmental assumptions, and
provide regional bone and soft-tissue composition with
exactitude. These dual-energy methods also overcome the
relatively large errors of prediction equations. In concert
with other methods DPA and DEXA can improve our understanding of
body compartments in health and disease. In particular, DEXA
can be valuable because of its speed (10 minutes), low radiation
dose (<0.02 mrem), and excellent precision (1%). The better
spatial resolution, compared to DPA, allows more exact
delineation of regional composition, and application to both
pediatric cases and animal models.

REFERENCES

Baker, P.T., 1961, Human bone mineral variability and body
        composition estimates, _in_: "Techniques for Measuring
        Body Composition," J. Brozek, ed., National Academy of
        Sciences, Washington, DC.
Barden, H.S., and Mazess, R.B., 1989, Bone densitometry of the
        appendicular and axial skeleton. _Top Geriatr Rehab_., 4:1.
Gallagher, J.C., Goldgar, D., and Moy, A., 1987, Total body
        calcium in normal women: effect of age and menopause
        status. _J Bone Miner Res_., 2:491.
Gotfredsen, A., Borg, J., Christiansen, C., and Mazess, R.B.,
        1984a, Total body bone mineral _in vivo_ by dual photon
        absorptiometry. I. Measurement procedures, _Clin Physiol_.,
        4:343.
Gotfredsen, A., Borg, J., Christiansen, C., and Mazess, R.B.,
        1984b, Total body bone mineral _in vivo_ by dual photon
        absorptiometry. II. Accuracy, _Clin Physiol_., 4:357.
Gotfredsen, A., Jensen, J., Borg, J., and Christiansen, C.,
        1986, Measurement of lean body mass and total body fat
        using dual photon absorptiometry. _Metabolism_. 35:88.
Heymsfield, S.B., Wang, J., Kehayias, J.J., and Pierson, R.N.,
        1989, Dual photon absorptiometry: comparison of bone
        mineral and soft-tissue mass measurement _in vivo_ with
        established methods. _Am J Clin Nutr_., 49:1283.
Mazess, R.B., Cameron, J.R., and Sorenson, J.A., 1970,
        Determining body composition by radiation absorption
        spectrometry. _Nature_. 228:771.
Mazess, R.B., Peppler, W.W., Chestnut, C.H., Nelp, W.B.,
        Cohn, S.H., and Zanzi, I., 1981, Total body bone mineral

and lean body mass by dual-photon absorptiometry. II. Comparison with total body calcium by neutron activation analysis. <u>Calcif Tissue Int.</u>, 33:361.

Mazess, R.B., Peppler, W.W., and Gibbons, M., 1984a, Total body composition by dual-photon 153-Gd absorptiometry. <u>Am. J Clin Nutr.</u>, 40:834.

Mazess, R.B., Peppler, W.W., Chesney, R.W., Lange, T.W., Lindgren, U., and Smith, E., 1984a, Total body and regional bone mineral by dual-photon absorptiometry in metabolic bone disease. <u>Calcif Tissue Int.</u>, 36:8.

Mazess, R.B., Hanson, J.A., Sorenson, J.A., and Barden, H.S., 1988, Accuracy and precision of dual-photon absorptiometry, <u>in</u>: "Bone Mineral Measurements by Photon Absorptiometry: Methodological Problems," J.V. Dequeker, P. Geusens, and H.W. Wahner, eds., Leuven, Leuven University Press.

Mazess, R.B., Collick, B., Trempe, J. Barden, H.S., and Hanson, J.A., 1989, Performance evaluation of a dual-energy x-ray bone densitometer. <u>Calcif Tissue Int.</u>, 44:228.

Nijs, J., Geusens, P., Dequeker, J., and Verstraeten, A., 1988, Reproducibility and intercorrelation of total bone mineral and dissected regional BMC measurements, <u>in</u>: "Bone Mineral Measurements by Photon Absorptiometry: Methodological Problems," J.V. Dequeker, P. Geusens, and H.W. Wahner, eds., Leuven, Belgium: Leuven University Press.

Nuti, R., Righi, G., Martini, G., Turchetti, V., Lepore, C., and Caniggia, A., 1987, Methods and clinical applications of total body absorptiometry. <u>J Nucl Med Allied Sci.</u> 31:213.

Peppler, W.W., and Mazess, R.B., 1981, Total body bone mineral and lean body mass by dual-photon absorptiometry. I. Theory and measurement procedure. <u>Calcif Tissue Int.</u>, 3:353.

Sartoris, D.J., and Resnick, D., 1989, Dual-energy radiographic absorptiometry for bone densitometry: current status and perspective. <u>AJR.</u>, 152:241.

Sorenson, J.A., and Cameron, J.R., 1964, Body composition determination by differential absorption of monochromatic x-rays. Proceedings of Symposium on Low-energy x- and gamma sources and applications. Illinois Institute of Technology Research Institute, Chicago, IL. AEC report ORNL-IIC.

THE USE OF ELECTRICAL IMPEDANCE AND INFRA-RED INTERACTANCE TO
DETERMINE NON WATER LEAN COMPONENT OF FAT FREE MASS

S.N. Kreitzman and A. Coxon

Department of Physical Education
Cambridge University, England, U.K.

INTRODUCTION

New information on the extent of water shifts has questioned the
interpretation of the results of indirect measurements of body composition
(Brodie et al, 1987; Kreitzman, 1988a). This is true not only of those
methods relying on direct body water determination (isotope dilution and
electrical impedance), but also the traditional methods of hydrodensitometry, total body potassium (TBK) and skinfolds. While all methods can
detect shifts of the total fat free mass (FFM) compartment, none are able
to distinguish the components of FFM, particularly the vital protein
element of the non-water lean component (Kreitzman, 1988b).

The advent of infra-red interactance may provide an answer to this
dilemma. Measurement of body composition assumes that the total body
weight is composed of fat and FFM. FFM is generally calculated from
estimates of body water assuming a constant hydration of 73%. Body fat is
calculated by subtracting this figure for FFM from body weight. An error
in the assumption of hydration therefore influences the calculated value
of non-water lean and body fat. By using two direct measures this
constant can be eliminated.

Impedance is well recognized to measure total body water with high
degree of reliability and reproducibility (Kushner and Schoeller, 1986).
Near infra-red interactance is based upon a principle of differential
absorption capacity of fat, protein or water, and a practical instrument,
the FUTREX-5000, has been designed to measure fat (Rosenthal, 1988). If
body water is measured by impedance and body fat measured by infra-red
interactance, the non-water component of FFM can be easily calculated.

With the changes in body composition associated with weight loss there
are major fluid shifts which make it important to determine body composition by methods which do not assume constant hydration. These investigations are relevant because of concern about the effect of hypocaloric
regimes of different protein and calorie composition on protein sparing in
non-water FFM.

These concepts were evaluated in a weight loss study demonstrating the
preservation of FFM by a ketogenic diet (Cambridge Diet Extra) in

Table 1.   Non-water Lean Percentage of the Weight Lost

	Group A	Group B	Group C
Weight loss	13.0 ± 1.0	9.6 ± 0.9*	7.1 ± 0.8**
NWL	3.9 ± 1.2	4.3 ± 0.3	3.3 ± 0.5
NWL % wt loss	28% ± 7.6	44% ± 1.5°	51% ± 7.4***

Weight loss: significance from Group A p=<0.001 ** p=<0.01*
NWL% weight loss: significance from Group A *** p=<0.05 °p=ns

comparison with higher calorie, non-ketogenic weight reduction diets. It
is well recognized that ketosis reduces the demand for brain glucose
during calorie restriction, and thus, by reducing gluconeogenesis,
preserves protein (Blackburn, 1973; Cahill, 1970). Recent studies have
also demonstrated the direct action of betahydroxybutyric acid in ketosis
on increased protein synthesis in starvation (Nair et al, 1988).

MATERIALS AND METHODS

Subjects

Three groups of 5 females of matched BMI (Group A 30.7 ± 1.3: Group B
31.1 ± 2.0: Group C 30.5 ± 1.6) were assigned to differing hypolcaloric
regimes for eight weeks. Strict compliance was ensured by daily
supervision including measurement of weight and urine ketones. Weekly
measurements of body composition by impedance and infra-red interactance
were obtained and validated at the start and end of the study against
skinfolds and hydrodensitometry.

Diets

Group A were given a 402kcal/day (42g protein) ketogenic diet formula
(Cambridge Diet Extra). This diet was supplemented in Group B to 800kcal
by preparing the formula with milk instead of water. In Group C the
supplement to 1200kcal was achieved by adding a 400kcal formula meal.

Body Composition Studies

Weekly measures of body composition were obtained with bioelectrical
impedance (RJL systems) and infra-red interactance (FUTREX-5000).
Measurements were made concurrently in subjects seen at the same time of
day, one hour after any fluid intake, after voiding urine, so as to
standardise for hydration, food intake, and diurnal fluctuations in water.

Calculations

Non water lean mass was calculated as:

$$NWL = (TBW - FAT) - WATER$$

NWL = Non water lean mass
TBW = Total body weight
FAT = as measured by infra-red interactance (FUTREX 5000)
WATER = as measured by bioelectrical impedance (RJL)

434

RESULTS

After eight weeks of strict adherence to a 405kcal ketogenic diet
(Group A), an 800kcal non-ketogenic liquid diet (Group B) and a 1200kcal
non-ketogenic diet (Group C), weight losses were 13.0kg ± 1.0. 9.6kg ±
0.4, and 6.6kg ± 0.8, respectively.

There was close correlation of FFM as determined by the two methods in
each subject on each occasion (r=0.98). There was also close correlation
between these measures of FFM and those obtained by skinfolds (r=0.96;
r=0.94) and hydrodensitometry (r=0.98: r=0.97) at the start and end of the
study.

The non water lean mass losses averaged 3.9 ± 1.2, 4.2 ± 0.2, and 3.3
± 0.5 for the three groups. These values amount to 28% ± 7.6 of the lost
weight in the ketogenic diet, vs 44% ± 1.5 in the 800kcal diet, and 51% ±
7.4 in the 1200kcal non-ketogenic diet.

DISCUSSION

Near infra-red interactance may provide an important tool when used in
conjunction with impedance, to assess changes in body composition under
conditions where tissue hydration cannot be assumed to be constant.
Further evidence of the specificity of near infra-red analysis for fat in
the presence of other tissue components should be sought. In this study
the combination of these methodologies has provided a simple technique to
assess changes in non-water lean tissue associated with weight loss. The
study appears to conform the protective value of ketosis on non-water lean
tissue during calorie restriction.

REFERENCES

Blackburn, G.L., 1973, "A new concept and its application for protein
    sparing therapies during semistarvation", PhD thesis MIT.
Brodie, D., Coxon, A., Eston, R., and Kreitzman, S., 1987, Body
    composition changes during diuresis, Int J Obesity, 11:306A.
Cahill, G.F., Jr. 1970, Starvation in man, New Eng J Med, 282:668.
Kreitzman, S., 1988a, Total body water and very low calorie diets,
    Lancet, 1:111.
Kreitzman, S., 1988b, Bioelectrical impedance and its role in body
    composition analysis, Int J Obesity, 13:1.17A.
Kushner, R.F., and Schoeller, D.A., 1986, Estimation of total body water
    by bioelectrical impedance analysis, Am J Clin Nut, 44:417.
Nair, K.S., Welle, S.L., Halliday, D. and Campbell, R.G., 1988, Effect of
    B-hydroxybutyrate on whole body leucine kinetics and fractional
    mixed skeletal muscle protein synthesis in humans, J Clin Invest,
    82:198.
Rosental, R.D., 1988, "Futrex-5000 Research Manual", Futrex Inc.
    Gaithersburg, MD.

DEVELOPMENT OF A TECHNIQUE TO MEASURE BONE ALUMINIUM

IN VIVO USING A CF-252 NEUTRON SOURCE

W.D. Morgan, E.A. McNeil, R.M. Wyatt, S.J.S. Ryde,
C.J. Evans*, J. Dutton*, A. Sivyer and A.J. Williams+

Swansea In Vivo Analysis Research Group
Department of Medical Physics and Clinical
Engineering, Singleton Hospital, Swansea SA2 8QA,
*University College Swansea and +Morriston Hospital

INTRODUCTION

Aluminium toxicity still presents a potential hazard to
patients with chronic renal failure through the continued use
of aluminum-based phosphate binders to prevent hyperphospha-
taemia. Consequently, there is a need to measure aluminium
accumulation in these patients to detect those at risk and to
monitor the effectiveness of subsequent treatment. Aluminium in
biological fluids, including serum, can be measured by flameless
AAS, but there are limitations in extrapolating these findings
to the assessment of tissue content (Charhon et al., 1985).

The technique of in-vivo neutron activation analysis has
been used to obtain an indirect assessment of total body aluminium by analysis of the apparent excess of phosphorus in the body
(Williams et al., 1980). Ellis and Kelleher (1987) have
measured bone aluminium directly by using a thermal/epithermal
reactor beam which greatly reduces the fast neutron interfering
reaction on phosphorus. These authors reported a detection
limit of 0.4 mg aluminium in the hand for a dose equivalent of
20 mSv. The aim of this study was to determine the extent to
which an existing californium-252 based instrument (Ryde et al.,
1987) could be adapted for this purpose, so that the method
need not be restricted to the availability of a nuclear reactor.

THEORETICAL ANALYSIS

The limitations of a Cf-252 source compared with a reactor
are (i) a lower neutron output, and (ii) a higher energy spectrum. The first difficulty may be partially overcome by using a
cyclic activation technique, and the second by filtering and/or
moderating the neutron spectrum (McNeil, 1990).

By assuming a thermal fluence rate 50 times smaller than
Ellis and Kelleher (1987) (i.e.~2×10^5 ncm^{-2}s^{-1}) but the
and detection efficiency as these authors, it can be shown that
a 5-cycle sequence of activation (240s) and counting (240s) for
a total time of 2400s can induce 34% of the activity of the

reactor method. Depending on the actual detection efficiency
available and the background level achieved in Swansea, this
implies a detection limit of Al in the hand of 1.2 to 2.4 mg.
Analyses (by AAS) of iliac crest bone biopsies from a group of
ten local dialysis patients revealed concentrations of aluminium
in bone that would imply amounts in the hand of 0.4 to 6.4 mg.

EXPERIMENTAL APPROACH

The neutron beam was modified by a short (~26cm) composite
filter of sulphur and aluminium in an attempt to transmit
selectively only epithermal and intermediate energy neutrons.
Irradiation of a phantom containing, successively phosphorus
and aluminium, showed that filtering approximately doubled the
ratio of thermal to fast fluence, such that the ratio of Al:P cts
g^{-1} improved to 600:1 compared with Williams et al.'s (1980) 8:1
and Ellis and Kelleher's (1987) 12500:1. However, the absolute
thermal fluence rate was only $\sim 10^4$ ncm^{-2}s^{-1}. Therefore we decided
to moderate the neutron beam by filling the collimator aperture
with water and measuring with activation foils the fast and
thermal neutron fluence profiles within the water volume. The
results showed that the required thermal fluence rate could be
obtained at a depth of 15cm from the source. At this position,
the ratio of thermal to fast fluence was sufficiently high to
maintain the ratio of Al:P cts g^{-1} at about 600:1.

CONCLUSIONS

A high efficiency detection system, consisting of opposed
large volume NaI detectors, separated by a 70mm gap, is under
construction. We expect that this will be suitable for diagno-
sing raised levels of aluminium in the hand (e.g.~2 to >10mg). The
fast neutron interference on phosphorus will produce a comparable
response to the activation of aluminium, but the former can be
adequately predicted by measuring simultaneously the calcium
activity and assuming a constant ratio of P:Ca in bone.

ACKNOWLEDGEMENTS

This work is supported by a grant from the Welsh Scheme for
the Development of Health and Social Research.

REFERENCES

Charhon, S.A., Chavassieux, P.M., Neunier, P.J., and Accominotti,
    M., 1985, Serum aluminium concentration and aluminium
    deposits in bone in patients receiving haemodialysis, Br.
    Med. J., 290 : 1613.
Ellis, K.J., and Kelleher, S.P., 1987, In vivo bone aluminium
    measurements in patients with renal disease, in : "In-Vivo
    Body Composition Studies", K.J. Ellis, S. Yasumura, W.D.
    Morgan, eds, Inst. of Physical Sciences in Medicine, London.
McNeil, E.A., 1990, M.Phil. Dissertation, University of Wales.
Ryde, S.J.S., Morgan, W.D., Sivyer, A., Evans, C.J., and Dutton
    J., 1987, A clinical instrument for multi-element in vivo
    analysis by prompt, delayed and cyclic neutron activation
    using Cf-252, Phys. Med. Biol., 32 : 1257.
Williams, E.D., Elliott, H.L., Boddy, K., Haywood, J.K.,
    Henderson, I.S., Harvey, I., and Kennedy, A.C., 1980, Whole
    body aluminium in chronic renal failure and dialysis
    encephalopathy, Clin. Nephrol., 14 : 198.